Origin of Mass and Strong Coupling Gauge Theories (SCGT15)

Proceedings of the Sakata Memorial KMI Workshop

SCGT15
Sakata Memorial KMI Workshop on
Origin of Mass and Strong Coupling Gauge Theories
March 3-6, 2015
Nagoya University

Origin of Mass and Strong Coupling Gauge Theories (SCGT15)

Proceedings of the Sakata Memorial KMI Workshop

Editors

Yasumichi Aoki
KEK, Japan & RIKEN BNL Research Center, USA

Toshihide Maskawa
Shin'ichi Nojiri
Masaharu Tanabashi
Koichi Yamawaki
Nagoya University, Japan

World Scientific

NEW JERSEY • LONDON • SINGAPORE • BEIJING • SHANGHAI • HONG KONG • TAIPEI • CHENNAI • TOKYO

Published by

World Scientific Publishing Co. Pte. Ltd.
5 Toh Tuck Link, Singapore 596224
USA office: 27 Warren Street, Suite 401-402, Hackensack, NJ 07601
UK office: 57 Shelton Street, Covent Garden, London WC2H 9HE

British Library Cataloguing-in-Publication Data
A catalogue record for this book is available from the British Library.

ORIGIN OF MASS AND STRONG COUPLING GAUGE THEORIES (SCGT15)
Proceedings of the Sakata Memorial KMI Workshop

ISBN 978-981-3231-45-0

For any available supplementary material, please visit
http://www.worldscientific.com/worldscibooks/10.1142/10749#t=suppl

Printed in Singapore

Preface

Nagoya SCGT workshop has a long history, starting back in 1988: "New Trends in Strong Coupling Gauge Theories (SCGT 88)". The SCGT 88 was largely focusing on a couple of our new proposals in mid 80's: The "Hidden Local Symmetries" formulating massive composite vector bosons as gauge bosons, and the "Walking Technicolor" which, having a large anomalous dimension near unity, predicted a "Technidilaton" as a light composite Higgs. The latter model was based on the scale-invariant/conformal strong gauge dynamics, with the coupling being almost constant near criticality on the fixed point. Following the line we propose yet another model "Top Quark Condensate", a composite Higgs model having even larger anomalous dimension, which was a main topic together with the above two at the second meeting SCGT 90 in 1990.

The Nagoya SCGT series then continued in 1996, 2002, 2006, 2009, 2012 (SCGT12Mini and SCGT12), 2014 (SCGT14Mini and SCGT14) and the present one, SCGT15, all discussing the development of the relevant new strong gauge dynamics. For some information on these workshops see http://www.kmi.nagoya-u.ac.jp/workshop/SCGT15/.

In March 3-6, 2015, we organized the "Sakata Memorial KMI Workshop on Origin of Mass and Strong Coupling Gauge Theories (SCGT15)" at the Sakata-Hirata Hall of Nagoya University. Shoichi Sakata at Nagoya University proposed in 1956 the Sakata model, a composite model for hadrons, which paved a way to the quark model and eventually to the Standard Model today. We are inspired by his never-ending enthusiasm seeking the deeper level of matter.

Following the tradition of previous SCGT workshops, the purpose of the SCGT15 was to discuss theoretical and phenomenological issues of strong coupling gauge theories, particularly in view of the new phase of the LHC experiments and the conformal fixed point for the gauge/gravity. Synergy of the lattice, string and phenomenological studies at the meeting was extremely important in this phase of the particle physics.

During the 4-day sessions we had 85 participants (21 from abroad), with 39 talks in the main program and 23 talks at poster sessions. Topics were broad, not just conformal strong gauge dynamics but various aspects of strong coupling gauge theories, in a wide variety of approaches covering lattice studies, holography, Schwinger-Dyson/Bethe-Salpeter equation, effective theories, and so on. We do believe that the presentations included in the Proceedings are very useful for those who are interested in the strong coupling gauge theories in the revolutionary era of LHC.

The workshop was sponsored by Kobayashi-Maskawa Institute for the Origin of Particles and the Universe (KMI), Nagoya University. It was financially supported by the JSPS Grant-in-Aid for Scientific Research (S), No.2224003; (C) No.23540300, Daiko Foundation and Nagoya University. We would like to express our sincere thanks to these organizations for generous supports.

We would also like to thank all the members of the International Advisory Committee, the Organizing Committee, the Local Organizing Committee, and the SCGT15 Secretariat and the GCOE support stuff office, and the graduate students at Nagoya University for their devoted contributions to make the workshop successful.

Editors
Yasumichi Aoki
Toshihide Maskawa
Shin'ichi Nojiri
Masaharu Tanabashi
Koichi Yamawaki

Workshop Organization

President: T. Maskawa (KMI)

International Advisory Committee:

T.W. Appelquist (Yale)	W.A. Bardeen (FNAL)
R.S. Chivukula (Michigan State)	H. Georgi (Harvard)
A. Hasenfratz (Colorado)	J.B. Kogut (DOE/Maryland)
Y. Iwasaki (KEK/Tsukuba)	M. Kobayashi (KEK/JSPS/Nagoya)
J. Kuti (UC San Diego)	T. Kugo (Kyoto Sangyo)
V.A. Miransky (Western)	M. Rho (Saclay)
E.H. Simmons (Michigan State))	G. 't Hooft (Utrecht)

Organizing Committee:

Chairperson K. Yamawaki (KMI)

Co-chairpersons

Y. Aoki (KMI) S. Nojiri (KMI/Nagoya) M. Tanabashi (KMI/Nagoya)

K.-I. Aoki (Kanazawa)	M. Harada (Nagoya)	Y. Kikukawa (Tokyo-Komaba)
R. Kitano (KEK)	K.-I. Kondo (Chiba)	T. Onogi (Osaka)
T. Sakai (Nagoya/KMI)	A. Shibata (KEK)	T. Yamazaki (Tsukuba)

Local Organizing Committee:

Y. Akamatsu	Y. Aoki	T. Aoyama	H.S. Fukano	M. Harada
M. Kurachi	N. Maekawa	S. Matsuzaki	K. Miura	K-i. Nagai
S. Nojiri	C. Nonaka	H. Ohki	T. Sakai	M. Tanabashi
K. Tobe	K. Yamawaki			

Sponsorship

Kobayashi-Maskawa Institute for the Origin of Particles and the Universe (KMI), Nagoya University

Financial support also from
JSPS Grant-in-Aid for Scientific Research (S): No.22224003
Daiko Foundation
Nagoya University Foundation
JSPS Grant-in-Aid for Scientific Research (C): No.23540300

Contents

IR Fixed Points in SU(3) Gauge Theories

Yoichi Iwasaki

Center for Computational Sciences, University of Tsukuba, Tsukuba, Ibaraki 305-8577, Japan

The main objective in the talk is to identify the location of the IR fixed point in the SU(3) gauge theories with N_f fundamental fermions within the conformal window $N_f^c \leq N_f \leq 16$. We propose a novel and simple RG method to specify the location and the anomalous mass dimension, and further the method to derive the propagator of a meson in the continuum limit on the IR fixed point.

Status of a Minimal Composite Higgs Theory

Zoltan Fodor

University of Wuppertal, Department of Physics, Wuppertal D-42097, Germany
Juelich Supercomputing Center, Forschungszentrum Juelich, Juelich D-52425, Germany
E-mail: fodor@bodri.elte.hu

Kieran Holland

University of the Pacific, 3601 Pacific Ave, Stockton CA 95211, USA
Albert Einstein Center for Fundamental Physics, Bern University, Bern, Switzerland
E-mail: kholland@pacific.edu

Julius Kuti*

University of California, San Diego, 9500 Gilman Drive, La Jolla, CA 92093, USA
E-mail: jkuti@ucsd.edu
** Presenter*

Santanu Mondal

Eotvos University, MTA-ELTE Lendulet Lattice Gauge Theory Group, Budapest, Hungary
E-mail: santanu@bodri.elte.hu

Daniel Nogradi

Eotvos University, MTA-ELTE Lendulet Lattice Gauge Theory Group, Budapest, Hungary
E-mail: nogradi@bodri.elte.hu

Chik Him Wong*

University of Wuppertal, Department of Physics, Wuppertal D-42097, Germany
E-mail: cwong@uni-wuppertal.de
** Presenter*

We analyze three sets of gauge ensembles in our extended physics program of a particularly important BSM gauge theory with a fermion doublet in the two-index symmetric (sextet) representation of the SU(3) BSM color gauge group. Our investigations include chiral symmetry breaking (χSB) in the p-regime and ϵ-regime, the mass of the composite 0^{++} scalar, resonance spectroscopy, new physics from gauge anomaly constraints, and the role of stable sextet BSM baryons with Electroweak interactions in dark matter searches. Important new goals include studies of the 0^{++} scalar entangled with Goldstone dynamics in the p-regime and the ϵ-regime, the resonance spectrum with particular attention to emerging LHC signals, like recent hints for diphoton excess at 750 GeV or diboson anomalies in the 2 TeV range. All results reported here are preliminary before journal publication including some post-conference material for the discussion.

1. Introduction

An important strongly coupled near-conformal gauge theory built on the minimally required SU(2) bsm-flavor doublet of two massless fermions, with a confining gauge force at the TeV scale in the sextet representation of the new SU(3) BSM color gauge group is an intriguing possibility for the minimal realization of the composite Higgs mechanism. Early discussions of the model as a BSM candidate were initiated in systematic explorations of higher fermion representations of color gauge groups[1–3] for extensions of the original Higgsless Technicolor paradigm[4,5]. In fact, the first appearance of the particular two-index symmetric SU(3) fermion representation can be traced even further back to Quantum Chromodynamics (QCD) where a doublet of sextet quarks was proposed as a mechanism for Electroweak symmetry breaking (EWSB) without an elementary Higgs field[6]. This idea had to be replaced by a new gauge force at the TeV scale, orders of magnitude stronger than in QCD, to facilitate the dynamics of EWSB just below the lower edge of the conformal window in the new BSM paradigm[1–3]. It should be noted that throughout its early history the important near-conformal behavior of the model was not known and definitive results had to wait for recent non-perturbative investigations with lattice gauge theory methods as used in our work.

Near-conformal BSM theories raise the possibility of a light composite scalar, perhaps a Higgs impostor, to emerge from new strong dynamics, far separated from the associated composite resonance spectrum in the few TeV mass range with interesting and testable predictions for the Large Hadron Collider (LHC). This scenario is very different from what was expected from QCD when scaled up to the Electroweak scale, as illustrated by the failure of the Higgsless Technicolor paradigm. Given the discovery of the 125 GeV Higgs particle at the LHC, any realistic BSM theory must contain a Higgs-like state, perhaps with some hidden composite structure.

Based on our *ab initio* non-perturbative lattice calculations we find accumulating evidence for near-conformal behavior in the sextet theory with the emergent low mass 0^{++} scalar state far separated from the composite resonance spectrum of bosonic and baryonic excitations in the 2-3 TeV energy range[7–10]. The identification of the light scalar state is numerically challenging since it requires the evaluation of disconnected fermion loop contributions to correlators with vacuum quantum numbers in the range of light fermion masses we explore. The evidence to date is very promising that the 0^{++} scalar is light in the chiral limit and that the model at this stage remains an important BSM candidate. This report presents new results and outlines the need for important further work needed for definitive results.

In Section 2 for the first time we describe the anomaly-free Electroweak embedding of the strongly coupled sextet gauge theory in a more comprehensive way than before and present the need for more extensive baryon analysis of our gauge ensembles with relevance for dark matter searches and model viability. In Section 3 new results are reported and analysis tools are proposed for mass-deformed chiral perturbation theory and its chiral limit using p-regime and ϵ-regime technologies including the improved chiral properties of mixed actions in the valence sector. This is critically important for accurate BSM scale setting from the chiral limit of the Goldstone decay constant F_π and the chiral condensate . In Section 4 the status of the light 0^{++} scalar and the resonance spectrum is discussed with new plans briefly outlined. Section 5 describes efforts to understand physics at several scales in nearly conformal gauge theories from the asymptotically free UV regime with an almost conformal crossover to the infrared with spontaneous chiral symmetry breaking. Flavor dependence of the running coupling and beta function in the sextet and fundamental fermion representations illustrates this problem. In Section 6 we provide a brief summary of the computational aspects of our work describing the staggered fermion action we use in the simulations with well known taste breaking effects on the exact chiral symmetries of the continuum theory. All figures are attached at the end of the report without interrupting the flow of the narrative.

2. Electroweak multiplet structure, gauge anomalies, and baryons

As in the minimal scheme of Susskind[4] and Weinberg[5], the gauge group of the theory is $SU(3)_{bsm} \otimes SU(3)_c \otimes SU(2)_w \otimes U(1)_Y$ where $SU(3)_c$ designates the QCD color gauge group and $SU(3)_{bsm}$ represents the BSM color gauge group of the new strong gauge force. In addition to quarks and leptons of the Standard Model, we include one SU(2) bsm-flavor doublet (u, d) of fermions which are $SU(3)_c$ singlets and transform in the six-dimensional sextet representation of BSM color, distinct from the fundamental color representation of fermions in the original Technicolor scheme[4,5]. The formal designation (u, d) for the bsm-flavor doublet of sextet fermions uses a similar notation to the two light quarks of QCD but describes completely different physics. The massless sextet fermions form two chiral doublets $(u,d)_L$ and $(u,d)_R$ under the global symmetry group $SU(2)_L \otimes SU(2)_R \otimes U(1)_B$. Baryon number is conserved for quarks of the Standard Model separate from baryon number conservation for sextet fermions which carry 1/3 of BSM baryon charge associated with the BSM sector of the global $U(1)_B$ symmetry group.

2.1. *Electroweak multiplet structure*

It is straightforward to define consistent multiplets for the sextet fermion flavor doublet under the $SU(2)_w \otimes U(1)_Y$ Electroweak gauge group with hypercharge assignments for left- and right-handed fermions transforming under the $SU(2)_w$ weak isospin group. The two fermion flavors u^{ab} and d^{ab} of the strongly coupled sector carry six colors in two-index symmetric tensor notation, $a, b = 1, 2, 3$, associated with the gauge force of the $SU(3)_{bsm}$ group. This is equivalent to a six-dimensional vector notation in the sextet representation. The fermions transform as left-handed weak isospin doublets and right-handed weak isospin singlets for each color,

$$\psi_L^{ab} = \begin{pmatrix} u_L^{ab} \\ d_L^{ab} \end{pmatrix}, \qquad \psi_R^{ab} = (u_R^{ab},\ d_R^{ab}). \tag{1}$$

With this choice of representations, the normalization for the hypercharge Y of the $U(1)_Y$ gauge group is defined by the relation $Y = 2(Q - T_3)$, with T_3 designating the third component of weak isospin.

Once Electroweak gauge interactions are turned on, the chiral symmetry breaking pattern $SU(2)_L \otimes SU(2)_R \to SU(2)_V$ of strong dynamics breaks Electroweak symmetry in the expected pattern, $SU(2)_w \times U(1)_Y \to U(1)_{em}$, and with the simultaneous dynamical realization of the composite Higgs mechanism. It is important to note that the dynamical Higgs mechanism is facilitated through the electroweak gauge couplings of the sextet fermions and does not depend on the hypercharge assignments of the multiplets[4]. Recently, we presented a detailed analysis on anomaly constraints of hypercharge assignments[11]. In this report we summarize what is relevant for the sextet baryon analysis of our existing gauge ensembles with dark matter and model viability implications[11].

2.2. *Anomaly conditions*

Anomaly constraints have a long history in Technicolor motivated BSM model building with representative examples in[1,12–15]. The first condition for model construction with left-handed doublets is the global Witten anomaly constraint which requires an even number of left-handed SU(2) multiplets to avoid inconsistency in the theory from a vanishing fermion determinant of the partition function[16]. In addition, gauge anomaly constraints also have to be satisfied[17]. With vector current $V^i_\mu(x) = \overline{\psi} T^i \gamma_\mu \psi(x)$ and axial current $A^i_\mu(x) = \overline{\psi} T^i \gamma_\mu \gamma_5 \psi(x)$ constructed from fermion fields and internal symmetry matrices T^i in some group representation R for fermions, the anomaly in the axial vector Ward identity is proportional to $\mathrm{tr}(\{T^i(R), T^j(R)\}T^k(R))$ and must vanish. In the sextet theory fermions are either left-handed doublets or right-handed singlets under the $SU(2)_w$ gauge group. The matrices T^i will be either the τ^i Pauli matrices or the diagonal U(1) hypercharge Y. Since the SU(2) group is anomaly free, $\mathrm{tr}(\{\tau^i, \tau^j\}\tau^k) = 0$, we only need to consider anomalies where at least one T^i is the hypercharge Y. The non-trivial constraints come from two conditions on hypercharge traces,

$$\mathrm{tr}(Y) = 0, \quad \mathrm{tr}(Y^3) \propto \mathrm{tr}(Q^2 T_3 - Q T_3^2) = 0\,, \tag{2}$$

where $Y = 2(Q - T_3)$ with electric charge Q, and T_3 as the third component of weak isospin. There are two simple solutions for BSM model building with sextet fermions to satisfy the Witten anomaly condition and gauge anomaly constraints on $\mathrm{tr}(Y)$ and $\mathrm{tr}(Y^3)$ in Eq. (2). The first solution with the choice $Y(f_L) = 0$ for doublets of left-handed sextet fermions (f_L) leads to half-integer electric charges for composite baryons. The second solution with the choice $Y(f_L) = 1/3$ for doublets of left-handed sextet fermions leads to integer electric charges for composite baryons. The hypercharges of right-handed singlets are automatically set from consistent electric charge assignments in both cases. The two choices have very different implications for sextet baryons.

2.3. *Sextet baryons and their early Universe*

In the sextet BSM theory we do not have direct observations of new heavy baryons to set unique hypercharge assignments for left-handed doublets and right-handed singlets of sextet fermions from two alternate solutions to the anomaly conditions. Viability of the choices $Y(f_L) = 0$, or $Y(f_L) = 1/3$, is affected by the different electric charge assignments they imply. With heavy sextet baryon masses in the 3 TeV range, as determined from our recent lattice simulations[11], the seemingly minimal solution with $Y = 0$ for left-handed doublets would lead to intriguing predictions of baryon states with half-integer electric charges for future accelerator searches and relics with fractional electric charges from the early Universe with observable consequences. Problems with half-integer electric charges, from the choice $Y(f_L) = 0$ in our case, were anticipated earlier from strong observational limits on stable fractional charges in the early Universe and their terrestrial relics[18,19]. The non-controversial $Y(f_L) = 1/3$ anomaly solution for the sextet model has new dark matter implications[11] which require new lattice calculations proposed here.

2.4. *Ongoing lattice work on sextet baryons and future plans*

The lightest baryons in the strongly coupled sextet gauge sector are expected to form isospin flavor doublets (uud, udd), similar to the pattern in QCD. As we noted earlier, baryons in the sextet model should carry integer multiples of electric charges if $Y(f_L) \neq 0$ to avoid problems with the relics of the early Universe. This leads to the simplest choice $Y(f_L) = 1/3$ with gauge anomalies to be compensated. A new pair of left-handed lepton doublets emerged from this choice as the simplest manifestation of the anomalies and the Electroweak extension of the strongly coupled sextet gauge sector[11].

Neutron-like udd sextet model baryons (n_6) will carry no electric charge and proton-like uud sextet model baryons (p_6) have one unit of positive electric charge from the choice $Y(f_L) = 1/3$. The two baryon masses are split by electromagnetic interactions. The ordering of the two baryon masses in the chiral limit of massless sextet fermions will require non-perturbative *ab initio* lattice calculations of the electromagnetic mass shifts to confirm intuitive expectations that the neutron-like n_6 baryon has lower mass than the proton-like p_6 baryon. In QCD this pattern was confirmed by recent lattice calculations[20]. We expect the same ordering in the sextet model so that the proton-like p_6 baryon will decay very fast, $p_6 \to n_6 + ...$, with a lifetime $\tau \ll 1$ second. It is unlikely for rapidly decaying p_6 baryons to leave any relic footprints from dark nucleosynthesis before they decay.

With BSM baryon number conservation the neutral n_6 baryon is stable and observational limits on its direct detection from experiments like XENON100[21] and LUX2013[22] have to be estimated. In charge symmetric thermal evolution sextet model baryons are produced with relic number density ratio $n_{B_6}/n_B \approx 3 \cdot 10^{-7}$. For 3 TeV sextet model baryon masses we can estimate the detectable dark matter ratio of respective mass densities ρ_{B_6} and ρ_B as $\rho_{B_6}/\rho_B \approx 10^{-4}$, about $5 \cdot 10^4$ times less than the full amount of unaccounted dark mass, $\rho_{\rm dark} \approx 5 \cdot \rho_{\rm B}$. We will use this mass density estimate to guide observational limits on relic sextet model baryons emerging from charge symmetric thermal evolution where tests of dark baryon detection come from elastic collisions with nuclei in dark matter detectors[11]. The neutral and stable n_6 baryon can interact several different ways with heavy nuclei in direct detection experiments including (a) magnetic dipole interaction, (b) Z-boson exchange, (c) Higgs boson exchange, and (d) electric polarizability. It turns out that cross sections from (a) and (b) can be parametrized and estimated even without lattice simulations. Cross sections from (c) and (d) require lattice calculations using our existing gauge ensembles and capacity computing from new allocation we request for gpu capacity computing. Based on these estimates we expect to show that the sextet BSM model is consistent with observational limits and stable baryons will contribute a small fraction to the missing dark matter content. New physics implied by gauge anomaly constraints, like new lepton generations with neutrinos[11], can also contribute to the relic abundance of dark matter. These are important and interesting issue for future investigations.

3. Mass-deformed chiral perturbation theory and the chiral condensate

One of the most important goals of lattice BSM models is to accurately set the Electroweak scale as a function of the lattice spacing. This allows control on the continuum limit when the cutoff is removed and phenomenologically relevant BSM predictions are made. The chiral $\rm SU(2)_L \times SU(2)_R$ symmetry of the model is dynamically broken to the diagonal vector symmetry $\rm SU(2)_V$ and three associated Goldstone pions facilitate the minimal realization of the Higgs mechanism after the Electroweak interactions are turned on. The Electroweak scale in finite lattice spacing units is set from the decay constant $\rm F_\pi$ of the Goldstone pion in the chiral limit with $\rm F = 250$ GeV in continuum physics notation. It can be identified as the fundamental scale of the theory related to the chiral (Higgs) condensate through the GMOR relation.

3.1. *Taste breaking cutoff effects in the staggered pion spectrum*

Since the determination of the Goldstone decay constant F in the chiral limit is critically important for the location of the light scalar mass and the well-separated resonance spectrum in the 2-3 TeV range, we carefully monitor taste breaking effects in the pion spectrum with the goal of removing cutoff effects from physics predictions. This also serves as guidance for our choice of lattice spacings for new configuration generation. To illustrate cutoff dependent taste breaking effects, spectra of mass-deformed non-Goldstone pion states are shown in Figure 1 from our newest data with the definition of the relevant correlators and quantum numbers given in[7,30]. In the fermion mass range of our data set the taste breaking pattern

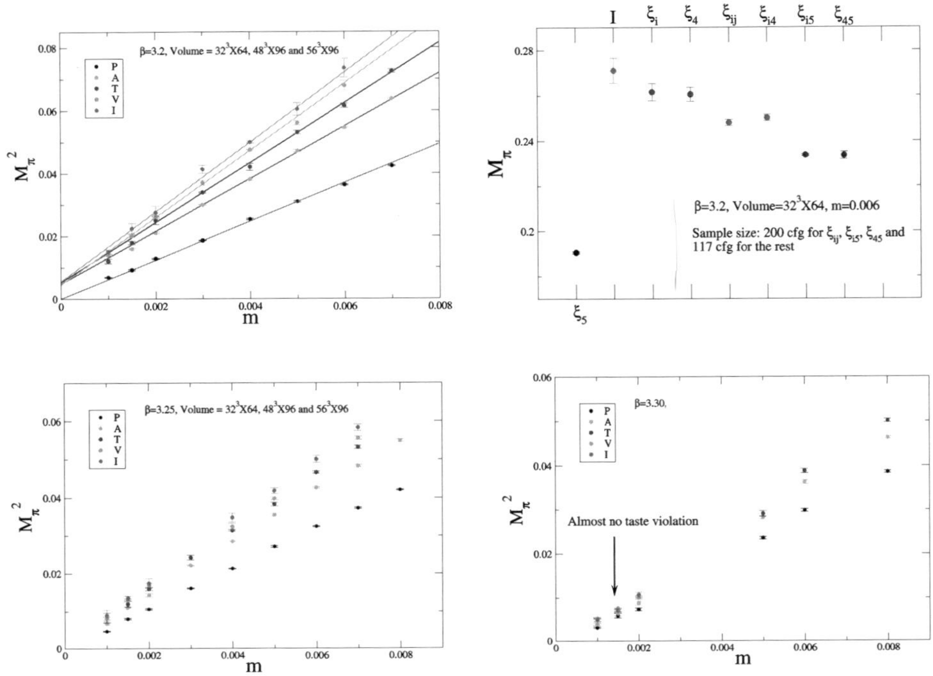

Fig. 1. To illustrate cutoff dependent taste breaking effects, spectra of mass-deformed non-Goldstone pion states are shown from our newest data with the definition of the relevant correlators and quantum numbers given in [7,30]. In the fermion mass range of our data set the taste breaking pattern is different from QCD where the residual Δ mass shifts of the non-Goldstone pions are equispaced in the chiral limit with approximately degenerate SO(4) taste multiplets and with parallel slopes for finite fermion mass deformations of Goldstone and non-Goldstone pion states [29]. For example, as part of the equispaced split of degenerate SO(4) multiplets, the observed approximate split $\Delta_{\rm ij} \sim 2\Delta_{\rm sc}$ of two multiplets in QCD appears to have collapsed in the sextet model. The other distinct difference from QCD is the non-parallel slopes which fan out in Goldstone and non-Goldstone mass deformations of the pion spectrum as shown. While the Δ additive mass shifts are LO taste breaking effects in the chiral Lagrangian [29,31], the taste breaking slope corrections δ can plausibly be identified with NLO analytic terms in the chiral analysis [32]. The corrected mass relation is $M^2_{\rm NLO} = M^2_{\rm LO}(1+\delta)$ where δ depends on the taste quantum number of the pion state. Several relations constrain the δ taste breaking corrections [32].

is different from QCD where the residual Δ mass shifts of the non-Goldstome pions are equispaced in the chiral limit with approximately degenerate SO(4) taste multiplets and with parallel slopes for finite fermion mass deformations of Goldstone and non-Goldstone pion states [29]. For example, as part of the equispaced split of degenerate SO(4) multiplets, the observed approximate split $\Delta_{\rm ij} \sim 2\Delta_{\rm sc}$ of two multiplets in QCD appears to have collapsed in the sextet model. The other distinct difference from QCD is the non-parallel slopes which fan out in Goldstone and non-Goldstone mass deformations of the pion spectrum as shown in Figure 1. While the Δ additive mass shifts are LO taste breaking effects in the chiral Lagrangian [29,31], the taste breaking slope corrections δ can plausibly be identified with NLO analytic terms in rooted staggered chiral perturbation theory (rsχPT)[32]. The corrected mass relation is $M^2_{NLO} = M^2_{LO}(1 + \delta)$ where δ depends on the taste quantum number of the pion state. Several relations constrain the δ taste breaking corrections[32]. The pion spectrum with taste breaking cutoff effects is the input to analyze the fundamental parameters of rsχPT as worked out for the SU(3) group in[31]. Our adaptation to the SU(2) group of rsχPT in the sextet model is straightforward.

3.2. *Fundamental parameters from rooted staggered chiral perturbation theory (p-regime)*

For the SU(2) analysis we adapted the procedure from[31]. There are two fundamental parameters F and B in the SU(2) chiral Lagrangian. The fundamental parameter F of χPT, defined as the chiral limit of the pion decay constant F_π, sets the Electroweak scale and the fundamental parameter B sets the fermion mass deformation of the Goldstone spectrum. With bare fermion mass m, the RG invariant combination $m \cdot BF^2$ is related to the chiral condensate via the GMOR relation.

We apply rooted staggered chiral perturbation theory to the mass-deformed pion spectrum and F_π. The fitting procedure in the p-regime proceeds in several steps. In the first step finite volume correction is applied to the M_π and F_π data from 1-loop continuum χPT. This is sufficient to assure that in the next step the fitting procedure is applied to data free from volume dependence. A linear fit is applied to the quadratic masses of the non-Goldstone pion spectrum to determine their mass shifts and slopes. In the final analysis of rooted chiral perturbation theory, non-Goldstone pion states run in the chiral loops including their mass splittings and fan-out slope structure from taste breaking as determined from the linear fits to the non-Goldstone spectrum. We applied this analysis at two values of the gauge coupling where we have extensive ensembles.

For illustration, preliminary results from rsχPT are shown in Figure 2 from fits at gauge coupling $\beta = 3.20$ which corresponds to our coarser lattice of the two extended sets of gauge ensembles. The upper left panel shows the linear fits to the quadratic masses of the non-Goldstone pions to determine their mass shifts and slopes as input. The upper right panel shows the rsχPT fit to F_π as a function of fermion mass deformations away from the chiral limit. The two lower panels show rsχPT fits to M_π as a function of fermion mass deformations away from the chiral limit. Fits at the finer lattice spacing $\beta = 3.25$ are quite similar in quality but with lower confidence level. The unambiguous determination of the cutoff dependent F and B parameters and their continuum limit from rsχPT will require extended analysis. Partial quenching with valence fermions is the first added step to make the ongoing analysis more robust.

Although our results are consistent with chiral symmetry breaking and rsχPT, ongoing work will require considerable extensions for definitive results. Important new work, besides partial quenching includes a solution to the entanglement problem of the light scalar with the low pion spectrum in perturbation theory with

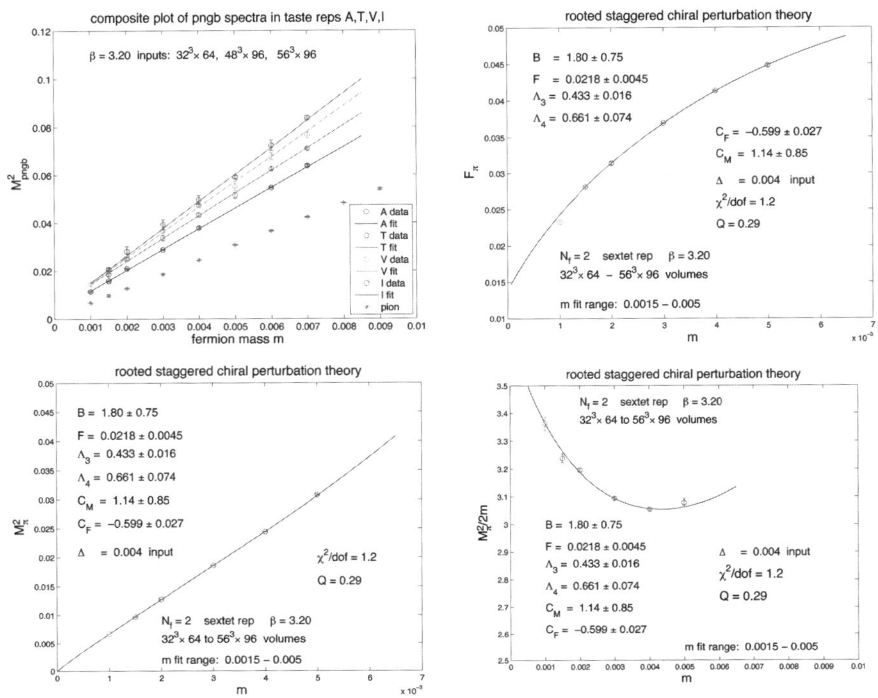

Fig. 2. Preliminary results from rooted χPT are shown from fits at gauge coupling $\beta = 3.20$ which corresponds to our coarser lattice of the two extended sets of gauge ensembles. The upper left panel shows the linear fits to the quadratic masses of the non-Goldstone pions to determine their mass shifts and slopes as input. The upper right panel shows the rooted χPT fit to F_π as a function of fermion mass deformations away from the chiral limit. The two lower panels show rooted χPT fits to M_π as a function of fermion mass deformations away from the chiral limit. We have similar analysis for M_π^2 and F_π at $\beta = 3.25$.

comparable masses in the rsχPT regime, crossover analysis from the p-regime to the ϵ-regime and applications of Random Matrix Theory (RMT) in the ϵ-regime. Coupled chiral dynamics of the low mass scalar 0^{++} state with the pions requires new analysis based on an extended effective theory. We are using the modified effective field theory of χPT on existing gauge ensembles but new gauge configuration generation is also needed in the crossover to the ϵ regime. For independent control on the results for the fundamental parameters F and B we developed and apply now mixed actions with improved chiral symmetry without taste breaking in the valence sector of the analysis. Ongoing new efforts in the p-regime and RMT based ϵ-regime in mixed action setting will resolve important aspects of rsχPT with better determination of F and B.

3.3. *Epsilon-regime, RMT, and mixed action in the valence sector*

Safe extrapolation from the entangled regime of the low mass 0^{++} scalar with pions to the massless fermion limit is enabled by crossover to the ϵ-regime of χPT at low enough scales λ where Goldstone dynamics begins to decouple from the scalar state. This is demanding and requires significant resources. To control taste breaking we cannot go to lattice spacings coarser than the one set by $\beta = 3.20$. The uncertainties in the value of $F \sim 0.018 - 0.025$ with limitations from rooted chiral perturbation theory at this lattice spacing requires large $V = 56^3 \times 96$ and $V = 48^3 \times 96$ lattice volumes to control the $F \cdot L \geq 1$ condition which is necessary for convergent expansion in all regimes of χPT, including the ϵ-regime. Even for our largest $V = 56^3 \times 96$ and $V = 48^3 \times 96$ lattice volumes control with $F \cdot L \sim 1$ is just barely sufficient. For the lowest fermion mass $m = 0.0010$, we have now at these volumes and at this lattice spacing, the scaling variable $m\Sigma V \sim 80$ is very large and more appropriate for the p-regime analysis of χPT. Reaching the ϵ-regime requires substantial decrease in the scaling variable $m\Sigma V$ targeting $m = 0.0003$ which presents considerable algorithmic challenge for accelerated inversion methods and also calls for mixed action innovation. We deploy accelerated inverters in configuration generation to the $m = 0.0010 - 0.0003$ range and analyze these configurations with mixed valence actions of good chiral properties as described below. Our limited resources this year allowed us to test these methods without comprehensive deployment for phenomenologically relevant results like the $M_{0^{++}}/F$ ratio in the continuum limit.

The $m = 0.0010 - 0.0003$ range is in the crossover from the p-regime to the ϵ-regime where known methods of χPT are based on partial quenching and mixed action analysis. For reliable testing, we performed χPT analysis in the crossover to the ϵ-regime with partial quenching and a mixed valence action with improved chiral symmetries. We take the p-regime gauge configurations of the lowest fermion masses on the largest lattice volumes and analyze the fermion condensate and the Dirac spectrum with valence fermion action where the original gauge link variables are replaced with the ones with a fixed number of small stout steps which corresponds to fixed gradient flow time t in lattice spacing units at each gauge coupling. This strategy can be viewed as a mixed action based analysis with very good chiral properties of the fermion valence action.

The newest test results are shown in Figure 3. The valence action is defined with a large number of very small stout steps which corresponds to gradient flow time $t = 3$ in cutoff units. We checked the eigenvalues of the Dirac operator which order into nearly degenerate quartets with the smeared gauge links of the gradient flow. The degenerate eigenvalues follow the index theorem count matching the topology of each gauge configuration as measured from the topological charge operator on the gradient flow. The upper left panel shows the RMT distribution of four degenerate quartets showing that the ϵ-regime is reached with the scaling variable $\lambda\Sigma_{\rm mixed}V \sim 10$ where the fermion mass is replaced by the scale of the gradient

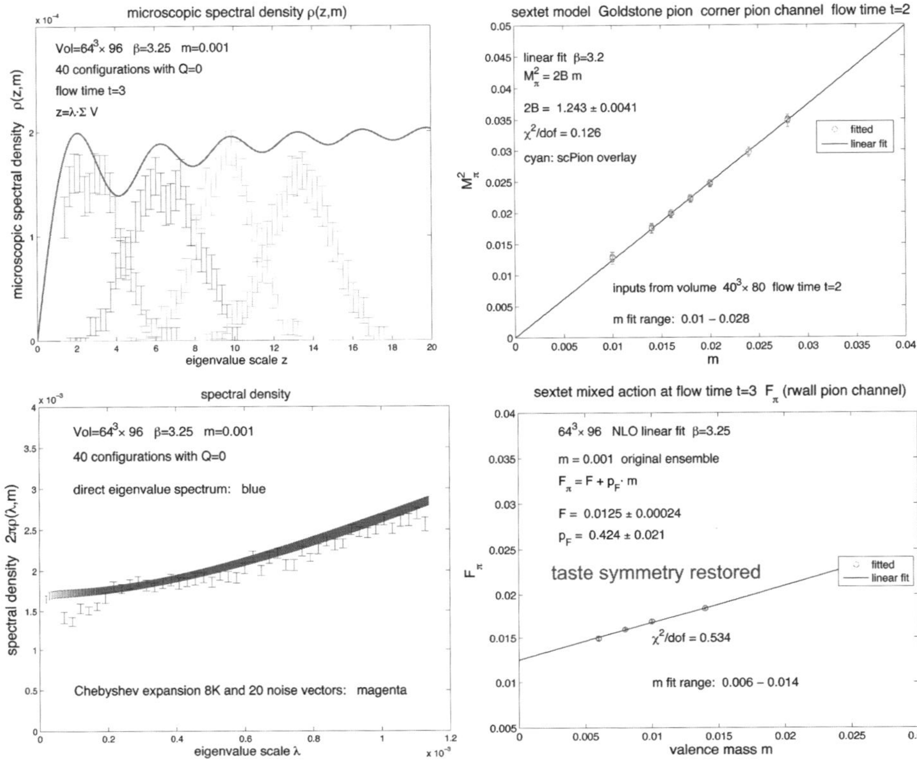

Fig. 3. Results from the mixed action are explained in the text.

flow defined valence Dirac spectrum (m $\to \lambda$). The fermion condensate Σ_{mixed}, not RG invariant itself, is consistently determined from the gradient flow defined valence Dirac operator. The upper right panel illustrates the perfect degeneracy of the Goldstone pion with one selected non-Goldstone pion (scPion in the plot). We checked that the degeneracy holds for all non-Goldstone pion states. The lower left panel shows the infrared part of the directly calculated Dirac spectral density on the gauge configurations and its Chebyshev expansion based approximation. The lower right panel shows the pion decay constant F_π fitted with the mixed action for fixed sea mass as a function of valence masses. Continued future work is needed for definitive results of BSM phenomenology.

4. The light 0^{++} scalar and the resonance spectrum

The most important goals of our lattice Higgs project are to establish the emergence of the light scalar state with 0^{++} quantum numbers and the resonance spectrum far separated from the light composite scalar.

4.1. *The light scalar state*

The f_0 meson (in QCD terminology) has 0^{++} quantum numbers and acts as the scalar state in the sextet model (σ particle in QCD). Close to the conformal window, the f_0 meson of the sextet model is not expected to be similar to its counterpart in QCD. If it turns out to be light, it can replace the elementary Higgs particle and pose as the Higgs impostor. Two types of different 0^{++} operators, the fermionic one and the gluonic one (0^{++} glueball), are expected to mix in the relevant correlation functions for mass determination. Such mixing was not included in the pilot study[9] but becomes an important goal of our ongoing effort. We will report our new results without including these mixing effects.

A particular flavor-singlet correlator is needed to capture the 0^{++} scalar state with vacuum quantum numbers. It requires connected and disconnected diagrams of fermion loop propagators on ensemble gauge configurations. The connected diagram corresponds to the non-singlet correlator $C_{non-singlet}(t)$. The correlator of the disconnected diagram is $D(t)$ at time separation t. The f_0 correlator $C_{singlet}(t)$ is defined as $C_{singlet}(t) \equiv C_{non-singlet}(t) + D(t)$. The transfer matrix has the spectral decomposition of the $C_{singlet}(t)$ correlator in terms of the sum of all energy levels $E_i(0^{++}), i = 0, 1, 2, ...$ and their parity partners $E_j(0^{-+}), j = 0, 1, 2, ...$ but at large time separation t the lowest states $E_0(0^{++})$ and $E_0(0^{-+})$ dominate. They correspond to m_{f_0} and $m_{\eta_{sc}}$. The relevant non-singlet staggered correlator can be fitted well with non-oscillating a_0 contribution and oscillating π_{sc} contribution, with the non-Goldstone pion π_{sc} discussed in Section 3. One of the most important new developments in our analysis is to use correlators which project out non-zero momentum states of the scalar. This projection eliminates the vacuum contribution in the disconnected part and improves the mass extraction procedure.

We estimate the connected and disconnected diagrams with stochastic source vectors of fermion propagators. To evaluate the disconnected diagram, we need to calculate closed loops of quark propagators. We introduce Z_2 noise sources on the lattice where each source is defined on individual time-slice t_0 for color a. The scheme can be viewed as a "dilution" scheme which is fully diluted in time and color and even/odd diluted in space. Results from the original pilot study[9] on $32^3 \times 64$ lattice volumes at $\beta = 3.20$ could only extend down to the lowest fermion mass at $m = 0.003$. From our new analysis some representative examples of 0^{++} effective mass fits are shown in Figure 4 probing the light scalar closer to the chiral limit than before at fermion mass $m = 0.0015$. The upper left panel is at $\beta = 3.20$ with $48^3 \times 96$ lattice volume and the upper right panel is at $\beta = 3.20$ with $56^3 \times 96$ lattice volume to check against finite volume dependence in this low fermion mass range. The two lower panels of the plot show results at $\beta = 3.25$.

Although our original estimate $M_{0^{++}}/F \sim 1 - 3$ for the chiral limit remains consistent with the ongoing new analysis, important further work is needed on the light f_0 scalar with 0^{++} quantum numbers. We want better control on the slowly changing topology of the RHMC algorithm and the related dependence of the extracted

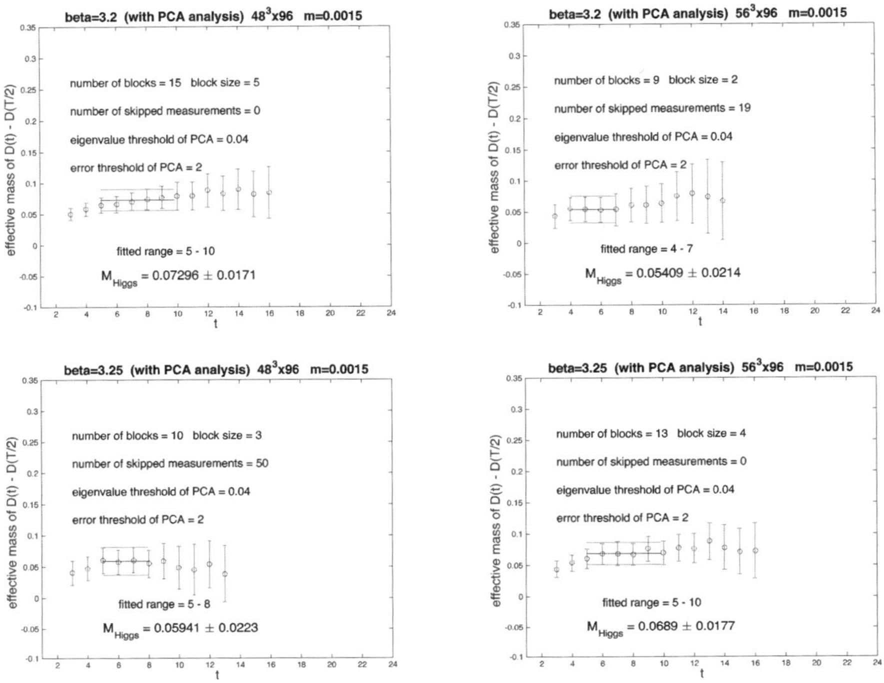

Fig. 4. Representative fits of the low mass scalar from two ensembles using double Jackknife procedure on the covariance matrix with Principal Component Analysis (PCA).

masses on the topological quantum numbers of the gauge configurations. We are also in the process of a closely related study of the η' problem which is particularly interesting and important in the staggered fermion formulation. Fermion mass deformations of the low-lying f_0 state and the Goldstone pion are expected to be entangled which requires extended χPT analysis. Our ongoing work will have to address these issues.

4.2. *The emerging resonance spectroscopy*

It is important to investigate the chiral limit of composite hadron states separated from the Goldstones and the light scalar by finite mass gaps. The baryon mass gap in the chiral limit, for example, provides further evidence for χSB with preliminary results reported earlier[24]. Resonance masses of parity partners provide important additional information with split parity masses in the chiral limit. This is particularly important for consistency with χSB and for a first estimate of the S parameter when probing the model via Electroweak precision tests[37].

A remarkable resonance spectrum is emerging in our new analysis which is sketched in Figure 5 for illustration only. The scale is set by F in TeV units at both lattice spacings with caveats from discussions in Section 3 of the report. Any

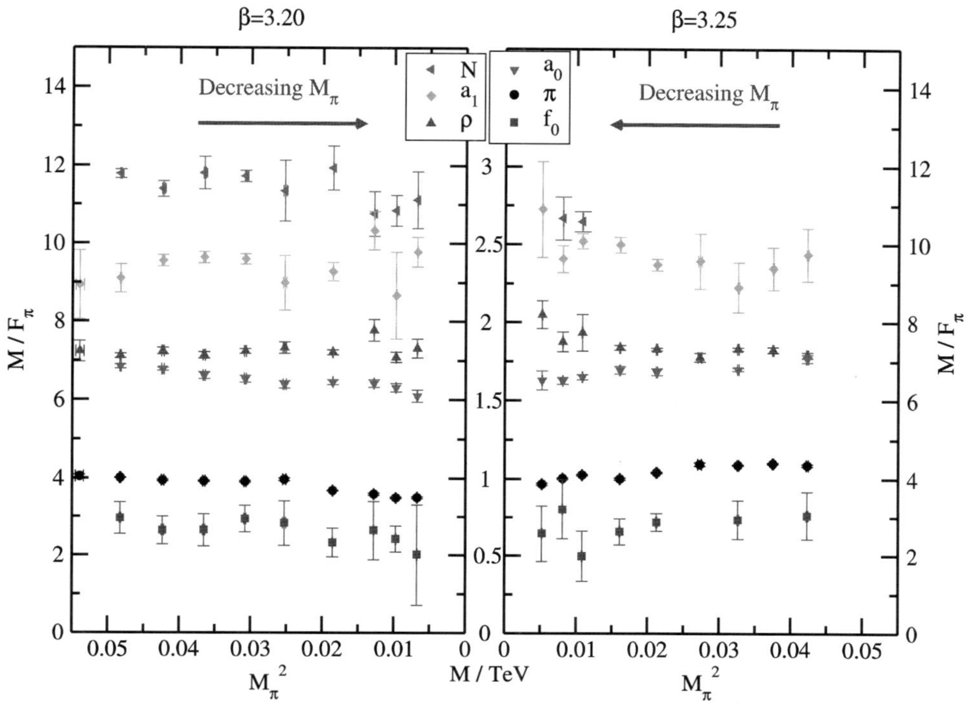

Fig. 5. New resonance spectroscopy results are shown in the plot for illustration only. The scale is set by F=250 GeV at both lattice spacings with caveats from discussions in Section 3 of the report. Any conclusion about χSB or conformal behavior from eyeballed inspection of the data would be inappropriate and misleading.

conclusion about χSB or conformal behavior from eyeballed inspection of the data would be inappropriate and misleading. Although with more work needed for confirmation, the sextet model appears to be close to the conformal window and due to χSB exhibits the right Goldstone spectrum for the minimal realization of the composite Higgs mechanism with a light scalar separated from the associated resonance spectrum in the 2-3 TeV region. Chiral symmetry breaking and a very small beta function are not sufficient to guarantee a light dilaton-like state as the natural interpretation for the emergence of the light scalar. Consistent with our observations, a light Higgs-like scalar is still expected to emerge near the conformal window as a composite state with 0^{++} quantum numbers, but not necessarily with a dilaton interpretation. This scalar state has to be light but is not required to match exactly the observed 125 GeV mass. The light scalar from composite strong dynamics gets lighter from electroweak loop corrections, dominated by the large negative mass shift from the top quark loop[38–40].

5. The scale dependent renormalized coupling and beta function

An important and independent consistency condition of the model would be provided by matching the scale dependent renormalized coupling of the perturbative regime to the scale dependent coupling of the non-perturbative phase associated with χSB. We proposed a gauge coupling earlier $g(\mu = 1/L)$, running with the scale set by the finite volume[41] and defined on the gradient flow of the gauge field[42]. Since the gradient flow at flow time t probes the gauge field at the scale $\sqrt{8t}$, the running coupling can be defined as a function of L in finite volume $V = L^4$ while holding $c = (8t)^{1/2}/L$ fixed with $\alpha_c(\mathrm{L}) = 4\pi\langle \mathrm{t}^2\mathrm{E}(\mathrm{t})\rangle/[3(1+\delta(\mathrm{c})]$ with details of the definition given in [41]. This volume dependent coupling is particularly suitable to study the perturbative regime and departures from it. The measured renormalized couplings are very accurate and the scheme defines a one-parameter family when c is varied and adjusted for different goals[43,44].

For illustration, published results are presented in Figure 6 to show the efficiency of the method. For comparison, red color marks the step β-function of four massless fermion flavors in the fundamental representation from an earlier publication[43]. The new published result for the step β-function of eight massless fermion flavors in the fundamental representation is marked with green color in the plot[45]. The step β-function of the sextet model of this report is marked with blue color in the plot from a recent publication[46]. This result ruled out an earlier claim of an infrared fixed point (IRFP) in the $\mathrm{g}^2 = 6$ range of the renormalized coupling which would have implied a conformal phase for the sextet model[47].

It is important to note that an IRFP was reported in the $\mathrm{g}^2 = 6.0 - 6.4$ range of the renormalized coupling with twelve massless fermions in the fundamental representation[48]. This IRFP is based on the study of the scale-dependent running coupling from the gradient flow with the choice $c = 0.2$ and has been generally accepted as the most convincing evidence that the model with twelve massless flavors in the fundamental representation is inside the conformal window. Recent work using larger c values in the gradient flow scheme does not find an IRFP in the reported range[49]. Further work would be required to resolve the tension between these two findings.

To establish the volume dependent running coupling without an IRFP in the range of renormalized couplings which can be reached within limitations of the method should be only one important part of the analysis. Equally important is to find independent and direct evidence for χSB to rule out the conformal phase. In this report we presented a preponderance of evidence for χSB in the sextet model. For added consistency, in the non-perturbative phase with χSB we are interested in a scale-dependent and volume independent renormalized coupling. At fixed lattice size, bare coupling, and fermion mass m we determine the appropriate flow time $\mathrm{t}(\mathrm{g}^2, \mathrm{m})$ to match any targeted flow-dependent renormalized coupling g^2 calculated from $\langle \mathrm{t}^2\mathrm{E}(\mathrm{t})\rangle$. Assuming that the footprint of the operator on the gradient flow is sufficiently small compared to the Compton wavelength of the pion for p-regime

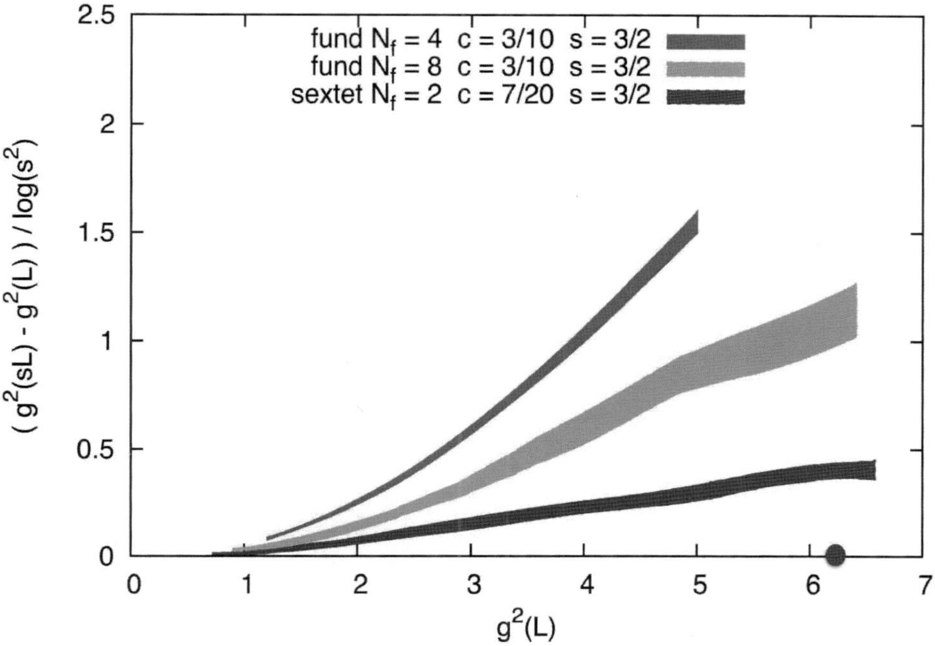

Fig. 6. Comparative results are shown in the plot to illustrate the efficiency of the gradient flow based running coupling and β-function. For comparison, red color marks the step β-function of four massless fermion flavors in the fundamental representation from an earlier publication[43]. The new published result for the step β-function of eight massless fermion flavors in the fundamental representation is marked with green color in the plot[45]. The step β-function of the sextet model of this proposal is marked with blue color in the plot from a recent publication[46]. This result ruled out an earlier claim of an infrared fixed point (IRFP) in the $g^2 = 6$ range of the renormalized coupling which would have implied conformal phase for the sextet model. The purple point in the plot marks the IRFP reported in the renormalized coupling with twelve massless fermions in the fundamental representation[48].

analysis, the dependence of $t(g^2, m)$ on m can be replaced by $t(g^2, M_\pi^2)$ in χPT of pion dynamics with linear dependence of $t(g^2, M_\pi^2)$ on M_π^2 in leading order[50]. Any residual finite volume dependence can be corrected in χPT. A scale dependent and volume independent step beta function can be determined from this procedure in the phase with χSB. The analysis is far from complete in the sextet model. Plans for similar calculations in the fundamental rep with twelve flavors, motivated by our newest results, would be interesting to consider.

6. Computational framework

We developed promising and well-tested new methods to overcome some limitations of more conventional lattice BSM methods. They include the analysis of Goldstone dynamics with mixed actions, the chiral condensate, and spectroscopy in crossover from the p-regime to the ϵ-regime, close to the chiral limit of the theory. As a candidate theory of the composite Higgs particle the new approach requires new runs and a mixed valence action with excellent chiral properties of valence fermions. The required fermion mass range of the gauge configurations is below the range of our existing run set. Runs in the range from $m = 0.0010$ to $m = 0.0003$ operate in the crossover region from the p-regime to the ϵ-regime and runs with valence fermions in the $m = 0.0003 - 0.0001$ range reach further down in the fermion mass range of the RMT analysis. Three gauge couplings are needed to control the continuum limit.

Our physics analysis includes spectroscopy, the Dirac spectra, and the running coupling in capacity computing mode with new configuration generation plans. The disconnected diagrams of the low mass 0^{++} state are extremely demanding on gpu resources. The conventional spectroscopy including analysis with a series of valence fermion masses in mixed actions, Dirac spectra including RMT, and configuration generation for the volume dependent running coupling also require large gpu resources for every given run parameter set in large volumes.

We use the tree-level Symanzik-improved gauge action for all simulations reported here. The conventional $\beta = 6/g^2$ lattice gauge coupling is defined as the overall factor in front of the well-known terms of the Symanzik lattice action. Its values are in the $\beta = 3.20 - 3.30$ range of the proposed ensemble generation. The link variables in the staggered fermion matrix are exponentially smeared with two stout steps[25]; the precise definition of the staggered stout action was given in[26]. The RHMC algorithm is deployed in all runs. The fermion flavor doublet requires rooting in the algorithm. For molecular dynamics time evolution we apply multiple time scales[27] and the Omelyan integrator[28]. Our error analysis of hadron masses uses correlated fitting with double jackknife procedure on the covariance matrices. The time histories of the fermion condensate, the plaquette, the topological charge, and correlators are used to monitor autocorrelation times in the simulations.

The composite Higgs theory of this report requires a code structure which is distinct from lattice QCD in one important way. The fermions in QCD are in the fundamental representation of the SU(3) gauge group and represented in the code by a complex vector q_i, with color index $i = 1, 2, 3$. In the sextet model the fermions are represented as a two-index symmetric tensor T_{ij} which is constructed as the symmetric part of the tensor $q_i q_j$.

The staggered Dirac operator, represented in matrix form and built from the SU(3) gauge link variables, has to be applied to the fermions field during the CG iteration steps consuming the dominant part of the ensemble generating evolution code. We developed two versions of the code. In the vector code, using the 6-dimensional vector representation of the fermions, the SU(3) gauge link matrices U_{ij} are brought in from main memory and from the product $U_{ij} \cdot U_{kl}$ on each link a new 6x6 matrix V_{mn}, $m, n = 1, 2, ...6$ is constructed which acts on the Q_n vectors as a $V_{mn} \cdot Q_n$ matrix-vector product. It is possible, however, to keep the Dirac matrix-vector algebra in the original two-index symmetric representation which entirely eliminates the overhead required to reconstruct the 6x6 matrix V_{mn} and directly use the product of SU(3) link matrices on each link in tensor representation. The load on the memory bandwidth remains unchanged and on the BG/Q platform with excellent balance of the network bandwidth and memory bandwidth this leads to a factor of two reduction in flops and to a corresponding speedup factor in performance.

The code, with high utilization of the fast BG/Q network and with register level optimization of the local code on the nodes, was thoroughly tested for weak scaling and strong scaling across a large number of nodes on the BG/Q platform. The efficiency of the code in CG performance is 29 percent of the peak speed of the BG/Q platform on $48^3 \times 96$ lattices on one rack with little loss of efficiency on half-rack. The numbers illustrate our efficient BG/Q code in lattice gauge theory applications of staggered fermions.

Acknowledgements

Julius Kuti and Ricky Wong are grateful to Koichi Yamawaki and the organizing committee for their kind hospitality and for the most interesting SCGT 2015 workshop at KMI. We acknowledge support by the DOE under grant DE-SC0009919, by the NSF under grants 0970137 and 1318220, by the DOE ALCC award for the BG/Q Mira platform of Argonne National Laboratory, by OTKA under the grant OTKA-NF-104034, and by the Deutsche Forschungsgemeinschaft grant SFB-TR 55. Computational resources were provided by the Argonne Leadership Computing Facility under an ALCC award, by USQCD at Fermilab, by the University of Wuppertal, by The Juelich Supercomputing Center on Juqueen and by the Institute for Theoretical Physics, Eotvos University. We are grateful to Szabolcs Borsanyi for his code development for the BG/Q platform. We are also grateful to Sandor Katz and Kalman Szabo for their code development for the CUDA platform[51]. KH wishes to thank the Institute for Theoretical Physics and the Albert Einstein Center for Fundamental Physics at the University of Bern for their support.

References

1. D. D. Dietrich, F. Sannino and K. Tuominen, *Light composite Higgs from higher representations versus electroweak precision measurements: Predictions for CERN LHC*, Phys. Rev. D **72**, 055001 (2005) [hep-ph/0505059].
2. F. Sannino and K. Tuominen, *Orientifold theory dynamics and symmetry breaking*, Phys. Rev. D **71**, 051901 (2005) [hep-ph/0405209].
3. D. K. Hong, S. D. H. Hsu and F. Sannino, *Composite Higgs from higher representations*, Phys. Lett. B **597**, 89 (2004) [hep-ph/0406200].
4. L. Susskind, *Dynamics Of Spontaneous Symmetry Breaking In The Weinberg-Salam Theory*, Phys. Rev. D **20**, 2619 (1979).
5. S. Weinberg, *Implications Of Dynamical Symmetry Breaking: An Addendum*, Phys. Rev. D **19**, 1277 (1979).

6. W. J. Marciano, *Exotic New Quarks and Dynamical Symmetry Breaking*, Phys. Rev. D **21**, 2425 (1980).
7. Z. Fodor, K. Holland, J. Kuti, D. Nogradi, C. Schroeder and C. H. Wong, *Can the nearly conformal sextet gauge model hide the Higgs impostor?*, Phys. Lett. B **718**, 657 (2012).
8. Z. Fodor, K. Holland, J. Kuti, D. Nogradi, C. Schroeder and C. H. Wong, *The sextet gauge model, light Higgs, and the dilaton*, PoS LATTICE **2012**, 024 (2012).
9. Z. Fodor, K. Holland, J. Kuti, D. Nogradi and C. H. Wong, *Can a light Higgs impostor hide in composite gauge models?*, PoS LATTICE **2013**, 062 (2014) [arXiv:1401.2176 [hep-lat]].
10. Z. Fodor, K. Holland, J. Kuti, S. Mondal, D. Nogradi and C. H. Wong, *Toward the minimal realization of a light composite Higgs*, PoS LATTICE **2014**, 244 (2015).
11. Z. Fodor, K. Holland, J. Kuti, S. Mondal, D. Nogradi and C. H. Wong, *Electroweak interactions and dark baryons in the sextet BSM model with a composite Higgs particle*, [arXiv:1601.03302 [hep-lat]].
12. K. Kainulainen, K. Tuominen and J. Virkajarvi,*The WIMP of a Minimal Technicolor Theory*, Phys. Rev. D **75**, 085003 (2007) [hep-ph/0612247].
13. R. Foadi, M. T. Frandsen, T. A. Ryttov and F. Sannino, *Minimal Walking Technicolor: Set Up for Collider Physics*, Phys. Rev. D **76**, 055005 (2007) [arXiv:0706.1696 [hep-ph]].
14. M. Antola, M. Heikinheimo, F. Sannino and K. Tuominen, *Unnatural Origin of Fermion Masses for Technicolor*, JHEP **1003**, 050 (2010) [arXiv:0910.3681 [hep-ph]].
15. K. Kainulainen, J. Virkajarvi and K. Tuominen, *Superweakly interacting dark matter from the Minimal Walking Technicolor*, JCAP **1002**, 029 (2010) [arXiv:0912.2295 [astro-ph.CO]].
16. E. Witten, *An SU(2) Anomaly*, Phys. Lett. B **117**, 324 (1982).
17. S. L. Adler and W. A. Bardeen, *Absence of higher order corrections in the anomalous axial vector divergence equation*, Phys. Rev. **182**, 1517 (1969).
18. R. S. Chivukula and T. P. Walker, *Technicolor Cosmology*, Nucl. Phys. B **329**, 445 (1990).
19. P. Langacker and G. Steigman, *Requiem for an FCHAMP? Fractionally CHArged, Massive Particle*, Phys. Rev. D **84**, 065040 (2011) [arXiv:1107.3131 [hep-ph]].
20. S. Borsanyi *et al.*, *Ab initio calculation of the neutron-proton mass difference*, Science **347**, 1452 (2015) [arXiv:1406.4088 [hep-lat]].
21. E. Aprile *et al.* [XENON100 Collaboration],*Dark Matter Results from 225 Live Days of XENON100 Data*, Phys. Rev. Lett. **109**, 181301 (2012) [arXiv:1207.5988 [astro-ph.CO]].

22. D. S. Akerib *et al.* [LUX Collaboration], *First results from the LUX dark matter experiment at the Sanford Underground Research Facility*, Phys. Rev. Lett. **112**, 091303 (2014) [arXiv:1310.8214 [astro-ph.CO]].
23. Z. Fodor, K. Holland, J. Kuti, D. Nogradi and C. H. Wong, *The chiral condensate from the Dirac spectrum in BSM gauge theories*, PoS LATTICE **2013**, 089 (2014) [arXiv:1402.6029 [hep-lat]].
24. Z. Fodor, K. Holland, J. Kuti, S. Mondal, D. Nogradi and C. H. Wong, PoS LATTICE **2014**, 270 (2015).
25. C. Morningstar and M. J. Peardon, *Analytic smearing of SU(3) link variables in lattice QCD*, Phys. Rev. D **69**, 054501 (2004).
26. Y. Aoki, Z. Fodor, S. D. Katz, K. K. Szabo, *The Equation of state in lattice QCD: With physical quark masses towards the continuum limit*, JHEP **0601**, 089 (2006).
27. C. Urbach, K. Jansen, A. Shindler, U. Wenger,*HMC algorithm with multiple time scale integration and mass preconditioning*, Comput. Phys. Commun. **174**, 87-98 (2006).
28. T. Takaishi, P. de Forcrand,*Testing and tuning new symplectic integrators for hybrid Monte Carlo algorithm in lattice QCD*, Phys. Rev. **E73**, 036706 (2006).
29. W. J. Lee and S. R. Sharpe, *Partial Flavor Symmetry Restoration for Chiral Staggered Fermions*, Phys. Rev. D **60**, 114503 (1999)
30. Z. Fodor, K. Holland, J. Kuti, D. Nogradi, C. Schroeder, *Twelve massless flavors and three colors below the conformal window*, Phys. Lett. B **703**, 348 (2011) [arXiv:1104.3124 [hep-lat].
31. C. Aubin and C. Bernard, *Pion and Kaon masses in Staggered Chiral Perturbation Theory*, Phys. Rev. D **68**, 034014 (2003)
32. S. R. Sharpe and R. S. Van de Water, *Staggered chiral perturbation theory at next-to-leading order*, Phys. Rev. D **71**, 114505 (2005) [hep-lat/0409018].
33. M. Gell-Mann, R. J. Oakes and B. Renner, *Behavior of current divergences under SU(3) x SU(3)*, Phys. Rev. **175**, 2195 (1968).
34. T. Banks and A. Casher,*Chiral Symmetry Breaking In Confining Theories*, Nucl. Phys. B **169**, 103 (1980).
35. Z. Fodor, K. Holland, J. Kuti, S. Mondal, D. Nogradi and C. H. Wong, in preparation.
36. L. Giusti and M. Luscher,*Chiral symmetry breaking and the Banks-Casher relation in lattice QCD with Wilson quarks*, JHEP **0903**, 013 (2009) [arXiv:0812.3638 [hep-lat]].
37. M. E. Peskin and T. Takeuchi,*Estimation of oblique electroweak corrections*, Phys. Rev. D **46**, 381 (1992).
38. R. Foadi, M. T. Frandsen and F. Sannino,*125 GeV Higgs from a not so light Technicolor Scalar*, Phys. Rev. D **87**, 095001 (2013) [arXiv:1211.1083 [hep-ph]].
39. G. Cacciapaglia and F. Sannino,*Fundamental Composite (Goldstone) Higgs Dynamics*, JHEP **1404**, 111 (2014) [arXiv:1402.0233 [hep-ph]].

40. S. Di Chiara, R. Foadi, K. Tuominen and S. Tahtinen,*Dynamical Origin of the Electroweak Scale and the 125 GeV Scalar*, arXiv:1412.7835 [hep-ph].
41. Z. Fodor, K. Holland, J. Kuti, D. Nogradi and C. H. Wong, *The Yang-Mills gradient flow in finite volume*, JHEP **1211**, 007 (2012) [arXiv:1208.1051 [hep-lat]].
42. M. Luscher, *Properties and uses of the Wilson flow in lattice QCD*, JHEP **1008**, 071 (2010) [arXiv:1006.4518 [hep-lat]].
43. Z. Fodor, K. Holland, J. Kuti, S. Mondal, D. Nogradi and C. H. Wong, *The lattice gradient flow at tree-level and its improvement*, JHEP **1409**, 018 (2014).
44. Z. Fodor, K. Holland, J. Kuti, S. Mondal, D. Nogradi and C. H. Wong, *The lattice gradient flow at tree level*, arXiv:1410.8801 [hep-lat].
45. Z. Fodor, K. Holland, J. Kuti, S. Mondal, D. Nogradi and C. H. Wong,*The running coupling of 8 flavors and 3 colors*, JHEP **1506**, 019 (2015).
46. Z. Fodor, K. Holland, J. Kuti, S. Mondal, D. Nogradi and C. H. Wong,*The running coupling of the minimal sextet composite Higgs model*, JHEP **1509**, 039 (2015).
47. A. Hasenfratz, Y. Liu and C. Y. H. Huang, *The renormalization group step scaling function of the 2-flavor SU(3) sextet model*, arXiv:1507.08260 [hep-lat].
48. A. Cheng, A. Hasenfratz, Y. Liu, G. Petropoulos and D. Schaich, *Improving the continuum limit of gradient flow step scaling*, JHEP **1405**, 137 (2014).
49. C.-J. D. Lin, K. Ogawa and A. Ramos, *The Yang-Mills gradient flow and SU(3) gauge theory with 12 massless fundamental fermions in a colour-twisted box*, JHEP **1512**, 103 (2015) doi:10.1007/JHEP12(2015)103 [arXiv:1510.05755 [hep-lat]].
50. O. Bar and M. Golterman, *Chiral perturbation theory for gradient flow observables*, Phys. Rev. D **89**, no. 3, 034505 (2014) [Erratum-ibid. D **89**, no. 9, 099905 (2014)].
51. G. I. Egri, Z. Fodor, C. Hoelbling, S. D. Katz, D. Nogradi and K. K. Szabo, *Lattice QCD as a video game*, Comput. Phys. Commun. **177**, 631 (2007) [hep-lat/0611022].

Spectral Properties of $N_f = 8$ SU(3) Gauge Theory

Y. Aoki[a], T. Aoyama[a], E. Bennett[b], M. Kurachi[c], T. Maskawa[a], K. Miura[ad], K.-i. Nagai[a], H. Ohki[e], E. Rinaldi[f], A. Shibata[g], K. Yamawaki[a] and T. Yamazaki[h] (LatKMI Collaboration)

[a] *Kobayashi–Maskawa Institute for the Origins of Particles and the Universe (KMI), Nagoya University, Nagoya, Aichi, 464-8602 Japan*

[b] *Department of Physics, Swansea University, Singleton Park, Swansea SA2 8PP UK*

[c] *Institute of Particle and Nuclear Studies, High Energy Accelerator Research Organization (KEK), Tsukuba, Ibaraki 305-0801 Japan*

[d] *Centre de Physique Theoretique (CPT), Aix-Marseille University, Campus de Luminy, Case 907, 163 Avenue de Luminy, 13288 Marseille cedex 9, France*

[e] *RIKEN/BNL Research Center, Brookhaven National Laboratory, Upton, NY 11973-5000 USA*

[f] *Lawrence Livermore National Laboratory, 7000 East Ave., Livermore, CA 94550-9234 USA*

[g] *Computing Research Center, High Energy Accelerator Research Organization (KEK), Tsukuba, Ibaraki 305-0801 Japan*

[h] *Graduate School of Pure and Applied Sciences, University of Tsukuba, Tsukuba, Ibaraki 305-8571, Japan*

QCD with 8 flavors of fundamental fermions has been investigated in LatKMI collaboration. In the series of our studies, we first pointed out the possibility of this theory having walking dynamics. Further, the lowest flavor singlet scalar meson was found as light as pions in the range of quark mass we investigated. In this talk we present an update of the spectral properties of this theory.

Results are later published in Ref. [1].

References

1. Y. Aoki, T. Aoyama, E. Bennett, M. Kurachi, T. Maskawa, K.-i. Nagai, H. Ohki, E. Rinaldi, A. Shibata, K. Yamawaki and T. Yamazaki, Light flavor-singlet scalars and walking signals in $N_f = 8$ QCD on the lattice, *Phys. Rev.* **D96**, p. 014508 (2017).

Investigation of the Scalar Spectrum in SU(3) with Eight Degenerate Flavors

E. Rinaldi
for the Lattice Strong Dynamics (LSD) Collaboration
Nuclear and Chemical Sciences Division, Lawrence Livermore National Laboratory, Livermore, California 94550, USA
E-mail: rinaldi2@llnl.gov

The Lattice Strong Dynamics collaboration is investigating the properties of a SU(3) gauge theory with $N_f = 8$ light fermions on the lattice. We measure the masses of the lightest pseudoscalar, scalar and vector states using simulations with the nHYP staggered-fermion action on large volumes and at small fermion masses, reaching $M_\rho/M_\pi \approx 2.2$. The axial-vector meson and the nucleon are also studied for the same range of fermion masses. One of the interesting features of this theory is the dynamical presence of a light flavor-singlet scalar state with 0^{++} quantum numbers that is lighter than the vector resonance and has a mass consistent with the one of the pseudoscalar state for the whole fermion mass range explored. We comment on the existence of such state emerging from our lattice simulations and on the challenges of its analysis. Moreover we highlight the difficulties in pursuing simulations in the chiral regime of this theory using large volumes.

1. Introduction

There is increasing evidence that non-abelian gauge theories with a large number of flavors display interesting infrared features that are qualitatively different from QCD. While experimental results can help our understanding of QCD, it is much harder to study SU(3) gauge theories with many light flavors. Lattice simulations offer a full non-perturbative description of the low energy regime for these models with controllable and improvable uncertainties.

The Lattice Strong Dynamics (LSD) collaboration is currently performing a high-quality lattice computation of many physical observables that can help our understanding of the infrared dynamics in a SU(3) gauge theory with $N_f = 8$ massless fermions. Such a theory is one of the plausible candidates for a composite Higgs model, based on the walking technicolor scenario[1]. Moreover, it is a strongly-coupled system which differs considerably from QCD, but that we have not completely theoretically understood.

With our investigation of the $N_f = 8$ theory, we are pursuing two main goals. On one hand, we want to use numerical lattice simulations to determine the presence of a light flavor-singlet scalar state in the massless limit of the theory, since this is a major experimental constraint for the theory to be a viable composite Higgs model. On the other hand, we want to study this theory *per se* in order to gain insights on the dynamics of strongly-coupled theories near the edge of the conformal window.

So far there have been a handful of lattice results for the SU(3) gauge theory with $N_f = 8$ massless fermions and they will be discussed and referred to in the next sections. We will show that we are able to improve over existing studies by using lighter fermion masses and larger volumes. This is a first step towards exploring the limit of massless fermions with controllable systematic and statistical errors. We remind the reader that all results included in the following come from a preliminary study, presented at the Sakata memorial workshop on strongly-coupled gauge theories (SCGT15) in March 2015.

2. Lattice setup and scale setting

For our investigation we use the same quark and gauge action of previous USBSM studies[2] focused on the spectrum, and previous works of the Boulder group[3]. The quark action includes two species of degenerate nHYP-smeared staggered quarks with smearing parameter $\alpha = \{0.5, 0.5, 0.4\}$ and bare mass m_f to describe eight degenerate flavors in the continuum theory. The SU(3) gauge action includes a fundamental and an adjoint plaquette term with lattice couplings related by $\beta_F = -4\beta_A$. The main advantages of this action are the cheaper computational cost with respect to the HISQ action while maintaining a good control over taste-splitting effects, and the possibility of exploring stronger couplings.

The strongest coupling that can be simulated with this action is however limited by the presence of a lattice bulk phase transition. We choose a lattice gauge coupling $\beta_F = 4.8$ that is stronger than the one used in a previous USBSM study[2] ($\beta_F = 5.0$) but still outside the bulk lattice phase. We simulate hyper-cubic lattices $L^3 \times T$ with $L = 24,\ 32,\ 48,\ 64$ and $T = 2L$ at small masses $0.00889 \geq m_f \geq 0.00125$. The numerical simulations of this large-volume and small-mass study are handled by the *QHMC* code that is part of the FUEL project[4] and optimized for the BlueGene/Q architecture.

As a first step we monitor the energy E and the topological charge Q evolution at different Wilson flow times[5]. These gauge observables are used to get a quantitative idea of the quality of the ensembles in terms, for example, of topological tunneling and autocorrelation times. We observe frequent tunnelling between different topological sectors, even at our lightest quark mass, which is a reassuring sign of ergodicity in the HMC dynamics. This is shown in the left panel of Fig. 1.

We use the energy measurements $E(t)$ along the Wilson flow to define a lattice scale t_0 which is the Wilson flow time satisfying the following equation: $t^2\langle E(t)\rangle = c$, where $c = 0.3$ is chosen for reference in the following. The choice of c is arbitrary and it depends on the system under scrutiny. One interesting observation is that in this system the scale t_0 is heavily dependent on the fermion mass m_f, contrary to the behavior observed at lower number of flavors.

The same lattice scale t_0 is defined on the USBSM ensembles at $\beta_F = 5.0$ and plotted in the right panel of Fig. 1. By defining a reference length $a_{\mathrm{ref}} = \sqrt{8t_0}$

and using dimensionless observables like $a_{\text{ref}}M_\pi$ we will compare the spectrum from different ensembles. Moreover we will add results obtained with the HISQ action[6] to the comparison.

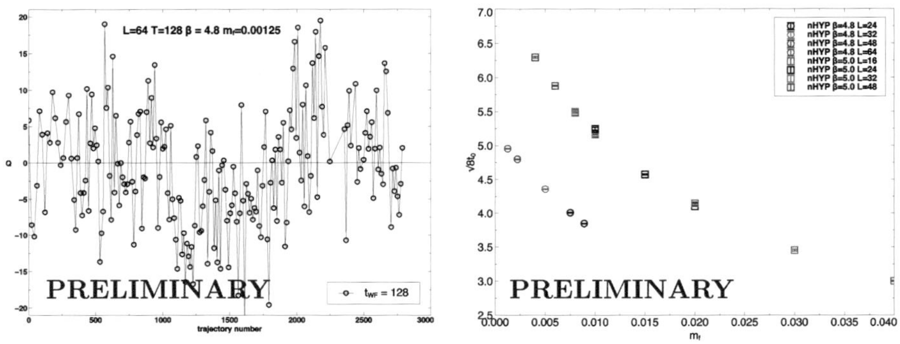

Fig. 1. Left panel: topological charge evolution on the $L = 64$ $T = 128$ $m_f = 0.00125$ ensemble measured at Wilson flow time $t_{\text{WF}} = 128$ corresponding to a *smearing* radius of $\sqrt{8t_{\text{WF}}} = L/2$. Right panel: lattice scale t_0 for our $\beta_F = 4.8$ ensembles and for the USBSM ensembles at $\beta_F = 5.0$. There is clear evidence of a strong fermion mass dependence, which is absent in simulations with $N_f = 2$ or 4.

3. Exploring the connected spectrum towards the chiral limit

For all fermion masses at $\beta_F = 4.8$ we collect more than $\mathcal{O}(10000)$ HMC trajectories, except for $m_f = 0.00125$ at $L = 64$, where we are only able to generate ~ 2000 trajectories due to limited computational resources at this time. The lattice volumes and masses are chosen with $M_\pi L \geq 5.3$ to keep volume effects under control and still be able to reach very light quark masses with available resources.

For the first time we are able to explore, on large volumes, a region of fermion masses where the vector meson (ρ) is heavier than twice the pseudoscalar meson (π) while previous state-of-the-art simulations with the HISQ staggered action[7] and more costly domain-wall-fermion simulations[8] could only achieve $M_\rho/M_\pi \lesssim 1.5$.

This is a needed and very important step forward in order to understand the long-distance behavior of this gauge theory towards the chiral limit, where $M_\rho/M_\pi \to \infty$ if chiral symmetry is spontaneously broken as in QCD. A clear picture of the progress in this direction is reported in the left panel of Fig. 2, where we collect our results at $\beta_F = 4.8$, together with the ones from the nHYP study at $\beta_F = 5.0$[2] and from a HISQ study[7]. We are able to qualitatively compare on the same plot different coupling constants for the nHYP action, as well as different lattice discretization, thanks to a common reference scale a_{ref} (see previous section) which we use to rescale the bare fermion mass m_f. The same plot includes the value of M_ρ/M_π for a SU(3) gauge theory with $N_f = 12$[9] which is known to have an infrared-conformal fixed point[10] and displays hyperscaling when it is mass-deformed such that all mass ratios are constant in the massless limit where all scales disappear.

The right panel of Fig. 2 shows a summary of the connected spectrum on all our ensembles. The lightest state is the pseudoscalar one, followed by the non-singlet scalar (a_0) which becomes degenerate with the vector at large fermion masses. Heavier states include the axial-vector meson (a_1) and the nucleon (N).

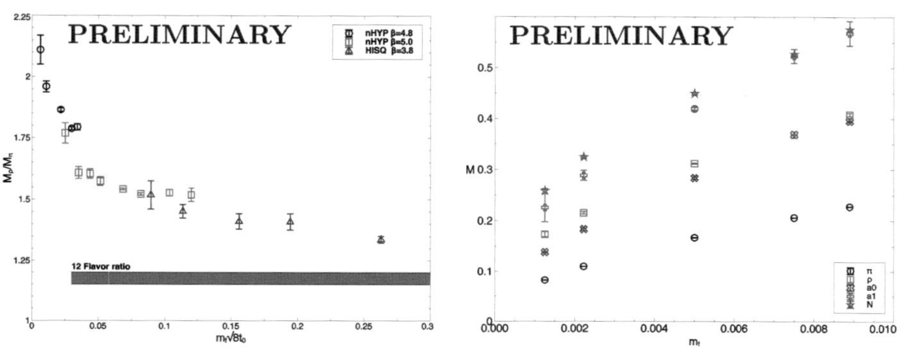

Fig. 2. Left panel: The ratio M_ρ/M_π would diverge in the massless limit of a chirally broken theory. Here we collect this ratio for different gauge couplings and lattice actions, using a rescaled bare fermion mass $a_{\text{ref}}m_f$ that allows for a direct comparison. The value of the same ratio which is constant for a $N_f = 12$ theory is also shown for comparison. Right panel: The spectrum of connected two-point functions for all the fermion masses and volumes investigated in this study includes the pseudoscalar, vector, scalar, axial-vector and nucleon states.

4. The flavor-singlet scalar channel

One of the most interesting spectroscopic observables in the SU(3) gauge theory with $N_f = 8$ flavors is the study of the flavor-singlet scalar channel, where a particle with the same quantum numbers of the Higgs boson (0^{++}) has been found to be as light as the pseudoscalar π state[6]. Such a 0^{++} state happens to be dynamically light in other gauge theories with a variety of different number of flavors[11,12] and even fermion representations[13–15] for SU(3) and SU(2): the common feature among these theories is that they are all close or inside the conformal window, where their beta function would have a non-trivial infrared conformal fixed point.

Studying the 0^{++} channel requires extra care due to the non-zero overlap with the gauge vacuum fluctuations and the presence of disconnected contributions. From previous QCD studies, it is known that a larger amount of gauge configurations is needed compared to the connected correlation functions. Moreover, specific techniques to reduce the computational cost associated with disconnected diagrams have to be employed. There has been a great deal of development in computing all-to-all fermionic propagators needed for this type of calculations and in particular we use U(1) full-volume stochastic sources with full dilution[16] in time and color, and with even-odd dilution in space. The scalar interpolating operator

$\mathcal{O}_S(t) = \bar{\psi}\psi(t)$ used in our measurements and the technical details of constructing its correlators are the same used for a companion project[12].

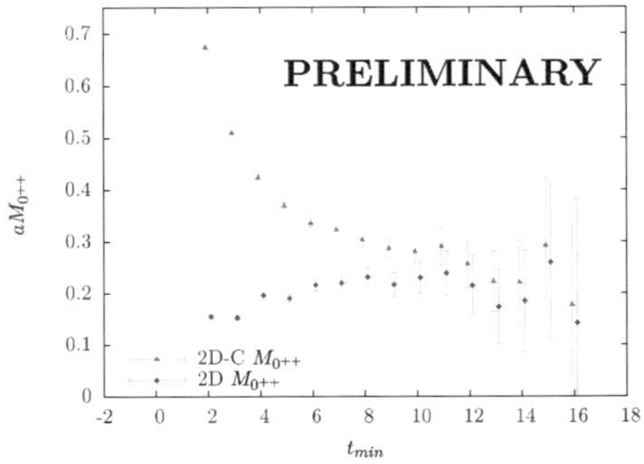

Fig. 3. A fit to the $2D(t) - C(t)$ and $2D(t)$ correlators for $L = 24$ and $m_f = 0.00889$ using the function in Eq. (3) and for different fitting windows $[t_{\rm min}, T/2]$. At large $t_{\rm min}$ both fits give a consistent answer, albeit with large statistical errors.

Let us define two primary correlation functions, $C(t)$ and $D(t)$, which represent the fully-connected correlator and the vacuum-subtracted disconnected correlator, respectively. While the connected $C(t)$ correlator contains contributions from non-singlet states, the disconnected $D(t)$ one contains only singlet contributions. Due to the staggered nature of the fermionic operators, both scalar correlation functions couple to the respective pseudoscalar parity-partner as well. However, if we call σ the lightest state in the scalar channel, the following two combinations will feature an asymptotic large-time behavior dominated by M_σ:

$$S(t) \equiv 2D(t) - C(t) \approx c_\sigma e^{-M_\sigma t} \qquad \text{when } t \to \infty \tag{1}$$

$$2D(t) \approx c_\sigma e^{-M_\sigma t} \qquad \text{when } t \to \infty \,, \tag{2}$$

where the factor of two in front of $D(t)$ represents the number of staggered species. We fit both $S(t)$ and $2D(t)$ with the following fit function

$$F(t) = c_{0^{++}} \cosh M_{0^{++}}(t - T/2) + (-1)^t \, c_1 \cosh M_1(t - T/2) + v_0 \,, \tag{3}$$

that includes a primary state with mass $M_{0^{++}}$, a parity partner state with mass M_1. We also include a free constant parameter v_0 representing our inability to precisely resolve the vacuum contribution in $D(t)$ with our limited statistics, in particular at large temporal distance. The plot in Fig. 3 shows how $M_{0^{++}}$ depends on the window of time-slices $[t_{\rm min}, T/2]$ considered in the fit for both correlators. When contributions from excited states at low $t_{\rm min}$ are removed, we note that $S(t)$ and $2D(t)$ contain a propagating ground state with the same mass: this mass is identified as M_σ. This comparison is reassuring and it provides, *a posteriori*, a justification for using, in the following, the results coming from $2D(t)$ given that this correlation function appears to be less contaminated by excited states and with a smaller statistical error.

In Fig. 4 we show a summary of M_π, M_σ and M_ρ on our ensembles. In the left panel, these masses are compared to the $2M_\pi$ threshold: for all masses but the lightest two, the ρ is below threshold, while σ is always below threshold. Moreover, at light fermion masses, M_σ is compatible with M_π, except for the lightest one where calculations have not yet been performed, even though statistical errors are large. This particular result confirms previous observations by the LatKMI collaboration[6] and a comparison is shown in the right panel of Fig. 4, using a common reference scale.

It is clear that this work explores a regime with considerably lighter fermions. Nonetheless a light flavor-singlet 0^{++} scalar continues to be a prominent feature of the spectrum. Whether this feature survives in the massless limit is one of the major questions we are addressing using the large computational resources at our disposal. However, going to lighter fermion masses requires a significant increase in CPU time and might not be feasible.

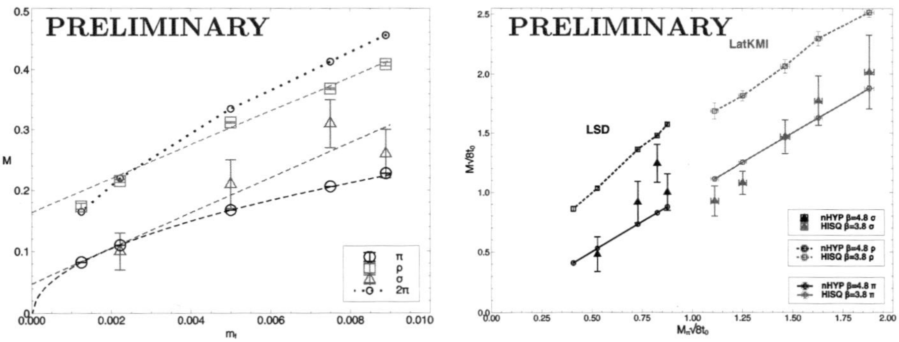

Fig. 4. Left panel: The low-lying spectrum of the eight-flavor theory, including the pseudoscalar π, the vector ρ and the 0^{++} flavor-singlet scalar σ state. The dotted line represents the 2π threshold. Dashed lines are for illustration purpose only and guide the eye toward the massless limit of the theory. Right panel: Comparison, in a_{ref} units, between the π, ρ and σ masses in this work with nHYP fermions (LSD) and in the work with HISQ fermions[6] (LatKMI). The pseudoscalar mass is used as an indication of the distance from the massless limit. Solid lines are π, dotted lines are ρ and triangles are σ.

5. Conclusions

With the goal to shed light on the $N_f = 8$ SU(3) theory as a phenomenologically viable composite Higgs model and to understand its low-energy dynamics, the LSD collaboration is pursuing large scale simulations of the zero-temperature hadron spectrum with emphasis on the flavor-singlet scalar particle. While the results presented in the previous sections are still preliminary, it is outstanding that a unified picture is beginning to emerge by combining efforts from different collaborations.

There are still unanswered questions ranging from the behavior of the non-perturbative beta function with massless quark, to the fate of the 0^{++} state in the chiral limit. Moreover, the existence of a finite-temperature phase transition is still an open question[17]. Our future plans, beside completing the analyses presented here, include a careful comparison of the scalar spectrum and of the beta function with $N_f = 4$, which is our template for QCD. Moreover, we want to expand our previous study[8] of the S parameter to investigate the electroweak constraints on this theory in the massless limit.

Acknowledgments

We are grateful to the LatKMI collaboration for sharing their unpublished results on the Wilson flow scale t_0. We thank Anna Hasenfratz and all the members of the LSD collaboration who contributed to the studies discussed in this proceeding. We acknowledge the Lawrence Livermore National Laboratory (LLNL) Multiprogrammatic and Institutional Computing program for Grand Challenge allocations on the LLNL BlueGene/Q supercomputer. Additional numerical analyses were carried out on clusters at LLNL, Boston University and DOE-funded USQCD resources at Fermilab. ER was supported by LLNL LDRD 13-ERD-023 "Illuminating the Dark Universe with PetaFlops Supercomputing" and acknowledges the support of DOE under Contract DE-AC52-07NA27344 (LLNL).

References

1. K. Yamawaki, M. Bando and K. I. Matumoto, Scale-invariant hypercolor model and a dilaton, *Physical Review Letters* **56**, 1335 (1986).
2. D. Schaich, Eight light flavors on large lattice volumes, *PoS* **LATTICE2013**, p. 072 (2014).
3. A. Cheng, A. Hasenfratz and D. Schaich, Novel phase in SU(3) lattice gauge theory with 12 light fermions, *Physical Review D* **85**, p. 94509 (May 2012).
4. J. Osborn, The FUEL code project, *PoS* **LATTICE2014**, p. 028 (2014).
5. M. Lüscher, Properties and uses of the Wilson flow in lattice QCD, *JHEP* **08**, p. 071 (2010), [Erratum: JHEP03,092(2014)].
6. Y. Aoki, T. Aoyama, M. Kurachi, T. Maskawa, K. Miura, K.-i. Nagai, H. Ohki, E. Rinaldi, A. Shibata, K. Yamawaki and T. Yamazaki, Light composite scalar in eight-flavor QCD on the lattice, *Physical Review D* **89**, p. 111502 (June 2014).
7. Y. Aoki, T. Aoyama, M. Kurachi, T. Maskawa, K.-i. Nagai, H. Ohki, A. Shibata, K. Yamawaki and T. Yamazaki, Walking signals in N_f=8 QCD on the lattice, *Phys. Rev. D* **87**, p. 94511 (May 2013).
8. T. Appelquist, R. C. Brower, G. T. Fleming, J. Kiskis, M. F. Lin, E. T. Neil, J. C. Osborn, C. Rebbi, E. Rinaldi, D. Schaich, C. Schroeder, S. Syritsyn, G. Voronov, P. Vranas, E. Weinberg and O. Witzel, Lattice simulations with eight flavors of domain wall fermions in SU(3) gauge theory, *Physical Review D* **10**, 1 (May 2014).

9. A. Cheng, A. Hasenfratz, Y. Liu, G. Petropoulos and D. Schaich, Finite size scaling of conformal theories in the presence of a near-marginal operator, *Physical Review D* **90**, p. 014509 (July 2014).
10. A. Cheng, A. Hasenfratz, Y. Liu, G. Petropoulos and D. Schaich, Improving the continuum limit of gradient flow step scaling, *Journal of High Energy Physics* **2014**, p. 137 (May 2014).
11. Y. Aoki, T. Aoyama, M. Kurachi, T. Maskawa, K.-i. Nagai, H. Ohki, E. Rinaldi, A. Shibata, K. Yamawaki and T. Yamazaki, Light composite scalar in twelve-flavor QCD on the lattice, *Phys. Rev. Lett.* **111**, p. 162001 (2013).
12. R. Brower, A. Hasenfratz, C. Rebbi, E. Weinberg and O. Witzel, Targeting the conformal window with 4+8 flavors, *PoS* **LATTICE2014**, p. 254 (2014).
13. Z. Fodor, K. Holland, J. Kuti, D. Nogradi and C. H. Wong, Can a light Higgs impostor hide in composite gauge models?, *PoS* **LATTICE2013**, p. 062 (2014).
14. L. Del Debbio, B. Lucini, A. Patella, C. Pica and A. Rago, The infrared dynamics of Minimal Walking Technicolor, *Phys. Rev. D* **82**, p. 14510 (2010).
15. A. Athenodorou, E. Bennett, G. Bergner and B. Lucini, Infrared regime of SU(2) with one adjoint Dirac flavor, *Phys. Rev.* **D91**, p. 114508 (2015).
16. J. Foley, K. Jimmy Juge, A. O'Cais, M. Peardon, S. M. Ryan and J.-I. Skullerud, Practical all-to-all propagators for lattice QCD, *Comput. Phys. Commun.* **172**, 145 (2005).
17. D. Schaich, A. Hasenfratz and E. Rinaldi, Finite-temperature study of eight-flavor SU(3) gauge theory, in *Sakata Memorial KMI Workshop on Origin of Mass and Strong Coupling Gauge Theories (SCGT15) Nagoya, Japan, March 3-6, 2015*, 2015.

Strongly Coupled Gauge Theories: What Can Lattice Calculations Teach Us?

A. Hasenfratz

Department of Physics, University of Colorado,
Boulder, CO 80309, USA
E-mail: Anna.Hasenfratz@colorado.edu

R. C. Brower, C. Rebbi and E. Weinberg

Department of Physics and Center for Computational Science, Boston University,
Boston, MA 02215, USA

O. Witzel

Higgs Centre for Theoretical Physics, School of Physics & Astronomy,
The University of Edinburgh, Edinburgh, EH9 3FD, UK

The dynamical origin of electroweak symmetry breaking is an open question with many possible theoretical explanations. Strongly coupled systems predicting the Higgs boson as a bound state of a new gauge-fermion interaction form one class of candidate models. Due to increased statistics, LHC run II will further constrain the phenomenologically viable models in the near future. In the meanwhile it is important to understand the general properties and specific features of the different competing models.

In this work we discuss many-flavor gauge-fermion systems that contain both massless (light) and massive fermions. The former provide Goldstone bosons and trigger electroweak symmetry breaking, while the latter indirectly influence the infrared dynamics. Numerical results reveal that such systems can exhibit a light 0^{++} isosinglet scalar, well separated from the rest of the spectrum. Further, when we set the scale via the *vev* of electroweak symmetry breaking, we predict a 2 TeV vector resonance which could be a generic feature of SU(3) gauge theories.

Keywords: Composite Higgs, 2 TeV resonance, Lattice field theory, 4+8 fundamental flavors, SU(3) gauge theory.

1. Introduction

The Standard Model is a very successful description of electromagnetic, weak and strong interactions, yet the dynamical nature of its central feature, electroweak symmetry breaking (EWSB), has remained elusive. While experimental data strongly constrains physics Beyond the Standard Model (BSM), many options are still viable. New LHC results, such as the recently published di-boson resonance around 2 TeV, hint at new interactions[1,2]. Increased statistics of LHC run II might finally reveal the origin of the Higgs boson and give insight into BSM physics.

Strongly coupled gauge-fermion systems are among the major contenders to describe BSM dynamics. In these models a new strongly interacting sector gives rise to a new set of hadronic states that lead to experimentally verifiable predictions. While weakly coupled "QCD-like" systems are not compatible with electroweak precision measurements, recent non-perturbative lattice calculations show that strongly coupled systems, especially those near the conformal window, have properties that are very different from the weakly coupled models, making them good BSM candidates. There are several promising systems with different gauge and fermion content that have received significant attention lately, see e.g.[3–8]. In this paper we describe a simple model that, while maybe not phenomenologically viable, can serve as basis of future, more sophisticated, and phenomenologically better motivated investigations. Preliminary results of our study have been presented in[9,10].

The strongly coupled BSM candidate models we are interested in must exhibit spontaneous chiral symmetry breaking. Their infrared (IR) spectrum contains three massless Goldstone pions and a relatively light 0^{++} scalar, in addition to a tower of additional massive states. When coupled to the electroweak sector the Goldstone pions become the longitudinal components of the W and Z bosons, triggering electroweak symmetry breaking. In this scenario the pseudoscalar (pion) decay constant F_π sets the scale of the BSM model, $F_\pi \approx 250$ GeV, the *vev* of EWSB. Among the massive states the 0^{++} meson plays the role of the Higgs boson, while the other states predicted by the model appear as heavy resonances at higher energies. Even though radiative corrections can significantly alter the mass of the 0^{++} scalar, for phenomenological applications it is desirable to have the isosinglet scalar well separated from the higher excitations.

If the model has only two massless fermions, there are only three Goldstone pions in the IR. However many models rely on more fermion flavors to push the system closer to the conformal window where non-perturbative effects can change the IR dynamics. In this case the IR spectrum contains many more Goldstone pion states than are required for EWSB. Those massless states have to acquire mass via a sequence of symmetry breaking steps between the UV and IR energy regions. While the explicit mechanism is model dependent and can be complicated, we can approximate the steps by lifting the mass of all but two fermions with explicit mass terms allowing them to decouple in the IR limit.

Theories that are either chirally broken or conformal but close to the conformal window can look very similar in a finite volume lattice study. Nevertheless, it is essential to ensure that the BSM candidate model breaks chiral symmetry spontaneously. If a gauge-fermion system turns out to be conformal, it has to be driven below the conformal window. This can easily be done by lifting the fermion masses in many-fermion systems as we do in this work. Alternatively, other interactions, possibly 4-fermion terms, may be added to ensure spontaneous chiral symmetry breaking.

2. The 4+8 flavor model

The model we consider in this work is based on the SU(3) gauge theory with $N_f = 12$ fundamental flavors. This system is conformal when all 12 flavors are massless[11–15]. Ideally we would like to lift the mass of 10 of the flavors to end up with a system that contains only 3 Goldstone pions in the IR. In this pilot study, however, we give mass only to 8 fermion species, keeping 4 in the chiral limit. The reason for this choice is technical. We use staggered fermions that naturally come in multiplets of four. It is possible to split the masses of the multiplets by taking the fourth root of the fermion determinant but taking the continuum limit requires special care, especially in strongly coupled systems. To avoid any potentially uncontrolled lattice artifacts we decided to consider a system with 4 massless (or at least light) and 8 heavy flavors.

Even if we kept only two fermions light, the $N_f = 12$ model is not ideal as it has a relatively small mass anomalous dimension. As we will argue in the next section, the anomalous dimension dictates scaling in the mass-split system as well. Systems closer to the conformal boundary, like $N_f = 10$ or perhaps 8 flavors could lead to phenomenologically more interesting systems. We are considering to investigate systems with 2+6 or 2+8 flavors in the future.

3. Wilson renormalization group description

The Wilson renormalization group (RG) approach applied for conformal and mass deformed conformal systems predicts the general scaling behavior of all dimensional quantities. In the following brief review we follow the notation of Ref.[16], but extend that approach by allowing only a subset of the fermions to become massive while keeping the rest in the massless chiral limit.

Consider a 4-dimensional gauge-fermion system characterized by a set of gauge couplings g_i and masses m_i. Perturbatively, the gauge couplings are dimensionless while the dimension of the fermion mass is carried by the lattice cutoff. For convenience we introduce

$$\widehat{m}_i = am_i \propto m_i/\Lambda_a, \tag{1}$$

with $\widehat{m}_i$ the dimensionless "lattice mass" and $\Lambda_a = \pi/a$ the lattice cutoff.

At the perturbative UVFP one of the gauge couplings, g_1, is marginally relevant (the usual asymptotically free gauge coupling), while all other gauge couplings are irrelevant. The $\widehat{m}_i$ mass couplings are relevant at the perturbative UVFP and we assume they will stay relevant everywhere in the parameter space considered.

We choose the number of flavors such that the system is still asymptotically free but inside the conformal window, i.e. it has an IRFP at ($g_i = g_i^\star$, $\widehat{m}_i = 0$) where all of the gauge couplings are irrelevant. If we tune the system to the conformal FP, it will be conformal at any energy range. Away but still close to the IRFP in the gauge couplings, at ($g_i \approx g_i^\star$, $\widehat{m}_i = 0$), there are corrections to conformal behavior.

In the IR those die out and the system is conformal everywhere. Finite mass breaks conformality. If $\widehat{m}_i \gtrsim 0$ we have a mass-deformed system.

For the remainder of this section we will assume that the system is in the vicinity of the IRFP, i.e. the bare parameters as defined at the cutoff scale Λ_a are ($g_i \approx g_i^\star$, $\widehat{m}_i \gtrsim 0$) and we can use Wilsonian RG equations to describe the parameter and energy scale dependence of any 2-point correlation function

$$C_H(t; g_i, \widehat{m}_i, \mu = \Lambda_a) = \int d^3x \langle H(t,x) H(0)^\dagger \rangle \Big|_{g,\widehat{m},\Lambda_a} . \tag{2}$$

In the vicinity of the IRFP an RG transformation that changes the scale

$$\mu \to \mu' = \mu/b, \quad b > 1 \tag{3}$$

transforms the parameters as

$$\begin{aligned}(g_i(\mu) - g_i^\star) \to (g_i(\mu') - g_i^\star) &= b^{y_{g_i}} (g_i(\mu) - g_i^\star), \\ \widehat{m}_i(\mu) \to \widehat{m}_i(\mu') &= b^{y_{m_i}} \widehat{m}_i(\mu)\end{aligned} \tag{4}$$

while the correlation function changes as

$$C_H(t; g_i, \widehat{m}_i, \mu) = b^{-2\gamma_H} C_H(t; g_i', \widehat{m}_i', \mu'). \tag{5}$$

Here γ_i denotes the anomalous dimension while $y_i = d_i + \gamma_i$ is the scaling dimension of the corresponding operator at the IRFP. Finally we rescale all dimensional quantities in Eq. (5) by b and obtain

$$C_H(t; g_i, \widehat{m}_i, \mu) = b^{-2y_H} C_H(t/b; g_i', \widehat{m}_i', \mu), \tag{6}$$

where for simplicity we do not indicate the explicit μ dependence of the couplings.

For the irrelevant couplings the scaling dimensions $y_i < 0$, while for relevant couplings $y_i > 0$. With repeated RG steps b increases and the irrelevant couplings g_i approach $g_i^\star$. The fermion mass parameters, on the other hand, increase as $\widehat{m}_i \to b^{y_{m_i}} \widehat{m}_i$. The fermions decouple from the IR dynamics when $\widehat{m}_i' = \mathcal{O}(1)$ because the masses are now above the cutoff. For concreteness we choose $b = \widehat{m}_i^{-1/y_m}$, so $\widehat{m}(b^{y_m}\mu) = 1$ is the scale where the massive fermions no longer affect the IR dynamics.

It is worth noting that we can define b uniquely only if all massive fermions start with the same bare mass, $\widehat{m}_i(\Lambda_a) = \widehat{m}$. The RG equations become considerably more complicated if the fermions have different bare mass values. There is no complication however if we keep some fermions massless. These fermions will remain massless and present in the IR theory.

If b is large enough we can neglect any dependence on $g' - g_\star$ and the RG equation of Eq. (6) reduces to

$$C_H(t; \widehat{m}_i(\mu), \mu) = \mathcal{C}_H F(t(\widehat{m})^{1/y_m}, \mu), \tag{7}$$

where F is some function that, for fixed μ, depends only on the rescaled variable $t(\widehat{m})^{1/y_m}$. Any 2-point correlation function is expected to show exponential behavior at large distances,

$$C_H(t; \widehat{m}_i, \mu) \propto e^{-M_H t}, \qquad t \to \infty. \tag{8}$$

Combining equations (7) and (8) leads to the well-known (hyper)scaling relation

$$aM_H \propto (\widehat{m})^{1/y_m}. \tag{9}$$

If all fermions are massive, Eqs. (7) and (9) describe the scaling behavior of correlation functions of degenerate fermions. Equations (7) and (9) also describe correlation functions of pure gauge quantities and the scaling of glueballs, the square root of the string tension, etc. In the deep infrared limit the system is an SU(N_c) gauge theory, no dynamical fermions are present. However this system is not identical to what we usually refer to as pure-gauge or quenched SU(N_c) theory. The fermions, while decoupled in the IR, do influence the IR dynamics. As long as the system evolves from the scaling regime of the IRFP, the scaling relations of Eq. (9) hold for any bound states. The dimensionless ratios of physical quantities are expected to be independent of $\widehat{m}$, up to corrections to scaling due to both the irrelevant and relevant couplings.

The situation becomes more complicated if we keep some of the fermions massless, i.e. we have N_l flavors in the $\widehat{m}_\ell = 0$ chiral limit, while we give mass to N_h fermion flavors, $\widehat{m}_h > 0$. Equations (7) and (9) are still valid, but the correlation function $C_H(t)$ can contain operators constructed from massless fermions, massive fermions, or a combination of both. If the number of massless fermions, N_ℓ, is below the conformal window, the system is chirally broken and contains $N_\ell^2 - 1$ massless Goldstone pions in the IR. It also contains light hadronic states, similar to an N_l flavor SU(N_c) theory in the chiral limit. States that contain at least one heavy flavor decouple and are not present in the deep IR limit, nevertheless their influence on the light hadron spectrum remains. All massive states, even those that contain only massless fermions, show hyperscaling with respect to the heavy flavor mass $\widehat{m}_h$. In particular, ratios of masses are expected to be independent of m_h (up to corrections due to scaling violations). While these ratios are governed by the conformal IRFP, their value can depend on N_ℓ, in addition to $N_\ell + N_h$ that determines the scaling dimensions at the IRFP.

3.1. *The running gauge coupling*

We consider first the running coupling of conformal and mass-deformed systems based on the Wilsonian RG discussion of the preceding section.

a) Conformal systems The solid curve in Fig. 1 depicts the running gauge coupling of a conformal system. All fermion masses are $\widehat{m}_i = 0$ and the quantity $\Lambda_a = \pi/a$ denotes the lattice scale where the bare parameter g_0 is set. If we want to define a continuum theory at the perturbative UVFP the bare gauge coupling $g(\Lambda_a)$ has to be tuned to $g(\Lambda_a) \to 0$. As we lower the energy scale from Λ_a towards the IR, the gauge coupling changes according to Eq. (4), i.e. it approaches the IRFP. We denote the scale where the gauge coupling gets close enough to the IRFP such that $g(\mu) - g^\star$ can be neglected by Λ_U, the UV scale where the dynamics governed by the IRFP sets in. The running gauge coupling at this scale takes the value of the IRFP, $g(\mu) \approx g^\star$ and approaches $g^\star$ as $\mu \to 0$.

b) Mass deformed system Next we consider the case where all fermions acquire a small bare mass at the cutoff scale Λ_a. The running of the gauge coupling is only minimally affected until $\widehat{m}_i(\mu) = \mathcal{O}(1)$. We denote the scale where $\widehat{m}_i(\mu) = 1$ as Λ_{IR}, the IR scale where the fermions decouple. As the system runs deeper in the IR, the gauge coupling follows the running coupling of a pure gauge system. This behavior is denoted by the blue dashed curve in Fig. 1. Unless the bare fermion mass is tuned $\widehat{m}(\Lambda_a) \to 0$, in the deep IR a mass deformed system has only gauge degrees of freedom. That does not mean, however, that the system is equivalent to a pure gauge system, as we have discussed already in section 3.

c) Mass deformed systems far from the IRFP If the fermion mass at the cutoff level is too large, it is possible that $\Lambda_{IR} > \Lambda_U$. In that case the fermions decouple before the gauge coupling reaches the IRFP. This scenario is depicted by the red long-dashed lines in Fig. 1. In this case the IR dynamics is close to the pure gauge system with heavy fermions, and is not of interest for us.

d) Mass deformed system with split fermion masses In our final case we keep N_ℓ fermions massless and assign finite, degenerate bare mass $\widehat{m}(\Lambda_a)$ to the others. We will assume that $\widehat{m}(\Lambda_a)$ is small enough that $\Lambda_{IR} \ll \Lambda_U$ and that in the deep IR, where only the N_ℓ massless fermions contribute to the dynamics the system is chirally broken, i.e. N_ℓ is below the conformal window. Qualitatively the running coupling is very similar to that of the mass deformed system as discussed in case *b)*. While the non-zero fermion masses are small the gauge coupling behaves like in a conformal system. At the Λ_{IR} scale the massive fermions decouple, chiral symmetry breaks and the IR system has N_ℓ massless fermions. The running gauge coupling is driven by this dynamics, similarly to the dashed blue line in Fig. 1. Just like in case *b)* this does not mean that the deep IR dynamics is equivalent to an SU(N_c) gauge system with N_ℓ massless fermions. Before decoupling, the additional fermions have influenced the IR dynamics, leading to specific hyper scaling relations as we discussed in section 3.

3.2. *The running coupling of the 4+8 flavor system*

The gradient flow (GF) transformation provides a natural definition of the running coupling that is easy to implement numerically[17–19]. The GF running coupling at

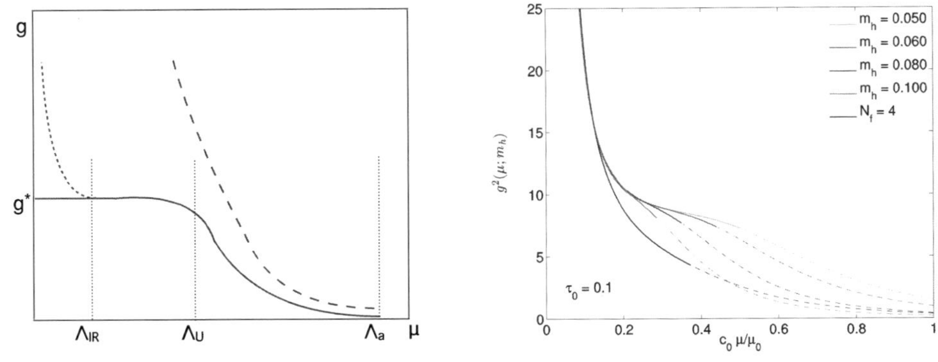

Fig. 1. Left: The expected running gauge coupling of conformal and mass-deformed system. The solid blue curve sketches the evolution of the gauge coupling with the energy scale in a conformal system. The dashed blue curve shows the modification in a mass deformed system while the red long-dashed curve corresponds to a situation where the fermions decouple before the gauge coupling could approach the conformal IRFP. Right: The running coupling constant $\widetilde{g}^2_{\rm GF}$ at the mass scale μ for different values of m_h with m_ℓ extrapolated to the chiral limit. μ_0 and $c_0 = \mu_0^{-1}|_{m_h=0.050}$ serve as normalization constants that ensure that the different systems are compared at matching energy scales and τ_0 is the shift parameter to remove discretization errors. The dashed sections of the lines indicate where we suspect cutoff effects may be significant. The running gauge coupling in our model at four different m_h mass values and in the $m_h = \infty$ 4-flavor limit. The emergence of the walking regime is evident as $m_h \to 0$.

energy scale μ is defined as

$$g^2_{GF}(\mu) = \frac{1}{\mathcal{N}}\langle t^2 E(t)\rangle, \tag{10}$$

where $t = a^2 t_{\rm lat}$ ($t_{\rm lat} \gg 1$) is the flow time that is related to the energy scale as $\mu^{-1} = \sqrt{8t}$. The energy density

$$E(t) = -\frac{1}{2}{\rm ReTr}[G_{\mu\nu}(t)G^{\mu\nu}(t)] \tag{11}$$

can be evaluated by any appropriate lattice operator while the constant $\mathcal{N} = 3(N^2-1)/128\pi^2$ is chosen such that g^2_{GF} matches the traditional $\overline{MS}$ coupling in perturbation theory[19].

We show $g^2_{\rm GF}(\mu)$ as function of the energy μ from our numerical simulation of the 4+8 flavor system in the right panel of Fig. 1. We consider four different values for the heavy flavors and extrapolate to the chiral limit in the light quark masses. The right plot in Fig. 1 also shows the running coupling of the 4-flavor ($m_h = \infty$) system. We rescale μ by the lattice scale $\mu_0^{-1} = \sqrt{8t_0}$ (see[19]) in each case such that the running couplings match in the IR. The dashed curves correspond to flow time where $\sqrt{8t} \leq 2.0$ and cutoff effects could be important. While the choice $\sqrt{8t} = 2.0$ to separate discretization effects from physical behavior is rather arbitrary, the trend as the mass of the 8 heavy flavors decrease from ∞ towards the chiral limit is clear. The resemblance between the RG inspired behavior depicted on the left panel and the numerically observed prediction of the right panel is striking though

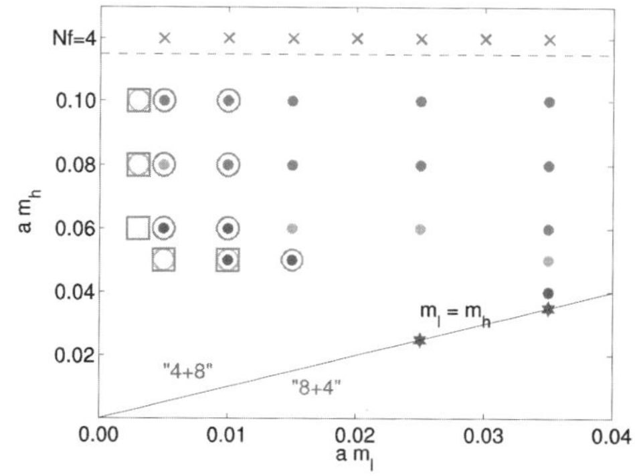

Fig. 2. Light (m_ℓ) and heavy (m_h) mass values for the simulations carried out on $24^3 \times 48$ lattices (filled symbols) and $32^3 \times 64$ lattice (open circles). The colors are meant to caution about finite size effects, likely negligible for green, but of increasing importance as the color turns to orange and red.

not unexpected: both show the theoretically well motivated running coupling of mass deformed systems near the conformal IRFP.

4. Numerical simulations

Numerical simulations of our model with four light and eight heavy flavors are carried out using nHYP-smeared[20] staggered fermions and the plaquette gauge action with fundamental and adjoint terms[12,21]. In previous works[12,21] with eight and twelve fundamental flavors, these actions demonstrated small taste breaking artifacts and with our choice of smearing parameters the numerical simulations are stable everywhere. We generated a large set of gauge field configurations with different bare input masses for the light and the heavy fermions at a single gauge coupling, $\beta = 4.0$. This gauge coupling is close enough to the $N_f = 12$ IRFP that the simulations can be considered to be in the mass deformed region. Starting with small volumes, $L^3 \times T = 24^3 \times 48$, we eventually increased the lattice sizes to $32^3 \times 64$, $36^3 \times 64$, and even $48^3 \times 96$ to study and account for finite volume effects. We present a graphical overview of our set of ensembles in Fig. 2 and indicate by the color the significance of finite volume effects: green (likely negligible), orange (moderate), red (severe). All gauge fields were generated by the hybrid Monte Carlo (HMC) update algorithm[22] using the implementation in the FUEL software package[23].

As usual we monitor algorithmic parameters and physical observables during the evolution of gauge field configurations maintaining e.g. an average acceptance rate of greater 65% on all ensembles. Likewise we measure fermionic ($\langle \psi\bar{\psi} \rangle$) and gluonic (average plaquette) quantities and their dependence on the four space-time coordinates to detect artificial lattice phases. So far all our simulations are consistent with chiral symmetry breaking and do not exhibit any signs of the S4-broken lattice phase[24].

5. Spectrum

We have investigated the light fermion spectrum and the dependence of "hadronic" masses on the heavy input quark mass m_h. As we discussed in Sec. 3 we expect hyperscaling in the heavy mass m_h when the light mass m_ℓ is extrapolated to the chiral limit. Consequently dimensionless ratios should show independence on m_h up to lattice corrections.

In analogy to QCD we investigate "mesonic" quark-antiquark pairs, studying in particular the pseudoscalar (pion, π), vector (rho, ϱ), axial-vector (a_1), and isotriplet scalar (a_0) bound states. Moreover, we determine the lowest mass baryon (three quark bound state), the nucleon (n). These quantities are described by quark-line connected diagrams, thus the numerical determination is straightforward and we obtain our results using propagators with wall-sources. In addition we determine the mass of the isosinglet scalar (0^{++}) meson bound state which has the same quantum numbers as the Higgs boson. Determining the isosinglet scalar is much more challenging because quark-line disconnected diagrams also contribute. These disconnected diagrams are numerically much noisier and require different numerical methods.

5.1. *Connected light fermion spectrum*

Using ensembles with sufficient number of thermalized trajectories and sufficiently large volumes, the left panel of Fig. 3 shows our results for the pion and the rho. We present the masses in units of a lattice reference scale $a_\bigstar$ which we define through the gradient flow $\sqrt{8t_0}$ scale on our $36^3 \times 64$, $m_h = 0.080$, $m_\ell = 0.003$ reference ensemble. Values in lattice units obtained on other ensembles are converted by multiplying appropriate ratios of the gradient flow scale[17,19] which we determine on all ensembles.

As one expects based on the arguments in Sec. 3, the dependence on the input quark mass m_h is weak. Except for small finite volume effects, our measurements of the rho mass shows no dependence on m_h. The pion mass, M_π shows the same overall trend; however a small dependence on m_h, in particular for the heavier m_ℓ data sets is visible. Hyperscaling in m_h is only expected in the $m_\ell = 0$ limit. Our data are consistent with this expectation.

In the right panel of Fig. 3 we show the ratio M_ϱ/M_π as function of the light quark mass. In contrast to a mass deformed conformal system (e.g. 12 flavors[12]), where all hadrons scale the same way with respect to the mass, we observe that for all our choices of m_h the ratio M_ϱ/M_π is consistent with a divergent behavior for $m_\ell \to 0$. We take this as a certain indicator that in our simulations chiral symmetry is broken spontaneously in the limit of vanishing light quark mass.

In addition to the hadronic masses we also determine the pseudoscalar decay constant F_π which is the preferred choice for setting the scale in relation to EWSB. The quantity $F_\pi L$ controlling the chiral perturbative expansion is $1.0 \lesssim F_\pi L \lesssim 1.35$ for all our relevant data sets. However the smallness of this parameter is not the only

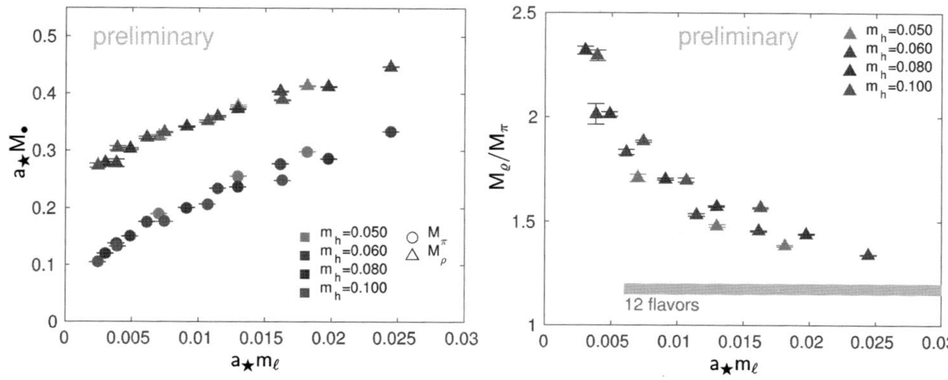

Fig. 3. Left: Pion and rho vs. light input quark mass shown in lattice units $a_\bigstar$ defined on our $36^3 \times 64$, $m_h = 0.080$, $m_\ell = 0.003$ reference ensemble. The rho shows (except for small finite volume effects) no dependence on m_h and the pion exhibits only a very weak dependence, most pronounced at larger m_ℓ values. Right: plotting the ratio M_ϱ/M_π vs. the light input quark mass confirms that for $m_\ell \to 0$ chiral symmetry is broken for all our choices of m_h. As reference the value of M_ϱ/M_π found in the corresponding 12 flavor system[12] is indicated by the grey band.

problem in performing a chiral extrapolation. The 0^{++} state, as will be discussed in the next section, is nearly degenerate with the pion, rendering standard chiral perturbative expansions and consequently chiral extrapolations questionable.

5.2. *Determination of the isosinglet* 0^{++} *scalar*

The determination of the isosinglet 0^{++} meson is more complicated and requires the evaluation of disconnected diagrams. We construct the disconnected operator from the vacuum subtracted operator

$$\mathcal{O}_{\rm disc} = \langle\bar{\psi}\psi\rangle(t) - \langle\langle\bar{\psi}\psi\rangle\rangle_e, \tag{12}$$

and determine the vacuum contribution via the ensemble average $\langle\langle\bar{\psi}\psi\rangle\rangle_e$. The operator $\langle\bar{\psi}\psi\rangle(t)$ is measured using $N_r = 6$ full volume noise sources diluted in time, color, as well as spatially even/odd in order to reduce the stochastic noise.[25]

In systems with degenerate flavors the 0^{++} state is the ground state of the correlator $ND(t) - C(t)$ where $D(t)$ and $C(t)$ are the disconnected and connected correlators of the scalar operator $\langle\bar{\psi}\psi\rangle(t)\,\langle\bar{\psi}\psi\rangle(0)$ and N denotes the number of flavors. When using staggered fermions $N = N_f/4$. In our case the physical 0^{++} correlator is more complicated as it is a mixture of light-light, light-heavy, and heavy-heavy scalar states. In addition the connected correlator couples strongly to excited states, making the determination of the 0^{++} numerically challenging. However, as was pointed out in Ref.[26], the ground state mass can be determined from the disconnected correlator alone. $D(t)$ couples to the 0^{++} state, and since that state is the lightest, it dominates the asymptotic behavior. Using only the disconnected correlator provides a numerically more stable determination of the lowest isosinglet scalar mass and avoids the cumbersome mixing of the light and

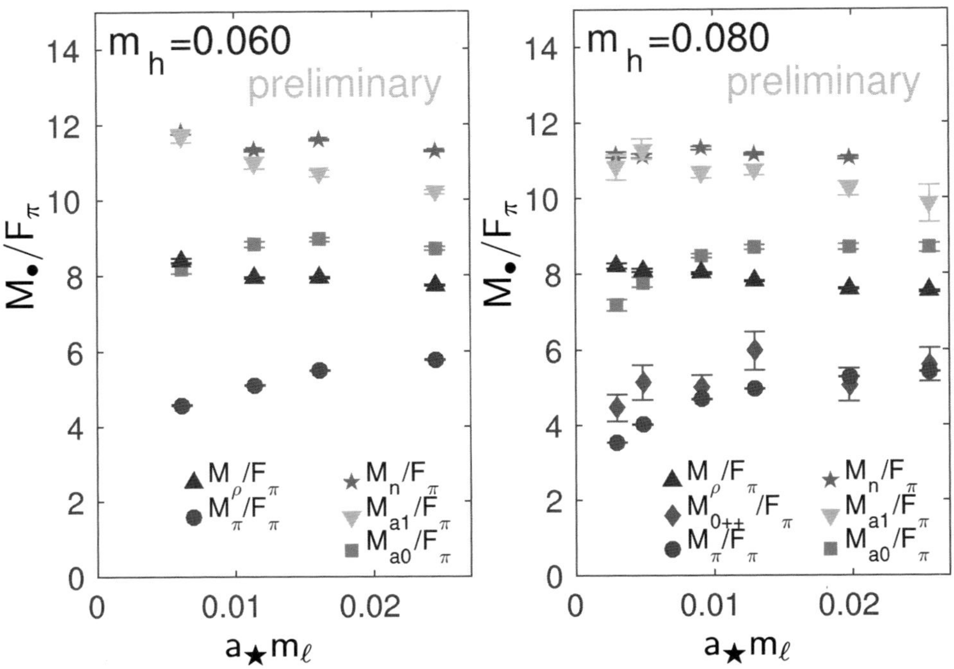

Fig. 4. Pion, rho, scalar 0^{++}, axial, a_0, and nucleon mass of the light flavor spectrum in units of F_π as function of the light quark mass $a_\bigstar m_\ell$ for $m_h = 0.060$ and 0.080. If a chirally broken system triggered EWSB, $F_\pi \approx 250$ GeV would set the correct electroweak scale.

heavy states as well. The vacuum subtraction in Eq. (12) is the source of large statistical fluctuations. In practice we find it is better to fit the finite difference $D(t+1) - D(t)$ of the disconnected correlator and avoid the vacuum subtraction altogether.

We perform measurements every 20 MDTU and have extended almost all of our Monte Carlo evolutions to at least 20,000 MDTU, in some cases even twice as much. Nevertheless the determination of the mass from the correlators is particularly difficult for larger values of m_ℓ. We therefore restrict the presentation of our 0^{++} results to the subset of our ensembles with $m_h = 0.080$. We expect hyperscaling in the $m_\ell = 0$ chiral limit, therefore the 0^{++} mass at any fixed m_h is indicative of its value everywhere in the scaling region of m_h.

6. Summary

The right panel of Fig. 4 summarizes our results for the 0^{++} isosinglet and several other isomultiplet states, normalized by the pion decay constant F_π, as the function of the light fermion mass m_ℓ at $m_h = 0.080$. The left panel shows the isomultiplet states at $m_h = 0.060$ where our statistics is not yet sufficient to determine the mass of the 0^{++} scalar. Hyperscaling, as we argued in Sec. 3, implies that all dimensionless ratios are identical, up to lattice corrections, in the $m_\ell = 0$ chiral limit. Comparing the two panels of Fig. 4 shows that this expectation is satisfied by our data for the connected spectrum. There is no reason to believe that the 0^{++} scalar state would be any different.

Unlike in QCD, we observe a 0^{++} state at $m_h = 0.080$ that is nearly degenerate with the pion and much lighter than the rho. We have not attempted any kind of chiral extrapolation — standard chiral perturbation theory is based on the pion being the only light particle in the spectrum which is clearly not true in our case. Moreover we observe that M_ϱ/F_π, M_{a_1}/F_π, and M_n/F_π have a fairly linear dependence on m_ℓ. In particular the ratio M_ϱ/F_π appears nearly independent of both m_ℓ and m_h with a value around 8. Coincidentially if $F_\pi \approx 250$ GeV, the mass of the vector resonance in our model is around 2 TeV, matching the recently reported signal by ATLAS and CMS collaborations[1,2]. It is a curiosity to note that $M_\varrho/F_\pi \approx 8$ appears to be a rather general feature of SU(3) gauge fermion systems: this ratio is 8.4 in QCD[27]; approximately 8.0 in the conformal SU(3) $N_f = 12$ model[28,29]; about 7.9 with 8 fundamental flavors[30,31]; approximately 8 in $N_f = 2$ flavor SU(3) sextet model[32]. Various theoretical models and approaches also predict that this ratio is independent of the specifics of the IR dynamics, see for example Refs.[33,34].

While we expect the dimensionless ratios to be independent of m_h, these ratios can depend on the details of the system, like the IRFP of the underlying conformal massless model and the number of light vs. heavy flavors. Comparing different systems could reveal which, if any, of these models are potentially compatible with experimental observations.

Our work is just the first step in this direction. Choosing a system with two (or four) massless and many massive fermions ensure that in the IR the model is chirally broken while choosing the total number of fermions such that in the massless limit the system is conformal ensures hyper scaling. While a similar construction is possible even if the underlying mass-degenerate system is not conformal, it is not obvious what kind of scaling relations one can obtain in that case.

In the future we would like to extend these investigation to systems with 2 massless fermions with $N_f = N_\ell + N_h$ closer to the conformal window. Numerical simulations indicate that systems with $N_f = 8$ or 10 fundamental fermions exhibit large mass anomalous dimensions, making them better candidates to satisfy electroweak constraints.

Acknowledgments

Computations for this work were carried out in part on facilities of the USQCD Collaboration, which are funded by the Office of Science of the U.S. Department of Energy, on computers at the MGHPCC, in part funded by the National Science Foundation, and on computers allocated under the NSF Xsede program to the project TG-PHY120002. We thank Boston University, Fermilab, the NSF and the U.S. DOE for providing the facilities essential for the completion of this work. R.C.B., C.R. and E.W. were supported by DOE grant DE-SC0010025 and in addition acknowledge the support of NSF grant OCI-0749300. A.H. acknowledges support by the DOE grant DE-SC0010005. O.W. is supported by STFC, grant ST/L000458/1. R.C.B., A.H. and C.R. thank the Aspen Center for Physics, which is supported by National Science Foundation grant PHY-1066293. R.C.B., A.H., and O.W. thank the KITP, Santa Barbara, supported in part by the National Science Foundation under Grant No. NSF PHY11-25915.

References

1. ATLAS Collaboration, Search for high-mass diboson resonances with boson-tagged jets in proton-proton collisions at $\sqrt{s} = 8$ TeV with the ATLAS detector (2015).
2. CMS Collaboration, Search for massive WH resonances decaying to $\ell\nu b\bar{b}$ final state in the boosted regime at $\sqrt{s} = 8\,\text{TeV}$ (2015).
3. ATLAS Collaboration, Search for high-mass dilepton resonances in pp collisions at $\sqrt{s} = 8$TeV with the ATLAS detector, *Phys. Rev.* **D90**, p. 052005 (2014).
4. A. Arbey, G. Cacciapaglia, H. Cai, A. Deandrea, S. Le Corre and F. Sannino, Fundamental Composite Electroweak Dynamics: Status at the LHC (2015).
5. Y. Aoki *et al.*, Light composite scalar in eight-flavor QCD on the lattice, *Phys.Rev.* **D89**, p. 111502 (2014).
6. A. Hietanen, R. Lewis, C. Pica and F. Sannino, Fundamental Composite Higgs Dynamics on the Lattice: SU(2) with Two Flavors, *JHEP* **07**, p. 116 (2014).
7. Z. Fodor, K. Holland, J. Kuti, S. Mondal, D. Nogradi and C. H. Wong, Toward the minimal realization of a light composite Higgs, *PoS* **LATTICE2014**, p. 244 (2015).
8. D. Schaich, A. Hasenfratz and E. Rinaldi, Finite-temperature study of eight-flavor SU(3) gauge theory (2015).
9. R. Brower, A. Hasenfratz, C. Rebbi, E. Weinberg and O. Witzel, A novel approach to the study of conformality in the SU(3) theory with multiple flavors, *J. Exp. Theor. Phys.* **120**, 423 (2015).
10. R. Brower, A. Hasenfratz, C. Rebbi, E. Weinberg and O. Witzel, Targeting the conformal window with 4+8 flavors, *PoS* **LATTICE2014**, p. 254 (2014).
11. A. Cheng, A. Hasenfratz, G. Petropoulos and D. Schaich, Scale-dependent mass anomalous dimension from Dirac eigenmodes, *JHEP* **1307**, p. 061 (2013).
12. A. Cheng, A. Hasenfratz, Y. Liu, G. Petropoulos and D. Schaich, Finite size scaling of conformal theories in the presence of a near-marginal operator, *Phys.Rev.* **D90**, p. 014509 (2014).
13. M. P. Lombardo, K. Miura, T. J. N. da Silva and E. Pallante, On the particle spectrum and the conformal window, *JHEP* **12**, p. 183 (2014).
14. E. Itou, The anomalous dimension at the infrared fixed point of Nf = 12 SU(3) theory, *PoS* **LATTICE2013**, p. 481 (2013).
15. E. Itou and A. Tomiya, Determination of the mass anomalous dimension for

$N_f = 12$ and $N_f = 9$ SU(3) gauge theories, *PoS* **LATTICE2014**, p. 252 (2014).
16. L. Del Debbio and R. Zwicky, Hyperscaling relations in mass-deformed conformal gauge theories, *Phys. Rev.* **D82**, p. 014502 (2010).
17. R. Narayanan and H. Neuberger, Infinite N phase transitions in continuum Wilson loop operators, *JHEP* **0603**, p. 064 (2006).
18. M. Lüscher, Trivializing maps, the Wilson flow and the HMC algorithm, *Commun.Math.Phys.* **293**, 899 (2010).
19. M. Lüscher, Properties and uses of the Wilson flow in lattice QCD, *JHEP* **1008**, p. 071 (2010).
20. A. Hasenfratz, R. Hoffmann and S. Schaefer, Hypercubic smeared links for dynamical fermions, *JHEP* **0705**, p. 029 (2007).
21. A. Cheng, A. Hasenfratz, G. Petropoulos and D. Schaich, Determining the mass anomalous dimension through the eigenmodes of Dirac operator, *PoS* **LATTICE2013**, p. 088 (2013).
22. S. Duane, A. Kennedy, B. Pendleton and D. Roweth, Hybrid Monte Carlo, *Phys.Lett.* **B195**, 216 (1987).
23. J. Osborn *et al.*, Framework for unified evolution of lattices (FUEL).
24. A. Cheng, A. Hasenfratz and D. Schaich, Novel phase in SU(3) lattice gauge theory with 12 light fermions, *Phys.Rev.* **D85**, p. 094509 (2012).
25. J. Foley, K. Jimmy Juge, A. Ó Cais, M. Peardon, S. M. Ryan and J.-I. Skullerud, Practical all-to-all propagators for lattice QCD, *Comput. Phys. Commun.* **172**, 145 (2005).
26. Y. Aoki *et al.*, Light composite scalar in twelve-flavor QCD on the lattice, *Phys.Rev.Lett.* **111**, p. 162001 (2013).
27. K. A. Olive *et al.*, Review of Particle Physics, *Chin. Phys.* **C38**, p. 090001 (2014).
28. Z. Fodor *et al.*, Twelve massless flavors and three colors below the conformal window, *Phys.Lett.* **B703**, 348 (2011).
29. Y. Aoki, T. Aoyama, M. Kurachi, T. Maskawa, K.-i. Nagai *et al.*, Lattice study of conformality in twelve-flavor QCD, *Phys.Rev.* **D86**, p. 054506 (2012).
30. Y. Aoki *et al.*, Walking signals in $N_f = 8$ QCD on the lattice, *Phys.Rev.* **D87**, p. 094511 (2013).
31. LSD collaboration, in preparation (2015).
32. J. Kuti, The simplest composite higgs candidate with resonance spectrum in the 2-tev range talk at KITP, Santa Barbara "Lattice Gauge Theory for the LHC and Beyond" on Aug. 5, 2015.
33. H. Pagels and S. Stokar, The Pion Decay Constant, Electromagnetic Form-Factor and Quark Electromagnetic Selfenergy in QCD, *Phys. Rev.* **D20**, p. 2947 (1979).
34. T. Appelquist and L. C. R. Wijewardhana, Chiral Hierarchies and Chiral Perturbations in Technicolor, *Phys. Rev.* **D35**, p. 774 (1987).

One-Family Walking Technicolor in Light of LHC Run-II

Shinya Matsuzaki

Department of Physics, Nagoya University, Nagoya 464-8602, Japan

Institute for Advanced Research, Nagoya University, Nagoya 464-8602, Japan

The LHC Higgs can be identified as the technidilaton, a composite scalar, arising as a pseudo Nambu-Goldstone boson for the spontaneous breaking of scale symmetry in walking technicolor. One interesting candidate for the walking technicolor is the QCD with the large number of fermion flavors, involving the one-family model having the eight-fermion flavors. The smallness of the technidilaton mass can be ensured by the generic walking feature, Miransky scaling, and the presence of the "anti-Veneziano limit" characteristic to the large-flavor walking scenario. To tell the standard-model Higgs from the technidilaton, one needs to wait for the precise estimate of the Higgs couplings to the standard model particles, which is expected at the ongoing LHC-Run II. In this talk the technidilaton phenomenology in comparison with the LHC Run-I data is summarized with the special emphasis placed on the presence of the anti-Veneziano limit supporting the lightness of technidilaton. Besides the technidilaton, the walking technicolor predicts the rich particle spectrum such as technipions and technirho mesons, arising as composite particles formed by technifermions. The LHC phenomenology of those technihadrons and the discovery channels are also discussed, which are smoking-guns of the walking technicolor, to be accessible at the LHC-Run II.

1. Introduction

A Higgs boson with the mass about 125 GeV was discovered at the LHC Run I[1] and its coupling property has so far been almost consistent with the Higgs boson predicted in the standard model (SM). Yet the dynamical origin of the Higgs, related to the issue left in the SM (such as the naturalness problem) has been uncovered, which is of great importance to be explored at the ongoing LHC Run II.

The dynamical origin of the electroweak symmetry breaking and the Higgs can elegantly be supplied by so-called technicolor (TC)[2,3]. However, the technicolor based on the naive-scale up version of QCD (QCD-like technicolor) has severely been disfavored by several inconsistencies with experiments. Most dramatically, it was ruled out by the recent discovery of the Higgs at LHC, because of the absence of the light Higgs at around 125 GeV in the QCD-like technicolor.

In sharp contrast, the walking technicolor (WTC)[4,5] predicts a light composite Higgs, which we call the technidilaton (TD). The TD arises as a pseudo Nambu-Goldstone (NG) boson for the spontaneous breaking of the approximate scale symmetry of the WTC, triggered by technifermion condensation. Thus its lightness can potentially be ensured by the approximate scale symmetry inherent to the WTC.

One interesting candidate to realize the walking theory is one-family model with the number of techniflavors $N_F = 8$[3,6], classified into QCD with the large number of flavors. As it will turn out, actually, the large flavor nature of the walking provides us with a limit where the TD can be regarded as an exactly massless NG boson, analogously to the η' in the Veneziano limit of QCD. That is what will be called "anti-Veneziano limit".

The LHC signatures of the TD in the one-family model of WTC were studied[7–13]. It has been shown[9,12,13] that when the walking theory has the number of technicolor $N_C = 4$, the coupling property of the 125 GeV TD is consistent with the LHC Higgs at the almost same level as the SM Higgs.

This talk summarizes the 125 GeV TD phenomenology at the LHC Run I and discusses future prospect in the Run II. The emphasis is also placed on several supports on the theoretical ground that the TD can indeed be as light as the 125 GeV boson due to the intrinsic feature of large flavor WTC. Besides the TD, the walking technicolor predicts the rich particle spectrum such as technipions and technirho mesons, arising as composite particles formed by technifermions, The LHC phenomenology of those technihadrons and the discovery channels are also discussed, which are smoking-guns of the walking technicolor, to be accessible at the LHC-Run II.

2. Characteristic Features of Walking Technicolor

In Fig. 1 a schematic view of the WTC is depicted in terms of the gauge coupling α (left panel) and its beta function $\beta(\alpha)$ (right panel). The walking, almost nonrunning region is literally defined as the domain where the running behavior of α looks like almost constant in scale (or the beta function β looks almost zero in terms of the beta function β). The presence of walking region implies a pseudo infrared fixed point (α_*), a la Caswell-Banks-Zack, based on the two-loop beta function in the large N_f QCD[14]. During the walking region, the gauge coupling slowly reaches the critical coupling, which is slightly off from an infrared fixed point α_*, $\alpha_{\rm cr}(< \alpha_*)$, where the chiral/electroweak symmetry is dynamically broken by technifermion condensation $\langle \bar{F}F \rangle \neq 0$ and hence technifermions get the dynamical mass m_F on the order of TeV, where F_π denotes the technipion decay constant associated with the chiral symmetry breaking. The walking regime ends at two edges: below the infrared scale $m_F \sim \mathcal{O}(\text{TeV})$, technifermions decouple and hence the balance with technigluon contributions gets lost, leading to the one edge of walking. The other edge is above the ultraviolet scale $\Lambda_{\rm TC}(\sim 10^3 - 10^4$ TeV) the theory will be embedded into an extended technicolor (ETC)[15].

The dynamical mass generation at $\alpha = \alpha_{\rm cr}$ is characterized by so-called Miransky scaling[16], tied with the conformal phase transition[17],

$$m_F \sim \Lambda_{\rm TC} e^{-\frac{\pi}{\sqrt{\alpha/\alpha_{\rm cr} - 1}}}, \qquad \text{for} \qquad \alpha > \alpha_{\rm cr} \,. \tag{1}$$

This scaling property, i.e. the criticality actually supports the existence of the wide-range walking regime above $\alpha = \alpha_{\rm cr}$ in Fig. 1, i.e., the large scale hierarchy

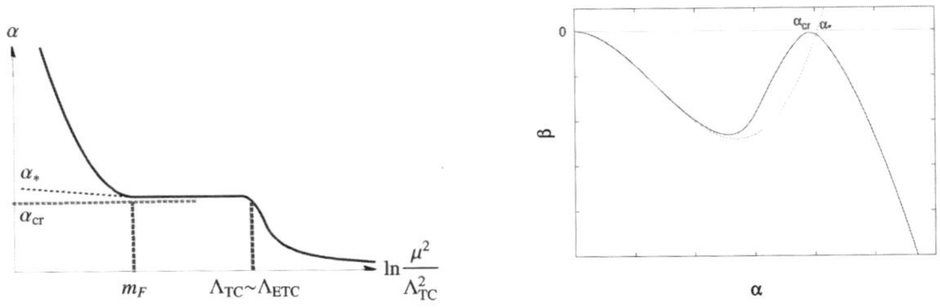

Fig. 1. A schematic picture of possible perturbative running coupling (left) and the beta function (right) in the region $\alpha < \alpha_{\rm cr}$, in comparison with the nonperturbative region $\alpha > \alpha_{\rm cr}$.

$m_F \ll \Lambda_{\rm TC}$. Thus, the WTC realizes a typical technihadron mass scale $M_{\rm Had}$ on the order of electroweak/TeV scale, much smaller than $\Lambda_{\rm TC}$:

$$M_{\rm Had} = \mathcal{O}({\rm EW}) \ll \Lambda_{\rm TC}\,, \tag{2}$$

in sharp contrast to QCD where $M_{\rm Had} \sim \Lambda_{\rm QCD}$.

As seen from Fig. 1, the WTC possesses the (approximate) scale invariance ($\beta(\alpha) \simeq 0$ for $m_F < \mu < \Lambda_{\rm TC}$), which is spontaneously broken by the technifermion condensation/mass generation. This implies the presence of a (pseudo) Nambu-Goldstone boson ("dilaton") for the scale symmetry. which arises as a flavor-singlet composite scalar $\sim \bar{F}F$. This composite scalar is what we call the TD, technidilaton.

The mass of the TD is essentially provided by the "nonperturbative" scale anomaly of the WTC: according to the dynamical mass generation in Eq.(1), the gauge coupling α is renormalized, starts running and hence the "nonperturbative" beta function $\beta_{\rm NP}(\alpha)$ is generated[18]:

$$\beta_{\rm NP}(\alpha) = \Lambda_{\rm TC}\frac{\partial\alpha}{\partial\Lambda_{\rm TC}} = -\frac{2\alpha_{\rm cr}}{\pi}\left(\frac{\alpha}{\alpha_{\rm cr}} - 1\right)^{3/2}\,. \tag{3}$$

This induces the "nonperturbative" scale anomaly:

$$\partial_\mu D^\mu = \frac{\beta_{\rm NP}(\alpha)}{4\alpha^2}\left(\alpha G_{\mu\nu}^2\right) \neq 0\,, \tag{4}$$

where D_μ denotes the dilatation current and $G_{\mu\nu}$ the field strength of technigluon field. Thus the TD becomes massive due to the nonperturbative scale anomaly[4,5]. Note that this scale anomaly is induced by the fermion mass generation itself: the scale symmetry is spontaneously broken by the technifermion condensation, at the same time it is explicitly broken by the nonperturbative running, as in Eq.(3), triggered by the technifermion mass generation itself.

The TD mass generation can be dictated by the partially conserved dilation current (PCDC) relation,

$$\langle 0|\theta^\mu_\mu|0\rangle = \frac{F_\phi^2 M_\phi^2}{4}\,, \tag{5}$$

where $\theta_\mu^\mu = \partial_\mu D^\mu$ is the trace of the energy-momentum tensor as in Eq.(4) and F_ϕ is the TD decay constant defined as

$$\langle 0|D_\mu(x)|\phi(p)\rangle = -ip_\mu F_\phi e^{-ipx} \,. \tag{6}$$

(The F_π is not necessarily equal to F_π, can rather be larger, as it will turn out below.) Since the scale anomaly is generated by the technifermion mass generation, one can find that the $\langle 0|\theta_\mu^\mu|0\rangle$ generically scales like

$$\langle 0|\theta_\mu^\mu|0\rangle \sim N_C N_F m_F^4 \,, \tag{7}$$

up to some loop factor dependent on details of nonperturbative computations. Combining this with the PCDC in Eq.(15), one arrives at a generic scaling law of the TD mass,

$$M_\phi \sim \sqrt{N_C N_F}\frac{m_F^2}{F_\phi} \,. \tag{8}$$

There have been much progress of the WTC particularly on the light TD, not just in the ladder Schwinger-Dyson (SD) equation, but also in a variety of approaches such as the ladder Bethe-Salpeter equation combined with the ladder SD equation[19,20], the effective theory based on the scale-invariant chiral perturbation theory[9,21], holographic method[11,22,23], and eventually, the first-principle calculation of the flavor-singlet scalar meson in the large N_F QCD on the lattice[24–27]. In particular, it is remarkable that such a light flavor-singlet scalar meson as a candidate for the TD was observed in the lattice $N_F = 8$ QCD[24], the theory shown to have signatures of the lattice walking theory including the mass anomalous dimension $\gamma_m \simeq 1$[28–30]. Note that $N_F = 8$ (four weak-doublets) corresponds to the "one-family model"[3,6] which is the most straightforward model building of the ETC[15] as a standard way to give masses to the quarks and leptons. The one-family model of the WTC with $N_C = 4$ is in fact best fit to the 125 GeV Higgs data[8–11], which has been updated recently[13] and will be summarized later, and is shown to be most natural for the ETC model building[31].

Below we shall propose a novel understanding for the realization of the light TD: it is the presence of the "anti-Veneziano limit", intrinsic to the WTC based on large N_F QCD. This observation has separately been described in the recent paper[13].

3. Parametrically Light Technidilaton in Large N_F Walking Technicolor

The key observation is that as long as the PCDC relation is satisfied, the TD as a pseudo NG boson has a vanishing mass in the anti-Veneziano limit, quite independently of the numerical details of nonperturbative calculations. To see this one should note from the large N_F and N_C scalings that

$$m_F \sim \frac{F_\pi}{\sqrt{N_C}} \,, \qquad F_\phi \sim \sqrt{N_C N_F} m_F \,, \tag{9}$$

and the F_π is related to the electroweak scale $v_{\rm EW} \simeq 246$ GeV as $F_\pi = 1/\sqrt{N_F/2}\, v_{\rm EW}$. Then one immediately finds that the TD mass scaling in Eq.(8) implies

$$\frac{M_\phi}{v_{\rm EW}} \sim \frac{1}{\sqrt{N_F N_C}}\,. \tag{10}$$

Now consider the anti-Veneziano limit where

$$N_C \to \infty\,, \qquad N_F \to \infty\,, \qquad \text{with} \qquad r \equiv \frac{N_F}{N_C} \gg 1 \qquad \text{fixed}\,. \tag{11}$$

In this limit one readily gets $M_\phi/v_{\rm EW} \to 0$ in the chiral/electroweak broken phase with $v_{\rm EW} \simeq 246$ GeV fixed. Thus the TD as the pseudo NG boson has a vanishing mass limit, though not exact massless point, in the anti-Veneziano limit, where the nonperturbative scale anomaly vanishes in units of F_ϕ via the PCDC in Eq.(15) as a measure of the spontaneous symmetry breaking of the scale symmetry. This is similar to the η' meson in QCD, which is regarded as a pseudo NG boson whose mass, evaluated through the anomalous WT identity with the $U(1)_A$ anomaly, does vanish in the large N_F and N_C limit with $r = N_F/N_C$ fixed ($\ll 1$) (Veneziano limit): $M^2_{\eta'}/F^2_\pi \sim N_F/N^2_C \to 0$, without the exact massless point. In this sense, the TD can be a vanishingly light pseudo NG boson at the same level as the η' in QCD.

Actually, more suprising thing will happen when one employs the ladder approximation for the WTC with the nonrunning gauge coupling. (The details has been given in the full paper[13].) Then the fermion dynamical mass m_F is related to the electroweak scale $v_{\rm EW}$ precisely through the Pagels-Stokar formula:

$$v^2_{\rm EW} = (246\,{\rm GeV})^2 \simeq \frac{N_F N_C}{4\pi^2}\, m_F^2 \simeq m_F^2 \left[\frac{N_F}{8}\frac{N_C}{4}\right]. \tag{12}$$

From this and Eq.(10) one can find a natural estimate of the TD mass for the one-family model with $N_F = 8$ and $N_C = 4$ to be

$$M_\phi = \mathcal{O}(m_F/2) = \mathcal{O}(v_{\rm EW}/2) = \mathcal{O}(125\,{\rm GeV})\,, \tag{13}$$

in agreement with the LHC Higgs as the TD.

More precisely, one has

$$M^2_\phi \simeq \left(\frac{v_{\rm EW}}{2}\right)^2 \cdot \left(\frac{5\, v_{\rm EW}}{F_\phi}\right)^2 \cdot \left[\frac{8}{N_F}\frac{4}{N_C}\right]. \tag{14}$$

Remarkablly, it was shown[8–11] that the TD best fit to the current LHC data is realized when

$$F_\phi \simeq 5\, v_{\rm EW} \simeq 1.25\,{\rm TeV} \quad \text{for} \quad M_\phi = 125\,{\rm GeV} \quad (N_F = 8,\, N_C = 4) \tag{15}$$

(See also the later discussions). With the fact that $v^2_{\rm EW} \propto N_F N_C m_F^2 \sim F^2_\phi$, the result reflects the generic scaling:

$$\frac{M_\phi}{v_{\rm EW}} \sim \frac{M_\phi}{F_\phi} \sim \frac{m_F}{F_\phi} \sim \frac{1}{\sqrt{N_F N_C}} \to 0, \tag{16}$$

in the anti-Veneziano limit.

4. The Low-Energy Effective Model for the TD

The effective theory for the TD involving the SM particles are constructed based on the scale-invariant chiral perturbation theory[9,21] with the scale anomaly terms supplied properly to the underlying WTC.

The chiral/electroweak and scale invariant Lagrangian thus takes the form[9]

$$\mathcal{L}_{\rm inv} = \frac{v_{\rm EW}^2}{4}\chi^2 {\rm tr}[D_\mu U^\dagger D^\mu U] + \mathcal{L}_{\rm kin}(\chi)\,, \tag{17}$$

where $\chi = e^{\phi/F_\phi}$ is a nonlinear base of the scale symmetry, which parametrizes the TD field ϕ with the decay constant F_ϕ and has the scale dimension 1; $D_\mu U = \partial_\mu U - iW_\mu U + iUB_\mu$ with the $SU(2)_W$ and $U(1)_Y$ gauge fields W and B; $\mathcal{L}_{\rm kin}(\chi)$ denotes the scale invariant kinetic term of TD and U the usual chiral field parameterizing the (eaten) Nambu-Goldstone boson fields π as $U = e^{2i\pi/v_{\rm EW}}$; The $|D_\mu U|^2$ term gives the TD couplings to massive weak bosons:

$$g_{\phi WW/ZZ} = \frac{2m_{W/Z}^2}{F_\phi}\,. \tag{18}$$

As seen in Eqs.(3) and (4), the scale symmetry is actually broken explicitly as well as spontaneously by dynamical mass generation of technifermions, which has to be respected also in the nonlinear realization[9]. Such explicit breaking effects arise in the TD Yukawa couplings to the SM fermions, which reflect underlying ETC-induced four-fermion terms, and couplings to QCD gluons and photons related to the scale anomaly in the SM gauge sector. In order to incorporate these effects into the scale-invariant Lagrangian, we introduce a spurion field S having the scale dimension 1 coupled to the SM fermions, digluon gg and diphoton $\gamma\gamma$ in such a way that[9]

$$\begin{aligned}\mathcal{L}_S = &-m_f\left(\left(\frac{\chi}{S}\right)^{2-\gamma_m}\cdot\chi\right)\bar{f}f \\ &+\log\left(\frac{\chi}{S}\right)\left\{\frac{\beta_F(g_s)}{2g_s}G_{\mu\nu}^2 + \frac{\beta_F(e)}{2e}F_{\mu\nu}^2\right\}\,,\end{aligned} \tag{19}$$

where $G_{\mu\nu}$ and $F_{\mu\nu}$ respectively denote the field strengths for QCD gluon and photon fields; g_s and e are the QCD gauge and electromagnetic couplings, respectively; β_Fs are the beta functions only including the technifermion loop contributions.

The TD Yukawa coupling to the SM f-fermion arises from the first line of Eq.(19) as[5]

$$g_{\phi ff} = \frac{(3-\gamma_m)m_f}{F_\phi}\,, \tag{20}$$

along with scale dimension of technifermion bilinear operator $(3-\gamma_m)$. The anomalous dimension $\gamma_m \simeq 1$ in WTC, which is crucial to obtain the realistic mass of the SM fermions of the first and the second generations without suffering from the flavor-changing neutral current problems. However it was known for long time that it is not enough for the mass of the third-generation SM f-fermions like t, b, τ: A simplest resolution would be the strong ETC model[32] having much larger anomalous dimension $1 < \gamma_m < 2$ due to the strong effective four-fermion coupling from the ETC dynamics in addition to the walking gauge coupling. Here we take $\gamma_m \simeq 2$, i.e., $(3 - \gamma_m) \simeq 1$, as in the strong ETC model for the third-generation SM f-fermions like t, b, τ which are relevant to the current LHC data.

In addition, the TD potential should be included so as to reproduce the PCDC relation Eq.(15) in the underlying walking theory:

$$V(\chi) = \frac{F_\phi^2}{4} m_\phi^2 \chi^4 \left(\log \frac{\chi}{S} - \frac{1}{4} \right) . \tag{21}$$

Here we stress remark on stability of the light TD mass against radiative corrections. As a pseudo Nambu-Goldstone boson of scale invariance, the quadratic divergence is suppressed by the scale invariance for the walking regime $m_F (\simeq v_{\rm EW}) < \mu < \Lambda_{\rm TC} (\sim \Lambda_{\rm ETC})$. The scale symmetry breaking in the ultraviolet region $\mu > \Lambda_{\rm TC}$ has no problem thanks to the naturalness as usual just as in QCD where the theory has only logarithmic divergences. Only possible source of the scale symmetry violation is from an effective theory for $\mu < m_F (\simeq v_{\rm EW})$. See Fig. 1.

Now it turns out that the TD mass is stable against the feedback effects of the ETC through particularly the top quark loop, because of the large $F_\phi \simeq 5 v_{\rm EW}$: below $\mu = m_F (\simeq v_{\rm EW})$ the dominant corrections to the TD mass M_ϕ come from the SM top quark and TD self-loops. These can be estimated from the effective Lagrangian in Eqs.(17) and (19) including the SM sector and ETC effects, which are estimated to be[9,13]

$$\begin{aligned} \frac{\delta M_\phi^2|_{\phi^4}}{M_\phi^2} &\simeq 24 \frac{m_F^2}{(4\pi F_\phi)^2} \simeq 6 \times 10^{-3} , \\ \frac{\delta M_\phi^2|_{\rm ETC/Yukawa}}{M_\phi^2} &\simeq 12 (3 - \gamma_m)^2 \frac{m_F^2}{(4\pi F_\phi)^2} \frac{m_t^2}{M_\phi^2} \\ &\simeq (3 - \gamma_m)^2 \frac{\delta M_\phi^2|_{\phi^4}}{M_\phi^2} , \end{aligned} \tag{22}$$

where the cutoff has been set to $m_F \simeq v_{\rm EW}$. These yield

$$\delta M_\phi^2 / M_\phi^2 = 0.01 \, (\gamma_m = 2) \, - 0.03 \, (\gamma_m = 1) . \tag{23}$$

Thus the 125 GeV TD mass is fully stable against the radiative corrections, in contrast to the unnatural SM Higgs case.

In passing, the TD potential in Eq.(21) written in terms $\chi = e^{\phi / F_\phi}$ is rewritten in the TD field ϕ as[9]

$$V(\phi) = -\mathcal{L}^S_{(2)\rm anomaly} = -\frac{M_\phi^2 F_\phi^2}{16} + \frac{1}{2} M_\phi^2 \phi^2 + \frac{4}{3} \frac{M_\phi^2}{F_\phi} \phi^3 + 2 \frac{M_\phi^2}{F_\phi^2} \phi^4 + \cdots . \tag{24}$$

It is remarkable to notice that in the anti-Veneziano limit Eq.(11) the TD self couplings (trilinear and quartic couplings) are highly suppressed:

$$\frac{4}{3}\frac{M_\phi^2}{F_\phi} \sim \frac{1}{\sqrt{N_F N_C}}\,, \qquad 2\frac{M_\phi^2}{F_\phi^2} \sim \frac{1}{N_F N_C} \tag{25}$$

by $M_\phi/F_\phi \sim 1/\sqrt{N_F N_C}$ and $M_\phi \sim N_F^0 N_C^0$. It is also interesting to numerically compare the TD self couplings for the one-family model ($N_F = 8, N_C = 4$) having $v_{\rm EW}/F_\phi \simeq 1/5$ with the self couplings of the SM Higgs with $m_h = M_\phi$, by making the ratios:

$$\begin{aligned} \left.\frac{g_{\phi^3}}{g_{h^3_{\rm SM}}}\right|_{M_\phi=m_h} &= \left.\frac{\frac{4M_\phi^2}{3F_\phi}}{\frac{m_h^2}{2v_{\rm EW}}}\right|_{M_\phi=m_h} \simeq \frac{8}{3}\left(\frac{v_{\rm EW}}{F_\phi}\right) \simeq 0.5\,, \\ \left.\frac{g_{\phi^4}}{g_{h^4_{\rm SM}}}\right|_{M_\phi=m_h} &= \left.\frac{\frac{2M_\phi^2}{F_\phi^2}}{\frac{m_h^2}{8v_{\rm EW}^2}}\right|_{M_\phi=m_h} = 16\left(\frac{v_{\rm EW}}{F_\phi}\right)^2 \simeq 0.6\,. \end{aligned} \tag{26}$$

This shows that the TD self couplings, although generated by the strongly coupled interactions, are even smaller than those of the SM Higgs, a salient feature of the approximate scale symmetry in the ant-Veneziano limit. This is in sharp contrast to the widely-believed folklore, "Strong coupling solutions like Technicolor tend to lead to a strongly coupled Higgs" [33].

5. LHC Higgs vs. Technidilaton in One-Family Model with $N_C = 4$ and $N_F = 8$

One finds from the effective Lagrangian that the TD couplings to the SM gauge bosons and the SM fermions can just be obtained by scaling from the SM Higgs as $v_{\rm EW} \to F_\phi$ [9,10]:

$$\begin{aligned} \frac{g_{\phi WW/ZZ}}{g_{h_{\rm SM}WW/ZZ}} &= \frac{g_{\phi ff}}{g_{h_{\rm SM}ff}} \quad (\text{for} \quad f = t, b, \tau) \\ &= \frac{v_{\rm EW}}{F_\phi} \quad \left[\simeq \frac{1}{5} \ll 1 \quad (N_F = 8\,, N_C = 4)\right]\,. \end{aligned} \tag{27}$$

On the other hand, in the one-family model with $N_F = 8$ the couplings to digluon and diphoton include the colored/charged techni-fermion loop contributions along with a factor N_C [9,10],

$$\begin{aligned} \mathcal{L}_{\rm eff}^{\gamma\gamma,gg} &= \frac{\phi}{F_\phi}\left\{\frac{\beta_F(g_s)}{2g_s}G_{\mu\nu}^2 + \frac{\beta_F(e)}{e}F_{\mu\nu}^2\right\}\,, \\ \beta_F(g_s) &= \frac{g_s^3}{(4\pi)^2}\frac{4}{3}N_C\,, \quad \beta_F(e) = \frac{e^3}{(4\pi)^2}\frac{16}{9}N_C\,, \end{aligned} \tag{28}$$

where the beta functions have been evaluated at the one-loop level. Thus one finds the scaling from the SM Higgs [10],

$$\begin{aligned} \frac{g_{\phi gg}}{g_{h_{\rm SM}gg}} &\simeq \frac{v_{\rm EW}}{F_\phi}\cdot(1+2N_C)\ , \\ \frac{g_{\phi\gamma\gamma}}{g_{h_{\rm SM}\gamma\gamma}} &\simeq \frac{v_{\rm EW}}{F_\phi}\cdot\left(\frac{63-16}{47}-\frac{32}{47}N_C\right)\ , \end{aligned} \tag{29}$$

where in estimating the SM contributions we have incorporated only the top (the terms of 1 and 16/47 for gg and $\gamma\gamma$ rates, respectively) and the W boson (the term of 63/47 for $\gamma\gamma$ rate) loop contributions. In Table 1 the branching fractions for relevant decay channels of the TD at 125 GeV are listed in the case of the one-family model with $N_C = 4$. Note that the total width $\Gamma_{\rm tot} = 1.15\,{\rm MeV}$ is smaller than the SM Higgs, which reflects the weaker couplings than the those of the latter, in contrast to the widely spread folklore.

Table 1. The TD branching ratios at 125 GeV in the one-family model with $N_C = 4$. The total width is also given.

BR[%]	gg	bb	WW	ZZ	$\tau\tau$	$\gamma\gamma$	$Z\gamma$	$\mu\mu$
$\Gamma_{\rm tot} = 1.15\,{\rm MeV}$	75.1	19.6	3.56	0.38	1.19	0.068	0.0048	0.0042

Calculating the signal strengths for the LHC production categories (gluon gluon fusion (ggF), vector boson fusion (VBF), vector boson associate production (VH) and top associate production (ttH)),

$$\mu^i_{X_1X_2} = \frac{\sigma^i_\phi \times {\rm BR}(\phi\to X_1X_2)}{\sigma^i_{h_{\rm SM}} \times {\rm BR}(h\to X_1X_2)}\ , \tag{30}$$

as a function of the overall coupling $v_{\rm EW}/F_\phi$ for given the number of N_C, we may fit the $\mu^i_{X_1X_2}$ to the latest data on the Higgs coupling measurements [34]. to determine the best-fit value of $v_{\rm EW}/F_\phi$. The result of the goodness of fit is shown in Table 2, which updates the previous analysis [10]. The Table 2 shows that the TD in the one-family model with $N_C = 4$ is favored by the current LHC Higgs data as much the same level as the SM Higgs. Remarkably, the best fit value $[v_{\rm EW}/F_\phi]_{\rm best} \simeq 0.2$, i.e. $F_\phi \simeq 5v_{\rm EW}$ for $N_C = 4$ is in excellent agreement with the ladder estimate of the TD mass $\simeq 125$ GeV in Eq.(15)!

In Table 3 we also make a list of the predicted signal strengths for each production category for the best fit value of $v_{\rm EW}/F_\phi \simeq 0.23$ in the case with $N_C = 4$, along with the latest result reported from the ATLAS and CMS experiments [34]. Note the TD signal strengths in the dijet category (VBF), which involves the contamination by about 30% from the ggF + gluon jets, $gg \to \phi + gg$. The contribution from the ggF is highly enhanced compared to the SM Higgs case, due to the extra techniquark loop contribution, which compensates the overall suppression by the direct VBF coupling $v_{\rm EW}/F_\phi \simeq 0.2$ to lift the event rate up to be comparable to the SM Higgs case. Note also the suppression of the VH-$b\bar{b}$-channel, which would be the characteristic signature of the TD to be distinguishable from the SM Higgs. More data from the upcoming LHC Run-II will draw a conclusive answer to whether or not the LHC Higgs is the SM Higgs, or the TD.

Table 2. The best fit values of $v_{\rm EW}/F_\phi$ for the one-family model with $N_C = 3, 4, 5$ displayed together with the minimum of the χ^2 ($\chi^2_{\rm min}$) normalized by the degree of freedom. Also has been shown in the last column the case of the SM Higgs corresponding to $N_C = 0$ and $v_{\rm EW}/F_\phi = 1$.

N_C	$[v_{\rm EW}/F_\phi]_{\rm best}$	$\chi^2_{\rm min}$/d.o.f.
3	0.27	$25/17 \simeq 1.5$
4	0.23	$16/17 \simeq 0.92$
5	0.17	$32/17 \simeq 2.0$
0 [SM Higgs]	1	$8.0/18 \simeq 0.44$

Table 3. The predicted signal strengths of the TD with $v_{\rm EW}/F_\phi = 0.23$ in the case of the one-family model with $N_C = 4$. The numbers in the parentheses correspond to the amount estimated without contamination from the ggF process. Also have been displayed the latest data on the Higgs coupling measurements reported from the ATLAS and CMS experiments[34].

TD signal strengths ($v_{\rm EW}/F_\phi = 0.23, N_C = 4$)	ATLAS	CMS
$\mu^{\rm ggF}_{\gamma\gamma} \simeq 1.4$	1.32 ± 0.38	1.13 ± 0.35
$\mu^{\rm ggF}_{ZZ} \simeq 1.0$	1.7 ± 0.5	0.83 ± 0.28
a $\mu^{\rm ggF}_{WW} \simeq 1.0$	0.98 ± 0.28	0.72 ± 0.37
$\mu^{\rm ggF}_{\tau\tau} \simeq 1.0$	2.0 ± 1.4	1.1 ± 0.46
$\mu^{\rm VBF}_{\gamma\gamma} \simeq 0.87$ (0.019)	0.8 ± 0.7	1.16 ± 0.59
$\mu^{\rm VBF}_{ZZ} \simeq 0.61$ (0.014)	0.3 ± 1.3	1.45 ± 0.76
$\mu^{\rm VBF}_{WW} \simeq 0.61$ (0.014)	1.28 ± 0.51	0.62 ± 0.53
$\mu^{\rm VBF}_{\tau\tau} \simeq 0.61$ (0.014)	1.24 ± 0.57	0.94 ± 0.41
$\mu^{\rm VH}_{bb} \simeq 0.014$	0.52 ± 0.40	1.0 ± 0.50

6. Other Walking Technihadrons: Technipions and Technirhos

In addition to the TD, the WTC predicts the rich spectrum. Such typical technihadrons involve the technipions, technirhos (and technia_1s.) In the case of the one-family model having the chiral $SU(8)_L \times SU(8)_R$ symmetry broken down to the vectorial $SU(8)_V$, the technipion spectra can be classified by the SM gauge charges, as well as the technirhos and techni-a_1s.

6.1. *Walking Technipions*

Technipion masses are all from explicit breaking outside of the WTC sector, i.e, SM gauge interactions and ETC gauge interactions. One can estimate the technipion masses in the WTC, based on the first order perturbation of the explicit chiral symmetry breaking by the "weak gauge couplings" of SM gauge interactions and the ETC gauge interactions (Dashen's formula). This is the same strategy as the QCD estimate of the $\pi^+ - \pi^0$ mass difference. It turns out [23,35,36] that the technipion masses are enhanced through the chiral condensate by the anomalous dimension as $(Z_m^{-1})^2 \sim (\Lambda/m_F)^{2\gamma_m}$ [4,37,38], to be on the order of TeV scale:

$$M_\pi \sim \mathcal{O}(\text{TeV})\,. \tag{31}$$

It is striking that although the explicit chiral symmetry breakings are formally very small due to the "weak gauge couplings", the nonperturbative contributions from the WTC sector lift all the technipions masses to the TeV region so that they all lose the nature of the "pseudo NG bosons". This is actually a universal feature of the dynamics with large anomalous dimension, "amplification of the symmetry violation" [39]. This amplification effect should not be confused with that of the pseudo NG boson mass due to the technifermion bare mass effects, like the pion mass due to the current quark mass, $F_\pi^2 m_\pi^2 = 2m\langle\bar\psi\psi\rangle$, which are not amplified by the large anomalous dimension, since the mass operator times the mass is totally invariant against the renormalization. (In the actual technicolor model, all the technifermions are set to be exactly massless and such a type of explicit breaking is not considered anyway.)

Note that although the left-over light spectra are just three exact NG bosons absorbed into W/Z bosons, our theory with $N_F \gg 2$ in the anti-Veneziano limit is completely different from the model with massless flavors $N_f = 2$ where the symmetry breaking is $SU(2)_L \times SU(2)_R/SU(2)_V$. In fact, even though all the NG bosons, other than the three exact NG bosons to be absorbed into W, Z bosons, are massive and decoupled from the low energy physics, they are composite of the linear combinations of all the N_F technifermions, not just 2 of them.

Including the walking technipions one can discuss the LHC phenomenology based on the scale-invariant chiral perturbation theory described by the Lagrangian in Eq.(17) [36]. The LHC-Run I limits on the 60 massive technipions were evaluated to be [23,36]

$$\begin{array}{lll} \text{color-octet technipions } (\theta): & M_\theta > 1.5 - 1.6\,\text{TeV} & \text{from } \theta \to t\bar{t} \\ \text{color-triplet technipions } (T): & M_T > 1.0 - 1.1\,\text{TeV} & \text{from leptoquark search}\,. \\ \text{color-singlet technipions } (P): & M_P > 850\,\text{GeV} & \text{from } \theta \to t\bar{t} \end{array}$$

More data from the Run II will thus reveal if walking technipions with the mass as large as $\mathcal{O}(\text{TeV})$ are present.

6.2. *Non-Pseudo NG-Technihadrons: Walking Technirhos*

In contrast to the TD and technipions, all the non-NG boson technihadrons, such as the technirho, techni-a_1, etc., have no constraints from the PCDC as the explicit breaking of the scale symmetry but do have constraints from the spontaneous breaking of the scale symmetry, so that they should have masses on the scale of spontaneous breaking of the scale symmetry, characterized by F_ϕ much larger than $2m_F$ of the naive nonrelativistic quark model picture:

$$M_\rho, M_{a_1}, \cdots = \mathcal{O}(\mathrm{TeV's}) > \mathcal{O}(F_\phi) \gg 2m_F \gg M_\phi . \tag{32}$$

In fact, the infrared conformal physics of the WTC should be described by the low-lying composite fields as effective fields, in a way to realize all the symmetry structure of the underlying theory.

Such an effective theory of WTC is constructed as a straightforward extension of scale-invariant chiral perturbation theory[9,21], i.e,, the scale-invariant version[40] of the Hidden Local Symmetry (HLS) model[41,42], (the "s-HLS model"), where the technirho mass terms have the *scale-invariance nonlinearly realized* by the TD field $\chi = e^{\phi/F_\phi}$, with the spontaneous breaking of the scale invariance characterized by the scale of F_ϕ, while the Higgs (TD) mass term in the TD potential, on the order of $m_F (\ll F_\phi)$, is the only source of the explicit breaking of the scale symmetry related (via PCDC) to the nonperturbative scale anomaly of the underlying theory.

One interesting candidate for such technihadrons may be a resonance behind the diboson excess recently observed at the LHC at 2 TeV[43,44], which can be identified with the walking technirho[45]. The excesses suggest a characteristically small width $\Gamma_{\mathrm{total}} < 100$ GeV[43], which can be naturally realized in the anti-Veneziano limit in Eq.(11):

$$\frac{\Gamma_{\mathrm{total}}}{M_\rho} \simeq \frac{\Gamma(\rho \to WW/WZ)}{M_\rho} \simeq \frac{1}{48\pi}\frac{g^2_{\rho\pi\pi}}{N_D} \sim \frac{1}{N_F N_C} \to 0 , \tag{33}$$

where $N_D = N_F/2$ is the number of the weak-doublets. In fact our one-family model $N_F = 8, N_C = 4$ can reproduces the features of the excesses very well[45]. A smoking gun of the walking technirho is the absence of the decay to the 125 GeV Higgs (TD), which is forbidden by the scale symmetry explicitly broken only by the Higgs (TD) mass term (corresponding to the nonperturbative scale anomaly in the underlying WTC)[46]. Actually, the salient feature of the scale symmetry of the generic effective theory containing the SM gauge bosons and the Higgs plus new vector bosons (any other massive particles as well), is the absence of the decay of the new vector bosons such as the technirho (and also other higher resonances) into the 125 GeV Higgs plus the SM gauge bosons[46]. If such decays of new particles are not found at LHC Run II, then the 125 GeV Higgs is nothing but the dilaton (TD in the case of the WTC) responsible for the nonlinearly realized scale symmetry, i.e., the spontaneous breaking of the scale symmetry, no matter what underlying theory may be beyond the SM. This should be tested in the ongoing LHC Run-II.

7. Conclusion

The technidilaton, arising as a pseudo NG boson for spontaneous breaking of the scale symmetry in the walking techniicolor, can be identified with the Higgs discovered at the LHC. The smallness of the TD mass can be ensured by the characteristic feature of the walking, and the presence of the anti-Veneziano limit in the case of the large-flavor walking gauge theory including the one-family model. The couplings of the technidilaton to the standard model particles have so far been consistent with those of the LHC Higgs. Crucial deviation from the standard model Higgs will be found in the decay channel to bottom quark pair, which is significantly suppressed by the large decay constant compared to the electroweak scale, to be tested in the ongoing Run II or other future colliders.

Walking technipions, pseudo NG bosons for the chiral symmetry, get heavy due to the salient feature of the walking dynamics having the large anomalous dimension. The masses are lifted to be on the order of a few TeV scale, which is consistent with the current LHC-Run I limits, so the walking technipions may be uncovered soon, or more severely constrained at the ongoing Run II through each discovery channel.

Non-pseudo technihadrons in the walking technicolor cannot be so light, in contrast to the technidilaton. The masses are thus expected to be on the order of the scale of the scale symmetry breaking, TeV scale, slightly higher than the electro weak scale. Among those Non-pseudos, the walking technirhos at around 2 TeV may be responsible for the recent excesses in the diboson channel, reported from the ATLAS group. The ongoing Run II will give the definite answer if it is the signal of the walking technirho.

Acknowledgments

I would like to thank Hidenori S. Fukano, Masafumi Kurachi and Koichi Yamawaki for useful discussions and collaborations. This work was supported by the JSPS Grant-in-Aid for Scientific Research (S) #22224003 and the JSPS Grant-in-Aid for Young Scientists (B) #15K17645.

References

1. G. Aad *et al.* [ATLAS Collaboration], Phys. Lett. B **716**, 1 (2012); S. Chatrchyan *et al.* [CMS Collaboration], Phys. Lett. B **716**, 30 (2012).
2. S. Weinberg, Phys. Rev. D **13**, 974 (1976); L. Susskind, Phys. Rev. D **20**, 2619 (1979).
3. For a review see E. Farhi and L. Susskind, Phys. Rept. **74**, 277 (1981).
4. K. Yamawaki, M. Bando and K. Matumoto, Phys. Rev. Lett. **56**, 1335 (1986); M. Bando, T. Morozumi, H. So and K. Yamawaki, Phys. Rev. Lett. **59**, 389 (1987).
5. M. Bando, K. Matumoto and K. Yamawaki, Phys. Lett. B **178**, 308 (1986).
6. S. Dimopoulos, Nucl. Phys. B **168**, 69 (1980).
7. S. Matsuzaki and K. Yamawaki, Prog. Theor. Phys. **127**, 209 (2012);
8. Phys. Rev. D **85**, 095020 (2012).
9. S. Matsuzaki and K. Yamawaki, Phys. Rev. D **86**, 035025 (2012).
10. Phys. Lett. B **719**, 378 (2013).

11. S. Matsuzaki and K. Yamawaki, Phys. Rev. D **86**, 115004 (2012).
12. S. Matsuzaki, arXiv:1304.4882 [hep-ph].
13. S. Matsuzaki and K. Yamawaki, arXiv:1508.07688 [hep-ph].
14. W. E. Caswell, Phys. Rev. Lett. **33**, 244 (1974); T. Banks and A. Zaks Nucl. Phys. B **196**, 189 (1982).
15. S. Dimopoulos and L. Susskind, Nucl. Phys. B **155**, 237 (1979); E. Eichten and K. D. Lane, Phys. Lett. B **90**, 125 (1980).
16. V. A. Miransky, Nuovo Cim. A **90**, 149 (1985).
17. V. A. Miransky and K. Yamawaki, Phys. Rev. D **55**, 5051 (1997); Errata, **56**, 3768 (1997).
18. W. A. Bardeen, C. N. Leung and S. T. Love, Phys. Rev. Lett. **56**, 1230 (1986); Nucl. Phys. B **273**, 649 (1986).
19. M. Harada, M. Kurachi and K. Yamawaki, Phys. Rev. D **68**, 076001 (2003).
20. M. Kurachi and R. Shrock, JHEP **0612**, 034 (2006).
21. S. Matsuzaki and K. Yamawaki, Phys. Rev. Lett. **113**, 082002 (2014).
22. K. Haba, S. Matsuzaki and K. Yamawaki, Phys. Rev. D **82**, 055007 (2010).
23. M. Kurachi, S. Matsuzaki and K. Yamawaki, Phys. Rev. D **90**, no. 9, 095013 (2014).
24. Y. Aoki, T. Aoyama, M. Kurachi, T. Maskawa, K. Miura, K. -i. Nagai, H. Ohki and E. Rinaldi, A. Shibata, K. Yamawaki and T. Yamazaki (the LatKMI Collaboration), Phys. Rev. D **89**, 111502(R) (2014); PoS LATTICE 2013 (2013) 070.
25. Y. Aoki, T. Aoyama, M. Kurachi, T. Maskawa, K. -i. Nagai, H. Ohki and E. Rinaldi, A. Shibata, K. Yamawaki and T. Yamazaki (the LatKMI Collaboration), "Light composite scalar in twelve-flavor QCD on the lattice," Phys. Rev. Lett. **111**, no. 16, 162001 (2013).
26. Z. Fodor, K. Holland, J. Kuti, D. Nogradi and C. H. Wong, PoS LATTICE **2013**, 062 (2014) [arXiv:1401.2176 [hep-lat]].
27. R. Brower, A. Hasenfratz, C. Rebbi, E. Weinberg and O. Witzel, arXiv:1411.3243 [hep-lat].
28. Y. Aoki, T. Aoyama, M. Kurachi, T. Maskawa, K. -i. Nagai, H. Ohki, A. Shibata, K. Yamawaki and T. Yamazaki (The LatKMI Collaboration), Phys. Rev. D **87**, 094511 (2013).
29. T. Appelquist *et al.* [LSD Collaboration], Phys. Rev. D **90**, 114502 (2014).
30. A. Hasenfratz, D. Schaich and A. Veernala, arXiv:1410.5886 [hep-lat].
31. M. Kurachi, R. Shrock and K. Yamawaki, Phys. Rev. D **91**, no. 5, 055032 (2015).
32. V. A. Miransky, K. Yamawaki, Mod. Phys. Lett. **A4**, 129-135 (1989); K. Matumoto, Prog. Theor. Phys. **81**, 277-279 (1989); T. Appelquist, M. Einhorn, T. Takeuchi, L. C. R. Wijewardhana, Phys. Lett. **B220**, 223 (1989).
33. see, e.g., Talk given at Aspen Winter, https://indico.cern.ch/event/202554/session/3/material/slides/0?contribId=58

34. The ATLAS collaboration, ATLAS-CONF-2015-007, ATLAS-COM-CONF-2015-011; ATLAS-CONF-2013-012, ATLAS-COM-CONF-2013-015; G. Aad *et al.* [ATLAS Collaboration], Phys. Lett. B **726**, 88 (2013) [Erratum-ibid. B **734**, 406 (2014)]; ATLAS-CONF-2013-030, ATLAS-COM-CONF-2013-028; ATLAS-CONF-2013-108, ATLAS-COM-CONF-2013-095; ATLAS-CONF-2013-079, ATLAS-COM-CONF-2013-080; CMS-PAS-HIG-13-001; CMS-PAS-HIG-13-002; S. Chatrchyan *et al.* [CMS Collaboration], JHEP **1401**, 096 (2014); CMS Collaboration [CMS Collaboration], CMS-PAS-HIG-13-022; S. Chatrchyan *et al.* [CMS Collaboration], JHEP **1405**, 104 (2014); S. Chatrchyan *et al.* [CMS Collaboration], Phys. Rev. D **89**, no. 1, 012003 (2014).
35. M. Harada, M. Kurachi and K. Yamawaki, Prog. Theor. Phys. **115**, 765 (2006).
36. J. Jia, S. Matsuzaki and K. Yamawaki, Phys. Rev. D **87**, 016006 (2013).
37. B. Holdom, Phys. Rev. D **24**, 1441 (1981).
38. K. Yamawaki and T. Yokota, Phys. Lett. B **113**, 293 (1982); Nucl. Phys. B **223**, 144 (1983).
39. For reviews, see, e.g., K. Yamawaki, Lecture at 14th Symposium on Theoretical Physics, Cheju, Korea, July 1995, arXiv:hep-ph/9603293; C. T. Hill and E. H. Simmons, Phys. Rept. **381**, 235 (2003) [Erratum-ibid. **390**, 553 (2004)]; F. Sannino, Acta Phys. Polon. **B40**, 3533-3743 (2009); K. Yamawaki, in Proceedings of Moriond EW Conference 2013, arXiv:1305.6352 [hep-ph].
40. M. Kurachi, S. Matsuzaki and K. Yamawaki, Phys. Rev. D **90**, 055028 (2014).
41. M. Bando, T. Kugo, S. Uehara, K. Yamawaki, and T. Yanagida, Phys. Rev. Lett. **54** (1985) 1215; M. Bando, T. Kugo, and K. Yamawaki, Nucl. Phys. **B259** (1985) 493; M. Bando, T. Fujiwara, and K. Yamawaki, Prog. Theor. Phys. **79** (1988) 1140; M. Bando, T. Kugo, and K. Yamawaki, Phys. Rept. **164** (1988) 217–314.
42. M. Harada and K. Yamawaki, Phys. Rept. **381** (2003) 1–233.
43. G. Aad *et al.* [ATLAS Collaboration], arXiv:1506.00962 [hep-ex].
44. V. Khachatryan *et al.* [CMS Collaboration], JHEP **1408**, 173 (2014).
45. H. S. Fukano, M. Kurachi, S. Matsuzaki, K. Terashi and K. Yamawaki, Phys. Lett. B **750**, 259 (2015).
46. H. S. Fukano, S. Matsuzaki and K. Yamawaki, arXiv:1507.03428 [hep-ph].

Topological Insights in Many-Flavor QCD on the Lattice

Yasumichi Aoki[a], Tatsumi Aoyama[a], Ed Bennett[b], Masafumi Kurachi[c], Toshihide Maskawa[a], Kohtaroh Miura[ad], Kei-ichi Nagai[a], Hiroshi Ohki[e], Enrico Rinaldi[f], Akihiro Shibata[g], Koichi Yamawaki[a] and Takeshi Yamazaki[h] (LatKMI Collaboration)

[a] *Kobayashi–Maskawa Institute for the Origins of Particles and the Universe (KMI), Nagoya University, Nagoya, Aichi, 464-8602 Japan*

[b] *Department of Physics, Swansea University, Singleton Park, Swansea SA2 8PP UK*

[c] *Institute of Particle and Nuclear Studies, High Energy Accelerator Research Organization (KEK), Tsukuba, Ibaraki 305-0801 Japan*

[d] *Centre de Physique Theoretique (CPT), Aix-Marseille University, Campus de Luminy, Case 907, 163 Avenue de Luminy, 13288 Marseille cedex 9, France*

[e] *RIKEN/BNL Research Center, Brookhaven National Laboratory, Upton, NY 11973-5000 USA*

[f] *Lawrence Livermore National Laboratory, 7000 East Ave., Livermore, CA 94550-9234 USA*

[g] *Computing Research Center, High Energy Accelerator Research Organization (KEK), Tsukuba, Ibaraki 305-0801 Japan*

[h] *Graduate School of Pure and Applied Sciences, University of Tsukuba, Tsukuba, Ibaraki 305-8571, Japan*

LatKMI Collaboration discusses the topological insights in many-flavor QCD on the lattice. We explore walking/conformal/confining phase in $N_\mathrm{f} = 4$, 8 and 12 (in particular $N_\mathrm{f} = 8$) lattice QCD via the topological charge and susceptibility, eigenvalues and anomalous dimension.

1. Introduction

For a number of years, the LatKMI Collaboration has performed an ongoing investigation[1,2] of QCD with 4, 8, 12, and 16 light flavors on the lattice, with the aim to classify them as conformal, near-conformal, or confining and chirally broken, and calculate quantities of phenomenological relevance. In particular, $N_\mathrm{f} = 8$ QCD is a candidate for a Walking Technicolor theory, with the potential to provide a mechanism for Electroweak Symmetry Breaking and at the same time produce a (composite) Higgs.

Thus far the study has largely been spectroscopic in nature; the aim of the work described in these proceedings was to consider orthogonal directions to gain a fuller picture of the theories. We present below first results of an investigation of the topological observables of these theories, and discuss both the effect these have on the interpretation of the spectroscopic results, and how these results may in themselves be interpreted to give signals of conformality or otherwise. Subsequently, we show some initial results of an investigation of the eigenvalue spectrum of these theories.

2. Topological charge and susceptibility

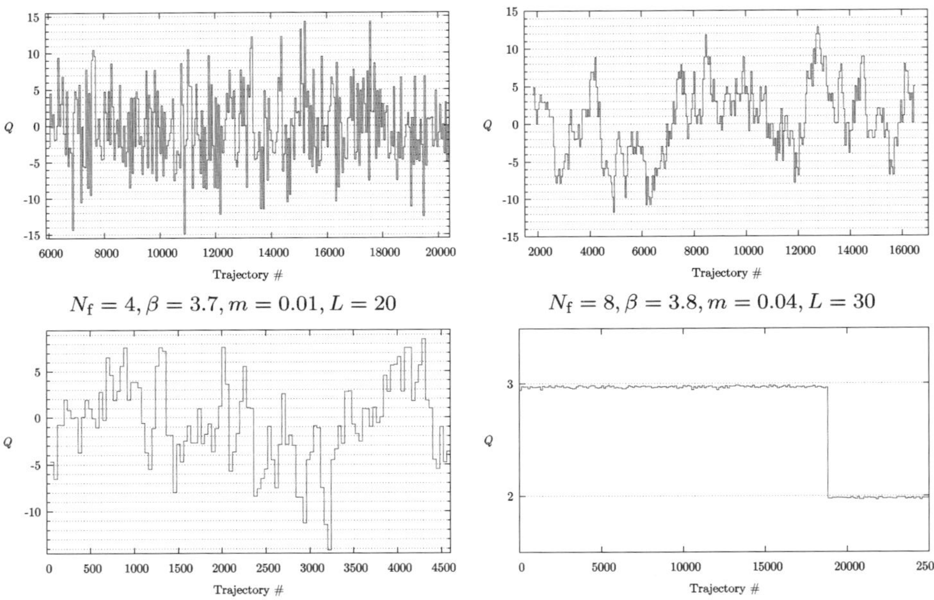

$N_{\mathrm{f}} = 4, \beta = 3.7, m = 0.01, L = 20$ $N_{\mathrm{f}} = 8, \beta = 3.8, m = 0.04, L = 30$

$N_{\mathrm{f}} = 12, \beta = 3.7, m = 0.16, L = 18$ $N_{\mathrm{f}} = 12, \beta = 4.0, m = 0.04, L = 36$

Fig. 1. Sample histories of the topological charge Q for the parameter sets shown. $N_{\mathrm{f}} = 12, \beta = 4.0$ shows the most substantial freezing observed outside of $N_{\mathrm{f}} = 16$.

In lattice gauge theories near the continuum limit, it is possible for the topological charge Q of a computation to become frozen, causing the simulation to become non-ergodic. It is thus necessary to compute the topological charge in order to verify whether or not our computations have sufficient ergodicity. Since the topological charge density operator is sensitive to UV fluctuations in the gauge field, it is necessary to smooth out the configuration. We use the gradient flow[3] for this purpose; specifically the Symanzik flow. We find that $N_{\mathrm{f}} = 4$ is highly ergodic, as are most parameters for $N_{\mathrm{f}} = 8$, with mild loss in ergodicity emerging at the lowest values of am. $N_{\mathrm{f}} = 12$ suffers from more ergodicity problems, with very long regions of frozen Q, particularly at low values of am. Example Q histories are shown in Fig. 1. Finally, the topology of $N_{\mathrm{f}} = 16$ is completely suppressed, with no local minima or maxima in the topological charge density distribution indicative of the presence of a topological excitation. (An example set of cooling histories is shown

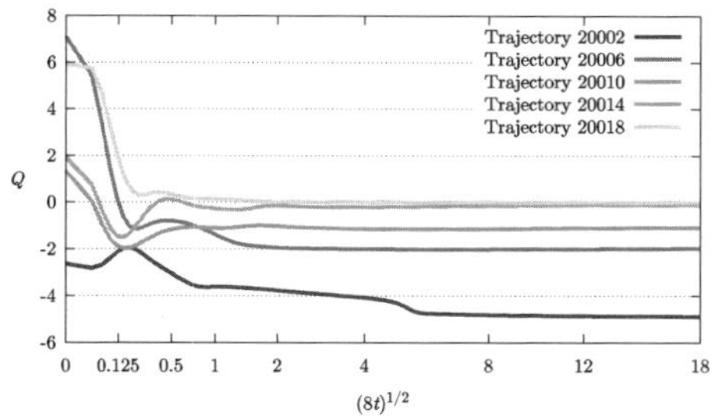

$N_f = 8, \beta = 3.7, m = 0.06, L = 24$

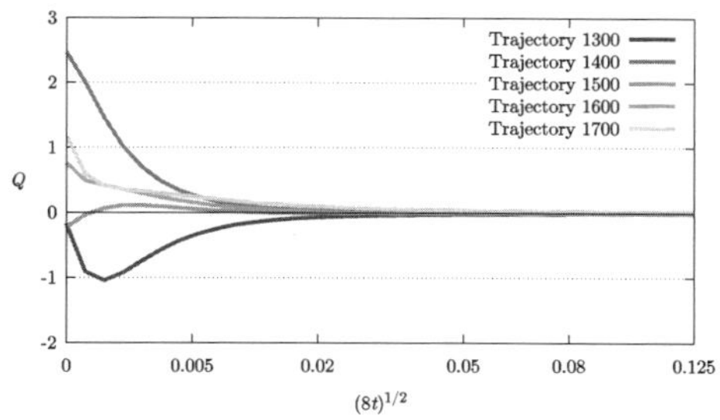

$N_f = 16, \beta = 12, m = 0.015, L = 48$

Fig. 2. History of the topological charge as a function of the gradient flow scale for selected parameter sets for $N_f = 8$ and $N_f = 16$. $N_f = 16$ is rapidly suppressed to zero, compared to $N_f = 8$ which takes longer to equilibrate and gives a variation in topological charge.

in Fig. 2. Owing to this lack of data, $N_f = 16$ will not be discussed further.

Having calculated these values, we would like to use them for more than just a check of ergodicity. In particular, it would be useful if they could contribute to our ongoing work to study the infrared regime of these theories, specifically whether they are conformal or confining and chirally broken, and if the latter, whether they are QCD-like or exhibit walking-type dynamics. EB and Lucini have previously performed a study of this kind[4] for SU(2) with $N_f = 2$ adjoint flavors, and we adopt a similar procedure here.

An IR-conformal theory becomes confining when deformed by a fermion mass. In the process, a scale is added (the fermion mass) which is intrinsically heavy—thus when considering gluonic IR observables, the theory should be indistinguishable from the equivalent quenched theory. One such observable is the topological susceptibility χ_{top}, which we calculate as

$$\chi_{\text{top}} = \frac{1}{V}\left(\langle Q^2\rangle - \langle Q\rangle^2\right)$$

where V is the lattice volume, and $\langle Q\rangle$ is expected to be zero. Since in IR-conformal theories, the deforming mass induces non-trivial RG corrections to the lattice spacing, we must consider only dimensionless ratios. Since we ultimately want to compare the data with SU(3) pure gauge theory, we scale with a gluonic observable.

To verify that our results for the topological susceptibility are stable, we perform a finite-volume study for the $N_f = 8$ data, which is shown in Fig. 3(a). In the cases suffering from topological freezing, it is obvious that the above definition of χ_{top} will give uncontrolled errors, as we are estimating the width of the distribution based on an inadequate sampling. Here, we use the subvolume method described by the LSD Collaboration[5], to whom we defer for details of the method.

The gradient flow, having been used to access the topological charge, may also be used at no extra cost to extract a scale[3] t_0, defined as the value of the flow time t for which $t^2\langle E\rangle = 0.3$, where E is a lattice discretized version of the continuum relation $E = \frac{1}{4}G_{\mu\nu}G_{\mu\nu}$. It is informative to see how this quantity varies as we move

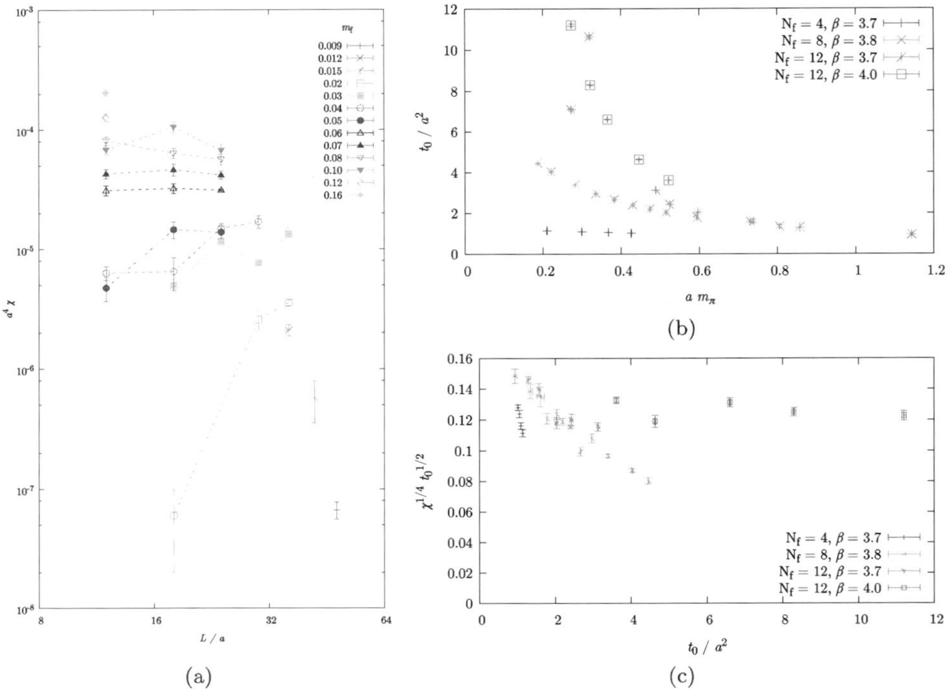

Fig. 3. (a) Finite-volume of all available values of χ_{top} for $N_f = 8$. (b) Comparison of the flow scale t_0 with the pion mass: in $N_f = 4$, the scale does not change significantly as the fermion mass changes, as we expect for a confining theory. $N_f = 12$ shows strong changes in scale as the deforming fermion mass is changed, while $N_f = 8$ lies between these two extremes. (c) Comparison of χ_{top} scaled by the gradient flow scale t_0 between theories: we see that in the heavy-fermion limit (that is, as $t_0 \to 0$), the theories match up. $N_f = 4$ rapidly diverges, with $N_f = 8, 12$ moving away more slowly. $N_f = 12$ turns over as t_0 increases, which is not seen in other theories.

towards the chiral limit; to this end, a plot of t_0 against m_π is shown in Fig. 3(b). As expected for a confining QCD-like theory, $N_f = 4$ shows very little deviation in scale as m_π varies. $N_f = 12$ varies strongly, with the scale blowing up as $m_\pi \to 0$, as we expect; we observe the limit of this behaviour in $N_f = 16$, with the quantity $t^2\langle E\rangle$ flattening well before we reach $t^2\langle E\rangle = 0.3$ (not plotted here). $N_f = 8$ interpolates between these two cases; with current data we cannot distinguish whether or not t_0/a^2 would be finite in the limit $m_\pi \to 0$.

If we scale χ_{top} using t_0, we obtain the results shown in Fig. 3(c). $N_f = 4$ and 8 show straight line behaviour as we head away from the heavy-fermion (small t_0 limit; $N_f = 12$ on the other hand shows a turnover. (Since the x-axis has a dependence on the lattice spacing, we cannot make more quantitative statements without some procedure to set the scale.)

Finally, we look to see whether the theories obey conformal hyperscaling relations, and if so, if the chiral condensate anomalous dimension γ_* extracted agrees with that found via other methods. In Fig. 4(a), we show the bare data for the

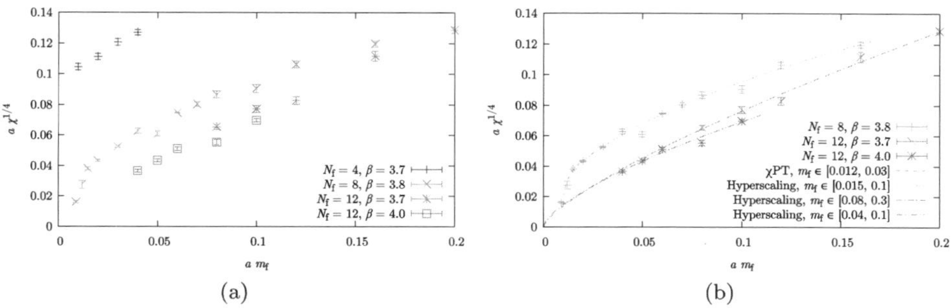

Fig. 4. Plots of $a\chi^{1/4}$ against m_f, showing (a) data for $N_\mathrm{f} = 4, 8, 12$, and (b) data for $N_\mathrm{f} = 8, 12$ and the results of fits performed upon these data.

(fourth root of) the susceptibility against the fermion mass, both in lattice units. Note that for the $N_\mathrm{f} = 4$ data, we do not see a susceptibility of zero in the chiral limit, as we would expect for a chirally broken theory. This is most likely due to staggered taste symmetry breaking effects, which will disappear as the continuum limit is approached. This will be investigated further in a future work.

In Fig. 4(b), we show the results of fits upon the $N_\mathrm{f} = 8$ and 12 data. For $N_\mathrm{f} = 8$ we attempt both chiral perturbation theory (χPT)

$$\chi_\mathrm{top} = Cm_\mathrm{f} + f(a) \qquad\qquad \Rightarrow \chi_\mathrm{top}^{1/4} = (Cm_\mathrm{f} + f(a))^{1/4}$$

(where $f(a)$ is a factor accounting for lattice artefacts) and hyperscaling

$$\chi_\mathrm{top}^{1/4} = Am_\mathrm{f}^{1/(1+\gamma_*)}$$

fits, while for $N_\mathrm{f} = 12$ we only fit with the hyperscaling form. For $N_\mathrm{f} = 8$, an anomalous dimension of $\gamma_* = 1.04(5)$ is extracted, while for $N_\mathrm{f} = 12$, values of 0.33(6) and 0.47(10) are found for $\beta = 3.7$ and 4.0 respectively. All values for γ_* found are consistent with those obtained from spectroscopic studies. Both hyperscaling and χPT fits are consistent with the data for $N_\mathrm{f} = 8$.

3. Eigenvalues

The second avenue we have explored is making use of the spectrum of eigenvalues of the Dirac operator. From the Banks–Casher relation[6]

$$\langle\overline{\psi}\psi\rangle = \lim_{m\to 0}\lim_{V\to\infty} \pi\rho(\lambda = 0)$$

we may investigate the chiral condensate of a theory via the density of eigenvalues (λ) of the Dirac operator, $\rho(\lambda)$. However, since the eigenvalue distribution is noisy, we can integrate it to find the mode number $\nu(\lambda)$, which smooths out some noise while carrying the same information. One such distribution is plotted in Fig. 5.

We would like to fit for the chiral condensate anomalous dimension. We attempt to fit using the most naïve form,

$$\rho(\lambda) = c\lambda^\alpha \qquad\qquad \Rightarrow \nu(\lambda) = d\lambda^{\alpha+1}$$

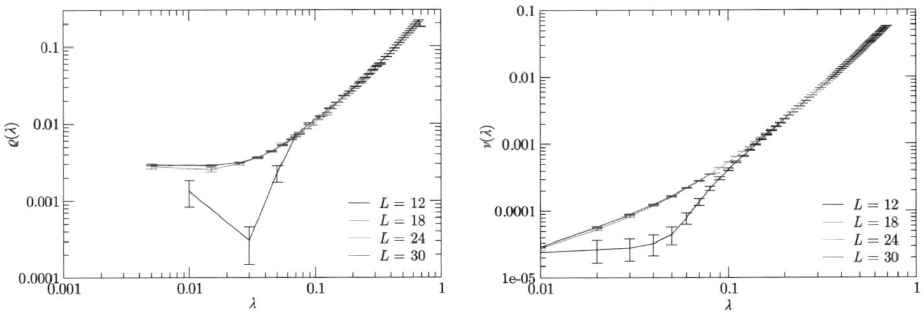

Fig. 5. Left: The spectral density $\rho(\lambda)$. Right: the mode number $\nu(\lambda)$, both plotted against the eigenvalue λ as log-log plots, for $N_f = 8, \beta = 3.8, m = 0.04, L = 30$.

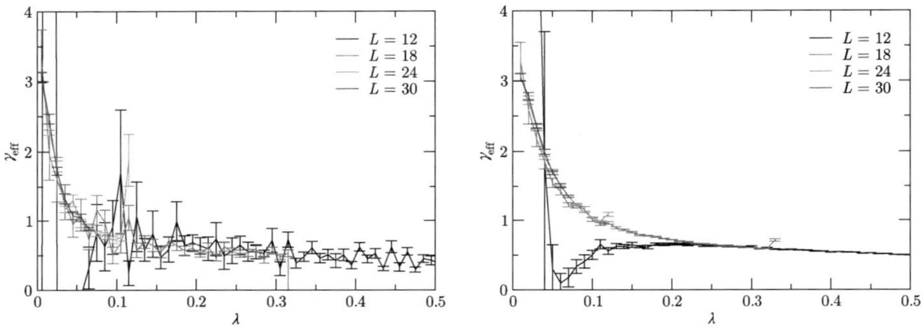

Fig. 6. The effective anomalous dimension $\gamma_{\rm eff}(\lambda)$, plotted against the eigenvalue λ, for the same ensemble as above. Left: via the spectral density ρ. Right: via the mode number ν.

where $\alpha + 1 = 4/(1 + \gamma)$.

Since we are not in the chiral limit, we cannot simply take the limit of smallest λ; instead we define an effective exponent across an interval of λ as

$$\alpha_{\rm eff} + 1 = \frac{\ln \nu_2 - \ln \nu_1}{\ln \lambda_2 - \ln \lambda_1}$$

where $\lambda_2 = \lambda_1 + \Delta$. From this we may calculate an effective anomalous dimension $\gamma_{\rm eff}$. We show the results of this fit in Fig. 6; here we see that $\gamma_{\rm eff}(\nu)$, while having the same overall shape as $\gamma_{\rm eff}(\rho)$, has significantly less noise. All volumes appear to lie on a universal curve, save for $L = 12$, which seems to suffer from finite-volume effects at small λ, while agreeing at larger values.

By choosing a region of small λ satisfying $\lambda \gg m$, we obtain an estimate $0.5 \lesssim \gamma_{\rm eff} \lesssim 0.7$, for the region $\lambda \gtrsim 0.2$. This value is smaller than that obtained from our spectroscopic studies, and the above fit of the topological susceptibility. The reason for this discrepancy is currently uncertain; it is possible that it is due to the finite fermion mass, or the lack of extrapolation to $\lambda = 0$.

4. Conclusions

We have shown some initial results of our study of topological observables in $N_{\rm f} = 4$, 8, and 12 QCD. We see that the topological susceptibility $\chi_{\rm top}$ may be used to give an estimate for the anomalous dimension of the chiral condensate γ_*. While whether $N_{\rm f} = 8$ QCD is conformal or confining is still an open question, assuming a conformal ansatz gives $\gamma_* = 1.04(5)$ for $N_{\rm f} = 8$ at $\beta = 3.8$, $\gamma_* = 0.33(6)$ for $N_{\rm f} = 12$ at $\beta = 3.7$, and $\gamma_* = 0.47(10)$ at $\beta = 4.0$—consistent with estimates from spectroscopic observables. We also showed initial results for the scaling of $\chi_{\rm top}$ as dimensionless products with the gradient flow scale t_0, which at this point are inconclusive, and observed that the behavior t_0 itself as a function of m_π gives some insight into the IR dynamics of the theories. The next steps of this study will be to obtain a comparable set of $N_{\rm f} = 0$ data to better interpret the behaviour of $\chi_{\rm top}\sqrt{t_0}$, and to consider dimensionless ratios of $\chi_{\rm top}$ with the string tension σ.

We have also begun to study the eigenvalue spectrum of the Dirac operator, obtaining a first estimate of the anomalous dimension $0.5 \lesssim \gamma_* \lesssim 0.7$. Since this is inconsistent with the results from the spectroscopy and topological susceptibility, further study is ongoing to determine why this is the case. The interaction of the gradient flow with the Dirac eigenvalues is also being investigated, as are the applications of the gradient flow to measuring the glueball spectrum and η' mass.

5. Acknowledgements

Computations have been carried out on the φ computer at KMI and the CX400 machine at the Information Technology Center in Nagoya University. This work is supported by the JSPS Grant-in-Aid for Scientific Research (S) No.22224003, (C) No.23540300 (K.Y.), for Young Scientists (B) No.25800139 (H.O.) and No.25800138 (T.Y.), and also by the MEXT Grants-in-Aid for Scientific Research on Innovative Areas No.23105708 (T.Y.), No. 25105011 (M.K.). E.R. acknowledges the support of the U.S. Department of Energy under Contract DE-AC52-07NA27344 (LLNL).

References

1. Y. Aoki, T. Aoyama, M. Kurachi, T. Maskawa, K.-i. Nagai, H. Ohki, E. Rinaldi, A. Shibata, K. Yamawaki and T. Yamazaki, Light composite scalar in twelve-flavor QCD on the lattice, *Phys. Rev. Lett.* **111**, p. 162001 (2013).
2. Y. Aoki *et al.*, Light composite scalar in eight-flavor QCD on the lattice, *Phys. Rev.* **D89**, p. 111502 (2014).
3. M. Lüscher, Properties and uses of the Wilson flow in lattice QCD, *JHEP* **08**, p. 071 (2010), [Erratum: JHEP03,092(2014)].
4. E. Bennett and B. Lucini, Topology of Minimal Walking Technicolor, *Eur. Phys. J.* **C73**, p. 2426 (2013).
5. R. C. Brower *et al.*, Maximum-Likelihood Approach to Topological Charge Fluctuations in Lattice Gauge Theory, *Phys. Rev.* **D90**, p. 014503 (2014).
6. T. Banks and A. Casher, Chiral Symmetry Breaking in Confining Theories, *Nucl. Phys.* **B169**, p. 103 (1980).

Recent Results from SU(2) with One Adjoint Dirac Fermion

A. Athenodorou[ab], E. Bennett[a*], G. Bergner[c] and B. Lucini[a]

[a] *Department of Physics, Swansea University, Singleton Park, Swansea, SA2 8PP UK*
** E-mail: E.J.Bennett@swansea.ac.uk*

[b] *Department of Physics, University of Cyprus, POB 20537, 1678 Nicosia, Cyprus*

[c] *Universität Bern, Institut für Theoretische Physik, Sidlerstr. 5, CH-3012 Bern, Switzerland*

We present some results for SU(2) with one adjoint Dirac flavour from the lattice studies. Data for the spectroscopy, the static potential, topological charge, and the anomalous dimension of the fermionic condensate are included. Our findings are found to be inconsistent with conventional confining behaviour, instead pointing tentatively towards a theory lying within or very near the onset of the conformal window, with an anomalous dimension of the fermionic condensate of almost 1. Implications of these findings on the building of models of strongly-interacting dynamics beyond the standard model are discussed.

1. Introduction

Novel strong interactions provide a promising explanation for the mechanism of electroweak symmetry breaking. Theories near the onset of the conformal window with an anomalous dimension of the chiral condensate of order one may in particular be relevant for phenomenology. It is important that these properties are verified for candidate models in a first-principles, fully non-perturbative fashion, and numerical lattice simulations provide an indispensable tool to this end. Recent lattice calculations[1,2] have demonstrated that the SU(2) gauge theory with two adjoint Dirac flavours is in the conformal window, while studies of many-flavour QCD[3–6] and of the SU(3) gauge theory with two Dirac flavours in the sextet representation[7–9] are ongoing.

In these proceedings we present results from an ongoing investigation of the SU(2) Yang–Mills theory with one adjoint Dirac flavour, using Monte Carlo studies of the theory discretised on a spacetime lattice. The chiral symmetry breaking of this model produces only two Goldstone bosons, making it insufficient to give mass to the $W^{\pm}$ and Z bosons. Thus the theory needs to be extended for phenomenological considerations. The main purposes of this study are firstly, to offer potential insights into the connection between near-conformal gauge theories and the presence of a large anomalous dimension. Secondly, to seek to pin down the lower end of the conformal window, which may give insight as to whether an interpolating theory between one and two adjoint Dirac flavours could be of phenomenological interest. Third, as an end rather than a means, as investigation and classification of gauge theories is also of theoretical interest.

2. Lattice computations

We simulate the theory on a 4-dimensional hypercubic lattice, using the Wilson gauge action and the Wilson fermionic action, giving

$$S = \beta \sum_{p} \mathrm{Tr}[1 - U(p)] + \sum_{x,y} \overline{\psi}(x) D(x,y) \psi(y) ,$$

where the first sum is over plaquettes, and $D(x,y)$ is the massive Dirac operator:

$$D(x,y) = \delta_{x,y} - \kappa \left[(1-\gamma_\mu) U_\mu(x) \delta_{y,x+\mu} + (1+\gamma_\mu) U^\dagger_\mu(x-\mu) \delta_{y,x-\mu} \right] .$$

κ parameterises the bare fermion mass am, being related to it as $\kappa = 1/(8+2am)$. $\beta = 2N/g^2 = 4/g^2$ is the inverse coupling, where $N = 2$ is the dimension of the gauge group.

Gauge ensembles were generated using the HiRep code suite, making use of the Rational Hybrid Monte Carlo (RHMC) algorithm. Further detail of the initial set of ensembles are given in[10]; the newer ensembles introduced in this work were generated using the same procedures.

Ensembles were generated at two values of β: at $\beta = 2.05$, masses of $-1.524 \leq am \leq -1.475$ were used, while at $\beta = 2.2$ the range $-1.378 \leq am \leq -1.280$. With the lattice volume fixed to $N_T \times N_L^3$, spatial lattice sizes of 12, 16, and 24 were chosen, in all cases being cautious to avoid the effects of finite volume artefacts in our observables.

Table 1. Parameter sets considered at $\beta = 2.05$ and 2.2, showing their volumes, bare masses, PCAC masses, and number of configurations.

(a) $\beta = 2.05$

	N_L	$-am$	am_{PCAC}	N_{conf}
B1	12	1.475	0.1493(5)	1500
B2	12	1.500	0.1113(8)	1500
B3	12	1.510	0.09226(92)	4000
C1	16	1.475	0.1485(4)	1500
C2	16	1.490	0.1279(2)	1500
C3	16	1.510	0.09111(31)	1500
C4	16	1.510	0.09048(52)	1500
C5	16	1.514	0.08223(34)	1500
C6	16	1.519	0.06587(37)	1500
D1	24	1.510	0.09130(27)	1534
D2	24	1.523	0.04722(43)	2168
D3	24	1.524	0.04081(71)	800*

(b) $\beta = 2.2$

	N_L	$-am$	am_{PCAC}	N_{conf}
2B1	12	1.280	0.2017(6)	4000
2B2	12	1.290	0.1891(5)	4000
2B3	12	1.300	0.1761(7)	4000
2B4	12	1.310	0.1633(7)	4000
2C1	16	1.320	0.1487(3)	4000
2C2	16	1.340	0.1189(4)	4000
2C3	16	1.350	0.1021(5)	4000
2C4	16	1.360	0.0844(5)	4000
2D1	24	1.360	0.0830(10)	1721
2D2	24	1.370	0.0652(2)	4000
2D3	24	1.378	0.0486(5)	1505

Note: * Run D3 used configurations saved every 5 trajectories rather than every one, giving equivalent statistics to 4000 trajectories.

Our aim in studying this theory is to probe its infrared regime—in particular, we would like to identify conformal properties and calculate the chiral condensate anomalous dimension. To this end, following[11,12], we have calculated the mass spectrum in the chiral regime, and tested finite-size scaling predictions. Additionally, we have performed fits of the Dirac mode number[13].

The massless fermionic action admits a global SU(2) chiral symmetry, which is broken by a non-zero chiral condensate to SO(2) $\equiv$ U(1). The generator of the unbroken subgroup is identified with the baryon number B. Since parity is unbroken, the states of the theory are labeled by B^P, where P is the parity, and by their spin. The spectrum comprises mesons, baryons, glueballs, and glue–fermion composite states[14]. In addition to these states, we also measure the string tension.

Mesonic states are accessed by the usual correlation functions of operators composed of fermion fields, which necessarily contain disconnected contributions due to the single flavour. Baryonic states meanwhile are accessed via the connected correlation functions of appropriately-chosen operators of the same form as for the mesons. Mesons have baryon number $B = 0$, while baryons have $B = \pm 2$.

Meanwhile glueball operators consist of linear combinations of ordered products of link matrices on closed path, chosen to transform irreducibly under spin transformations, parity, and charge conjugation.

The string tension is accessed via two methods: via operators of blocked smeared Polyakov loops, fitted to an effective string theory prediction, and via the temporal asymptotic behaviour of expectation values of Wilson loops, giving the static potential, which can be fitted to a Cornell-like form.

For further details of the methods used to generate the ensembles and to extract these observables, we defer to the more detailed presentation given in[10].

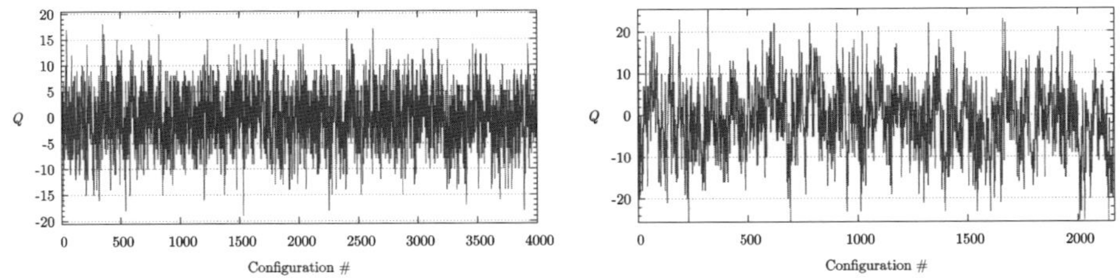

(a) (b)

Fig. 1. Sample topological charge histories for ensembles (a) C1, and (b) D2.

As a check of ergodicity, we look at the topological charge history, illustrative examples of which are shown in Fig. 1. Good ergodic movement is seen, with no sign of frozen regions. To verify that we are not in a region of lattice artefacts, the histogram of the average Polyakov loop in each direction is checked; in Fig. 2 we show illustrative examples showing an unbroken centre symmetry via the unimodal distribution—a broken centre would be indicated by a bimodal distribution.

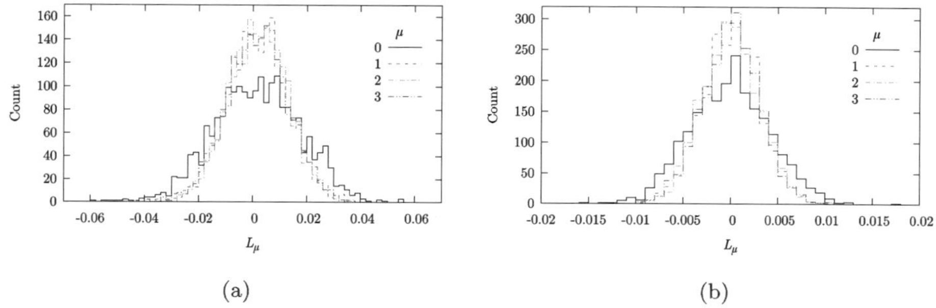

Fig. 2. Polyakov loop distribution for ensembles (a) C4, and (b) D2.

3. Spectrum

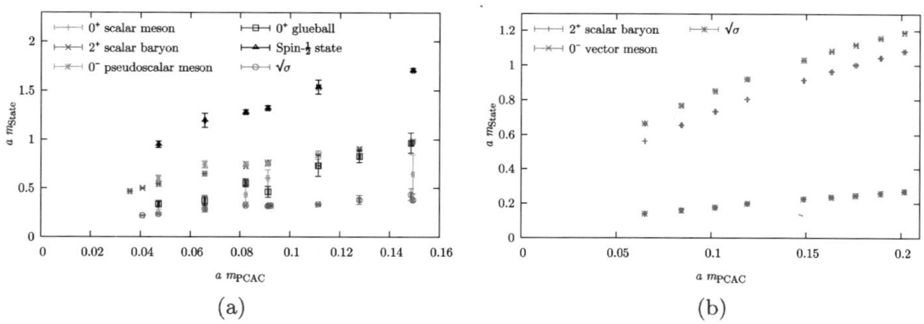

Fig. 3. Bare spectral quantities for (a) $\beta = 2.05$, and (b) $\beta = 2.2$, as a function of am_{PCAC}.

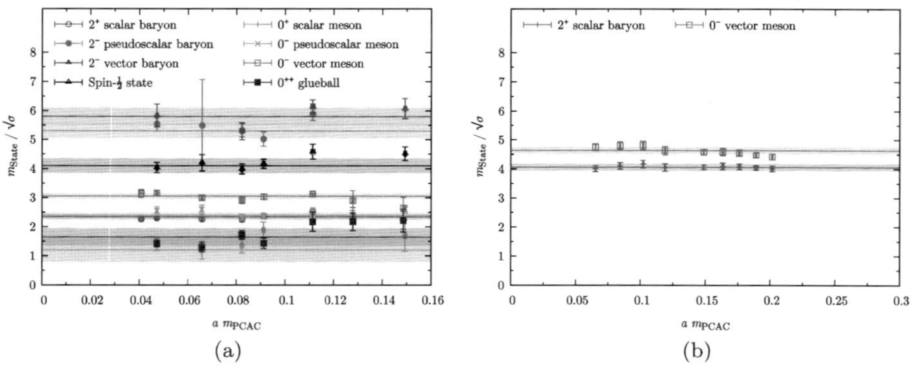

Fig. 4. Spectral quantities for (a) $\beta = 2.05$ and (b) $\beta = 2.2$, scaled by the string tension $\sqrt{\sigma}$, as a function of the PCAC mass am_{PCAC}. In all cases the behaviour observed is consistent with flat.

We now turn our attention to the spectrum of the theory. As seen in Fig. 3, all spectral observables of the theory, including meson, baryon, and glueball masses, and the string tension in lattice units, decrease monotonically towards zero as $am_{\text{PCAC}} \to 0$.

In Fig. 4 we show the spectrum of the theory, now as dimensionless ratios to the string tension, both for $\beta = 2.05$ and with available preliminary data for $\beta = 2.2$. All states show constant mass ratios to two standard deviations throughout the range in which they were observed. At $\beta = 2.05$, the 0^+ scalar meson is seen to be the lightest state in the scaling region, and appears to be degenerate with the 0^{++} glueball, with which it is anticipated to mix.

4. Chiral condensate anomalous dimension

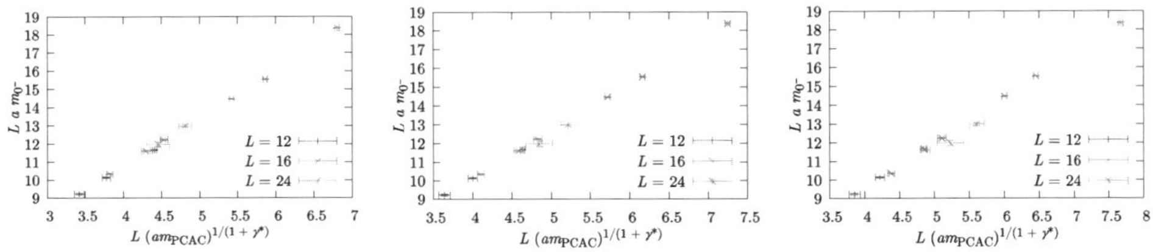

($\gamma_* = 0.9$) ($\gamma_* = 1.0$) ($\gamma_* = 1.1$)

Fig. 5. Plots of Lm_{0^-} as a function of $Lm_{\text{PCAC}}^{1/(1+\gamma_*)}$ for $\beta = 2.05$ for three values of γ_*. The results appear to identify a universal curve for $0.9 \le \gamma_* \le 1.0$.

We consider two independent methods to extract the anomalous dimension γ_*. The first gives an approximate value via finite size scaling predictions. For a conformal theory, a spectral quantity m_{X} of the system in a finite volume of spatial extent L, as $L \to \infty$ and the combination $Lm_{\text{PCAC}}^{1/(1+\gamma_*)}$ is kept constant, obeys the relation

$$Lm_{\text{X}} = f\left(Lm_{\text{PCAC}}^{\frac{1}{1+\gamma_*}}\right),$$

for some unknown function f. Thus by plotting Lm_{X} as a function of $Lm_{\text{PCAC}}^{1/(1+\gamma_*)}$ for our data at various volumes, we may extract γ_* by tuning it until the data lie on a single universal curve.

A sample of such plots are shown in figure 5; from this we can see a best fit in the range $0.9 \le \gamma_* \le 1.0$.

The second route we use to obtaining γ_* is by fitting the Dirac mode number $\overline{\nu}(\Omega)$ as a function of the Dirac eigenvalues Ω[13]. Since at low Ω the distribution is distorted by the finite fermion mass, while at large Ω the scaling region ends, we look for an ensemble with a large intermediary region to examine; thus a large lattice and small mass makes sense. We therefore examine the D2 ensemble, and

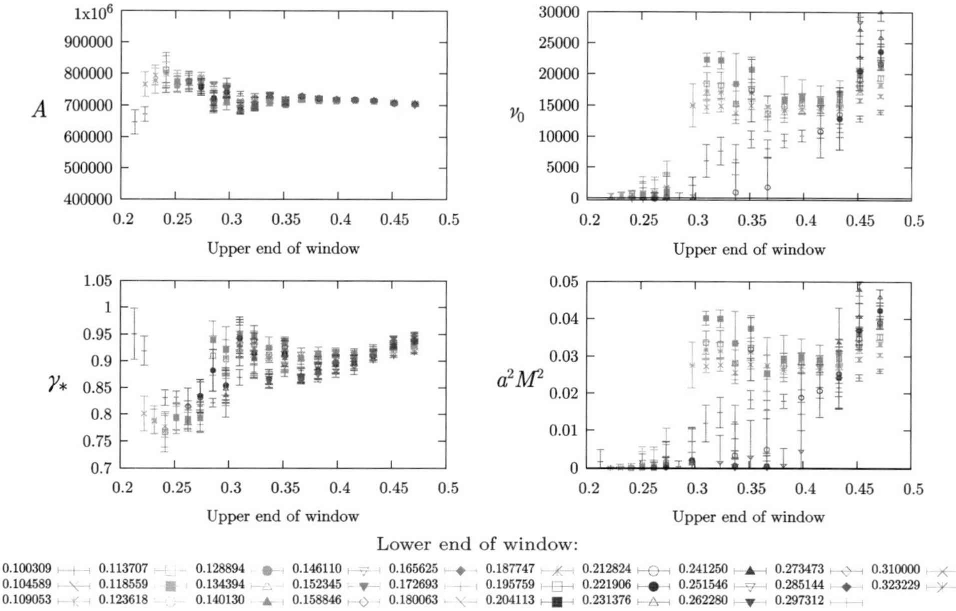

Fig. 6. Plots of the fitted parameters A, $\overline{\nu}_0$, γ_* and m^2 as a function of the window bounds. A region of stability is identified around $\gamma_* = 0.92(1)$.

look to fit

$$a^{-4}\overline{\nu}(\Omega) \approx a^{-4}\overline{\nu}_0(m) + A\left[(a\Omega)^2 - (aM)^2\right]^{\frac{2}{1+\gamma_*}} .$$

Since M here is a renormalised mass and not known *a priori*, we have four parameters to fit: $\overline{\nu}_0$, A, $(aM)^2$, and γ_*, However, since the scaling region is also not known *a priori*, an additional fitting degree of freedom is introduced in the choice of fitting window. Since the fit is highly sensitive to initial conditions, each possible window is fitted many times for varying starting parameters, and uncertainties estimated using the bootstrap method. We then look for a region where the fit is stable for a range of window lengths and positions, as can be seen in Fig. 6; we see this gives a best fit of the anomalous dimension of 0.92(1). Further details of the fitting procedure are given in[10].

5. Conclusions

We have extended our previous investigation of SU(2) with one adjoint Dirac flavour to a second value of β, where we have a preliminary set of observables available. At both values of β we see behaviour consistent with conformality and inconsistent with QCD-like confinement: namely, monotonic change of all spectral quantities, whose ratios are flat, consistent with hyperscaling behaviour. At one value of β, we find a large anomalous dimension of the chiral condensate, $0.9 \leq \gamma_* \leq 0.95$, with a best fit of 0.92(1). Work is ongoing to extend the study at $\beta = 2.2$ to the full set of observables studied for $\beta = 2.05$—including the anomalous dimension of the chiral condensate, which we currently have insufficient data to determine conclusively—after which we intend to study sufficiently many values of β to allow a continuum limit extrapolation.

Acknowledgements

Computational resources used in this work include the Dirac and Dirac2 facilities, and the Blue Joule machine at the Hartree Centre, supported by STFC, and the HPC Wales cluster supported through WEFO.

References

1. S. Catterall and F. Sannino, Minimal walking on the lattice, *Phys. Rev.* **D76**, p. 034504 (2007).
2. L. Del Debbio, B. Lucini, A. Patella, C. Pica and A. Rago, Conformal versus confining scenario in SU(2) with adjoint fermions, *Phys. Rev.* **D80**, p. 074507 (2009).
3. Y. Aoki, Spectral properties of $N_{\rm f} = 8$ SU(3) gauge theory, *these proceedings* .
4. A. Hasenfratz, Approaching conformality with 4+8 flavors, *these proceedings* .
5. E. Rinaldi, Investigation of the scalar spectrum in 8-flavor QCD, *these proceedings* .
6. D. Schaich, Finite-temperature study of eight-flavor SU(3) gauge theory, *these proceedings* .
7. A. Hasenfratz, The step scaling function of 2 flavor sextet SU(3) fermions, *these proceedings* .
8. J. Kuti, Toward the minimal realization of a light composite Higgs, *these proceedings* .
9. C. H. Wong, Baryon spectrum in the composite sextet model, *these proceedings.*
10. A. Athenodorou *et al.*, Infrared regime of SU(2) with one adjoint Dirac flavor, *Phys. Rev.* **D91**, p. 114508 (2015).
11. L. Del Debbio *et al.*, Mesonic spectroscopy of Minimal Walking Technicolor, *Phys. Rev.* **D82**, p. 014509 (2010).
12. L. Del Debbio *et al.*, The infrared dynamics of Minimal Walking Technicolor, *Phys. Rev.* **D82**, p. 014510 (2010).
13. A. Patella, A precise determination of the psibar-psi anomalous dimension in conformal gauge theories, *Phys. Rev.* **D86**, p. 025006 (2012).
14. G. Bergner *et al.*, Towards the spectrum of low-lying particles in supersymmetric Yang-Mills theory, *JHEP* **1311**, p. 061 (2013).

Lattice Study of the Higgs-Yukawa model with a Dimension-6 Operator

C.-J. David Lin

Institute of Physics, National Chiao-Tung University, Hsinchu 30010, Taiwan

We perform lattice investigation of the Higgs-Yukawa model with a dimension-6 operator as the prototype of physics beyond the standard model (BSM). By studying the Higgs boson mass, we can obtain bounds on the coupling for this BSM operator.

SU(2) Lattice Gauge Theory with Many Flavors of Domain-Wall Fermions

Hideo Matsufuru

KEK, High Energy Accelerator Research Organization,
1-1 Oho, Tsukuba, Ibaraki 305-0801, Japan

We numerically study the SU lattice gauge theory with many flavors. Dynamical simulation is performed with the standard domain-wall fermions in fundamental representation at numbers of flavors $N_f = 2, 4, 6$, and 8. N_f dependence of the static potential, meson spectrum, and the eigenvalue spectrum of the Dirac operator is investigated so as to determine the phase structure.

Recent Results on Renormalization-Group Evolution of Theories with Gauge, Fermion, and Scalar Fields

R. Shrock

C. N. Yang Institute for Theoretical Physics, Stony Brook University, Stony Brook, NY 11794, USA
E-mail: robert.shrock@stonybrook.edu

We discuss recent results on renormalization-group evolution of several types of theories. First, we consider asymptotically free vectorial gauge theories with various fermion contents and discuss higher-loop calculations of the UV to IR evolution in these theories, including an IR zero of the beta function and the value of the anomalous dimension γ_m at this point, together with comparisons with lattice measurements. Effects of scheme transformations are discussed. We then present a novel way to determine the value of N_{TC} in an SU(N_{TC}) technicolor model from a particular embedding in an extended technicolor theory. Finally, we analyze the renormalization-group behavior of several non-asymptotically free theories, including a U(1) gauge theory, a non-Abelian gauge theory with many fermions, an O(N) $\lambda|\vec{\phi}|^4$ scalar theory, and Yukawa theories.

Keywords: Renormalization group, IR, UV fixed points, Symmetry breaking.

1. Introduction

A subject of fundamental interest in quantum field theory is the dependence of the interaction coupling on the Euclidean momentum scale, μ, where it is measured. This dependence is described by the renormalization group (RG) and associated beta function of the theory. In this talk we discuss some of our recent results in this area and an application to physics beyond the Standard Model.

Since this conference on the Strong Coupling Gauge Theories (SCGT) and the Origin of Mass at the KMI Institute at Nagoya University has a dedication to Prof. Shoichi Sakata, it is appropriate to begin by recalling his pioneering work emphasizing the role of compositeness in particle physics. His idea of compositeness was confirmed with the understanding of hadrons as color-singlet bound states of quarks and gluons. The question of compositeness is still with us, in a different guise, namely whether the Higgs-like particle discovered at the Large Hadron Collider at CERN is composite or pointlike. In turn, this is connected with the continuing question of the origin of electroweak symmetry breaking (EWSB). The group of Prof. Koichi Yamawaki has made many important contributions to models of dynamical EWSB, and the SCGT conferences have been a key venue for discussion of progress in this area. It is also fitting to recall that Maki, Nakagawa, and Sakata (MNS) were the first to suggest that the neutrino weak eigenstates are linear combinations of mass eigenstates here at Nagoya University[1]. As is well known, this suggestion has been verified with the discovery of neutrino oscillations.

2. UV to IR Evolution of an Asymptotically Free Gauge Theory

We begin by considering an asymptotically free (AF) vectorial gauge theory with a (non-Abelian) gauge group G and N_f massless fermions in a representation R. We denote the gauge coupling as $g(\mu)$ and define $\alpha(\mu) = g(\mu)^2/(4\pi)$. The beta function is $\beta \equiv \beta_\alpha = d\alpha/dt$, where $dt = d\ln\mu$, with series expansion

$$\beta = -2\alpha \sum_{\ell=1}^{\infty} b_\ell \, a^\ell = -2\alpha \sum_{\ell=1}^{\infty} \bar{b}_\ell \, \alpha^\ell \ , \tag{1}$$

where $a = \alpha/(4\pi)$, ℓ denotes the number of loops, and $\bar{b}_\ell = b_\ell/(4\pi)^\ell$. The coefficients b_1 and b_2 are independent of the scheme used for regularization, while the b_ℓ with $\ell \geq 3$ are scheme-dependent. We have carried out higher-loop calculations of the evolution of this theory from the ultraviolet (UV) to the infrared (IR)[2-8], several papers being in collaboration with T. Ryttov. Some related work is[9-11]. We have also carried out corresponding studies for an asymptotically free vectorial $\mathcal{N} = 1$ supersymmetric gauge theory with N_f pairs of chiral superfields transforming in the R and $\bar{R}$ representations of the gauge group[3,4,12] (see also[13]).

Concerning the properties of the beta function away from the UV fixed point (UVFP) at the origin, there are two general possibilities. First, β may not have any IR zero (reached from evolution from weak coupling in the UV). In this case, as μ decreases, the gauge coupling increases, eventually beyond the perturbatively calculable region, as in QCD. The second type of situation, which we have studied, occurs if β does have a zero, at a value denoted α_{IR}, so that as μ decreases, $\alpha(\mu)$ increases from 0 toward α_{IR}. The infrared properties in this second case then depend on whether α_{IR} is larger or smaller than a critical minimal value, denoted α_{cr}, at which the gauge interaction causes the formation of a bilinear fermion condensate, with associated spontaneous chiral symmetry breaking (SχSB). If $\alpha_{IR} < \alpha_{cr}$, then the zero of β at α_{IR} is an exact IR fixed point (IRFP) of the RG. In this case, the IR behavior is that of a non-Abelian Coulomb phase (NACP) without any SχSB. As a function of N_f for a given fermion representation R, this NACP extends downward from $N_f = N_{f,b1z}$ to a value denoted $N_{f,cr}$.

If $\alpha_{IR} > \alpha_{cr}$, then the UV to IR evolution does not actually reach α_{IR}; instead, when μ decreases below a scale Λ such that $\alpha(\Lambda) > \alpha_{cr}$, there is S$\chi$SB, and the fermions pick up dynamical masses of order Λ. In this case, in the low-energy effective field theory applicable for $\mu < \Lambda$, one integrates these fermions out, and the β function becomes that of a pure gauge theory, which has no (perturbative) IR zero. We denote the IR zero of the n-loop beta function, $\beta_{n\ell}$, as $\alpha_{IR,n\ell}$.

The value of $N_{f,cr}$ is not known for this gauge theory, and one of the reasons for our higher-loop calculations is to gain insight into it. A second reason is that if α_{IR} is only slightly greater than α_{cr}, then, as $\alpha(\mu)$ approaches α_{IR}, the theory is approximately scale-invariant (also called "walking" or approximately dilatation-invariant), so when the SχSB occurs and breaks this approximate symmetry, it leads to a pseudo-Nambu-Goldstone boson (PNGB), the dilaton[14]. Continuum[15] and lattice[16] studies find that this dilaton PNGB can be light. A route by which dynamical EWSB models could remain viable is to have walking behavior, and a resultant dilaton PNGB that can match, to the requisite degree of accuracy, experimental measurements of the Higgs-like scalar.

For a given gauge group G and fermion representation R, β has an IR zero for N_f in the interval $I: \ N_{f,b2z} < N_f < N_{f,b1z}$, where

$$N_{f,b1z} = \frac{11C_A}{4T_f} \ , \quad N_{f,b2z} = \frac{34C_A^2}{4T_f(5C_A + 3C_f)} \ , \tag{2}$$

where C_A, C_f are Casimir invariants and T_f is the trace invariant (with $C_A = N$ and, e.g., $C_f = (N^2-1)/(2N)$, $T_f = 1/2$ for the fundamental representation in SU(N)). For example, for SU(3), this interval I is $8.05 < N_f < 16.5$ (integer values of N_f are understood implicitly). At the two-loop level, $\alpha_{IR,2\ell} = -4\pi b_1/b_2$, which is physical where $b_2 < 0$, i.e., in the interval I.

Using the three-loop and four-loop coefficients that had been calculated in the $\overline{\text{MS}}$ scheme[17], we found that

$$\alpha_{IR,3\ell} < \alpha_{IR,2\ell} \ . \tag{3}$$

In[3] we proved this inequality more generally. First, we proved that if $b_3 < 0$, then the inequality holds. Next, given that the two-loop beta function $\beta_{\alpha,2\ell}$ has an IR zero, since this is a scheme-independent property, one may require a physically acceptable scheme to maintain the existence of this IR zero at loop level $n \geq 3$. Imposing this condition and noting that $\alpha_{IR,3\ell}$ involves a square root $\sqrt{b_2^2 - 4b_1b_3}$, together with the fact that $b_1 > 0$ for asymptotic freedom and that $N_{f,b2z}$ occurs where b_2 passes through zero, it follows that the scheme must have $b_3 < 0$. The inequality (3) then follows.

Going from the three-loop to four-loop level, there is a slight change in the value of the IR zero, but this change is smaller than the decrease from the two-loop to three-loop level, so $\alpha_{IR,4\ell} < \alpha_{IR,2\ell}$. As an example, for SU(3) and $N_f = 12$ fermions in the fundamental represention, we find

$$\text{SU}(3), \ N_f = 12: \quad \alpha_{IR,2\ell} = 0.754, \quad \alpha_{IR,3\ell} = 0.435, \quad \gamma_{IR,4\ell} = 0.470 \tag{4}$$

Since $\alpha_{IR,n\ell}$ increases as N_f decreases from $N_{f,b1z}$ at the upper end of the interval I to $N_{f,b2z}$ at the lower end of this interval, and since the lower end of the NACP is determined by where α_{IR} increases through α_{cr}, our results suggest that this lower end and the corresponding value $N_{f,cr}$ is smaller than indicated by early estimates based on solutions of truncated Schwinger-Dyson equations (which gave $N_{f,cr} \simeq 4N$). This is in agreement with results from lattice studies. For example, for SU(3), most lattice groups find that for fermions in the fundamental representation, the $N_f = 12$ theory is in the NACP[16].

In[2] with Ryttov we also obtained the anomalous dimension of the fermion bilinear, $\gamma_m \equiv \gamma$ evaluated at the IR zero of the beta function, at n-loop order, with $n = 2,\ 3,\ 4$. In general, we found that $\gamma_{IR,n\ell}$ decreases as the loop order increases from 2 to 3 to 4. For example, for SU(3) and $N_f = 12$ fermions in the the fundamental representation, we obtained the results

$$\mathrm{SU}(3),\ N_f = 12: \quad \gamma_{IR,2\ell} = 0.773, \qquad \gamma_{IR,3\ell} = 0.312, \quad \gamma_{IR,4\ell} = 0.253 \tag{5}$$

Our four-loop result is in excellent agreement with several recent lattice measurements[18,19] (current reviews of lattice measurements are in[16]). Since there is a substantial decrease of γ_{IR} from two- to four-loop level, this shows the value of these higher-loop calculations. A necessary condition for a perturbative calculation to be reliable is that higher-order contributions do not modify the result too much. Our calculations satisfied this criterion, as is evident from the illustrative results for $\alpha_{IR,n\ell}$ and $\gamma_{IR,n\ell}$ in (4) and (5). Our higher-loop calculations of α_{IR} and γ enable one to explore the theory reliably down to smaller values of N_f and thus stronger couplings.

The values of $\alpha_{IR,n\ell}$ and $\gamma_{IR,n\ell}$ for fermions in the fundamental representation are similar in theories with different values of N and N_f but the same or similar values of the ratio $r = N_f/N$. To elucidate this behavior, we studied the limit $N \to \infty$, $N_f \to \infty$ with N_f/N fixed in[3,4] and showed that the behavior results because of the rapid approach of $\alpha_{IR,n\ell} N$ and $\gamma_{IR,n\ell}$ to their limiting values, with correction terms vanishing quite quickly, like $1/N^2$.

Theories with fermions in higher-dimensional representations of G are also of interest[20]. Higher-loop calculations have also been performed in[2,11] for SU(N) gauge theories with N_f fermions in the adjoint, symmetric and antisymmetric rank-2 tensor representations.

3. Scheme-Dependence in Calculation of IR Fixed Point

We have studied the scheme dependence of higher-loop calculations of RG evolution in a series of papers[5-8]. Since the beta function coefficients b_n with $n \geq 3$ are scheme-dependent, this is also true of $\alpha_{IR,n\ell}$ for $n \geq 3$. In order to assess the uncertainty in the higher-loop calculation of RG evolution due to this, we have constructed scheme transformations and determined how much of a change they induce in $\alpha_{IR,n\ell}$. A scheme transformation (ST) is a map between α and α' or equivalently, a and a', where $a = \alpha/(4\pi)$, which can be expressed as $a = a' f(a')$. The transformation function satisfies $f(0) = 1$, since the ST has no effect in the limit of zero coupling. The scheme transformation function has the series expansion

$$f(a') = 1 + \sum_{s=1}^{s_{max}} k_s (a')^s \ , \tag{6}$$

where s_{max} may be finite or infinite. In terms of the transformed coupling,

$$\beta_{\alpha'} = -2\alpha' \sum_{\ell=1}^{\infty} b'_\ell (a')^\ell \ . \tag{7}$$

We have calculated the b'_ℓ in terms of the b_ℓ and k_s [5]. In addition to reproducing the known results $b'_1 = b_1$ and $b'_2 = b_2$, we obtain $b'_3 = b_3 + k_1 b_2 + (k_1^2 - k_2) b_1$, and so forth for higher ℓ.

Physically acceptable scheme transformations must satisfy a number of conditions: (i) as condition C_1, the ST must map a real positive α to a real positive α'; (ii) C_2: the ST should not map a moderate value of α for which perturbative calculations may be reliable, to an excessively large value of α' where perturbative calculations are inapplicable; (iii) C_3: the Jacobian $J = d\alpha/d\alpha'$ should not vanish or diverge in the region of α and α' of interest, since otherwise the ST is singular; and (iv) C_4: since the existence of an IR zero of β is a scheme-independent property, the ST should satisfy the property that β_α has an IR zero if and only if $\beta_{\alpha'}$ has an IR zero. These four conditions can always be satisfied by STs near the UV fixed point at the origin, but they are a significant constraint on an ST applied in the vicinity of a zero of the beta function away from the origin. An example is provided by the following ST depending on a parameter r: $a = (1/r)\tanh(ra')$, with inverse $a' = (2r)^{-1} \ln[(1+ra)/(1-ra)]$. For $ra << 1$, as at a UVFP, this is acceptable, but as ra approaches 1 from below, as it could at a IRFP with $ra_{IR} \sim 1$, the ST fails to be acceptable.

We have studied the scheme dependence of the IR zero of the beta function by using several scheme transformations that are acceptable away from the origin[5-8]. One of these is $a = (1/r)\sinh(ra')$, which has the inverse transformation, $a' = (1/r)\ln[ra + \sqrt{1+(ra)^2}\ ]$[5]. Here, with no loss of generality, one can take $r > 0$. Two recent scheme transformations that we have constructed and applied are[8]

$$S_{L_r}: \quad f(a') = \frac{\ln(1+ra')}{ra'} \ , \tag{8}$$

with inverse

$$a' = \frac{e^{ra} - 1}{r} \tag{9}$$

and

$$S_{Q_r}: \quad a = \frac{a'}{1 - ra'} \ , \tag{10}$$

with inverse

$$a' = \frac{a}{1 + ra} \ . \tag{11}$$

In[8], continuing the work in[5-7], we have applied these transformations to the IR zero of the beta function in AF vectorial gauge theories. The results are presented in Tables 1-4 in[8]. As before, we find that for moderate r, the effects of scheme transformations are rather mild. Our results provide a quantitative measure of the scheme dependence of the location of an IR zero of β. In general, the effect of scheme dependence tends to be reduced (i) for a given N and N_f, as one calculates to higher-loop order, and (ii) for a given N, as $N_f \to N_{f,b1z}$, so $\alpha_{IR,n\ell} \to 0$.

Since the β function coefficients b_ℓ with $\ell \geq 3$ are scheme-dependent, one may investigate the extent to which it is possible to construct and apply scheme transformations that render these higher-loop coefficients zero. The possibility of this was remarked on by 't Hooft. We have constructed an explicit scheme transformation that can do this in the vicinity of zero coupling constant. However, we have also shown that it is much more difficult to try to do this at a zero of β away from the origin. Specifically, we construct a scheme transformation, denoted S_{R,m,k_1} with $s_{max} = m \geq 2$, that removes the terms in the beta function from loop order 3 up to $m+1$, inclusive, for small coupling. In the limit $m \to \infty$, this transforms to the 't Hooft scheme. Our solution for $S_{R,2,k_1}$ has[6,7].

$$k_2 = \frac{b_3}{b_1} + \frac{b_2}{b_1} k_1 + k_1^2 . \tag{12}$$

Our solution for $S_{R,3,k_1}$ has

$$k_3 = \frac{b_4}{2b_1} + \frac{3b_3}{b_1} k_1 + \frac{5b_2}{2b_1} k_1^2 + k_1^3 , \tag{13}$$

and so forth for S_{R,m,k_1} with higher m. The simplest case takes $k_1 = 0$[5]. In addition to the successful application near the origin, $\alpha = 0$, we have shown that this ST S_{R,m,k_1} can be applied over part, but not all, of the interval I where the 2-loop beta function has an IR zero. As shown in[6,7], the inclusion of a nonzero parameter k_1 can be used to optimize the fraction of the interval I over which the S_{R,m,k_1} transformation is physically acceptable, satisfying the conditions C_1-C_4 stipulated above.

4. Application to Models of Dynamical Electroweak Symmetry Breaking

In addition to studying the UV to IR evolution of AF vectorial gauge theories, we have also performed such studies for AF chiral gauge theories, including the recent works[21,22]. These have an important application to physics beyond the Standard Model. Although the Higgs-like scalar discovered at the LHC is consistent with being the SM Higgs, naturalness arguments still motivate studies of extensions of the SM, including possible models with dynamical electroweak symmetry breaking, such as technicolor (TC). A TC theory features an asymptotically free vectorial TC gauge symmetry and a set of TC-nonsinglet, SM-nonsinglet fermions. The TC theory becomes strongly coupled at the TeV scale, confining and producing technifermion condensates with associated spontaneous chiral symmetry breaking (SχSB) and dynamical EWSB. A crucial property of viable TC theories is walking behavior[14]. The Higgs-like scalar is then the technidilaton resulting from the SSB of the approximate scale invariance of the walking TC theory.

To give masses to SM fermions, one embeds the TC theory in a larger, extended technicolor (ETC) theory. With an SU(N_{TC}) TC gauge group, the SU(N_{ETC}) ETC theory gauges the SM fermion generation index and combines it with the TC gauge index, so

$$N_{ETC} = N_{TC} + N_{gen.} \ , \tag{14}$$

where $N_{gen.}$ denotes the number of SM fermion generations. The ETC theory is an asymptotically free chiral gauge theory that becomes strongly coupled and self-breaks in $N_{gen.}$ stages down to TC via the formation of various fermion condensates. Reasonably UV-complete ETC theories have been constructed that exhibit the requisite self-breaking[23]. TC/ETC theories face many challenges, including precision EW constraints, flavor-changing neutral current processes, t-b mass splitting, CKM mixing, ability to produce a light, Higgs-like scalar, ability to produce very small neutrino masses, etc.

In technicolor theories using an SU(N_{TC}) gauge group, the value of N_{TC} is not, *a priori*, determined and is typically chosen by phenomenological criteria. Recently, with Kurachi and Yamawaki, we have presented a novel way to determine N_{TC} from the embedding of a one-family technicolor model, with fermions in the fundamental represention of SU(N_{TC}), in an extended technicolor theory, and have used it to deduce that $N_{TC} = 4$ in this framework[24].

The SM-singlet part of this ETC theory has a chiral fermion in the antisymmetric rank-2 tensor rep. of SU(N_{ETC}), denoted A, together with ($N_{ETC} - 4$) copies ("flavors") of chiral fermions in the conjugate fundamental rep., denoted $\bar{F}$:

$$\psi^{ij}_R = \psi^{[ij]}_R \ , \qquad \chi_{i,s,R} \ , \quad 1 \le s \le N_{ETC} - 4 \ , \tag{15}$$

where i, j are ETC gauge indices and s is a flavor index. The ETC gauge interaction is asymptotically free, so as this ETC theory evolves from the UV to the IR, $\alpha_{ETC}(\mu)$ grows, and, at a scale denoted Λ_1, leads to fermion condensation in the channel $A \times \bar{F} \to F$, breaking SU(N_{ETC}) to SU($N_{ETC} - 1$). The associated condensate is

$$\langle \sum_{j=2}^{N_{ETC}} \psi^{1j\ T}_R C \chi_{j,1,R} \rangle \ , \tag{16}$$

where, by notation convention, we take the ETC gauge index $i = 1$ in ψ^{ij}_R and the flavor index $s = 1$ in $\chi_{j,s,R}$. The fermions ψ^{1j}_R and $\chi_{j,1,R}$ with $2 \le j \le N_{ETC}$ involved in this condensate gain dynamical masses of order Λ_1 and are integrated out of the SU($N_{ETC} - 1$) low-energy effective theory (LEET) applicable at scales $\mu < \Lambda_1$.

This SU($N_{ETC} - 1$) theory is again asymptotically free, with a gauge coupling that continues to grow, and we infer that at a lower scale, Λ_2, there is again condensation in the $A \times \bar{F} \to F$ channel, breaking SU($N_{ETC} - 1$) to SU($N_{ETC} - 2$). The associated condensate is $\langle \sum_{j=3}^{N_{ETC}} \psi^{2j\ T}_R C \chi_{j,2,R} \rangle$, where, by notation convention, we take the gauge index $i = 2$ in ψ_R^{ij} and the flavor index $s = 2$ in $\chi_{j,s,R}$. The fermions ψ_R^{2j} and $\chi_{j,2,R}$ with $3 \le j \le N_{ETC}$ involved in this condensate gain dynamical masses of order Λ_2 and are integrated out of the SU($N_{ETC} - 2$) LEET operative at $\mu < \Lambda_2$.

This sequential self-breaking of the SU(N_{ETC}) theory continues iteratively in $N_{ETC} - 4$ stages, using the $N_{ETC} - 4$ flavors of $\chi_{j,s,R}$ fermions, reducing the original SU(N_{ETC}) ETC gauge symmetry to the (vectorial) SU(N_{TC}) subgroup, with the broken indices being generation indices. Hence,

$$N_{gen.} = N_{ETC} - 4 \ . \tag{17}$$

Substituting this in the eq. $N_{ETC} = N_{gen.} + N_{TC}$, we get $N_{TC} = 4$. We have thus determined N_{TC} from the structure of the specific ETC theory in which our TC theory is embedded. (Note that $N_{gen.}$ cancels out in the algebra.) Setting $N_{gen.} = 3$, we thus get an SU(7) ETC gauge group. Interestingly, this value $N_{TC} = 4$ agrees with the preferred value obtained in [25] from a fit to the Higgs-like scalar as a technidilaton. In general, a technidilaton-like composite scalar in this theory contains $\bar{F}F$, AA, and techni-glue components.

This ETC model naturally accounts for the mass hierarchy in the SM fermion generations, since the SM fermion masses in the i'th generation result from exchange of ETC vector bosons with mass Λ_i and, in the ETC boson propagators, $\Lambda_1^{-2} \ll \Lambda_2^{-2} \ll \Lambda_3^{-2}$. The resultant running fermion masses $m_{f_i}(p)$ are constant up to Λ_i and have the power-law decay $m_{f_i}(p) \propto (\Lambda_i/p)^2$ for $p \gg \Lambda_i$ [26].

The SU$(4)_{TC}$ theory has one SM family of technifermions and the (SM-singlet) A fermion, which is self-conjugate in SU(4). At the TeV scale, the A fermion forms a condensate in the channel $A \times A \to 1$, which is invariant under TC and the SM. Technifermion condensates $\langle \bar{F}F \rangle$ cause EWSB in the usual way. In order for this model to be viable, it must exhibit walking behavior and must have quite strong suppression of the EW S parameter [27]. The 60 PNGBs and the techni-vector mesons must also also gain sufficiently large masses of order a few TeV to agree with current LHC bounds [25,28]. Further ingredients are needed to account for actual SM fermion masses and mixings, e.g., t-b mass splitting.

In addition to the application to dynamical EWSB, this model is of field-theoretic interest for the insight that it provides on how the structure of a low-energy effective field theory - here the TC theory - is determined by its embedding in an ultraviolet completion, the ETC theory.

5. Studies of RG Flows in Infrared-Free Gauge Theories

If the β function of a theory is positive near zero coupling, then this theory is IR-free; as μ decreases from the UV to the IR, the coupling decreases to zero (given that the theory starts in the basin of attraction of the UVFP at zero coupling). Put the other way, if one starts with a small coupling, then as μ increases, the coupling grows. Eventually (before reaching any Landau pole), the coupling would become too large for perturbative calculations to be applicable or it might approach a UV fixed point. There is thus interest in investigating whether an IR-free theory might have such a UV fixed point (UV zero of β). In[29] we investigated the possibility of such a UV zero in the n-loop beta functions of two IR-free vectorial gauge theories, namely a U(1) gauge theory and a non-Abelian gauge theory with sufficiently many fermions that the beta function is positive.

The U(1) gauge theory is taken to contain N_f fermions of charge q which can, without loss of generality, be set equal to 1. We write the beta function as in Eq. (1) but with the minus sign prefactor removed, so that $b_1 > 0$ corresponds to the IR free theory. The scheme-independent 1-loop and 2-loop coefficients are

$$b_1 = \frac{4N_f}{3} , \qquad b_2 = 4N_f . \tag{18}$$

Since these coefficients have the same sign, the two-loop beta function, $\beta_{\alpha,2\ell}$, does not have a UV zero, and this is the maximal scheme-independent information that one has about it. Using the coefficients calculated up to five loops in the $\overline{\text{MS}}$ scheme[30], we have investigated the beta function for a possible UV zero. We note that (for all N_f) b_3 and b_4 are negative, while b_5 is positive. We present some illustrative results in Table 1 a necessary condition for the perturbatively calculated

Table 1. Values of $\alpha_{UV,n\ell}$ for a U(N) gauge theory with N_f fermions at the n-loop level, for $2 \le n \le 5$.

N_f	$\alpha_{UV,2\ell}$	$\alpha_{UV,3\ell}$	$\alpha_{UV,4\ell}$	$\alpha_{UV,5\ell}$
1	–	10.2720	3.0400	–
2	–	6.8700	2.4239	–
3	–	5.3689	2.0776	–
4	–	4.5017	1.8463	–
5	–	3.9279	1.67685	2.5570
10	–	2.5871	1.2135	1.3120
20	–	1.7262	0.8483	–
100	–	0.7081	0.33265	–
500	–	0.3038	0.1203	–
10^3	–	0.2127	0.07678	–
10^4	–	0.016614	0.016965	–

β function to yield evidence for a stable UV zero is that it should remain present when one increases the loop order, and the fractional change in the value of the UV zero should decrease going from n to $n+1$ loops. As is evident from Table 1, we find that the UV zeros that we have calculated at $\ell = 3,\ 4,\ 5$ loop order for a large range of N_f values do not satisfy this necessary condition. Hence, our results give evidence against a UVFP in U(1) gauge theory for general N_f.

In addition to our analysis of the perturbative beta function up to five-loop order, we have carried out a nonperturbative analysis in the limit

$$N_f \to \infty \quad \text{with finite} \quad y(\mu) \equiv N_f\, a(\mu) = \frac{N_f\, \alpha(\mu)}{4\pi} \ . \tag{19}$$

We set $b_1 = b_{1,1} N_f$ with $b_{1,1} = 4/3$. Further,

$$b_\ell = \sum_{k=1}^{\ell-1} b_{\ell,k}\, N_f^k \quad \text{for } \ell \geq 2 \ , \tag{20}$$

where the $b_{\ell,k}$ are independent of N_f. Hence,

$$b_\ell \propto N_f^{\ell-1} \quad \text{for} \ \ \ell \geq 2 \quad \text{as} \ \ N_f \to \infty \ . \tag{21}$$

We call this the LNF (large-N_f) limit. We define the finite quantities

$$\check{b}_\ell \equiv \frac{b_\ell}{N_f^{\ell-1}} \quad \text{for } \ell \geq 2 \ , \tag{22}$$

so

$$\lim_{N_f \to \infty} \check{b}_\ell = b_{\ell,\ell-1} \quad \text{for } \ell \geq 2 \ . \tag{23}$$

We define a rescaled β function that is finite in the LNF limit as $\beta_y \equiv \beta_\alpha N_f$. Then

$$\beta_y = 8\pi b_{1,1}\, y^2 \Big[1 + \frac{1}{b_{1,1} N_f} \sum_{\ell=2}^{\infty} b_\ell\, y^{\ell-1} \Big] \ . \tag{24}$$

The condition that the the n-loop rescaled beta function $\beta_{y,n\ell}$ has a zero at $y \neq 0$ is the equation

$$1 + \frac{1}{b_{1,1} N_f} \sum_{\ell=2}^{n} b_\ell\, y^{\ell-1} = 0 \ . \tag{25}$$

In the LNF limit, of the $n-1$ roots of this equation, the relevant one has the approximate form

$$y_{_{UV,n\ell}} \sim \Big(- \frac{b_{1,1} N_f}{b_{n,n-1}} \Big)^{\frac{1}{n-1}} \ . \tag{26}$$

Hence, $\beta_{y,n\ell}$ has a zero for $y \neq 0$ in the LNF limit if and only if $b_{n,n-1}$ is negative, which is not, in general true. Further, even if it were true for a given loop order n, in the LNF limit, $\lim_{N_f \to \infty} y_{_{UV,n\ell}} = \infty$.

Moreover, one can reexpress β_y as a series in powers of $\nu \equiv 1/N_f$:

$$\beta_y = 8\pi b_{1,1}\, y^2 \Big[1 + \sum_{s=1}^{\infty} F_s(y) \nu^s \Big] \ . \tag{27}$$

An exact integral representation of $F_1(y)$ is known, which thus incorporates all-order nonperturbative information[31]. We have used this representation to determine the signs of $b_{n,n-1}$ up to $n = 24$ loops. We find that these signs are scattered, and show no indication of an onset of negative signs, further supporting our conclusion.

Thus, we do not find evidence of a UVFP in a U(1) gauge theory with N_f massless charged fermions for large N_f.

We have carried out a corresponding analysis of an SU(N) gauge theory with N_f larger than the AF range, so that the theory is IR-free. We find the same conclusion as for the U(1) theory, namely evidence against a UV zero. In both the U(1) and non-Abelian theories, we have studied the effect of scheme transformations on our results.

6. Studies of RG Flows in an O(N) $\lambda|\vec{\phi}|^4$ Theory

It is also of interest to consider an O(N) $\lambda|\vec{\phi}|^4$ theory (in $d = 4$ spacetime dimensions), which is IR-free, in order to investigate whether there is evidence for a UV fixed point of the renormalization group. In[32] we carried out this study up to five-loop order.

As background, it may be recalled that in 1976 we obtained an exact solution of another IR-free theory, namely the O(N) nonlinear σ model in $d = 2 + \epsilon$ spacetime dimensions, in the $N \to \infty$ limit[33]. From this solution, we calculated the resultant beta function and showed that it did have a UV zero. For small ϵ, this is[33]

$$\beta(\lambda) = \frac{d\lambda}{dt} = \epsilon\lambda\Big(1 - \frac{\lambda}{\lambda_c}\Big) , \tag{28}$$

i.e.,

$$\beta(x) = \frac{dx}{dt} = \epsilon x\Big(1 - \frac{x}{x_c}\Big) , \tag{29}$$

where λ is the effective coupling, $\lambda_c = 2\pi\epsilon/N$, $x = \lim_{N\to\infty} \lambda N$, and $x_c = 2\pi\epsilon$. Thus, this theory has a UVFP at x_c, so that if initial value of $x < x_c$, then $x \nearrow x_c$ as $\mu \to \infty$.

Our normalization of the interaction term in the O(N) $\lambda|\vec{\phi}|^4$ theory is

$$\mathcal{L}_{int} = -\frac{\lambda}{4!}|\vec{\phi}|^4 , \tag{30}$$

where $|\vec{\phi}| = \sqrt{\sum_{i=1}^N \phi_i^2}$. The beta function is $\beta_\lambda = d\lambda/dt$, or equivalently, in terms of $a_\lambda = \lambda/(16\pi^2)$,

$$\beta_{a_\lambda} = \frac{da_\lambda}{dt} = a\sum_{\ell=1}^{\infty} b_\ell \, (a_\lambda)^\ell . \tag{31}$$

The scheme-independent coefficients are

$$b_1 = \frac{1}{3}(N+8) \tag{32}$$

and

$$b_2 = -\frac{1}{3}(3N+14) . \tag{33}$$

Although the two-loop beta function of this theory has a UV zero, it occurs at too large a value of the coupling for the perturbative calculation to be reliable, as shown by the fact that when we examine the zeros of β_{a_λ} to higher-loop order, we find that the three-loop beta function $\beta_{a_\lambda,3\ell}$ has no UV zero, and the four-loop and five-loop beta functions $\beta_{a_\lambda,4\ell}$ and $\beta_{a_\lambda,5\ell}$ differ considerably from the two-loop and three-loop beta functions where the two-loop beta function has a zero. This is illustrated for the case $N = 1$ in Fig. 1. Some results on UV zeros are listed in Table 2. We have studied this further with scheme transformations and Padé approximants, confirming our findings with $\beta_{a_\lambda,n\ell}$. We thus conclude that in the range of λ where the perturbative calculation of the n-loop beta function is reliable, the O(N) $\lambda|\vec{\phi}|^4$ theory does not exhibit evidence of a UV zero up to the level of five loops.

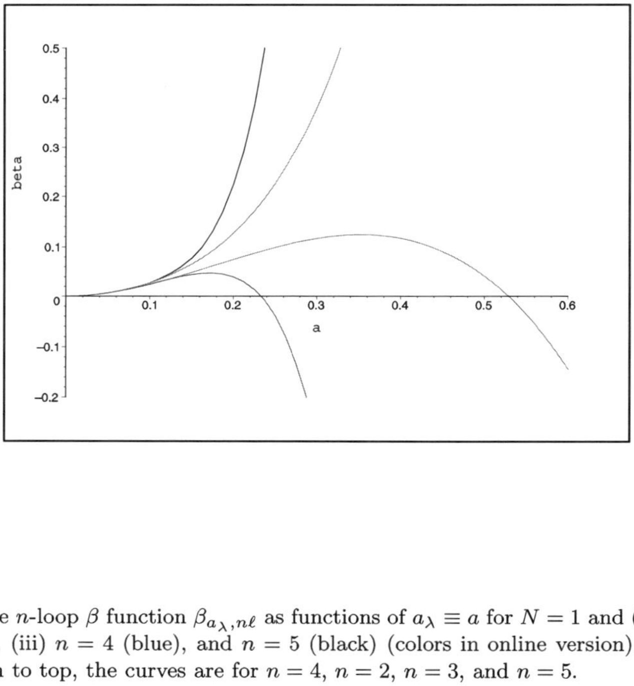

Fig. 1. Plot of the n-loop β function $\beta_{a_\lambda,n\ell}$ as functions of $a_\lambda \equiv a$ for $N = 1$ and (i) $n = 2$ (red), (ii) $n = 3$ (green), (iii) $n = 4$ (blue), and $n = 5$ (black) (colors in online version). At $a = 0.18$, going from bottom to top, the curves are for $n = 4$, $n = 2$, $n = 3$, and $n = 5$.

7. RG Flows in a Yukawa Theory

With E. Mølgaard, we have calculated RG flows for Yukawa theories in [34]. To study these flows in simple context, we first use the (one-generation) leptonic sector of the SM with the gauge fields turned off. This theory has a global chiral symmetry group: $SU(2)_L \otimes U(1)_Y$, forbidding bare fermion mass terms. This analysis is more complicated than the previous ones since the Yukawa theory involves two couplings

Table 2. Values of $a_{\lambda,UV,n\ell}$ for the O(N) $\lambda|\vec{\phi}|^4$ theory at the n-loop level, for $2 \le n \le 5$.

N	$a_{\lambda,UV,2\ell}$	$a_{\lambda,UV,3\ell}$	$a_{\lambda,UV,4\ell}$	$a_{\lambda,UV,5\ell}$
1	0.5294	–	0.2333	–
2	0.5000	–	0.2217	–
3	0.4783	–	0.2123	–
4	0.4615	–	0.2044	–
5	0.4483	–	0.1978	–
6	0.4375	–	0.1920	–
7	0.4286	–	0.1869	–
8	0.42105	–	0.1823	–
9	0.4146	–	0.1783	–
10	0.4091	–	0.1746	–
100	0.3439	–	0.1012	–
1000	0.3344	–	0.07241	0.02276
3000	0.3337	–	0.5475	0.008850
10^4	0.3334	–	–	0.003460

rather than one, so the RG flows depend on the Yukawa coupling y and the quartic scalar coupling λ. The normalizations of these couplings are those of the SM. We use the convenient variables $a_\lambda/(16\pi^2)$ as above and $a_y = y^2/(16\pi^2)$. The beta functions are $\beta_{a_\lambda} = da_\lambda/dt$ and $\beta_{a_y} = da_y/dt$. An important part of our analysis is a comparison of calculations obtained at different loop orders. In order for a perturbative result to be considered reliable, a necessary condition is that these must give similar results. Here we calculate the RG flows using β_{a_y} and β_{a_λ} computed to loop orders (1,1), (1,2), (2,1), (2,2), respectively, for this comparison. We show the results in Fig. 2. For small a_y and a_λ, where the perturbative calculations at different loop orders agree, the RG flow is to the IR-free zero of both beta functions at $a_y = a_\lambda = 0$, i.e., $y = \lambda = 0$. We have also analyzed a generalization of this Yukawa theory with a global $SU(N)_L \otimes U(1)_Y$ symmetry[34], with similar findings.

8. Conclusions

In conclusion, we have discussed our recent results on renormalization-group flows of various theories with gauge, fermion, and scalar fields. These include higher-loop calculations of the IR zero of the beta function and the anomalous dimension of the fermion bilinear at this zero. We have compared our calculation with lattice results for an intensively studied case, SU(3) with $N_f = 12$ fermions in the fundamental representation, finding excellent agreement with recent measurements. Higher-loop calculations involve scheme dependence, and we have presented a number of results on effects of scheme transformations. We have discussed interesting recent work with Kurachi and Yamawaki in which we determine the value of N_{TC} in an $SU(N_{TC})$ technicolor model from a particular embedding in an extended technicolor model. Finally, we have reported results on the RG flows of several IR-free theories, including a U(1) gauge theory, a non-Abelian gauge theory with many fermions, an

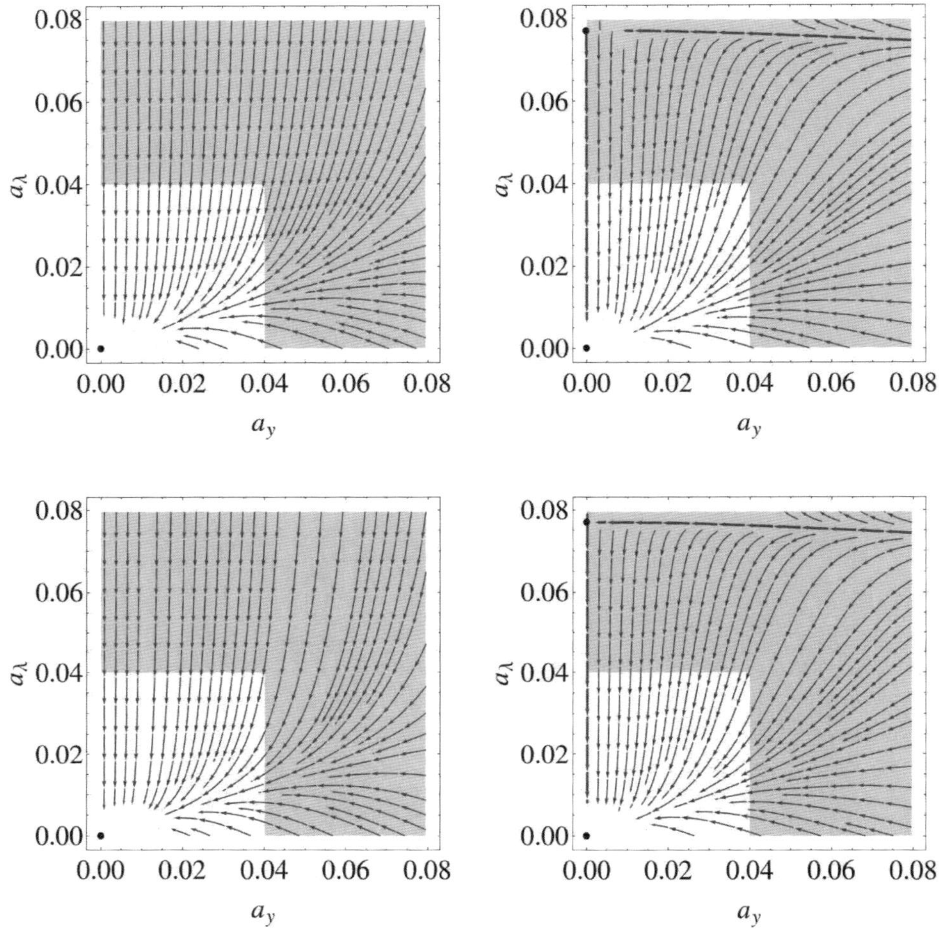

Fig. 2. RG flows obtained via integration of beta functions $\beta_{a_y,\ell}$ and $\beta_{a_\lambda,\ell'}$ for small a_y and a_λ, calculated for loop orders (ℓ,ℓ'): (1,1) (upper left); (1,2) (upper right); (2,1) (lower left); and (2,2) (lower right). Arrows are flows from UV to IR.

O(N) $\lambda|\vec{\phi}|^4$ theory, and Yukawa theories.

Acknowledgements

I thank K. Yamawaki and the other organizers for the invitation to speak at SCGT15 and for a very enjoyable and informative conference. Our research was partially supported by the grant NSF-PHY-13-16617

References

1. Z. Maki, M. Nakagawa, and S. Sakata, Prog. Theor. Phys. **28**, 243 (1962); see also M. Nakagawa, H. Okonogi, S. Sakata, and A. Toyoda, Prog. Theor. Phys. **30**, 727 (1963).
2. T. A. Ryttov and R. Shrock, Phys. Rev. D **83**, 056011 (2011) [arXiv:1011.4542].
3. R. Shrock, Phys. Rev. D **87**, 105005 (2013) [arXiv:1301.3209].
4. R. Shrock, Phys. Rev. D **87**, 116007 (2013) [arXiv:1302.5434].
5. T. A. Ryttov and R. Shrock, Phys. Rev. D **86**, 065032 (2012) [arXiv:1206.2366]; Phys. Rev. D **86**, 085005 (2012) [arXiv:1206.6895].
6. R. Shrock, Phys. Rev. D **88**, 036003 (2013) [arXiv:1305.6524].
7. R. Shrock, Phys. Rev. D **90**, 045011 (2014) [arXiv:1405.6244].
8. G. Choi and R. Shrock, Phys. Rev. D **90**, 125029 (2014) [arXiv:1411.6645].
9. T. Banks and A. Zaks, Nucl. Phys. B **196**, 189 (1982).
10. E. Gardi and M. Karliner, Nucl. Phys. B **529**, 383 (1998); E. Gardi and G. Grunberg, JHEP 03, 024 (1999).
11. C. Pica and R. Sannino, Phys. Rev. D **83**, 035013 (2011) [arXiv:1011.5917].
12. T. A. Ryttov and R. Shrock, Phys. Rev. D **85**, 076009 (2012), [arXiv:1202.1297].
13. R. Shrock, Phys. Rev. D **91**, 125039 (2015) [arXiv:1505.03588].
14. K. Yamawaki, M. Bando, and K. Matumoto, Phys. Rev. Lett. **56**, 1335 (1986); T. Appelquist, D. Karabali, and L. C. R. Wijewardhana, Phys. Rev. Lett. **57**, 957 (1986); T. Appelquist and L. C. R. Wijewardhana, Phys. Rev. D **35**, 774 (1987); B. Holdom, Phys. Rev. Lett. **60**, 1233 (1988).
15. M. Harada, M. Kurachi, and K. Yamawaki, Phys. Rev. D **70**, 033009 (2004); M. Kurachi and R. Shrock, JHEP **12**, 034 (2006).
16. Current reviews of lattice work are given at this SCGT15 conference.
17. T. van Ritbergen, J. A. M. Vermaseren, S. A. Larin, Phys. Lett. B **400**, 379 (1997).
18. A. Hasenfratz, A. Cheng, G. Petropoulos, and D. Schaich, PoS (Lattice 2012) 034 [arXiv:1207.7162] ($\gamma = 0.27(3)$); arXiv:1310.1124 ($\gamma \simeq 0.25$).
19. M.-P. Lombardo, K. Miura, T. J. Nunes, and E. Pallante, JHEP 12 (2014) 183 [arXiv:1410.0298] ($\gamma = 0.235(46)$).
20. See, e.g., D. D. Dietrich, F. Sannino, and K. Tuominen, Phys. Rev. D **72**, 055001 (2005); N. D. Christensen and R. Shrock, Phys. Lett. B **632**, 92 (2006); T. A. Ryttov and F. Sannino, Phys. Rev. D **76**, 105004 (2007); F. Sannino, Acta Phys. Polon. B **40**, 3533 (2009).
21. T. Appelquist and R. Shrock, Phys. Rev. D **88**, 105012 (2013) [arXiv:1310.6076].
22. Y. Shi and R. Shrock, Phys. Rev. D **91**, 045004 (2015) [arXiv:1411.2042].
23. T. Appelquist and R. Shrock, Phys. Lett. B **548**, 204 (2002); Phys. Rev. Lett. **90**, 201801 (2003); T. Appelquist, M. Piai, and R. Shrock, Phys. Rev. D **69**, 015002 (2004); T. A. Ryttov and R. Shrock, Phys. Rev. D **81**, 115013 (2010).

24. M. Kurachi, R. Shrock, and K. Yamawaki, Phys. Rev. D **91**, 055032 (2015) [arXiv:1501.06454].
25. S. Matsuzaki and K. Yamawaki, Phys. Rev. D **85**, 095020 (2012); Phys. Rev. D **86**, 115004 (2012); Phys. Lett. B **719**, 378 (2013).
26. N. D. Christenen and R. Shrock, Phys. Rev. Lett. **94**, 241801 (2005).
27. M. Kurachi, R. Shrock, and K. Yamawaki, Phys. Rev. D **76**, 035003 (2007).
28. R. S. Chivukula, M. Narain, and J. Womersley, in http://pdg.lbl.gov.
29. R. Shrock, Phys. Rev. D **89**, 045019 (2014) [arXiv:1311.5268].
30. A. L. Kataev and S. A. Larin, JETP Lett. **96**, 61 (2012); P. A. Baikov, K. G. Chetyrkin, J. H. Kühn, and J. Rittinger, JHEP 07, 017 (2012); P. A. Baikov, K. G. Chetyrkin, J. H. Kühn, and C. Sturm, Nucl. Phys. B **867**, 182 (2013).
31. B. Holdom, Phys. Lett. B **694**, 74 (2010).
32. R. Shrock, Phys. Rev. D **90**, 065023 (2014) [arXiv:1408.3141].
33. W. A. Bardeen, B. W. Lee, and R. E. Shrock, Phys. Rev. D **14**, 985 (1976). See also E. Brézin and J. Zinn-Justin, Phys. Rev. B **14**, 3110 (1976)).
34. E. Mølgaard and R. Shrock, Phys. Rev. D **89**, 105007 (2014) [arXiv:1403.3058].

Rethinking Naturalness: Can the Higgs be Elementary?

Francesco Sannino

CP3-Origins and the Danish IAS, University of Southern Denmark, 5230 Odense M, Denmark

I will critically introduce, classify and discuss the fundamental open issues related to either composite or elementary extensions of the standard model. As for the progress I will exhibit, among other things, the first proof of existence of nonsupersymmetric and non-asymptotically free 4D Gauge-Yukawa theories (structurally similar to the standard model) which are UV finite thanks to the existence of an exact interacting quantum UV fixed point in the gauge, Yukawa and scalar self-couplings. The quantum moduli space of the theory will also be precisely determined. Theories with this behaviour have been searched for on the lattice for the past several decades. Our results show the critical ingredients that are needed to construct new classes of fundamental theories featuring elementary scalars. I will then comment on the wide phenomenological impact of this discovery.

Dynamical Origin of the Electroweak Scale and a 125 GeV Boson

K. Tuominen

Department of Physics, University of Helsinki,
P.O. Box 64, FI-00014, Finland
E-mail: kimmo.i.tuominen@helsinki.fi
www.physics.helsinki.fi

A fully dynamical origin for the masses of weak gauge bosons and heavy quarks of the Standard Model is considered. Electroweak symmetry breaking and the gauge boson masses arise from new strong dynamics, which leads to the appearance of a composite scalar in the spectrum of excitations. In order to generate mass for the Standard Model fermions, we consider extended gauge dynamics, effectively represented by four fermion interactions at presently accessible energies. By systematically treating these interactions, we show that they lead to a large reduction of the mass of the scalar resonance. Therefore, interpreting the scalar as the recently observed 125 GeV state, implies that the mass originating solely from new strong dynamics can be much heavier, of the order of 1 TeV. The couplings of the scalar resonance with the Standard Model gauge bosons and fermions are evaluated, and found to be compatible with the current LHC results.

Keywords: Composite Higgs, Technicolor, Extended technicolor.

1. Introduction

The discovery of the Higgs boson at the Large Hadron Collider (LHC)[1,2] established the Standard Model (SM) as an accurate description of elementary particle interactions[3–5]. Nevertheless, understanding the hierarchical mass patterns of matter fields, origin of dark matter or the baryon-antibaryon asymmetry continue to provide motivation for searches of viable beyond-the-Standard-Model (BSM) scenarios.

In models of dynamical electroweak symmetry breaking[6,7] the lightest resonance is naturally expected to be much heavier than $M_H \simeq 125$ GeV[8]. It is known that a light scalar can arise from approximate global symmetries, as in models where the Higgs is a pseudo Goldstone boson associated with chiral symmetry[9–11] or scale invariance[12–16]. However, also with QCD-like new composite dynamics the scalar particle can become light because of loop corrections originating from extended sectors, anyway required in Technicolor (TC) models to account for the generation of fermion masses.

In this talk we review recent developments[17,18], where this effect was investigated in a fully dynamical model setup of simple extended technicolor (ETC). In the large-N limit, where N is the dimension of the technifermion representation under the TC gauge group, it is possible to rigorously demonstrate a large reduction of the scalar mass from the value arising solely from new strong dynamics. The amount of fine tuning involved is on the level of a few percent[17].

We consider TC theories in which the lightest resonances are composed of fermions belonging to one weak technidoublet $Q \equiv (U, D)$. Additional technifermions, heavy enough to be decoupled from the lightest resonances, may also exist similarly to the strange quark in QCD.

For phenomenology at the energies explored by the LHC, we consider the effective Lagrangian

$$\mathcal{L} = \mathcal{L}_{\overline{\mathrm{SM}}} + \mathcal{L}_{\mathrm{TC}} + \mathcal{L}_{\mathrm{ETC}} \ , \tag{1}$$

where $\mathcal{L}_{\overline{\mathrm{SM}}}$ is the SM Lagrangian without Higgs and Yukawa terms, $\mathcal{L}_{\mathrm{TC}}$ is a model Lagrangian accounting for the non-perturbative TC dynamics, and $\mathcal{L}_{\mathrm{ETC}}$ contains interactions mediated by exchanges of heavy ETC gauge bosons, which we assume to exceed the energy reach of the LHC. The full Lagrangian is invariant under the SM gauge group, and the electroweak symmetry is spontaneously broken by the TC force to electromagnetism.

The $\mathcal{L}_{\mathrm{TC}}$ part of the full Lagrangian is given by a symmetry-breaking chiral-techniquark model, containing both composite resonances and the techniquarks. Among the resonances, we assume the lightest ones to be the massless technipion isospin triplet Π^a, which becomes the longitudinal component of the W and Z bosons, and the scalar singlet H. Our main goal is to determine whether the latter can be the recently observed 125 GeV resonance. In order to account for compositeness, we take the kinetic terms for Π^a and H to be radiatively generated. Explicitly,

$$\begin{aligned} \mathcal{L}_{\mathrm{TC}} = & \ \overline{Q}_L \, i\not{D} Q_L + \overline{U}_R \, i\not{D} U_R + \overline{D}_R \, i\not{D} D_R \\ & - M\left(1 + \frac{y}{M} H + \cdots\right)\left(\overline{Q}_L \Sigma Q_R + \overline{Q}_R \Sigma^\dagger Q_L\right) - \frac{m^2}{2} H^2 + \cdots \ , \end{aligned} \tag{2}$$

where $\Sigma \equiv \exp\left(i\Pi^a \tau^a / F_\Pi\right)$, and τ^a, $a = 1, 2, 3$, are the Pauli matrices. The dynamical constituent mass of the techniquarks is denoted by M. The ellipses denote higher-order interactions, which we assume to give negligible contribution to the masses and decay constants. Note that we have suppressed the TC gauge index in the techniquarks, and that the covariant derivatives are with respect to the SM gauge interactions.

To illustrate the main feature, reduction of the scalar mass, we will first consider a simple model[17] for ETC interactions. Then we outline further development of the model framework[18]. The mechanism for the reduction of the scalar mass also works in this case. Furthermore, the effective couplings of the composite Higgs particle with the SM particles can be evaluated and they are very close to the SM-Higgs couplings, and hence compatible with the LHC data. The oblique corrections are computable and the model is viable with respect to the electroweak precision data.

2. A simple model example

We consider first the ETC sector consisting only of the four-fermion operator which allows the top quark to interact with the techniquark condensate and acquire mass,

$$\mathcal{L}_{\text{ETC}} = 2G\left(\overline{q}_L^i t_R \overline{U}_R Q_{iL} + \overline{Q}_L^i U_R \bar{t}_R q_{iL}\right) , \tag{3}$$

where $q \equiv (t, b)$ is the top-bottom doublet, and i is a weak isospin index. This is sufficient for the study of the generic effect of four-fermion interactions on the scalar mass. In a realistic ETC setup, additional four-fermion interactions coupling only technifermions or SM fermions among themselves are expected to arise. These will be discussed in section 4.

Using the Lagrangian of (1) we will compute several observables, such as the mass of the scalar singlet H and the technipion decay constant. We work in the large-N and large-N_c limit, with N/N_c finite: here N is the dimension of the techniquark representation under TC, whereas $N_c = 3$ is the dimension of the quark representation under QCD. Since we work with effective low energy Lagrangian, we use a physical cutoff. However, there are two relevant mass scales involved: the mass of the lightest ETC gauge boson, $\mathcal{M}$, and the scale of compositeness, Λ. It is not clear which of these should be used as a cutoff. If one used the smaller mass scale, Λ, this would imply loss of information from the dynamics occurring between Λ and $\mathcal{M}$. Furthermore, it is well know that making the techniquark loop integrals finite with a sharp cutoff does not account for confinement, as the fermion propagators go on-shell for sufficiently large external momenta. A solution to both problems is provided by models of confinement[19]: In each propagator the techniquark constituent mass is "smeared" by the distribution density $\rho(M)$, which plays a double role: First, it provides convergence of the integrals in momentum space, within a domain the size of which is determined by the mass scale Λ. Second, it prevents the techniquark propagators to go on-shell, thus removing unwanted production of free techniquarks.

If we use a distribution density $\rho(M)$ to smear the integrals over techniquarks, we may cutoff the $\mathcal{L}_{\text{TC}} + \mathcal{L}_{\text{ETC}}$ theory at $\mathcal{M}$. The integrals over SM quarks are cutoff at $\mathcal{M}$, whereas the integrals over techniquarks are naturally finite and of the order of Λ. No techniquark propagator can go on-shell, and confinement is therefore guaranteed. Clearly we must choose an appropriate function $\rho(M)$, and integrals are unavoidably more difficult to evaluate than the standard loop integrals. However in our analysis we are only interested in small external momenta, and no internal techniquark propagator can go on-shell, even without a distribution density.

Therefore, we approximate the distribution density by using a sharp cutoff Λ for the loop integrals over techniquark momenta, while still cutting off the SM-fermion loop integrals at $\mathcal{M}$. This approach makes the dynamics between Λ and $\mathcal{M}$ contribute to the low-energy observables, and preserves the global symmetries of the Lagrangian.

In the approximation (3) the top quark acquires mass, whereas the bottom quark

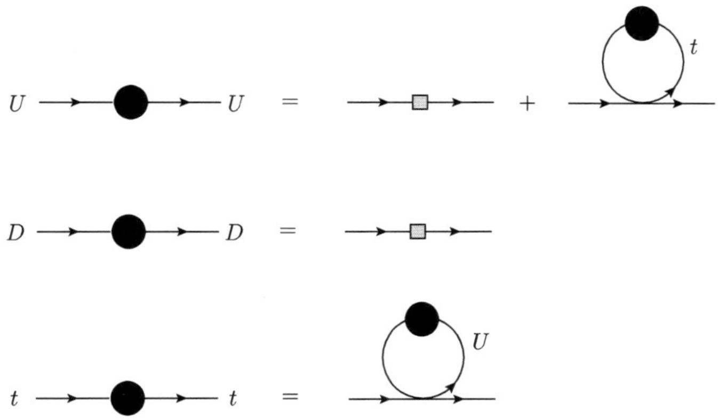

Fig. 1. Diagrams contributing to the fermion masses at leading order in N and N_c. M_D is entirely given by the tree-level term, whereas M_U receives both tree-level and loop contributions. Note that the bottom quark is massless in our approximation.

is massless. In the large-N, N_c limit, with N/N_c finite, the fermion masses are given by the diagrams of Fig. 1. These correspond to the coupled gap equations

$$\begin{aligned} M_U &= M + 4\,N_c\,G\,M_t\,I_1^{\mathcal{M}}(M_t)\ , \\ M_t &= 4\,N\,G\,M_U\,I_1^{\Lambda}(M_U)\ , \end{aligned} \tag{4}$$

and $M_D = M$, where $I_1^Q(M)$ is the standard integral

$$I_1^Q(M) \equiv i\int \frac{d^4k}{(2\pi)^4}\frac{1}{k^2 - M^2} = \frac{1}{16\pi^2}\left(Q^2 - M^2\log\frac{Q^2+M^2}{M^2}\right) \tag{5}$$

As explained above, here and in all computations below, the integrals over techniquarks are cutoff at Λ, whereas the integrals over SM fermions are cutoff at the ETC mass scale $\mathcal{M}$.

The charged and neutral technipion are the Goldstone bosons "eaten" by the longitudinal components of the W and Z boson, respectively. To ensure that the cutoff scheme respects chiral symmetry, we need to check that these absorbed Goldstone bosons are massless.

The self-energies of neutral and charged technicians are given by the infinite sums of diagrams shown in Fig. 2. Using the gap equations it is rather straightforward to show[17] that the expressions for the self-energies vanish as the external momentum goes to zero, proving that the technipions remain massless as they should.

Finally, in order to match this chiral-techniquark model to the electroweak theory, we need to impose that the charged-technipion decay constant is equal to $F_\Pi \equiv 246$ GeV. The decay constant can be extracted from the mixing term between the charged technipion and the charged component of the axial current. This is given by the sum of diagrams shown in Fig. 3.

With the help of the gap equations, the expression for $\mu_{\Pi\mathcal{A}}$ can be brought into

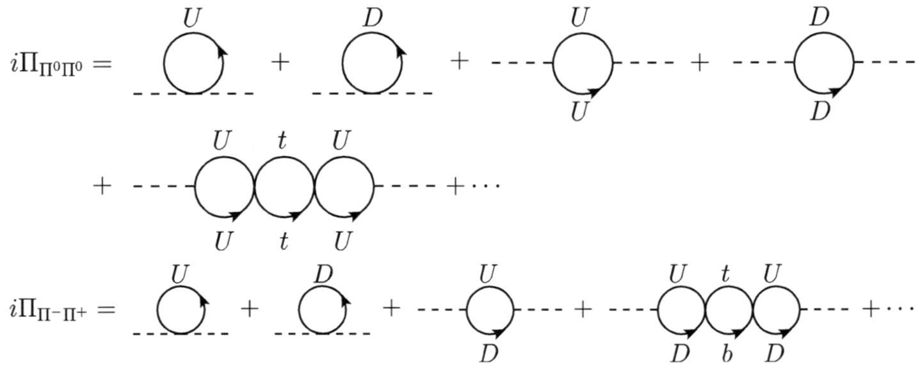

Fig. 2. Diagrams contributing to the neutral and charged technipion self-energies at leading order in N and N_c.

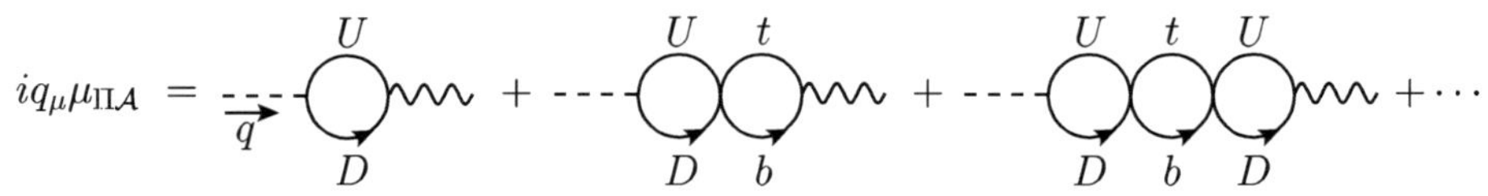

Fig. 3. Diagrams contributing to the mixing between charged technipion and axial current, at leading order in N and N_c. At zero momentum this gives F_Π times the square root of the charged-technipion wave-function renormalization.

the compact form

$$\mu_{\Pi\mathcal{A}} = 2\left[N\,(M_U + M)\,\mu^\Lambda(M_U, M) + N_c\,M_t\,\mu^{\mathcal{M}}(M_t, 0)\right] , \tag{6}$$

where

$$\begin{aligned}
\mu^Q(M_X, M_Y) &\equiv -i\int_0^1 dx \int \frac{d^4k}{(2\pi)^4}\frac{x\,M_X + (1-x)M_Y}{\left(k^2 - x\,M_X^2 - (1-x)M_Y^2\right)^2} \\
&= \frac{1}{32\pi^2(M_X^2 - M_Y^2)(M_X + M_Y)}\Bigg(Q^4 \log\frac{Q^2+M_X^2}{Q^2+M_Y^2} - (M_X^2 - M_Y^2)Q^2 \\
&\quad + M_X^3(M_X + 2M_Y)\log\frac{Q^2+M_X^2}{M_X^2} - M_Y^3(2M_X + M_Y)\log\frac{Q^2+M_Y^2}{M_Y^2}\Bigg) .
\end{aligned} \tag{7}$$

To obtain the decay constant we must divide this expression by the square root of the charged-technipion wave-function renormalization constant:

$$F_\Pi = \lim_{q^2\to 0}\frac{\mu_{\Pi\mathcal{A}}(q^2)}{\sqrt{\Sigma'_{\Pi^-\Pi^+}(q^2)}} . \tag{8}$$

In the limit $G \to 0$, this is nothing but the usual Pagels-Stokar relation:

$$F_\Pi^2 \underset{G\to 0}{=} 4\,N\,M^2\,I_2^\Lambda(M) , \tag{9}$$

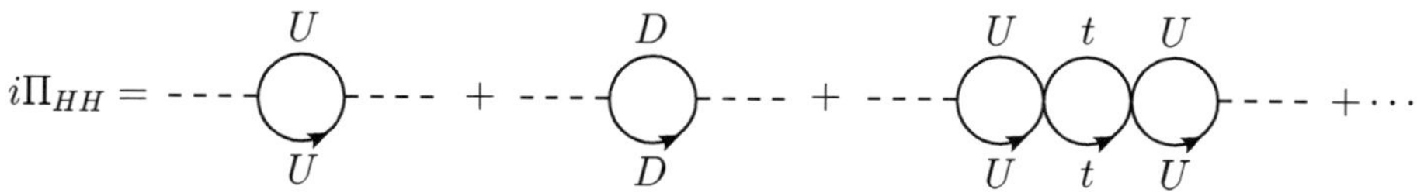

Fig. 4. Diagrams contributing to the TC-Higgs self-energy at leading order in N and N_c.

with the standard integral

$$I_2^Q(M) \equiv -i \int \frac{d^4k}{(2\pi)^4} \frac{1}{(k^2 - M^2)^2} = \frac{1}{16\pi^2} \left(\log \frac{Q^2 + M^2}{M^2} - \frac{Q^2}{Q^2 + M^2} \right) . \quad (10)$$

Once F_Π is fixed, this implies that small variations of M lead to exponentially large variations of Λ, as the Λ dependence in $I_2^\Lambda(M)$ is essentially logarithmic. Therefore, requiring Λ to be above 2-3 TeV and well below 10 TeV, as expected if Λ is of the order of the heavy technihadron masses, forces M to be within a very narrow interval around $\mathcal{O}(1 \text{ TeV})$.

We have five parameters: G, y, M, $\mathcal{M}$ and Λ. Using the gap equations and the values $M_t = 172$ GeV, $F_\Pi = 246$ GeV in Eq. (8) we can eliminate two of the five parameters. Requiring the physical mass of the Higgs particle to be 125 GeV imposes a constraint between the remaining three parameters and, consequently, everything is expressed in terms of $\mathcal{M}$ and Λ.

Next, we will consider the subject of our main interest in this section, which is the mass of the scalar singlet. Because of the interaction with the top quark, the *physical* mass M_H of the scalar singlet may be very different from its *dynamical* mass M_{H0}: the latter is defined as the mass solely due to the strong TC dynamics, and is obtained by setting $G = 0$. Taking also the iteraction with the top quark into account, the scalar-singlet self-energy is given by the diagrams of Fig. 4. Summing the series gives

$$\Sigma_{HH} = -m^2 + \frac{N\, y^2\, M^2}{v^2} (\xi_D + \xi_U) + \frac{N\, y^2\, M^2}{v^2} \frac{N\, N_c\, G^2\, \xi_U^2\, \xi_t}{1 - N\, N_c\, G^2\, \xi_U\, \xi_t} , \quad (11)$$

where

$$i\, \xi_X \equiv - \int \frac{d^4k}{(2\pi)^4} \text{Tr}\, \frac{\not{k} + M_X}{k^2 - M_X^2} \frac{\not{k} + \not{q} + M_X}{(k+q)^2 - M_X^2} . \quad (12)$$

Evaluating the integral gives

$$\xi_X = 2\left(q^2 - 4M_X^2\right) I_2^\Lambda(M_X) + 4 I_1^\Lambda(M_X) , \quad X = U, D \quad (13)$$

$$\xi_t = 2\left(q^2 - 4M_t^2\right) I_2^{\mathcal{M}}(M_t) + 4 I_1^{\mathcal{M}}(M_t) . \quad (14)$$

We eliminate m^2 by requiring the dynamical mass to be M_{H0}, i.e. we set Σ_{HH} of Eq. (11) equal to zero at $q^2 = M_{H0}$ at $G = 0$. This leads to the expression

$$\Sigma_{HH} = \frac{N y^2 M^2}{v^2} \left(4 I_2^{\Lambda_{TC}}(M) \left(q^2 - M_{H0}^2\right) - \xi_D + \frac{\xi_U}{1 - N N_c G^2 \xi_U \xi_t} \right) . \quad (15)$$

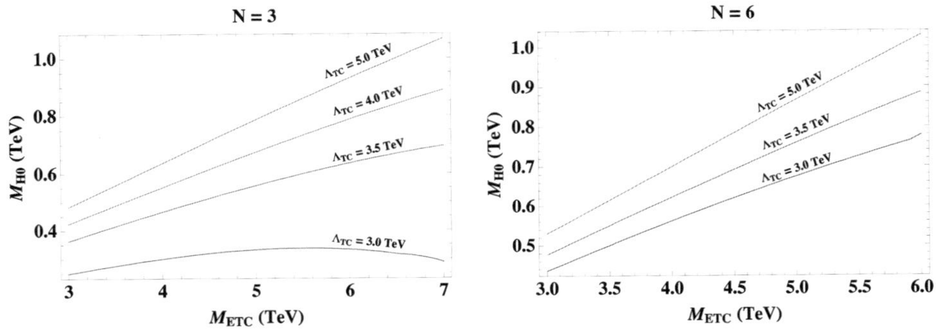

Fig. 5. The dynamical mass of the scalar singlet, M_{H0}, which is required to give a 125 GeV scalar, after the radiative corrections from the top quark are included. For large enough TC scale $\Lambda = \Lambda_{\rm TC}$, M_{H0} grows with the ETC mass $\mathcal{M} = M_{\rm ETC}$, as shown by Eq. (16). The left plot is for $N = 3$, whereas the right plot is for $N = 6$, where N is the dimension of the techniquark representation under the TC gauge group.

To test whether the scalar singlet H can be interpreted as the observed 125 GeV resonance, we set $\Sigma_{HH} = 0$ at $q^2 = M_H^2 = (125\ \text{GeV})^2$, and solve for M_{H0}. This can then be compared with independent estimates for the dynamical mass, such as those obtained from scaling up the QCD spectrum.

Let us first take the limit $M^2 \ll \Lambda_{\rm TC}^2 \ll M_{\rm ETC}^2$. Solving for M_{H0} gives, at leading logartithmic order,

$$M_{H0}^2 \simeq \frac{1}{\log \Lambda_{\rm TC}^2/M^2} \frac{\dfrac{N_c}{N}\dfrac{M_t^2}{M_U^2}}{1 - \dfrac{N_c}{N}\dfrac{M_t^2}{M_U^2}\dfrac{M_{\rm ETC}^2}{\Lambda_{\rm TC}^2}} M_{\rm ETC}^2 \,. \tag{16}$$

This shows that the dynamical mass required to give a 125 GeV physical mass grows with $M_{\rm ETC}$. Ignoring M_H^2 gives a linear growth, as shown in Fig. 5 (left) for $N = 3$, and (right) for $N = 6$. Note that M and M_{H0} are expected to scale like $1/\sqrt{N}$. We see that M_{H0} may indeed be much larger than 125 GeV.

3. A more realistic model example

In the previous section we only considered four-fermion operators containing the techniquark doublet and the top-bottom doublet $q \equiv (t, b)$. Now we consider a richer structure of four-fermion operators. We focus on ETC theories in which left-handed and right-handed fields belong to different representations, and classify the four-fermion operators according to the quantum numbers of the exchanged ETC gauge bosons. We assume that there is only one ETC gauge boson with a given set of quantum numbers, and that, under the TC and QCD gauge groups, the ETC bosons are either singlets or N- and N_c- multiplets, respectively. This categorises the ETC bosons into five distinct classes which we call A, B, C, D, and E: The classes A and B correspond to bosons which are TC and QCD singlet with hypercharge $Y = 0$ (for class A) and $Y = 1$ (for class B). The classes C, D and E consist of bosons which are N- and N_c-multiplets of TC and QCD, respectively, with hypercharge $Y = 1/6$ for class C, $Y = 5/6$ for class D and $Y = 7/6$ for class E.

3.1. *The ETC Lagrangian*

Below the ETC scale the ETC gauge bosons G_μ are heavy with masses $\mathcal{M}_G$. Integrating out the heavy bosons leads to effective four fermion interactions. Generally, the relevant terms in the fundamental Lagrangian are of the form

$$\mathcal{L}^G_{\rm ETC} \sim g_{XX'} \overline{X}\, \gamma_\mu\, X'\, G^\mu + \mathcal{M}_G^2 G_\mu G^{\mu *}, \tag{17}$$

where X and X' are any of the fermions Q, q, U, D, t, b, and all interaction terms allowed by the representation of the ETC boson under consideration should be taken into account. Integrating out the ETC boson G_μ at tree-level gives first

$$G^*_\mu \sim -\frac{g_{XX'}}{\mathcal{M}_G^2} \overline{X} \gamma_\mu X' \ , \tag{18}$$

and, after plugging back in $\mathcal{L}^G_{\rm ETC}$, one obtains the effective four fermion interaction

$$\mathcal{L}^G_{\rm ETC} \sim -\frac{|g_{XX'}|^2}{\mathcal{M}_G^2} \left|\overline{X}\, \gamma_\mu\, X'\right|^2 \ , \tag{19}$$

valid below the ETC scale. For example, for the class D and E bosons this procedure leads to

$$\mathcal{L}^{\rm D}_{\rm ETC} = -\frac{|g_{Ub}|^2}{\mathcal{M}_D^2} \left(\overline{U}_R \gamma_\mu b_R\right) \left(\overline{b}_R \gamma^\mu U_R\right) , \qquad \mathcal{L}^{\rm E}_{\rm ETC} = -\frac{|g_{Dt}|^2}{\mathcal{M}_E^2} \left(\overline{D}_R \gamma_\mu t_R\right) \left(\overline{t}_R \gamma^\mu D_R\right) , \tag{20}$$

where the quark color index $a = 1, 2, \ldots, N_c$ has been suppressed. The complete results of this classification are given in[18].

We assume that all ETC couplings are real,

$$g^*_{XY} = g_{XY} \ , \tag{21}$$

and that all ETC masses are identical:

$$\mathcal{M}_A = \mathcal{M}_B = \mathcal{M}_C = \mathcal{M}_D = \mathcal{M}_E \equiv \mathcal{M} \ . \tag{22}$$

Putting together all these ETC operators and using the Fierz rearrangement formulas under the above assumptions, gives the ETC Lagrangian

$$\begin{aligned}
\mathcal{L}_{\rm ETC} &= 2G_{QqUt} \left[\left(\overline{Q}_L U_R\right) \left(\overline{t}_R q_L\right) + \left(\overline{q}_L t_R\right) \left(\overline{U}_R Q_L\right)\right] \\
&\quad + 2G_{QqDb} \left[\left(\overline{Q}_L D_R\right) \left(\overline{b}_R q_L\right) + \left(\overline{q}_L b_R\right) \left(\overline{D}_R Q_L\right)\right] + 2G_{QQUU} \left(\overline{Q}_L U_R\right) \left(\overline{U}_R Q_L\right) \\
&\quad + 2G_{QQDD} \left(\overline{Q}_L D_R\right) \left(\overline{D}_R Q_L\right) + 2G_{qqtt} \left(\overline{q}_L t_R\right) \left(\overline{t}_R q_L\right) \\
&\quad + 2G_{qqbb} \left(\overline{q}_L b_R\right) \left(\overline{b}_R q_L\right) + \Delta\mathcal{L}_{\rm ETC} \ ,
\end{aligned} \tag{23}$$

where the couplings are defined as

$$G_{QqUt} \equiv \frac{g_{Qq}g_{Ut}}{\mathcal{M}^2} \; , \quad G_{QqDb} \equiv \frac{g_{Qq}g_{Db}}{\mathcal{M}^2} \; , \quad \text{etc.} \tag{24}$$

The contribution $\Delta\mathcal{L}_{\rm ETC}$ is more complicated and we do not reproduce it explicitly here. The operators not included in $\Delta\mathcal{L}_{\rm ETC}$ contribute both to scalar and fermion masses, whereas the operators included in $\Delta\mathcal{L}_{\rm ETC}$ contribute to neither.

Again, we compute observables in the large-N limit, with N/N_c finite. Proceeding as in the case of the simple model, we first set up the gap equations to determine fermion masses; now an equation also for the bottom mass appears. Then we evaluate the weak boson masses, and this provides the generalization of the Pagels-Stokar equation taking into account the ETC contributions. For detailed analysis, we refer to[18] and turn here now to the phenomenological details, in particular the Higgs mass.

3.2. *Mass of the TC Higgs*

The TC-Higgs self-energy is given by the chain of diagrams shown in Fig. 6. Including the tree-level mass M^2, we obtain

$$\Sigma_{HH} = -M^2 + Ny^2 \left[\frac{\mathcal{I}^{SS}_{UU}}{1 - N\,N_c\,G^2_{QqUt}\,\mathcal{I}^{SS}_{UU}\,\mathcal{I}^{SS}_{tt}} + \frac{\mathcal{I}^{SS}_{DD}}{1 - N\,N_c\,G^2_{QqDb}\,\mathcal{I}^{SS}_{DD}\,\mathcal{I}^{SS}_{bb}} \right] \tag{25}$$

where

$$\mathcal{I}^{SS}_{XX} \equiv \begin{cases} \dfrac{I^{SS}_{XX}}{1 - N\,G_{QQXX}\,I^{SS}_{XX}} \, , \; X = U, D \\[2ex] \dfrac{I^{SS}_{XX}}{1 - N_c\,G_{qqXX}\,I^{SS}_{XX}} \, , \quad X = t, b \, , \end{cases} \tag{26}$$

and

$$I^{SS}_{XY} \equiv -i \int \frac{d^4k}{(2\pi)^4}\, \mathrm{Tr}\, \frac{i\,(\not{k} + M_Y)}{k^2 - M^2_Y + i\,\varepsilon} \frac{i\,(\not{k} + \not{q} + M_X)}{(k+q)^2 - M^2_X + i\,\varepsilon} \tag{27}$$

Again, we trade the bare mass parameter M for the one-loop mass M_{H0} via

$$-M^2 + Ny^2 \left(I^{SS}_{UU} + I^{SS}_{DD}\right)_{q^2 = M^2_{H0}} = 0 \; . \tag{28}$$

Then we set the physical Higgs mass $M_H \simeq 125$ GeV and solve for M_{H0} from the equation

$$\Sigma_{HH}\left(M^2_H\right) = 0 \; . \tag{29}$$

To gain understanding on the dynamics here, assume for a moment that we can expand in the ETC couplings. Expanding up to the linear order in the four-fermion couplings, and ignoring corrections of the order of M^2_X/Λ^2, leads to the approximate relation

$$M^2_H \simeq M^2_{H0} - \frac{N\,(G_{QQUU} + G_{QQDD})\,\Lambda^2}{4\pi^2} \frac{\Lambda^2}{\langle \overline{\log}(\Lambda^2/M^2_Q) \rangle} \tag{30}$$

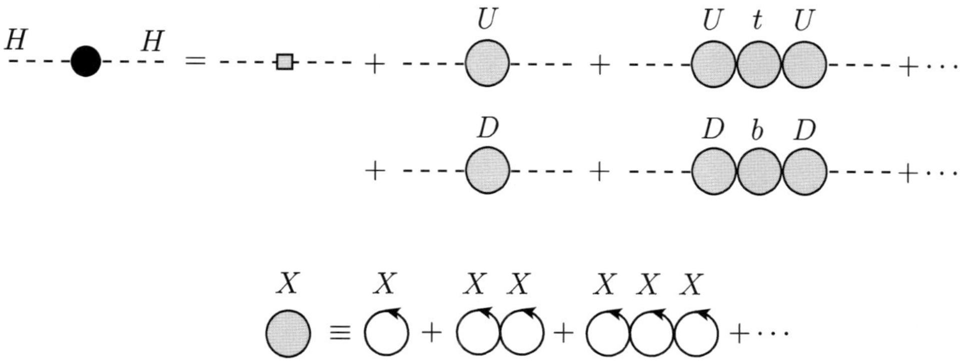

Fig. 6. The diagrams contributing to the TC-Higgs self-energy are shown in the top figure, including the tree-level mass. The shaded circles represent the sum of fermion-bubble chains, as shown in the bottom figure.

where we defined

$$\overline{\log}\frac{\Lambda^2}{M_X^2} \equiv \log\frac{\Lambda^2}{M_X^2} - 1 \; . \tag{31}$$

and

$$\langle\overline{\log}\frac{\Lambda^2}{M_Q^2}\rangle \equiv \frac{1}{2}\left(\overline{\log}\frac{\Lambda^2}{M_U^2} + \overline{\log}\frac{\Lambda^2}{M_D^2}\right) \tag{32}$$

The last factor in (30) is reasonably of the order of $(1\ \text{TeV})^2$. Estimates obtained by scaling up the mass of the σ meson from QCD to a TC theory suggest $M_{H0} \lesssim 1$ TeV[20]. Therefore, the factor containing the ETC couplings, in (30), must be $\mathcal{O}(1)$. Bounds on the ETC couplings can be derived by requiring that the series leading to (25) and (26) are convergent. This gives the constraints

$$N\, G_{QQUU}\, \Lambda^2 < 4\pi^2 \;, \quad N_c\, G_{qqtt}\, \mathcal{M}^2 < 4\pi^2 \;, \tag{33}$$

and similarly for G_{QQDD} and G_{qqbb} which guarantee, respectively, no $\overline{U}U$, $\overline{D}D$, $\bar{t}t$, and $\bar{b}b$ condensation from ETC, and the constraints

$$N\, N_c\, G_{QqUt}^2\, \Lambda^2\, \mathcal{M}^2 < 16\pi^4 \;, \quad N\, N_c\, G_{QqDb}^2\, \Lambda^2\, \mathcal{M}^2 < 16\pi^4 \;, \tag{34}$$

which prevent the ETC force from generating $\overline{U}t$ and $\overline{D}b$ condensates, respectively. If the ETC couplings are large enough, yet subcritical, we obtain $N\ (G_{QQUU} + G_{QQDD})\, \Lambda^2/4\pi^2 \sim 1$, which, according to the estimate (30), is the order of magnitude required for the ETC interactions to reduce the TC-Higgs mass from $\mathcal{O}(1\ \text{TeV})$ to $\mathcal{O}\ (100\ \text{GeV})$.

As in any BSM scenario which explains the relatively small mass of the observed Higgs boson, there is some amount of fine tuning involved in this scenario. The amount of fine tuning can be quantified in a simply as

$$\text{FT} = \frac{M_H^2}{M_{H0}^2}. \tag{35}$$

The concrete values are on the level of few percents. For example, for $M_{H0} \simeq 1$ TeV, FT $\simeq 1.6\%$, whereas for $M_{H0} \sim 600$ GeV, FT $\simeq 4.3\%$.

3.3. *Numerical results*

The model parameters are g, g', y, M_Q, M, Λ, $\mathcal{M}$, g_{QQ}, g_{qq}, g_{Qq}, g_{UU}, g_{DD}, g_{UD}, g_{tt}, g_{bb}, g_{tb}, g_{Ut}, g_{Dt}, g_{Ub}, g_{Db}, and N. The Yukawa coupling y disappears after renormalization, whereas M is exchanged for the dynamical mass M_{H0}. We trade g, g', M_Q, M_{H0}, g_{Ut}, and g_{Db} for the experimental values of α, G_F, M_Z, M_H, M_t, and M_b, respectively. Therefore, we end up with the free variables Λ, $\mathcal{M}$, g_{QQ}, g_{qq}, g_{Qq}, g_{UU}, g_{DD}, g_{UD}, g_{tt}, g_{bb}, g_{tb}, g_{Dt}, g_{Ub}, and N. We may obtain an estimate of Λ by scaling up the corresponding quantity from QCD using the scaling law

$$\Lambda = \sqrt{\frac{N_c}{N}} \frac{F_\Pi}{f_\pi} \Lambda_{\rm QCD} \ , \tag{36}$$

where $F_\Pi = (\sqrt{2}\, G_F)^{-1/2} \simeq 246$ GeV. Setting $f_\pi \simeq 93$ MeV, $m_\sigma \simeq 441$ MeV[21], and $N_c = 3$, gives $\Lambda = 2.2$ TeV for $N = 6$. We take this as an input to the numerical calculation. The ETC scale $\mathcal{M}$ remains as a free parameter, and we consider values increasing from $\mathcal{M} = \Lambda$. Then, we perform a random scan over the couplings g_{QQ}, g_{qq}, g_{Qq}, g_{UU}, g_{DD}, g_{UD}, g_{tt}, g_{bb}, g_{tb}, g_{Dt}, g_{Ub}.

The conditions of sub-criticality (33) and (34) imply direct bounds on the ETC couplings. Furthermore, to avoid obvious large corrections to T-parameter, we impose $g_{UU} = g_{DD}$, $g_{tt} = g_{bb}$ and $g_{UD} = g_{tb} = 0$.

Concretely, we scan the parameter space by assigning random values for the couplings within the following ranges:

$$\pi/3 \leq |g_{QQ}|, |g_{UU}|, |g_{Qq}|, |g_{Dt}|, |g_{Ub}| \leq 2\pi, \ -2\pi \leq |g_{qq}|, |g_{tt}| \leq 2\pi, \tag{37}$$

and we also set

$$g_{DD} = g_{UU} \ , \quad g_{bb} = g_{tt} \ , \quad g_{UD} = 0 \ , \quad g_{tb} = 0 \ . \tag{38}$$

For each generated data point we compute masses, the couplings of the Higgs particle, and evaluate the S and T parameters. For the Higgs couplings we find that they are within $\sim 10\%$ of the SM values for any value of the ETC couplings and mass. In particular, the Higgs coupling to the weak boson is always slightly larger than the SM value, the top Yukawa is slightly smaller, whereas the bottom Yukawa is standard within a few per cents.

For each considered value of Λ and $\mathcal{M}$, we sampled 25 000 points satisfying (37) and (38), and plotted these in the S, T plane, together with the experimentally viable 3σ contour. The dynamical mass M_{H0} is of course highly dependent on the ETC couplings. In Fig. 7, for $N = 6$, we only show those points for which M_{H0} is between the fractions 0.5 and 1.25 of the scaled up mass of the sigma meson:

$$0.5\, m_\sigma \frac{F_\Pi}{f_\pi} \sqrt{\frac{N_c}{N}} < M_{H0} < 1.25\, m_\sigma \frac{F_\Pi}{f_\pi} \sqrt{\frac{N_c}{N}} \ . \tag{39}$$

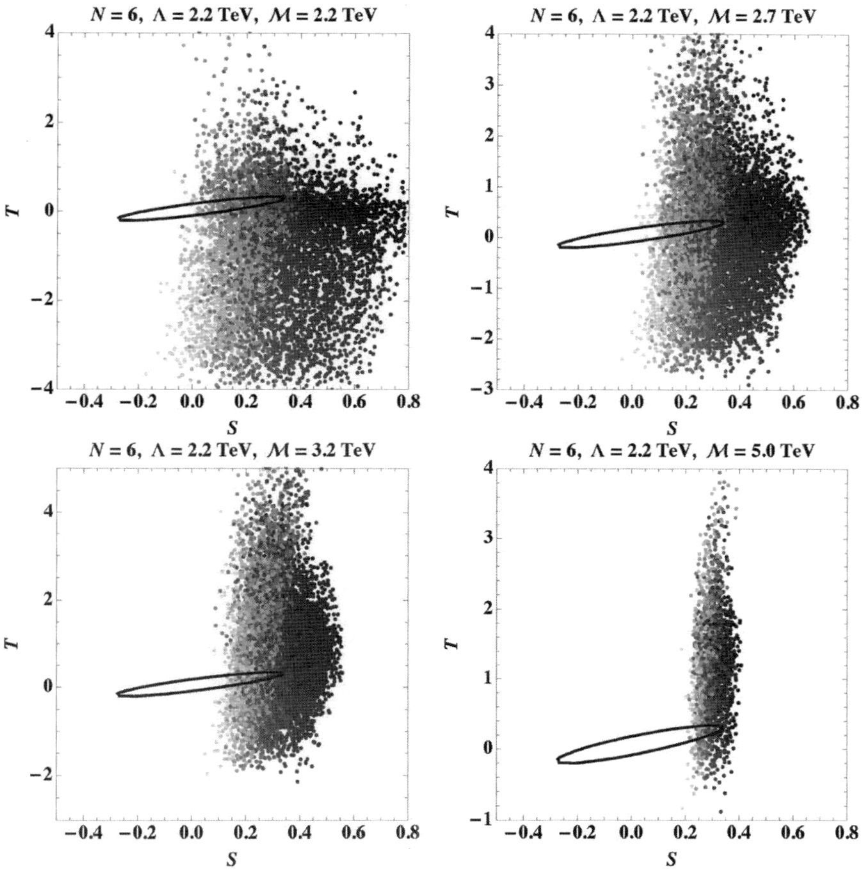

Fig. 7. Result over the random scan of the model parameter space for $N = 6$ and $\Lambda = 2.2$ TeV. The four figures correspond to $M = 2.2, 2,7, 3.2$ and 5 TeV, respectively.

The color coding of the points shown in the figure is related to the different couplings as follows: The bigger the value of $|g_{Dt}|$, the greener the point, and the larger the value of $|g_{Ub}|$, the redder the point. Furthermore, the larger the value of $|g_{Qq}|$ the darker the shade of the color, with a clear border at $|g_{Qq}| = 3$. Larger values of $|g_{Ub}|$ imply smaller S whereas larger values of $|g_{Dt}|$ imply smaller S and also smaller T. Therefore, as we can observe from the figure, the redder points are towards the top-left of the plot while the greener points are towards the bottom.

For small values of $\mathcal{M}$ the green points are disfavoured. This is because T is too large and negative. The favoured points on the other hand are orange, corresponding to approximately equal values of $|g_{Dt}|$ and $|g_{Ub}|$. Increasing the ETC scale $\mathcal{M}$ makes the mass splitting $M_U - M_D$ larger. In the gap equations this splitting originates from G_{QqUt}, the latter being fixed by the top quark mass. The large $M_U - M_D$ splitting implies a large positive one-loop contribution to the T parameter, and this needs to be balanced by a negative contribution. Such a balancing contribution can only arise from g_{Dt}, and therefore for large $\mathcal{M}$ the favoured points are greener.

Overall, for large values of $\mathcal{M}$, *e.g.*. 5 TeV, there are less points since larger values of $g_{QQ}g_{UU}$ and $g_{QQ}g_{DD}$ are required to have sufficient reduction for the Higgs mass. In other words, the portion of the sampled points shown in the figure shrinks as $\mathcal{M}$ increases. Finally, we note that varying Λ within, say, 20% range around the values we have used does not change the results qualitatively.

Our results can be considered as instructions on how to build an ETC theory satisfying the experimental constraints from precision electroweak observables. We leave for future work the more thorough analyses highlighting different domains and their relevance for microscopic ETC model building.

4. Conclusions and outlook

We have considered a realistic effective models where the electroweak scale and the masses of the weak gauge bosons arise dynamically from a TC force and the technifermions are coupled with the SM matter fields via four-fermion interactions, low-energy remnants of an ETC theory. We have demonstrated that due to these external couplings, the lightest scalar resonance can be as light as 125 GeV. Within a well defined ETC setup, we classified all four-fermion operators based on the quantum numbers of the exchanged ETC gauge bosons. We showed that the resulting light scalar has SM-like couplings with the particles of the SM. The contribution of the four fermion operators to the oblique electroweak parameters makes model is viable over large region of the parameter space.

Model building of the full gauge dynamics required by ETC theories is challenging[22]. Our effective theories, formulated in terms of four fermion couplings, and taking into account only the third generation quarks, can be seen as a stepping stone towards more complete dynamical theories of flavour. There exists lots of earlier work using NJL-like models to describe dynamical electroweak symmetry breaking and the associated Higgs physics[23–31]. There is also a large lattice program motivated by applications to BSM physics and aimed at studying strong dynamics in isolation[32–41]. Our analysis should be applicable in refining the phenomenological interpretation of the lattice results.

References

1. G. Aad *et al.*, Observation of a new particle in the search for the Standard Model Higgs boson with the ATLAS detector at the LHC, *Phys.Lett.* **B716**, 1 (2012).
2. S. Chatrchyan *et al.*, Observation of a new boson at a mass of 125 GeV with the CMS experiment at the LHC, *Phys.Lett.* **B716**, 30 (2012).
3. P. P. Giardino, K. Kannike, I. Masina, M. Raidal and A. Strumia, The universal Higgs fit, *JHEP* **1405**, p. 046 (2014).
4. A. Azatov, R. Contino and J. Galloway, Model-Independent Bounds on a Light Higgs, *JHEP* **1204**, p. 127 (2012).

5. T. Alanne, S. Di Chiara and K. Tuominen, LHC Data and Aspects of New Physics, *JHEP* **1401**, p. 041 (2014).
6. C. T. Hill and E. H. Simmons, Strong dynamics and electroweak symmetry breaking, *Phys.Rept.* **381**, 235 (2003).
7. F. Sannino, Conformal Dynamics for TeV Physics and Cosmology, *Acta Phys.Polon.* **B40**, 3533 (2009).
8. R. Foadi, M. T. Frandsen, T. A. Ryttov and F. Sannino, Minimal Walking Technicolor: Set Up for Collider Physics, *Phys.Rev.* **D76**, p. 055005 (2007).
9. D. B. Kaplan and H. Georgi, SU(2) x U(1) Breaking by Vacuum Misalignment, *Phys.Lett.* **B136**, p. 183 (1984).
10. D. B. Kaplan, H. Georgi and S. Dimopoulos, Composite Higgs Scalars, *Phys.Lett.* **B136**, p. 187 (1984).
11. G. Cacciapaglia and F. Sannino, Fundamental Composite (Goldstone) Higgs Dynamics, *JHEP* **1404**, p. 111 (2014).
12. K. Yamawaki, M. Bando and K.-i. Matumoto, Scale Invariant Technicolor Model and a Technidilaton, *Phys.Rev.Lett.* **56**, p. 1335 (1986).
13. M. Bando, K.-i. Matumoto and K. Yamawaki, TECHNIDILATON, *Phys.Lett.* **B178**, p. 308 (1986).
14. F. Sannino and K. Tuominen, Orientifold theory dynamics and symmetry breaking, *Phys.Rev.* **D71**, p. 051901 (2005).
15. D. D. Dietrich, F. Sannino and K. Tuominen, Light composite Higgs from higher representations versus electroweak precision measurements: Predictions for CERN LHC, *Phys.Rev.* **D72**, p. 055001 (2005).
16. Z. Chacko and R. K. Mishra, Effective Theory of a Light Dilaton, *Phys.Rev.* **D87**, p. 115006 (2013).
17. S. Di Chiara, R. Foadi and K. Tuominen, 125 GeV Higgs from a chiral-techniquark model (2014).
18. S. Di Chiara, R. Foadi, K. Tuominen and S. Thtinen, Dynamical Origin of the Electroweak Scale and the 125 GeV Scalar (2014).
19. G. Efimov and M. A. Ivanov, *The Quark confinement model of hadrons* (CRC press, 1993).
20. R. Foadi, M. T. Frandsen and F. Sannino, 125 GeV Higgs from a not so light Technicolor Scalar, *Phys.Rev.* **D87**, p. 095001 (2013).
21. I. Caprini, G. Colangelo and H. Leutwyler, Mass and width of the lowest resonance in QCD, *Phys.Rev.Lett.* **96**, p. 132001 (2006).
22. T. Appelquist, M. Piai and R. Shrock, Fermion masses and mixing in extended technicolor models, *Phys.Rev.* **D69**, p. 015002 (2004).
23. V. Miransky, M. Tanabashi and K. Yamawaki, Dynamical Electroweak Symmetry Breaking with Large Anomalous Dimension and t Quark Condensate, *Phys.Lett.* **B221**, p. 177 (1989).
24. V. Miransky, M. Tanabashi and K. Yamawaki, Is the t Quark Responsible for the Mass of W and Z Bosons?, *Mod.Phys.Lett.* **A4**, p. 1043 (1989).

25. W. A. Bardeen, C. T. Hill and M. Lindner, Minimal Dynamical Symmetry Breaking of the Standard Model, *Phys.Rev.* **D41**, p. 1647 (1990).
26. B. A. Dobrescu and C. T. Hill, Electroweak symmetry breaking via top condensation seesaw, *Phys.Rev.Lett.* **81**, 2634 (1998).
27. R. S. Chivukula, B. A. Dobrescu, H. Georgi and C. T. Hill, Top quark seesaw theory of electroweak symmetry breaking, *Phys.Rev.* **D59**, p. 075003 (1999).
28. H.-J. He, C. T. Hill and T. M. Tait, Top quark seesaw, vacuum structure and electroweak precision constraints, *Phys.Rev.* **D65**, p. 055006 (2002).
29. H. S. Fukano and K. Tuominen, A hybrid 4th generation: Technicolor with top-seesaw, *Phys.Rev.* **D85**, p. 095025 (2012).
30. H. S. Fukano and K. Tuominen, 126 GeV Higgs boson in the top-seesaw model, *JHEP* **1309**, p. 021 (2013).
31. M. Geller, S. Bar-Shalom and A. Soni, Hybrid dynamical electroweak symmetry breaking with heavy quarks and the 125 GeV Higgs boson, *Phys.Rev.* **D89**, p. 035012 (2014).
32. A. J. Hietanen, J. Rantaharju, K. Rummukainen and K. Tuominen, Spectrum of SU(2) lattice gauge theory with two adjoint Dirac flavours, *JHEP* **0905**, p. 025 (2009).
33. A. J. Hietanen, K. Rummukainen and K. Tuominen, Evolution of the coupling constant in SU(2) lattice gauge theory with two adjoint fermions, *Phys.Rev.* **D80**, p. 094504 (2009).
34. T. Karavirta, J. Rantaharju, K. Rummukainen and K. Tuominen, Determining the conformal window: SU(2) gauge theory with $N_f = 4$, 6 and 10 fermion flavours, *JHEP* **1205**, p. 003 (2012).
35. T. Appelquist, G. Fleming, M. Lin, E. Neil and D. Schaich, Lattice Simulations and Infrared Conformality, *Phys.Rev.* **D84**, p. 054501 (2011).
36. T. Appelquist, R. Brower, M. Buchoff, M. Cheng, G. Fleming *et al.*, Two-Color Theory with Novel Infrared Behavior, *Phys.Rev.Lett.* **112**, p. 111601 (2014).
37. T. DeGrand, Y. Shamir and B. Svetitsky, Near the sill of the conformal window: gauge theories with fermions in two-index representations, *Phys.Rev.* **D88**, p. 054505 (2013).
38. Z. Fodor, K. Holland, J. Kuti, D. Nogradi, C. Schroeder *et al.*, Can the nearly conformal sextet gauge model hide the Higgs impostor?, *Phys.Lett.* **B718**, 657 (2012).
39. Z. Fodor, K. Holland, J. Kuti, D. Nogradi and C. H. Wong, Can a light Higgs impostor hide in composite gauge models?, *PoS* **LATTICE2013**, p. 062 (2014).
40. Y. Aoki *et al.*, Light composite scalar in twelve-flavor QCD on the lattice, *Phys.Rev.Lett.* **111**, p. 162001 (2013).
41. Y. Aoki *et al.*, Light composite scalar in eight-flavor QCD on the lattice, *Phys.Rev.* **D89**, p. 111502 (2014).

Top-Mode Pseudo-Nambu-Goldstone Boson Higgs Model *

Hidenori S. Fukano†

Kobayashi-Maskawa Institute for the Origin of Particles and the Universe (KMI), Nagoya University, Nagoya 464-8602, Japan

We discuss the Top-Mode Pseudo-Nambu-Goldstone boson Higgs (TMpNGBH) model which has recently been proposed as a variant of the top quark condensate model in light of the 125 GeV Higgs boson discovered at the LHC. In this talk, we focus on the vacuum alignment and the phenomenologies of characteristic particles of the TMpNGBH model.

1. Introduction

The ATLAS[4] and CMS collaborations[5] have discovered a 125 GeV Standard Model (SM)-like Higgs boson. This implies that the era to reveal the origin of mass of the elementary particles has come. Preceding the discovery of the Higgs boson by about two decades the top quark has been discovered at the Tevatron[6,7]. The top quark is the heaviest particle among the observed particles and its mass is $m_t \simeq 173\,\text{GeV}$[8], which is coincidentally on the order of the Higgs mass and the electroweak symmetry breaking (EWSB) scale ($v_{\text{EW}} \simeq 246\,\text{GeV}$).

The top quark condensate model[9–11] is a scenario in which the top quark plays a crucial role to explain the dynamical origin for both the EWSB and the Higgs boson. However, the original top quark condensate model is somewhat far from a realistic situation, especially, a Higgs boson predicted as a $t\bar{t}$ bound state has the mass in a range of $m_t \lesssim m_H \lesssim 2m_t$, which cannot be identified with the 125 GeV Higgs boson at the LHC.

2. Top-Mode Pseudo-Nambu-Goldstone Boson Higgs (TMpNGBH) model

Recently, a variant class of the top quark condensate model, so-called Top-Mode pseudo-Nambu-Goldstone Boson Higgs (TMpNGBH) model, was proposed[1,12]. In these models a composite Higgs boson emerges as a pseudo-Nambu–Goldstone boson (pNGB) associated with the spontaneous breaking of a global symmetry, therefore it is light to be identified as the LHC Higgs boson.

The TMpNGBH model is constructed from the top and bottom quarks $q = (t, b)$ and a vectorlike χ quark, a flavor partner of the top quark having the same

---

*This talk is based on[1–3] and given at 2015 KMI workshop " Origin of Mass and Strong Coupling Gauge Theories" (SCGT 15), March 3-6, 2015

†fukano@kmi.nagoya-u.ac.jp

SM charges as those of the right-handed top quark, which form a four-fermion interaction:

$$\mathcal{L}_{4f} = G_{4f}(\bar{\psi}_L^i \chi_R)(\bar{\chi}_R \psi_L^i)\,, \tag{1}$$

where $\psi_L^i \equiv (t_L, b_L, \chi_L)^{T\,i}$ $(i = 1, 2, 3)$. This four-fermion interaction possesses the global symmetry $G = U(3)_L \times U(1)_R$. When the value of G_{4f} is large enough to form a fermion-bilinear condensate, namely $G_{4f} > G_{\rm crit} = 8\pi^2/(N_c\Lambda^2)$ with N_c being the number of QCD color and Λ the cutoff scale of the theory, the global symmetry is spontaneously broken down to $H = U(2)_L \times U(1)_V$. In association with the symmetry breaking, the five NGBs emerge as bound states of the t and χ quarks, in addition to a composite heavy scalar boson, corresponding to the σ mode of the usual Nambu-Jona-Lasinio (NJL) model[13]. Three of these five NGBs are eaten by the electroweak gauge bosons when the subgroup of G is gauged by the electroweak symmetry (and if the condensate is formed in a direction where the electroweak symmetry is broken). The other two remain as physical states, and they obtain their masses by additional interaction terms which explicitly break the global $G = U(3)_L \times U(1)_R$ symmetry:

$$\mathcal{L}_h = -\left[\Delta_{\chi\chi}\bar{\chi}_R\chi_L + \text{h.c.}\right] - G'\left(\bar{\chi}_L\chi_R\right)\left(\bar{\chi}_R\chi_L\right)\,. \tag{2}$$

Then two NGBs become pNGBs, dubbed as top-mode pNGBs (TMpNGBs). One of the TMpNGBs, which is the CP-even scalar (h_t^0), is identified as the 126 GeV Higgs boson discovered at the LHC, while the other is the CP-odd scalar (A_t^0). Furthermore, the model includes another four-fermion interaction term,

$$\mathcal{L}_t = G''\left(\bar{\chi}_L\chi_R\right)\left(\bar{t}_R\chi_L\right) + \text{h.c.}\,. \tag{3}$$

This, combined with Eq.(1), generates the top quark mass via the top-seesaw mechanism[14–18].

Note that Eq.(3) also explicitly breaks the G-symmetry, but does not contribute to the TMpNGBs' masses ($m_{h_t^0}$ and $m_{A_t^0}$) at the leading order. However, it was shown that at the next-to-leading order, the term in Eq.(3) gives large corrections to the masses of h_t^0 and A_t^0 via the top and χ-quark loops[1]. This, namely the fact that even a small explicit breaking term causes large correction to physical quantities at the loop level, poses a question: is the vacuum alignment stable at the loop level ? We address this question based on an effective Lagrangian described by the TMpNGBs (h_t^0 and A_t^0), the t' quark, the SM gauge bosons and fermions, including terms explicitly breaking the global $U(3)_L \times U(1)_R$ symmetry.

3. Vacuum alignment of TMpNGBH model

The effective Lagrangian relevant for the vacuum alignment is given by

$$\begin{aligned}\mathcal{L}_{\rm eff}(U) = {} & \frac{f^2}{2}\text{tr}\left[D_\mu U^\dagger D^\mu U \Sigma_0\right] - \tilde{m}_\chi\left[\bar{\psi}_L \mathcal{M}_f(U)\psi_R + \text{h.c.}\right] \\ & - c_1 f^2 \text{tr}\left[U^\dagger \Sigma_0 U \Sigma_0\right] + c_2 f^2 \text{tr}\left[U\Sigma_0 + \Sigma_0 U^\dagger\right]\,,\end{aligned} \tag{4}$$

where the unitary matrix U parameterizes the five NGBs and is given by

$$U = \exp\left[\frac{i}{f}\left(\sum_{a=4,5,6,7} \pi_t^a \lambda^a + \pi_t^A \Sigma_0\right)\right] . \tag{5}$$

Here, f is a decay constant, the Gell-Mann matrices λ^a are normalized as $\mathrm{tr}[\lambda^a \lambda^b] = 2\delta^{ab}$, and Σ_0 is defined as $\Sigma_0 \equiv \mathrm{diag}(0,0,1)$. $D_\mu U = \left(\partial_\mu - ig\hat{W}_\mu + ig'\hat{B}_\mu\right) U$, $\hat{W}_\mu = \sum_{\hat{a}=1}^{3} W_\mu^{\hat{a}}(\lambda^{\hat{a}}/2)$, $\hat{B}_\mu = B_\mu \cdot \mathrm{diag}(1/2, 1/2, 0)$, W_μ and B_μ are the usual $SU(2)_L$ and $U(1)_Y$ gauge fields with gauge couplings g and g', respectively. $\mathcal{M}_f(U), \tilde{m}_\chi$ are given by

$$\mathcal{M}_f(U) = U\Sigma_0 + \frac{G''}{G_{4f}}\Sigma_0 U \begin{pmatrix} 0 & 0 & 0 \\ 0 & 0 & 0 \\ 1 & 0 & 0 \end{pmatrix} , \tilde{m}_\chi = \frac{1}{\sqrt{2}} y f = \sqrt{\frac{8\pi^2}{N_c \ln(\Lambda^2/\Lambda_\chi^2)}} f , \tag{6}$$

where Λ is the cutoff scale of the ultraviolet theory and Λ_χ is an infrared scale corresponding to the cutoff scale of the effective theory Eq.(4). The coefficients c_1 and c_2 in Eq.(4) are given by

$$c_1 = \frac{y^2}{2}\frac{G'}{G_{4f}^2} , \quad c_2 = \frac{y}{\sqrt{2} f}\frac{\Delta_{\chi\chi}}{G_{4f}} . \tag{7}$$

At the tree level, the form of the potential term for NGBs, corresponding to the second line of Eq.(4), is determined solely by the $\mathcal{L}_h$. The effect of the explicit breaking terms in $\mathcal{L}_t$ and the electroweak sector appear only at loop level. Therefore, to see the effect of all the explicit breaking terms, we compute the effective Lagrangian at one-loop level by including all the contributions from the NGBs, electroweak gauge bosons, as well as fermions. The effective Lagrangian is calculated by keeping only the quadratic divergent terms, and the resultant expression becomes as follows (for the detail of the calculation, see[3]):

$$\mathcal{L}_{\mathrm{eff}}^{\text{1-loop}}(U) = \frac{F^2}{2}\mathrm{tr}\left[D_\mu U^\dagger D^\mu U \Sigma_0\right] - \tilde{m}_\chi \left[\bar{\psi}_L \mathcal{M}_f(U)\psi_R + \mathrm{h.c.}\right] - V_{\mathrm{eff}}(U) , \tag{8}$$

where the effective potential $V_{\mathrm{eff}}(U)$ is given by

$$V_{\mathrm{eff}}(U) = C_1 F^2 \mathrm{tr}\left[U^\dagger \Sigma_0 U \Sigma_0\right] - C_2 F^2 \mathrm{tr}\left[U\Sigma_0 + \Sigma_0 U^\dagger\right] . \tag{9}$$

The quadratic divergences can be absorbed by redefinitions of the bare coupling f, c_1 and c_2:

$$F^2 = f^2 - \frac{\Lambda_\chi^2}{4\pi^2} = \frac{N_c}{8\pi^2}\tilde{m}_\chi^2 \ln\frac{\Lambda^2}{\Lambda_\chi^2} - \frac{\Lambda_\chi^2}{4\pi^2} , \tag{10}$$

$$C_1 F^2 = c_1 f^2 \left(1 - \frac{3\Lambda_\chi^2}{8\pi^2 f^2}\right) - \frac{f^2\Lambda_\chi^2}{32\pi^2}\left(\frac{9}{4}g^2 + \frac{3}{4}g'^2 + 2N_c y^2 \left(\frac{G''}{G_{4f}}\right)^2\right) , \tag{11}$$

$$C_2 F^2 = c_2 f^2 \left(1 - \frac{5\Lambda_\chi^2}{32\pi^2 f^2}\right) . \tag{12}$$

Let us address the vacuum alignment of the TMpNGBH model based on the effective potential Eq.(9). First, with appropriate chiral $U(3)_{L,R}$ rotations of fermion fields $\psi_{L,R}$ and redefinition of the $\Delta_{\chi\chi}$, we parameterize the vacuum expectation value of U by a single angle parameter θ as

$$\langle U \rangle = \begin{pmatrix} \cos\theta & 0 & \sin\theta \\ 0 & 1 & 0 \\ -\sin\theta & 0 & \cos\theta \end{pmatrix} . \tag{13}$$

Taking $U = \langle U \rangle$, we have $V_{\text{eff}}(\langle U \rangle) = F^2 \left[C_1 \cdot \cos^2\theta - 2C_2 \cdot \cos\theta \right]$. It is possible to determine the vacuum alignment by minimizing the above potential energy with respect to the alignment parameter $\cos\theta$. In the present model, we find that the potential energy $V_{\text{eff}}(\langle U \rangle)$ is minimized at a nonzero $\theta = \theta_h$ with

$$\cos\theta \bigg|_{\theta=\theta_h} = \frac{C_2}{C_1} \qquad \text{only if} \quad C_1 > 0 \text{ and } \left| \frac{C_2}{C_1} \right| < 1 , \tag{14}$$

to realize the desired vacuum in which the electroweak symmetry is broken.

From the effective potential, we find the non-vanishing elements of the NGB mass-squared matrix take the following forms:

$$\begin{pmatrix} m_{44}^2 & m_{4A}^2 \\ m_{A4}^2 & m_{AA}^2 \end{pmatrix} = 2C_1 \times \begin{pmatrix} \cos\theta_h & -\sin\theta_h \\ \sin\theta_h & \cos\theta_h \end{pmatrix} \begin{pmatrix} 0 & 0 \\ 0 & 1 \end{pmatrix} \begin{pmatrix} \cos\theta_h & \sin\theta_h \\ -\sin\theta_h & \cos\theta_h \end{pmatrix} , \tag{15}$$

and

$$m_{55}^2 = 2C_1 \sin^2\theta_h \, . \tag{16}$$

Note that the stability of the effective potential requires $C_1 \geq 0^3$. The massive state in Eq.(15) is identified as the CP-odd scalar A_t^0 ($A_t^0 \equiv -\pi_t^4 \sin\theta_h + \pi_t^A \cos\theta_h$), while that in Eq.(16) is the CP-even scalar ($\pi_5 \equiv h_t^0$), dubbed as the "tHiggs". These masses are related by the alignment parameter θ_h:

$$m_{A_t^0}^2 = 2C_1 \, , \tag{17}$$

$$\begin{aligned} m_{h_t^0}^2 &= 2C_1 \sin^2\theta_h \\ &= m_{A_t^0}^2 \sin^2\theta_h \, . \end{aligned} \tag{18}$$

Other three eigenvalues of mass-squared matrix vanish, which corresponds to three massless NGBs ($\pi_t^{6,7}$, $\pi_t^4 \cos\theta_h + \pi_t^A \sin\theta_h$). These are the would-be NGBs to be eaten by the electroweak gauge bosons. It should be noted from Eqs.(17) and (18) that the quadratic divergent contributions to masses of TMpNGBs have been fully absorbed into the renormalization of the decay constant F, the coefficient C_1 and the alignment parameter θ_h (or the coefficient C_2).

4. Implications for collider physics

In this section, we discuss phenomenological implications for the TMpNGBH model. We take the alignment parameter $\cos\theta_h$ in the range of $0.97 \leq \cos\theta_h \lesssim 1$. This is the range where the coupling property of the tHiggs to SM particles is consistent with the LHC data at 95% C.L.[2]. For $0.97 \leq \cos\theta_h \lesssim 1$ the masses of A_t^0 and t' monotonically increase from $(m_{A_t^0}, m_{t'}) = (518\,\text{GeV}, 1.85\,\text{TeV})$ to infinity as $\cos\theta_h \to 1$. This value of $m_{t'}$ is consistent with the electroweak precision tests[19,20] as shown in[1]. We thus study the LHC phenomenologies of A_t^0 and t' with their masses from $(m_{A_t^0}, m_{t'}) = (518\,\text{GeV}, 1.85\,\text{TeV})$ to certain heavier mass regions which are considered to be relevant to the LHC.

The couplings of A_t^0 to the SM particles, the tHiggs (h_t^0) and the t' quark can be read off from the Lagrangian Eq.(8). The explicit expressions of the partial decay widths relevant to the LHC study can be found in[2] with the replacement, $f \to F$ and $\theta \to \theta_h$. In Fig. 1, we plot the branching ratio of A_t^0 as a function of $m_{A_t^0}$ in the range of $518\,\text{GeV} \leq m_{A_t^0} \leq 2\,\text{TeV}$ in the left panel of Fig. 1. In this plot, we also indicate the corresponding values of $\cos\theta_h$ in the upper horizontal axis. From the

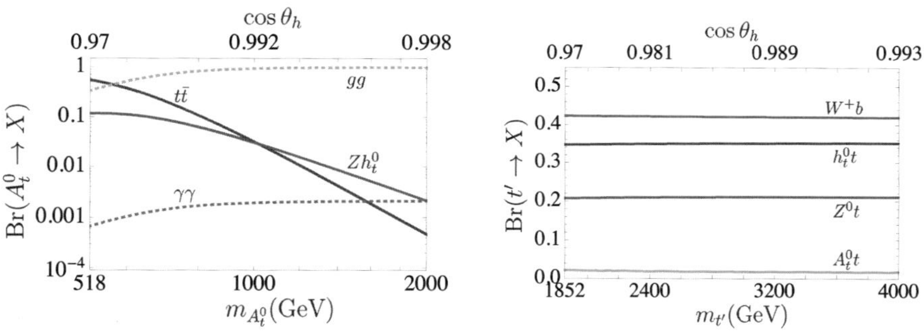

Fig. 1. The branching ratios of A_t^0 (left panel) and t' (right panel) as functions of $m_{A_t^0}$ and $m_{t'}$, respectively. Values of $\cos\theta_h$ are also shown in the upper horizontal axes.

plot we see that, in the smaller mass region, the $t\bar{t}$ and gg modes are the dominant decay channels, and therefore the main production process is the gluon-gluon fusion (ggF). The 8 TeV LHC cross sections $pp \to A_t^0 \to gg/tt$ for $m_{A_t^0} \geq 1\,\text{TeV}$ have not seriously been limited by the currently available data yet. It is therefore to be expected that more data from the upcoming Run-II would probe the A_t^0 through these channels. Another interesting channel would be $A_t^0 \to Zh_t^0$. However, with the updated branching ratio, this channel seems to be rather challenging even at the $\sqrt{s} = 14\,\text{TeV}$ LHC with 3000 fb^{-1} data due to the small branching ratio in the smaller mass region.

The t' quark arises as a mixture of the gauge-eigenstate top and χ-quarks through the diagonalization of the fermion mass matrix in the effective Lagrangian Eq.(8). The explicit expressions of the t' couplings and the partial decay widths relevant to the LHC study are listed in[3]. In the right panel of Fig. 1, we plot the branching ratios of the t' quark as a function of $m_{t'}$. In the same way as the plot for the branching ratios of A_t^0, the corresponding value of $\cos\theta_h$ is also shown in the upper horizontal axis. From the figure we read off $\mathrm{Br}(t' \to W^+b) \simeq \mathrm{Br}(t' \to h_t^0 t) \simeq 0.4$, $\mathrm{Br}(t' \to Zt) \simeq 0.21$ and $\mathrm{Br}(t' \to A_t^0 t) \simeq 0.02$. It is worth comparing these values with the branching ratios of the "singlet t' quark" in a benchmark model of t' quark[21], $\mathrm{Br}(t' \to W^+b) \simeq 0.5, \mathrm{Br}(t' \to Zt) \simeq 0.25, \mathrm{Br}(t' \to ht) \simeq 0.25$, for $m_{t'} \simeq 2\,\mathrm{TeV}$[22,23]. It is interesting to note that $\mathrm{Br}(t' \to h_t^0 t)$ in the present model is by about 40 % larger than that in the benchmark model. This is essentially due to the large $ht't$ coupling, which is the very consequence of the top quark condensate scenario.

5. Summary

We presented the Top-Mode Pseudo-Nambu-Goldstone boson Higgs (TMpNGBH) model which has recently been proposed as a variant of the top quark condensate model in light of the 125 GeV Higgs boson discovered at the LHC. We also discussed the vacuum alignment problem of TMpNGBH model based on the one-loop effective Lagrangian for the NGB sector, taking into account all the explicit breaking effects, including electroweak gauge interactions and four fermion interactions responsible for the top-seesaw mechanism. We found that the correct vacuum is determined by the configuration which minimizes the one-loop effective potential. It was found that the true vacuum is parameterized by $\cos\theta_h$, and a non-zero value of $\cos\theta_h$ realizes the EWSB phase with the appropriate breaking scale. Furthermore, we also discussed the phenomenological implications of the TMpNGBH model on the vacuum aligned at the one-loop level.

References

1. H. S. Fukano, M. Kurachi, S. Matsuzaki and K. Yamawaki, Higgs boson as a top-mode pseudo-Nambu-Goldstone boson, *Phys.Rev.* **D90**, p. 055009 (2014).
2. H. S. Fukano and S. Matsuzaki, Top-mode pseudo-Nambu-Goldstone bosona at the LHC, *Phys.Rev.* **D90**, p. 015005 (2014).
3. H. S. Fukano, M. Kurachi and S. Matsuzaki, Vacuum Alignment of the Top-Mode Pseudo-Nambu-Goldstone Boson Higgs Model, *Phys.Rev.* **D91**, p. 115005 (2015).
4. G. Aad *et al.*, Observation of a new particle in the search for the Standard Model Higgs boson with the ATLAS detector at the LHC, *Phys.Lett.* **B716**, 1 (2012).
5. S. Chatrchyan *et al.*, Observation of a new boson at a mass of 125 GeV with the CMS experiment at the LHC, *Phys.Lett.* **B716**, 30 (2012).
6. F. Abe *et al.*, Observation of top quark production in $\bar{p}p$ collisions, *Phys.Rev.Lett.* **74**, 2626 (1995).

7. S. Abachi *et al.*, Observation of the top quark, *Phys.Rev.Lett.* **74**, 2632 (1995).
8. K. Olive *et al.*, Review of Particle Physics, *Chin.Phys.* **C38**, p. 090001 (2014).
9. V. Miransky, M. Tanabashi and K. Yamawaki, Dynamical Electroweak Symmetry Breaking with Large Anomalous Dimension and t Quark Condensate, *Phys.Lett.* **B221**, p. 177 (1989).
10. V. Miransky, M. Tanabashi and K. Yamawaki, Is the t Quark Responsible for the Mass of W and Z Bosons?, *Mod.Phys.Lett.* **A04**, p. 1043 (1989).
11. W. A. Bardeen, C. T. Hill and M. Lindner, Minimal Dynamical Symmetry Breaking of the Standard Model, *Phys.Rev.* **D41**, p. 1647 (1990).
12. H.-C. Cheng, B. A. Dobrescu and J. Gu, Higgs mass from compositeness at a multi-TeV scale, *JHEP* **1408**, p. 095 (2014).
13. Y. Nambu and G. Jona-Lasinio, Dynamical Model of Elementary Particles Based on an Analogy with Superconductivity. 1., *Phys.Rev.* **122**, 345 (1961).
14. B. A. Dobrescu and C. T. Hill, Electroweak symmetry breaking via top condensation seesaw, *Phys.Rev.Lett.* **81**, 2634 (1998).
15. R. S. Chivukula, B. A. Dobrescu, H. Georgi and C. T. Hill, Top quark seesaw theory of electroweak symmetry breaking, *Phys.Rev.* **D59**, p. 075003 (1999).
16. H.-J. He, C. T. Hill and T. M. Tait, Top quark seesaw, vacuum structure and electroweak precision constraints, *Phys.Rev.* **D65**, p. 055006 (2002).
17. H. S. Fukano and K. Tuominen, A hybrid 4th generation: Technicolor with top-seesaw, *Phys.Rev.* **D85**, p. 095025 (2012).
18. H. S. Fukano and K. Tuominen, 126 GeV Higgs boson in the top-seesaw model, *JHEP* **1309**, p. 021 (2013).
19. M. E. Peskin and T. Takeuchi, A New constraint on a strongly interacting Higgs sector, *Phys.Rev.Lett.* **65**, 964 (1990).
20. M. E. Peskin and T. Takeuchi, Estimation of oblique electroweak corrections, *Phys.Rev.* **D46**, 381 (1992).
21. F. del Aguila, L. Ametller, G. L. Kane and J. Vidal, Vector Like Fermion and Standard Higgs Production at Hadron Colliders, *Nucl.Phys.* **B334**, p. 1 (1990).
22. J. Aguilar-Saavedra, Identifying top partners at LHC, *JHEP* **0911**, p. 030 (2009).
23. J. Aguilar-Saavedra, R. Benbrik, S. Heinemeyer and M. Prez-Victoria, Handbook of vectorlike quarks: Mixing and single production, *Phys.Rev.* **D88**, p. 094010 (2013).

Composite Dark Matter

G. T. Fleming†

Department of Physics, Sloane Laboratory, Yale University,
New Haven, Connecticut 06520, USA

Although dark matter has only been observed gravitationally, it would be an amazing coincidence if it did not interact at all with standard model particles. We will discuss how to sequester electroweak-charged dark constituents in electroweak-neutral stable composites. We present a model of "Stealth Dark Matter" where dark composites couple to the electroweak sector only via polarizability, rendering them nearly impossible to detect in present experiments.

†The presentation at the workshops was given by E. Rinaldi as a substitute.

Lattice Study of the Scalar and Baryon Spectra in Many-Flavor QCD

Yasumichi Aoki[a], Tatsumi Aoyama[a], Ed Bennett[b], Masafumi Kurachi[c], Toshihide Maskawa[a], Kohtaroh Miura[d], Kei-ichi Nagai[a], Hiroshi Ohki[e], Enrico Rinaldi[f], Akihiro Shibata[g], Koichi Yamawaki[a] and Takeshi Yamazaki[h]

(LatKMI Collaboration)

[a] *Kobayashi-Maskawa Institute for the Origin of Particles and the Universe (KMI), Nagoya University, Nagoya, 464-8602, Japan*

[b] *Department of Physics, Swansea University, Singleton Park, Swansea SA2 8PP, UK*

[c] *Institute of Particle and Nuclear Studies, High Energy Accelerator Research Organization (KEK), Tsukuba 305-0801, Japan*

[d] *Centre de Physique Theorique(CPT), Aix-Marseille Univerisity, Campus de Luminy, Case 907, 163 Avenue de Luminy, 13288 Marseille cedex 9, France*

[e] *RIKEN/BNL Research Center, Brookhaven National Laboratory, Upton, NY, 11973, USA*

[f] *Lawrence Livermore National Laboratory, Livermore, California, 94550, USA*

[g] *Computing Research Center, High Energy Accelerator Research Organization (KEK), Tsukuba, 305-0801, Japan*

[h] *Graduate School of Pure and Applied Sciences, University of Tsukuba, Tsukuba, Ibaraki, 305-8571, Japan*

In the search for a composite Higgs boson in walking technicolor models, many flavor QCD, in particular with $N_f = 8$, is an attractive candidate, and has been found to have a composite flavor-singlet scalar as light as the pion. Based on lattice simulations of this theory with the HISQ action, we will present our preliminary results on the scalar decay constant using the fermionic bilinear operator, and on the mass of the lightest baryon state which could be a dark matter candidate. Combining these two results, implications for dark matter direct detection are also discussed.

Keywords: Lattice gauge theory, Conformal/walking dynamics, Dark matter, composite Higgs particle.

1. Introduction

A new scalar particle with mass 125 GeV has been discovered at the LHC, and the current experimental data is consistent with the standard model (SM) of particle physics. However, the origin of mass and electroweak symmetry breaking remains unknown, and the SM does not have a viable dark matter (DM) candidate, which are reasons why we are seeking new physics beyond the SM.

It is possible that there is strong gauge dynamics beyond the TeV scale: one popular candidate is the Walking Technicolor model. This breaks the electroweak symmetry dynamically in an approximately scale-invariant dynamics, and predicts a large anomalous dimension $\gamma_m \simeq 1$ together with a "techni-dilaton" as a pseudo Nambu-Goldstone boson of the spontaneously broken approximate scale symmetry[1]. The Higgs scalar could be identified with the techni-dilaton, which is a composite bound-state of the flavor-singlet techni-fermion bilinear, parametrically lighter than other techni-hadrons. From a phenomenological point of view, the techni-dilaton decay constant (denoted here as F_σ) is a very important parameter in addition to its mass, since F_σ controls all the techni-dilaton's couplings to SM particles. In addition, the technicolor model has a rich hadron structure, and composite resonances like the techni-rho meson (ρ) may be in the discovery reach of run-II at the LHC. Furthermore the techni-baryon, its lightest neutral component, could be a stable particle due to techni-fermion number conservation. Thus it would be a candidate for DM.

In order to study the walking technicolor model, non-perturbative understanding of the strong gauge dynamics is necessary, and lattice numerical calculations are the most powerful tool for that purpose. In a previous lattice study, the LatKMI collaboration has shown that 8-flavor ($N_f = 8$) QCD could be a candidate for a walking gauge theory. We found an approximate hyperscaling relation for various hadron masses in a certain fermion mass range[2]. Remarkably, we also found that the flavor singlet-scalar (σ) mass is as light as the pseudoscalar (π), which might indicate that σ could be regarded as the techni-dilaton[3].

In this proceedings we investigate the decay constant of the flavor-singlet scalar ant its mass. Using the Ward-Takahashi (WT) relation of the scale symmetry in the continuum theory, we estimate the dilaton decay constant F_σ. As for the techni-baryon DM, it is important to investigate its direct detection, where the most dominant contribution of the DM scattering amplitude is the scalar mediated interaction. Based on a DM effective theory whose parameters include F_σ and the techni-baryon mass, we evaluate the leading order of the scattering cross section of the DM. Using the lattice input for the above parameters, we discuss experimental detectability and the possible realization of the techni-baryon DM scenario.

In the next section, we briefly explain how to calculate the mass of the flavor-singlet scalar and scalar decay constant on the lattice. We discuss a relation between the scalar and dilaton decay constants. In Section 3, we show our lattice result of the scalar mass and decay constants, including a chiral extrapolation based on a dilaton effective theory. In Section 4, we study the DM detection rate based on the dilaton chiral perturbation theory with baryon, and the summary is given in Section 5. Note that all the results shown in this proceeding are preliminary.

2. Scalar mass and decay constant

We investigate the mass of the scalar and its decay constant on the lattice. The mass of the flavor-singlet scalar, which we call σ, is calculated from the two-point correlation function ($C_\sigma(t)$) of the flavor-singlet scalar bilinear operator $\mathcal{O}_s$. By using the staggered lattice fermion field χ_i, $\mathcal{O}_s$ is written as $\mathcal{O}_s(x,t) = \sum_i \bar{\chi}_i(x,t)\chi_i(x,t)$, where i denotes the different staggered fermion species, namely $i = 1, 2$ for $N_f = 8$ QCD. The correlator is then given by

$$C_\sigma(t) = \frac{1}{V}\sum_x \langle \mathcal{O}_s(x,t)\mathcal{O}_s(0,0)\rangle , \tag{1}$$

where V is the spatial volume ($V = L^3$). The asymptotic behavior of $C_\sigma(t)$ is given by $C_\sigma(t) = A_\sigma(t) + (-1)^t A_{\pi_{\overline{SC}}}(t)$, where $A_{\pi_{\overline{SC}}}$ is a pseudoscalar correlator, and $A_H(t) = A_H(e^{-m_H t} + e^{-m_H(T-t)})$ for different hadrons H.

We define the scalar decay constant F_S as the scalar operator matrix element,

$$\langle 0|m_f O_s(0,0)|\sigma(0)\rangle = F_S m_\sigma^2. \tag{2}$$

The above matrix element can be calculated from $C_\sigma(t)$. Substituting the complete set, $\sum_n |n\rangle\langle n| = \int \frac{d^3p}{(2\pi)^3} \frac{|\sigma(p)\rangle\langle\sigma(p)|}{2E_p} + \cdots$ into Eq. 1, the correlator $C_\sigma(t)$ is given by

$$C_\sigma(t) = \frac{1}{V}|\langle 0|O_s(0,0)|\sigma(0)\rangle|^2 \frac{1}{2m_\sigma}\left(e^{-m_\sigma t} + e^{-m_\sigma(T-t)}\right). \tag{3}$$

Thus we calculate F_S as follows,

$$F_S = \frac{m_f\sqrt{2m_\sigma V A_\sigma}}{m_\sigma^2}. \tag{4}$$

We note that this quantity is renormalization group invariant and a physical quantity, and it is easy to measure on the lattice. It is also noted that F_S should obey a hyperscaling relation, $F_S \propto m_f^{1/(1+\gamma)}$ with $\gamma \sim \gamma_m$, if the theory is in the conformal window[4].

Let us now discuss the dilaton decay constant F_σ. The following discussion is based on the continuum theory. It is important to recall the basic definition of the dilaton decay constant, which is $\langle 0|\mathcal{D}^\mu(x)|\sigma(p)\rangle = -iF_\sigma p^\mu e^{-ipx}$. From this we obtain $\langle 0|\partial_\mu \mathcal{D}^\mu(0)|\sigma(0)\rangle = -F_\sigma m_\sigma^2$. Therefore the dilaton decay constant can be directly calculated from the matrix element of the dilatation current. However the dilatation current is rather difficult to construct on the lattice, since it contains a power divergence that needs to be subtracted.

Instead, we consider an alternative way to estimate it, which can be derived by using the WT relation for the dilatation current. Following the argument in[5], we consider the integrated WT relation for dilatation transformation. We use the scale transformation relation of an operator $\mathcal{O}$, $\delta_D \mathcal{O} = \Delta_\mathcal{O}\mathcal{O}$, where $\Delta_\mathcal{O}$ is the scaling dimension of $\mathcal{O}$. In the zero momentum transfer limit the WT relation leads to $\int d^4x T\langle \partial_\mu \mathcal{D}^\mu(x)\mathcal{O}(0)\rangle = \Delta_\mathcal{O}\langle\mathcal{O}\rangle$[a], from which, by assuming the σ pole dominance, we obtain $\Delta_\mathcal{O}\langle\mathcal{O}\rangle = (-F_\sigma m_\sigma^2)\frac{1}{m_\sigma^2}\langle\sigma|\mathcal{O}|0\rangle = -F_\sigma\langle\sigma|\mathcal{O}|0\rangle$.[b] In the case we take $\mathcal{O}(x) = m_f \sum_i^{N_F} \bar{\psi}_i\psi_i(x)$ (the flavor-singlet scalar operator), we obtain a relation,

$$F_S F_\sigma m_\sigma^2 = -\Delta_{\bar{\psi}\psi} m_f \sum_i^{N_F} \langle \bar{\psi}_i\psi_i\rangle. \tag{5}$$

(Dividing both sides by m_f leads to the relation obtained at $m_f = 0$[6].)

We note that this relation holds in the continuum theory with infrared conformality under the assumption of the σ pole dominance. While we do not know the

[a] In the formula, the vacuum contribution on the left hand side is absent in the WT relation.
[b] The so-called Partially Conserved Dilatation Current (PCDC) relation $F_\sigma^2 m_\sigma^2 = -4\langle\theta_\mu^\mu\rangle$ follows for $\mathcal{O} = \partial_\mu\mathcal{D}^\mu = \theta_\mu^\mu$ (nonperturbative trace anomaly).

real infrared behavior in $N_f = 8$ QCD towards the chiral limit, our recent lattice study shows that an approximate hyperscaling relation (walking behavior) is found for various hadron masses in a certain fermion mass region of m_f, so the above relation could be effective also in the walking region up to excited state contaminations and discretization effects. Therefore in the following preliminary lattice analysis, we shall use this relation for a semi-direct estimate of F_σ. In order to precisely calculate F_σ, however, a direct calculation of the dilatation current is needed.

3. Result

In this section we show our lattice results for $N_f = 8$ QCD. Details of the simulation and analysis for measurements can be found in[8]. The results for m_σ and F_S are summarized in Fig 1. It is found that the scalar is as light as π, and clearly lighter than ρ, and we obtain a good signal for F_S as well. The lightness of the scalar could be regarded as a reflection of a dilatonic nature of the scalar.

While our fermion mass is far from the chiral limit, we shall study the chiral extrapolation based on an effective theory, which is dilaton chiral perturbation theory (DChPT)[7]. At the leading order of the DChPT, the scalar mass is given by $m_\sigma^2 = d_0 + d_1 m_\pi^2$, where $d_1 = \frac{(3-\gamma_m)(1+\gamma_m)}{4}\frac{N_f F_\pi^2}{F_\sigma^2}$, m_π and F_π are the mass and decay constant of π. We also try a fit with a naive form of $m_\sigma = c_0 + c_1 m_f$. The fit results are shown in the left panel of Fig. 2, where a reasonable value of $\chi^2/\mathrm{dof} \sim \mathcal{O}(1)$ is obtained. The value of m_σ^2 in the chiral limit is $d_0 = -0.0028(98)$. Thus we obtain a very light σ. From the fit value of the slope $d_1 = 0.89(26)$ in DChPT, we also obtain $F_\sigma \sim \sqrt{N_f} F_\pi \sim 0.06$, with F_π being in the chiral limit[8] and $\gamma \sim \gamma_m \sim 1$.

On the right panel of Fig. 2, the result of F_σ from the semi-direct estimate (Eq. 5) is shown[c]. We also carry out the chiral extrapolation fits, whose results are also shown in the figure. In the chiral limit, we obtain $\frac{F_\sigma}{\Delta_{\bar\psi\psi}} \sim 0.03$. Given that $\Delta_{\bar\psi\psi} = 3 - \gamma$, we have shown how two different methods, DChPT and a semi-direct calculation, give a consistent result for F_σ.

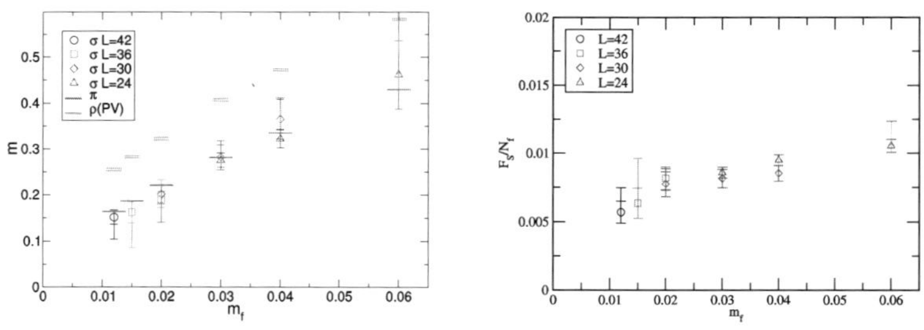

Fig. 1. (Left) Mass of the flavor-singlet scalar. Other hadron masses are also shown. Outer error represents the statistical and systematic uncertainties added in quadrature, while inner error is only statistical. (Right) Scalar decay constant of the flavor-singlet scalar.

[c]For the chiral condensate, we use its chiral limit value to avoid large lattice artefacts.

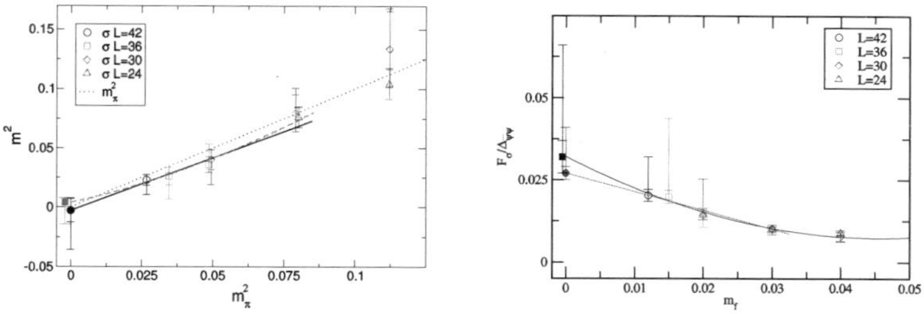

Fig. 2. (Left) m_σ^2 v.s. m_π^2. Black line shows the fit with DChPT formula. Dashed red curve shows the fit result with a naive form, $m_\sigma = c_0 + c_1 m_f$. (Right) Dilaton decay constant from a semi-direct method. Blue and Black lines show the chiral fits of the linear and quadratic polynomial in m_f with the lightest 4 and 5 data points used. Outer error represents the statistical and systematic uncertainties added in quadrature, while inner error is only statistical.

4. Dilaton ChPT with baryon and dark matter

The technicolor model may provide a good candidate for composite DM, in the form of a neutral baryonic bound state made of constituent (possibly charged) techni-fermions. In this case, the coupling between the SM particle and the DM as well as the mass of the DM are constrained by direct detection experiments. In most cases, the experiments determine the scattering rate of DM with heavy nuclei in the detector, one dominant contribution to which is the Higgs (scalar) mediated spin-independent process. In the following, we provide a low-energy effective theory of the walking gauge theory including the DM, whose low-energy constants can be determined from the lattice results of the baryon and scalar spectra. The result obtained in the previous section is useful, and it gives rise to information for the DM direct detection experiments.

We consider a DM effective theory including the dilaton based on the DChPT. Since the DM is the lightest techni-baryon, the extension to the baryon sector of the DChPT is straightforward[7]. As a result, in the leading order the dilaton field can only couple through the baryon mass term as

$$\mathcal{L} = \bar{B}(x)(i\gamma_\mu \partial^\mu - \chi(x) m_B)B(x), \tag{6}$$

where $\chi(x) = e^{\sigma(x)/F_\sigma}$, $B(x)$ is the baryon field, and m_B is its mass in the chiral limit. The parameter m_B explicity breaks the scale symmetry, and the (pseudo) dilaton acts on this term to make the action scale invariant. Actually this action is invariant under the scale transformation, $\delta B = (\frac{3}{2} + x_\nu \partial^\nu)B$, $\delta\chi = (1 + x_\nu \partial^\nu)\chi$.

From this effective theory, we can read off the dilaton-baryon effective coupling ($y_{\bar{B}B\sigma}$), which is uniquely determined as $y_{\bar{B}B\sigma} = m_B/F_\sigma$. Regarding to the SM sector, the dilaton-nucleon coupling is also determined within the framework of the dilaton effective theory[10], since the dilaton-quark coupling ($y_{\sigma\bar{f}f}$) could be related to the SM Yukawa coupling ($y_{h_{\rm SM}\bar{f}f}$) as $\frac{y_{\sigma\bar{f}f}}{y_{h_{\rm SM}\bar{f}f}} = \frac{(3-\gamma)v_{EW}}{F_\sigma}$. Combining both SM and technicolor sectors, the cross section with a target nucleus N is given as $\sigma_{SI} = \frac{M_R^2}{\pi}(Zf_p + (A-Z)f_n)^2$, where $M_R = (m_B m_N)/(m_B + m_N)$, and Z and

A are the total number of the protons (p) and neutrons (n) in the nucleus. The parameter $f_{(n,p)}$ is defined as $f_{(n,p)} = \frac{m_N}{\sqrt{2}m_\sigma^2}\frac{y_{\bar{B}B\sigma}}{F_\sigma}(3-\gamma)\left(\sum_{q=u,d,s} f_{T_q}^{(n,p)} + \frac{2}{9}f_{TG}^{(n,p)}\right)$, where $f_{T_q}^{(n,p)}$ is the nucleon σ-term of the light quarks ($q = u, d, s$), and $f_{TG}^{(n,p)}$ is that of the heavy quarks[d]. Thus lattice calculations are used in the technicolor theory as well as in QCD theory to obtain non-perturbative information about DM physics.

Here we show our numerical results of the DM cross section[e]. We use the lattice results of the dilaton decay constant (F_σ) obtained from the previous section and baryon mass, while the scalar mass m_σ is fixed to its experimental value (125 GeV) in this analysis. We use the values in[9] for $f_{T_q}^{(n,p)}$. To set the scale, we use the relation $\sqrt{N_f/2}F_\pi/\sqrt{2} = 246$ GeV. We again use the F_π in the chiral limit. To compare it with experiment, we use the cross section per nucleon (σ_0) instead of σ_{SI}. The result is shown in Fig. 3. According to DM direct detection experiments[f], our values for σ_0 are excluded, so that it may be difficult to explain the existence of DM as a techni-baryon. However we note that there exist other contributions to the DM cross section, e.g. gauge boson mediated interaction, and higher order terms, which might affect the DM cross section. Calculation of these contributions on the lattice is left for future investigations.

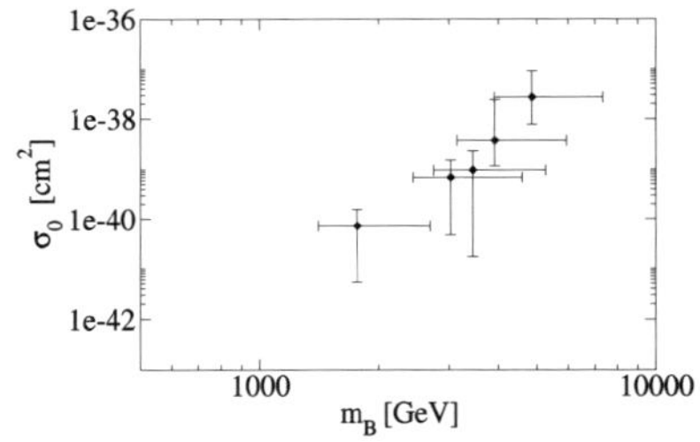

Fig. 3. σ_0 [cm^2] as a function of m_B [GeV]. The results for $m_f = 0.030, 0.020, 0.015, 0.012$, and the chiral limit from upper-right to lower-left. Both the statistical and systematic errors are included. Experimentally allowed region is below the plotted window.

5. Summary

The scalar mass and decay constant are very important parameters to probe a technicolor signature at the LHC. Based on the lattice theory, we derived a relation between the scalar decay constant and the (flavor-singlet) scalar correlation functions. Our numerical result shows that the signal of the decay constant is as good as that of the mass. Although the accuracy of our result is not enough to precisely extrapolate towards the chiral limit, we obtained a rough estimate of the

[d]In general, techni-fermions can be charged under the SM color, so there may exist additional contributions to the nucleon matrix elements from the techni-fermions. In this analysis, we omit these contributions for simplicity.
[e]A similar analysis on the lattice has been performed for a different composite DM model based on strong dynamics [11].
[f]For a recent experiment, see e.g. [12].

ratio $F_\sigma/F_\pi \sim 3$. This is useful for collider phenomenology. Besides it, F_σ can be also used for DM physics. We provided a DM effective theory based on the dilaton ChPT, where a dilaton-DM coupling is related to F_σ and the DM mass. This coupling is constrained by DM direct detection experiments, e.g. LUX. We then discussed a possible scenario for techni-baryon DM. We note that all the results shown here are preliminary. Simulation at lighter fermion mass, and the direct computations on the lattice are needed to precisely estimate the above quantities.

Acknowledgments – Numerical computations have been carried out on the high-performance computing systems at KMI (φ), at the Information Technology Center in Nagoya University (CX400), and at the Research Institute for Information Technology in Kyushu University (CX400 and HA8000). This work is supported by the JSPS Grant-in-Aid for Scientific Research (S) No.22224003, (C) No.23540300 (K.Y.), for Young Scientists (B) No.25800139 (H.O.) and No.25800138 (T.Y.), and also by the MEXT Grants-in-Aid for Scientific Research on Innovative Areas No.23105708 (T.Y.) and No.25105011 (M.K.). This work is supported by the JLDG constructed over the SINET of NII. The work of H.O. is supported by the RIKEN Special Postdoctoral Researcher program. E.R. acknowledges the support of the U.S. Department of Energy under Contract DE-AC52- 07NA27344 (LLNL).

References

1. K. Yamawaki, M. Bando and K. i. Matumoto, Phys. Rev. Lett. **56**, 1335 (1986).
2. Y. Aoki *et al.* [LatKMI Collaboration], Phys. Rev. D **87**, no. 9, 094511 (2013) [arXiv:1302.6859 [hep-lat]].
3. Y. Aoki *et al.* [LatKMI Collaboration], Phys. Rev. D **89**, 111502 (2014) [arXiv:1403.5000 [hep-lat]].
4. L. Del Debbio and R. Zwicky, Phys. Rev. D **82**, 014502 (2010) [arXiv:1005.2371 [hep-ph]].
5. M. Bando, K. i. Matumoto and K. Yamawaki, Phys. Lett. B **178**, 308 (1986).
6. S. Matsuzaki and K. Yamawaki, Phys. Rev. D **86**, 115004 (2012) [arXiv:1209.2017 [hep-ph]].
7. S. Matsuzaki and K. Yamawaki, Phys. Rev. Lett. **113**, no. 8, 082002 (2014) [arXiv:1311.3784 [hep-lat]].
8. LatKMI collaboration in preparation.
9. J. Hisano, K. Ishiwata and N. Nagata, JHEP **1506**, 097 (2015) [arXiv:1504.00915 [hep-ph]].
10. S. Matsuzaki and K. Yamawaki, Phys. Rev. D **86**, 035025 (2012) [arXiv:1206.6703 [hep-ph]].
11. T. Appelquist *et al.* [Lattice Strong Dynamics (LSD) Collaboration], Phys. Rev. D **89**, no. 9, 094508 (2014) [arXiv:1402.6656 [hep-lat]].
12. D. S. Akerib *et al.* [LUX Collaboration], Phys. Rev. Lett. **112**, 091303 (2014) [arXiv:1310.8214 [astro-ph.CO]].

Weak Renormalization Group Approach for Dynamical Chiral Symmetry Breaking

Ken-Ichi Aoki

Institute for Theoretical Physics, Kanazawa University, Kanazawa 920-1192, Japan

We set up weak renormalization group equation for dynamical chiral symmetry breaking systems which allows us to work with only fermionic degrees of freedom. So far the fermionic system encounters the divergence at the critical scale of the dynamical chiral symmetry breaking. We define the weak partial differential equation representing the scale dependent change of effective fermionic interactions. The weak renormalization group equation can be solved without any ambiguity and the solution perfectly describes the physically correct singularity due to the phase transition even in case of the first order transition having the finite jump of the dynamical mass at the critical scale.

Gauge Symmetry and the Functional Renormalization Group

Katsumi Itoh

Department of Education, Niigata University,
8050 Ikarashi-2, Nishi-ku, Niigata, 950-2181, Japan
E-mail: itoh@ed.niigata-u.ac.jp

In the functional renormalization group (FRG), the introduction of a momentum cutoff often breaks symmetry present in a theory. The most important example is the gauge symmetry. However a symmetry survives in the presence of a cutoff in a modified form. We apply our understanding to QED as the simplest case. A modified version of the Ward-Takahashi identity is solved partially to constrain the Wilson action. Furthermore, we study the flow equation for the photon 2-point function and find its analytical expression for the regulator function of the exponential type.

Keywords: Functional renormalization group; Gauge symmetry; QED.

1. Introduction

The functional renormalization group (FRG) has been successfully applied to various field theoretical problems as a robust non-perturbative method: e.g. QCD, quantum gravity, problems in condensed matter physics, non-equilibrium physics and others.

When applied to a gauge theory, however, the symmetry in its usual form is not compatible with the presence of the momentum cutoff Λ: it survives in a modified form that depends on a regularization function used to introduce the cutoff[1]. The path integration over the high momentum modes produces corrections to the gauge or BRST transformation at the scale Λ. The condition for the gauge invariance, described by the Ward-Takahashi (WT) identity[2 3 4] $\Sigma_\Lambda[\phi] = 0$, is the vanishing of the sum of the variation of the Wilson action, S_Λ, and the Jacobian factor, the contribution from the path integral measure. The symmetry, carrying all the quantum corrections above the cutoff, may be regarded as a truly quantum symmetry.

The WT identity $\Sigma_\Lambda[\phi] = 0$ and the flow equations for the couplings in the Wilson action, or equivalently the effective average action,[a] are two key equations to define a gauge theory with FRG. My talk[b] was consisted of two parts: 1) a study of the condition for the gauge invariance and a solution to it for QED; 2) to obtain the photon 2-point function by using the flow equation with its full momentum dependence for a finite momentum cutoff Λ.

The condition $\Sigma_\Lambda[\phi] = 0$ contains an infinite number of terms when expanded

[a]Two actions are related via a Legendre transformation.
[b]The talk was based on my work[5] in collaboration with Y. Igarashi and J. Pawlowski.

in terms of fields. The vanishing of the coefficient for each term gives us a relation among couplings in the Wilson action. Here, we consider first two relations from the WT identity, that correspond to gauge transformations of the gauge and fermion 2-point functions.

The Wilson action itself contains an infinite number of terms. For a practical calculation, we need to truncate the action. The WT identity tells us a necessary set of interaction terms to be included in a truncation scheme. Based on this observation, we include field bilinear terms of the gauge and fermion fields, their gauge interaction and four fermi interactions. For simplicity, we consider the massless fermion and assume the chiral symmetry.

Here, we solve two relations from the WT identity for the photon 2-point functions and 4-fermi interaction terms with their momentum dependence fully respected: we introduce form factors to the familiar 4-fermi couplings and yet another 4-fermi derivative couplings with the dimension eight. The first relation for the photon 2-point function determines its longitudinal part. We find that it can be evaluated exactly when we take a specific cutoff function of a gaussian type. The second relation consists of a tree and one loop contributions. It should be noted that even a tree part contains contributions from modes above the cutoff. As the first step, we assume that the tree and one-loop parts vanish separately. As results, we obtain: 1) the condition similar to the well-known "$Z_1 = Z_2$"; 2) two conditions on the 4-fermi form factors in terms of photon propagator.

Finally, we study the flow equation for the effective average action to obtain the photon 2-point function analytically. We find the anomalous dimension from the transverse part, while the longitudinal part agrees with our result from the WT identity.

Several different approaches to gauge symmetry in FRG have been proposed.[c] Among others, works based on the "modified Slavnov-Taylor identity (mST)" [8], the WT identity for average effective action Γ_Λ,[d] are closest to our present work. In these earlier studies, mST was solved with the derivative expansion.[9]

2. The Wilson action and the effective average action

This section provides a brief summary of some basic objects in the FRG and gauge symmetry, the Wilson action, the effective average action and the Ward-Takahashi (WT) identity.

We consider a theory in 4-dimensional Euclidean space, which is assumed to have some gauge symmetry written as a BRST symmetry. It is described by a gauge-fixed action $\mathcal{S}[\varphi]$, a functional of fields φ^A which collectively represent gauge and matter fields as well as ghosts and antighosts. The index A denotes the Lorentz indices of gauge fields, the spinor indices of the fermions, and indices distinguishing different

[c] Summaries and references will be found in review articles, e.g., Refs. 6 and 7.

[d] Strictly speaking, we need to introduce the antifields in order to discuss the mST.

types of generic fields. The Grassmann parity for φ^A is expressed as $\epsilon(\phi^A) = \epsilon_A$: $\epsilon_A = 0$ (1) if the field ϕ^A is Grassmann even (odd).

In FRG, we introduce an IR cutoff Λ through positive and monotonically decreasing functions of $x = p^2/\Lambda^2$ that behave as

$$K(x) \quad \sim \quad \begin{cases} 1 & (x \ll 1) \\ 0 & (x \gg 1) \end{cases} \tag{1}$$

and they go to zero sufficiently rapidly as $p^2/\Lambda^2 \to \infty$. For simplicity, we write the function as $K(p)$ in the rest of the paper. The action $\mathcal{S}$ defined at some UV scale Λ_0 is assumed to be decomposed into the kinetic and interaction terms,

$$\mathcal{S}[\varphi] = \frac{1}{2}\varphi \cdot D \cdot \varphi + \mathcal{S}_I[\varphi] \ . \tag{2}$$

In this paper we use matrix notation in the momentum space, for example,

$$\phi \cdot D \cdot \phi = \int \frac{d^4p}{(2\pi)^4}\phi^A(-p)D_{AB}(p)\phi^B(p). \tag{3}$$

We introduce another IR cutoff $\Lambda(< \Lambda_0)$ and integrate over the modes with the momenta between two cutoffs to give the Wilson action with the IR cutoff Λ.

$$S_\Lambda[\phi] = \frac{1}{2}\phi^A Z^A \ K^{-1} \ D_{AB}\phi^B + S_{I,\Lambda}[\phi] \ , \tag{4}$$

where the interaction part is given by a functional integral

$$\exp\Bigl(-S_{I,\Lambda}[\phi]\Bigr) = \int \mathcal{D}\chi \exp\Bigl(-\frac{1}{2}\chi \cdot (1-K)^{-1}D \cdot \chi - \mathcal{S}_I[Z^{1/2}\phi + \chi]\Bigr) \ . \tag{5}$$

The Wilson action S_Λ is the generating functional of the connected (amputated) cutoff Green functions. Another important object in the FRG is its 1PI part, the effective average action, Γ_Λ.

$$\Gamma_\Lambda[\Phi] = \frac{1}{2}Z\Phi \cdot D \cdot \Phi + \Gamma_{I,\Lambda}[\Phi] + \frac{1}{2}Z\Phi \cdot R \cdot \Phi \ , \tag{6}$$

where

$$R = \frac{K}{1-K}D \ . \tag{7}$$

The Wilson and the effective average actions are related to each other via the following two relations: 1) the interaction part of Γ_Λ is related to that of S_Λ as,

$$\Gamma_{I,\Lambda}[\Phi] = S_{I,\Lambda}[\phi] - \frac{1}{2}Z(\Phi - \phi) \cdot \frac{D}{1-K} \cdot (\Phi - \phi) \ ; \tag{8}$$

2) the fields Φ^A and ϕ^A are related as,

$$\left(Z^A\right)^{1/2}\Phi^A = \left(Z^A\right)^{1/2}\phi^A - \left(Z^A\right)^{-1/2}(\Delta_H)^{AB}\frac{\partial^l S_{I,\Lambda}[\phi]}{\partial \phi^B} \ . \tag{9}$$

By using Eqs. (8) and (9), the Wilson action $S_{I,\Lambda}$ is shown to be expanded in terms of the effective average action $\Gamma_{I,\Lambda}$, their field derivatives and the high energy propagators, $(\Delta_H)^{AB} = (1-K)\left(D^{-1}\right)^{AB}$.[10]

3. The WT identity for QED and its implications

Given the Wilson action $S_\Lambda[\phi]$ for the fields $\phi^A = (a_\mu,\ \bar{\psi}_{\hat{\alpha}},\ \psi_\alpha,\ c,\ \bar{c})$, the WT operator for QED is written as[2]

$$\begin{aligned}\Sigma_\Lambda[\phi] = & \int_p \Big\{ \frac{\partial S_\Lambda}{\partial a_\mu(p)}(-ip_\mu)c(p) + \frac{\partial^r S_\Lambda}{\partial \bar{c}(p)} \xi_0^{-1} p_\mu a_\mu(p) \Big\} \\ & -i\, e_0 \int_{p,\ q} \Big\{ \frac{\partial^r S_\Lambda}{\partial \psi_\alpha(q)} \frac{K(q)}{K(p)} \psi_\alpha(p) - \frac{K(p)}{K(q)} \bar{\psi}_{\hat{\alpha}}(-q) \frac{\partial^l S_\Lambda}{\partial \bar{\psi}_{\hat{\alpha}}(-p)} \Big\} c(q-p) \\ & -i\, e_0 \int_{p,\ q} U_{\beta\hat{\alpha}}(-q,p) \Big\{ \frac{\partial^l S_\Lambda}{\partial \bar{\psi}_{\hat{\alpha}}(-p)} \frac{\partial^r S_\Lambda}{\partial \psi_\beta(q)} - \frac{\partial^l \partial^r S_\Lambda}{\partial \bar{\psi}_{\hat{\alpha}}(-p) \partial \psi_\beta(q)} \Big\} c(q-p)\ ,\end{aligned} \tag{10}$$

where

$$U(-q,p) = Z_2^{-1} \left[K(q) \frac{1-K(p)}{\not{p}} - K(p) \frac{1-K(q)}{\not{q}} \right] .$$

To construct interaction part $S_{I,\Lambda}[\phi]$ using Eqs. (8) and (9), we first specify its 1PI part, namely effective average action $\Gamma_{I,\Lambda}[\Phi]$, imposing for simplicity chiral symmetry on the fermionic sector. We introduce form factors in interactions.

$$\begin{aligned}\Gamma_{I,\Lambda}[\Phi] = & \int_p \left[\frac{Z_3}{2} A_\mu(-p) \mathcal{M}_{\mu\nu}(p) A_\nu(p) + Z_2 \sigma(p) \bar{\Psi}(-p) \not{p} \Psi(p) \right] \\ & - e Z_2 Z_3^{1/2} \int_{p,q} \bar{\Psi}(-p) \not{A}(p-q) \Psi(q) + \frac{Z_2^2}{2\Lambda^2} \int_{p_1,\cdots,p_4} (2\pi)^4 \delta^4(p_1+p_2+p_3+p_4) \\ & \times \Big\{ h_S(s,u) \Big[\big(\bar{\Psi}\Psi\big)^2 - \big(\bar{\Psi}\gamma_5\Psi\big)^2 \Big] + h_V(s,u) \Big[\big(\bar{\Psi}\gamma_\mu\Psi\big)^2 + \big(\bar{\Psi}\gamma_5\gamma_\mu\Psi\big)^2 \Big] \\ & + \frac{1}{\Lambda^2}(p_1+p_4)_\mu (p_2+p_3)_\nu h_{V'}(s,u) \Big[\big(\bar{\Psi}\gamma_\mu\Psi\big)\big(\bar{\Psi}\gamma_\nu\Psi\big) + \big(\bar{\Psi}\gamma_5\gamma_\mu\Psi\big)\big(\bar{\Psi}\gamma_5\gamma_\nu\Psi\big) \Big] \Big\} .\end{aligned} \tag{11}$$

Here $\mathcal{M}_{\mu\nu}(p) = P^T_{\mu\nu}\mathcal{T}(p) + P^L_{\mu\nu}\mathcal{L}(p), P^T_{\mu\nu} = \delta_{\mu\nu} - p_\mu p_\nu/p^2, P^L_{\mu\nu} = p_\mu p_\nu/p^2$ and s, u are Mandelstam variables. Note that we introduced an unusual 4-fermi interaction in Eq. (11), that will presently be found necessary to satisfy the WT identity.

Using Eqs. (8) and (9), we obtain $S_{I,\Lambda}$ written in terms of the form factors appeared in Eq. (11).[10] We then substitute S_Λ into the WT identity $\Sigma_\Lambda[\phi] = 0$, which can be expanded in polynomial of ϕ^A. Consider terms that contain $a_\mu \times c$ or $\bar{\psi} \times \psi \times c$ in this expansion. For simplicity, we assume $\sigma(p) = 0$ for fermionic 2-point function. From $a_\mu(p) \times c(-p)$ terms, we have

$$Z_3 p_\nu \mathcal{L}(p) = -e_0\, e Z_2 Z_3^{1/2} \int_q \mathrm{Tr}[U(-p-q,q)\gamma_\nu]\ . \tag{12}$$

From $\bar{\psi}(p) \times \psi(-q) \times c(q-p)$ terms, we have the second WT relation

$$\left(e_0 Z_2 - e Z_2 Z_3^{1/2}\right)(\not{p} - \not{q}) - 2e_0 Z_2^2 \int_k \left[\frac{1}{\Lambda^2}\left\{\left(h_S - 2h_V\right)[k^2, (p+q)^2]\right.\right.$$
$$\left. -2h_V[(p+q)^2, k^2]\right\} + e^2 T(k^2) + \frac{1}{\Lambda^4}\left\{2(p-q)^2 h_{V'}[k^2, (p+q)^2]\right.$$
$$\left.\left. +k^2 h_{V'}[(p+q)^2, k^2]\right\}\right] U(-q-k, p+k) \tag{13}$$
$$-e_0 Z_2^2 \int_k \left[\frac{2}{\Lambda^4} h_{V'}[(p+q)^2, k^2] + e^2 \frac{1}{k^2}\{T(k^2) - L(k^2)\}\right] \not{k} U(-q-k, p+k) \not{k} = 0 \; .$$

where

$$T(p) = \frac{1-K}{p^2 + (1-K)\mathcal{T}(p)}, \qquad L(p) = \frac{\xi(1-K)}{p^2 + \xi(1-K)\mathcal{L}(p)}, \qquad \xi = Z_3 \xi_0 \; .$$

This constraint (13) splits into two parts: a constant $\times$ $(\not{p} - \not{q})$ and the one-loop part. Let us assume that they separately vanish. The tree part gives

$$e_0 = Z_3^{1/2} e \; . \tag{14}$$

This may be understood as the well-known WT relation in the standard realization of gauge symmetry in QED: $Z_1 = Z_2$. The one-loop part gives

$$\frac{1}{\Lambda^2}\left\{\left(h_S - 2h_V\right)[k^2, (p-q)^2] - 2h_V[(p-q)^2, k^2]\right\}$$
$$= e^2\left\{T[(p-q)^2] - L[(p-q)^2]\right\} - \frac{e^2}{2}\left\{T(k^2) + L(k^2)\right\} , \tag{15}$$
$$\frac{1}{\Lambda^4} h_{V'}[(p+q)^2, k^2] = -\frac{e^2}{2k^2}\left\{T(k^2) - L(k^2)\right\} .$$

These are relations between 4-fermi interactions and the photon propagator. We see here that the $h_{V'}$ coupling is necessary. Note that a derivative expansion of Eq. (13) gives us, $1 - Z_3^{1/2} e/e_0 \simeq [h_S(0,0) - 4h_V(0,0)]$, which should be compared with Eqs. (14) and (15).

Let us go back to Eq. (12). Remarkably, the longitudinal component of photon 2-point function $\mathcal{L}$ can be evaluated exactly for a specific cutoff function $K(p) = e^{-p^2/\Lambda^2}$ using formulae for the modified Bessel functions. We obtain

$$\mathcal{L}(p^2) = -e^2 \frac{\Lambda^2}{2\pi^2 \bar{p}^4}\left[1 - \exp(-\bar{p}^2/2) - \bar{p}^2\left(1 - \frac{1}{2}\exp(-\bar{p}^2/2)\right)\right] \tag{16}$$

where we have used $e_0 = Z_3^{1/2} e$ to eliminate e_0, and $\bar{p}^2 = p^2/\Lambda^2$. To find the transverse part $\mathcal{T}$, we need to use RG flow equations.

4. Momentum dependent flow equation

The flow equation for the 1PI action Γ_Λ is known as the Wetterich equation. Here we write it for the photon 2-point functions at $O(e^2)$:

$$\frac{Z_3}{2}\int_p A_\mu(-p)\left[P^T_{\mu\nu}\left\{2\eta_A p^2 - 2\mathcal{T}(p)\right\} + (-2)P^L_{\mu\nu}\mathcal{L}(p)\right]A_\nu(p)$$
$$= -e^2 Z_3 \int_{p,q} 2K'(q)\,(1-K(p+q))^2\,\frac{1}{(p+q)^2}\mathrm{Tr}\left[\not{A}(-p)(\not{p}+\not{q})\not{A}(p)\not{q}\right] . \quad (17)$$

The r.h.s. of Eq. (17) can be exactly evaluated with $K(p) = e^{-p^2/\Lambda^2}$ to give

$$-Z_3\frac{e^2}{2\pi^2}\int_p A_\mu(-p)\left[P^T_{\mu\nu}\frac{1}{4p^4}\left\{4-\left(4+2p^2-p^4\right)\exp(-p^2/2)\right\}\right.$$
$$\left.-P^L_{\mu\nu}\frac{1}{p^4}\left\{1-p^2-\left(1-\frac{p^2}{2}\right)\exp(-p^2/2)\right\}\right]A_\nu(p) .$$

The p^2 term in transverse part here generates the well-known anomalous dimension for the photon field $\eta_A = e^2/12\pi^2$ and the rest corresponds to $\mathcal{T}(p)$,

$$\mathcal{T}(p^2) = \frac{\Lambda^2 e^2}{8\pi^2\bar{p}^4}\left\{4+\frac{2\bar{p}^6}{3}-\left(4+2\bar{p}^2-\bar{p}^4\right)\exp(-\bar{p}^2/2)\right\} .$$

The longitudinal part, $\mathcal{L}$, obtained from Eq. (17) is exactly the same as Eq. (16) obtained by the WT identity. The result is a non-trivial check for the validity of our calculation.

Note that the same constant mass term appears in both $\mathcal{T}$ and $\mathcal{L}$

$$\mathcal{T} = \mathcal{L} = \frac{3e^2}{16\pi^2}\Lambda^2 + \mathcal{O}(\bar{p}^2)$$

5. Discussions

Here we presented a FRG approach to QED as a prototype for treating a gauge theory in FRG. We observed the roles of the WT identity in FRG approach. Our study certainly shows the importance to take account of the momentum dependence in the WT identity and flow equations.[e] Further study of the renormalization group flow is possible and in progress.

Let us close this article by pointing out several remaining problems in the approach presented here. Firstly, from the WT identity we found two relations among the three 4-fermi couplings. With another relation, we may be able to express them in terms of the photon 2-point function. A careful study of flow equations is needed. Secondly, in the present calculation, we assumed that the fermion sector takes a simple form, $\sigma(p) = 0$. With this assumption, we obtained the photon 2-point function analytically. Further study of the fermion sector should be possible and important.

---

[e]The momentum dependence of flow equations have been also studied in different approaches.[11] [12]

Acknowledgment

The work by Y. I. and K. I. had been supported in part by the Grants-in-Aid for Scientific Research nos. R2209 and 22540270 from Japan Society for the Promotion of Science.

References

1. Y. Igarashi, K. Itoh and H. So, Exact Symmetries realized on the Renormalization Group Flow, *Phys. Lett.* **B479**, 336 (2000).
2. H. Sonoda, On the construction of QED using ERG. *J. Phys.* **A40**, 9675 (2007).
3. Y. Igarashi, K. Itoh, and H. Sonoda, Quantum master equation for QED in exact renormalization group, *Prog. Theor. Phys.* 118, 121 (2007).
4. B. J. Warr, Renormalization of gauge theories using effective Lagrangians. 1, *Ann. Phys.* **183**, 1 (1988).
5. Y. Igarashi, K. Itoh and J. Pawlowski, in preparation.
6. J. M. Pawlowski, Aspects of the functional renormalisation group. *Annals Phys.* **322**, 2831 (2007).
7. K. Itoh Y. Igarashi and H. Sonoda, Realization of symmetry in the ERG approach to quantum field theory. *Prog. Theor. Phys. Suppl.* **No. 181**, (2009).
8. U. Ellwanger, Flow equations and BRS invariance for Yang-Mills theories, *Phys. Lett.* **B335**, 364 (1994).
9. U. Ellwanger, M. Hirsch, and A. Weber, Flow equations for the relevant part of the pure Yang- Mills action, *Z. Phys.* **C69**, 687 (1996).
10. M. Ishikake, Y. Igarashi, and N. Ukita, Tree expansion of the Wilson effective action and reduction of the Polchinski flow equations, *Prog. Theor. Phys.* **113** 229 (2005).
11. J. P. Blaizot, R. Mendez-Galain, and N. Wschebor, A new method to solve the non-perturbative renormalization group equations. *Phys. Lett.* **B632**, 571 (2006).
12. M. Mitter, J M. Pawlowski, and N. Strodthoff, Chiral symmetry breaking in continuum QCD. *Phys.Rev.* **D91**, 054035 (2015).

The Proton Mass and Scale-Invariant Hidden Local Symmetry for Compressed Baryonic Matter

Mannque Rho

Institut de Physique Théorique, CEA Saclay, 91191 Gif-sur-Yvette cédex, France
E-mail: mannque.rho@cea.fr

I discuss how to access dense baryonic matter of compact stars by combining hidden local symmetry (HLS) of light-quark vector mesons with spontaneously broken scale invariance of a (pseudo) Nambu-Goldstone boson, dilaton, in a description that parallels the approach to dilatonic Higgs. Some of the surprising observations are that the bulk of proton mass is not Nambu-Goldstonian, parity doubling emerges at high density and the EoS of baryonic matter can be soft enough for heavy-ion processes at low density and stiff enough at high density for ~ 2 solar mass neutron stars.

Keywords: Scale-chiral symmetry, Hidden local symmetry, Vector manifestation at high denstiy, Proton mass, Emergent parity doubling, Massive compact stars.

1. Introduction

Every speaker in this meeting, perhaps with one glaring exception, that is, myself, has so far been, and will be, talking about the Higgs boson or Higgs-like scalars and matters related to mass generation and above all what's beyond the Standard Model. But I am not in the field and that's not what I will talk about. So why am I invited by Koichi to give this talk and what is it all about ?

Here is why and what.

Whatever the true mechanism for the origin of the mass of the visible Universe, hotly debated in this meeting, may be, the Higgs boson may very well account for the mass of the constituents of the proton, say, quarks, but I will suggest that its presence *by itself* cannot explain the bulk of the proton mass. I will present this issue using what's being observed of massive compact stars.

Although the physics involved may look quite different, it seems that there are quite a few things in common between what's being discussed in this meeting and what I will talk about, i.e., dense baryonic matter that is on the verge of collapsing into a black hole. In my view it has a lot to do with light-quark vector bosons endowed with hidden local symmetry, the idea that initially germinated here in Nagoya and a scalar dilaton associated with spontaneously broken scale symmetry, which is one of the main goings-on in Nagoya, the techni-dilaton[1].

In describing this matter in my talk I will be a bit daring. I will do this without dragging in my close collaborators who have been working with me on the matter since some time because I don't want to involve them in my discussions, some of which could be too wild and above all I don't want them to be held responsible for some far-out ideas I will launch onto in this talk.

My talk is about the origin of the mass of the "ordinary matter" we see all around us. Now that the fundamental boson considered to be responsible for the origin of the mass of the Universe, the Higgs boson, is discovered, one would think that we now know where the mass of the proton, say, comes from. Given the precisely measured nucleon (say, proton) mass, the mass of all other visible matter around us – this building, molecules, atoms and nuclei – is almost completely accounted for: More than 99% of the mass is given by the simple sum of the masses of the nucleons involved. But this simple picture ends there. The mass of the nucleon is not just sum of "things." One misses more than 95% of the proton mass if one simply adds the masses of the QCD "things" of the nucleon, i.e., quarks, the mass of which is presumably given by the Higgs mechanism, and massless gluons. QCD on lattice, however, does give the proton mass correctly, hence QCD does have in it but it pops out of computers after highly complex and intricate manoeuvering[2]. But one cannot "see" it. The first question therefore is where does the proton mass come from? The Higgs mechanism does not provide an obvious answer to this question, so the statement that the origin of the SM mass is the Higgs mechanism cannot be taken to be the whole story. Now simple models like linear σ model and equivalently NJL model say that the mass can arise from spontaneous breaking of chiral symmetry – the Nambu-Goldstone (NG) mechanism – signalled by the presence of NG bosons, i.e., pions, with the nonzero order parameter, the quark condensate, $\langle\bar{q}q\rangle \neq 0$. Then the next question is, *if the mass is generated dynamically with a nonzero $\langle\bar{q}q\rangle$ from chiral symmetry, can one not unbreak the symmetry by tweaking the condensate to go to zero by, say, temperature and/or density?*

These are the questions that Gerry Brown and I raised in early 1980's. In the attempt to answer these questions – which do not require involving directly the microscopic (QCD) degrees of freedom, i.e., quarks and gluons, we came across the hidden local symmetry associated with the light-quark vector mesons $V \equiv (\rho, \omega)$ – the only effective field theory that we knew had the possibility of having a non-NG hadron with vanishing mass and the spontaneously broken scale symmetry associated with a dilaton, denoted σ, to be identified with $f_0(500)$ listed in the particle data booklet. The vectors V and the scalar σ have been around in the nuclear physics community since a very long time but what I will present here takes a totally different aspect that though quite different physics-wise, is closely along the line of the thinking currently prevalent in the particle physics community like in this meeting.

In this talk I will focus on density driving the quark condensate to vanish[a]. This is relevant to the interior of compact stars, where the density is expected to be high enough to be near chiral restoration. This issue became quite topical and

[a]For simplicity I will be thinking of the chiral limit in which the pion is massless although confrontation with Nature requires, sometimes quite crucially, the small pion mass to be taken into account.

particularly acute with the recent precise measurements of $\sim$ 2-solar mass neutron stars.

2. "Scalar Meson Conundrum" and Vector Mesons in Nuclear Physics

In the spirit of effective field theory – that I will adopt – a low-mass scalar meson and comparable-mass vector mesons turn out to be absolutely indispensable to understand what's going on both in nuclei and in dense nuclear matter. A scalar meson with a mass around 600 MeV has figured importantly in nuclear physics since a long time. The only scalar mesonic excitation with such a low mass currently listed in the particle data booklet is $f_0(500)$ with a broad width. Despite the observed large width, when taken, for no good reason other than convenience, as a local bosonic field, it has fairly successfully accounted for the attractive scalar channel in nucleon-nucleon potentials such as the well-known Bonn boson-exchange potential and, more significantly, in the highly popular relativistic mean field (RMF) theories for nuclei, nuclear matter and compact-star matter[b]. There has been a long-standing controversy as to whether the scalar excitation with such a large width can be depicted in terms of a local field in phenomenological or effective Lagrangian approaches. If it is a hadronic particle, then what is its QCD structure? Can it be described in quark models, as, say, quarkonium of $(q\bar{q})^m$ complex with $m \geq 1$, a gluonium (glueball) or a mixture thereof? In effective field theories of strong interactions, there is the fourth component of the chiral four-vector in linear sigma model. However it must have a mass $\gtrsim 1$ GeV to be compatible with the current algebras, so cannot be relevant far below the chiral scale $4\pi f_\pi \sim 1$ GeV. Furthermore what figures in relativistic mean field (RMF) theories – which are highly (perhaps too) successful in heavy nuclei and nuclear matter – must be a chiral-singlet scalar, not the scalar of linear sigma model. If it were the sigma-model scalar with a mass < 1 GeV as needed in nuclear phenomenology, it would destabilize nuclear matter due to strong attractive many-body forces. A possible way-out of this difficulty is that the scalar is a low-lying chiral singlet scalar at least in the vicinity of nuclear matter density. However at high density as the system approaches chiral restoration, it should transmute to a $\bar{q}q$ configuration that is expected to dominate at the large N_c limit[3].

I suggest that the idea recently put forward by Crewther and Tunstall[4] that the scalar is a NG boson or more precisely pseudo-NG (pNG) boson of spontaneously broken scale symmetry can provide a simple and appealing way to resolve the above "scalar conundrum" and offers a new perspective on various aspects of nuclear physics, including hadronic matter under extreme conditions that have remained

[b] This model consists of the vector mesons ρ, ω and a scalar of mass $\sim$ 600 MeV coupled to baryons treated in the mean-field. This model is presumably related to the Landau Fermi-liquid approach to nuclear matter, the reason why the model seems to work well in nuclear physics.

poorly understood. The power of the scheme is that the low-lying scalar, a chiral singlet, can be treated on the same footing as the pseudoscalar pseudo-NG bosons – the pions and kaons – in low-energy effective field theory. This is called "scale chiral perturbation theory ($s\chi$PT)"[4]. I don't think this idea has been confirmed rigorously, but I will take this as an *assumption that may be valid in medium* and see how it fares with Nature.

No less crucial for nuclear stability with the low-mass scalar providing the attractive interactions are vector-meson degrees of freedom. They provide the short-range repulsion that counters the scalar attraction at short distances to prevent nuclear matter from collapsing and in addition, control the nuclear tensor forces crucial in nuclear dynamics and, as it turns out, in compact stars. What figures in RMF theories for repulsive interaction is the isoscalar vector meson, ω, and for the tensor forces important for compact stars, it is the ρ meson. While both ρ and ω are essential for both nuclear matter and dense matter, their roles are basically different. Endowed with hidden local symmetry (HLS), the ρ meson has a fixed point called "vector manifestation (VM)" at which the HLS coupling g vanishes as the quark condensate – order parameter of chiral symmetry – is tuned to zero, as a consequence of which the ρ mass vainshes[5]. In this framework, it is the VM fixed point that baryonic matter will be driven to when the condensate $\langle\bar{q}q\rangle$ is dialled to zero. I will exploit this property in describing the EoS of star matter, even though the system may not reach the chiral restoration point.

As for the ω, it seems that the $U(2)$ symmetry that puts the ω in the same multiplet as the ρ works very well in matter-free space. Apart from near degeneracy in mass between them, the vector dominance picture holds very well for both the pion and nucleon EM form factors if one takes into account the infinite towers of the isovector ρ's and isoscalar ω's – embedded in 5-D local $U(2)$ symmetry – as in holographic QCD models[6,7]. For this, the infinite tower is essential. However it has been known since Sakurai proposed the idea in 1970's that while the vector dominance with the lowest ρ meson alone worked well for the pion EM form factor, it failed famously for the nucleon form factors. For the latter, other degrees of freedom such as quark bag[8] were required. This implied among others that the flavor (and local) $U(2)$ symmetry could breakdown, quite drastically, in a vacuum modified by dense matter as I will discuss below.

3. Implementing Scale Invariance to Hidden Local Symmetry

An early idea favored by many people in particle and nuclear physics, including Nambu himself (see[9] for a current review), for the origin of $\sim 99\%$ of the proton mass was the spontaneous breaking of chiral symmetry associated with the nearly zero-mass quarks. This idea provided a natural order parameter, the quark condensate $\langle\bar{q}q\rangle$, for the phase transition that could be probed by "unbreaking" chiral symmetry by external disturbances. "Seeing" dropping mass of the proton by experiments, say, by increased density, would provide a "smoking gun" signal for the origin of the proton mass. Unfortunately lattice gauge calculation cannot yet access high density due to the the famous sign problem. Furthermore, there are presently no reliable model-independent theoretical tools available. Experimental efforts are being, and will be, made in heavy-ion collisions to explore this uncharted area. Up to date there is no clear indication either for or against the notion that the proton mass can be unsheathed by density or temperature. The only laboratory available at present is the space laboratory and the object is the dense compact star matter.

As I will argue, the proton mass is not, it turned out, the right object to look at for chiral symmetry. In fact some $\gtrsim 70\%$ of the proton mass seems to remain unaffected by the quark condensate going to zero, thus, contrary to what was generally accepted, unconnected to spontaneously broken chiral symmetry (SBCS). The first suggestion on what to look for in Nature for a signature for chiral symmetry in hadronic matter was made by Gerry Brown with his collaborators[10]. His idea was to observe instead dropping ρ mass at high temperature (or density), which ultimately goes to zero at the chiral restoration point (that is, in the chiral limit). He was initially interested in high-temperature cases but I believe the suggestion is more appropriate for high density matter. Heavy-ion collisions producing high temperature matter are too swamped with mundane backgrounds to directly and cleanly probe the vacuum-change caused by temperature.

As mentioned, the only known way in which a zero-mass ρ meson figures *naturally* in nuclear dynamics under extreme conditions is in a hidden gauge symmetric setting[5]. It is also in this way that chiral symmetry and scale symmetry could be intricately locked to each other. This accounts for why Gerry and I advocated – since many years and more or less isolated from the rest of nuclear physics community – the approach based on HLS[5,11] – and dilaton. We will see that this feature plays an extremely important role in dense medium, particularly in what is called the symmetry energy E_{sym} that is crucial in the EoS of compact-star matter.

Together with the scalar degree of freedom treated below, the ω meson plays an equally indispensable role in nuclear physics as a whole and is crucial at high density. The $U(2)$ symmetry, while holding well in the matter-free space for the ω and ρ mesons, is most likely broken badly at high density with the ω insensitive to the VM fixed point of the ρ meson. I will therefore consider independently $SU(2)$ HLS for the ρ and $U(1)$ HLS for the ω in treating dense medium. I write the HLS Lagrangian to $\mathcal{O}(p^4)$ in the expansion in covariant derivatives [c] $D_\mu \xi_{R,L} = (\partial_\mu - iV_\mu)\,\xi_{R,L}$, where V_μ represents the gauge bosons of the HLS as $V_\mu = \frac{1}{2}\left(g_\omega \omega_\mu + g_\rho \rho_\mu\right)$ with $\rho_\mu = \vec{\tau}\cdot\vec{\rho}_\mu$. The Lagrangian is

$$\mathcal{L}_{\rm HLS} = \mathcal{L}_{(2)}^{\rm HLS} + \mathcal{L}_{(4)}^{\rm HLS} + \mathcal{L}_{(4){\rm anom}}^{\rm HLS} + \mathcal{L}_{\rm mass} + \cdots . \tag{1}$$

The numerical subscript stands for the power of covariant derivatives. Here $\mathcal{L}_{\rm mass}$ is the quark mass term and the anomalous Wess-Zumino term contains four independent terms. The V_μ kinetic terms are in the leading order term $\mathcal{L}_{(2)}^{\rm HLS}$.

[c] For convenience, I follow the notation of Ref. [12].

As stressed in Introduction, the scalar degree of freedom in nuclear interactions has been elusive since a long time. The recent idea, contested by some workers in the field, of Crewther and Tunstall[4] that the low mass scalar $f_0(500)$ is a pseudo-Nambu-Goldstone boson – with the scale symmetry explicitly broken by the trace anomaly – *and* the quark mass – arising from spontaneous breaking of scale invariance, i.e., dilaton, offers a simple and potentially elegant approach to the problem. In incorporating scale symmetry to HLS to dense nuclear matter, in the past, the dilaton scalar was introduced as a lower-mass component of two-component structure of gluons figuring in the trace anomaly of QCD[13,14]. The gluons that saturate the trace anomaly were decomposed into "soft" and "hard" components, with the soft-glue supposed to "melt" at the chiral restoration, leaving the "hard" component remaining above the chiral restoration point. The basic assumption was that the melting of the"soft-glue" can be tied to the bilinear quark condensate Σ going to zero. Based on the behavior of the gluon condensate in temperature wherein roughly half of the gluon condensate vanishes at the critical temperature, it was assumed that the same "melting" occurs in density at the critical density n_c. The CT approach does not rely on this separation. Instead it posits that there is an IR fixed point at large α, say, at α_{IR} – where $\alpha = g^2/4\pi$ with the g the QCD gauge coupling – at which the β function goes to zero. At that point the trace anomaly $\theta^\mu_\mu = \frac{\beta(\alpha_s)}{4\alpha_s} G^a_{\mu\nu} G^{a\mu\nu} + (1+\gamma_m(\alpha_s)) \sum_{q=u,d,s} m_q \bar{q} q \to \theta^\mu_\mu|_{IR} = \partial^\mu D_\mu = (1+\gamma_m(\alpha_s)) \sum_{q=u,d,s} m_q \bar{q} q$. At the IR fixed point and in the chiral limit with $m_q = 0$, the dilatation current is conserved $\partial^\mu D_\mu = 0$ even if the gluon condensate and the quark condensate are non-zero separately or together. This combined symmetry is "scale-chiral symmetry (SCS)." Spontaneous breaking of this SCS leads to a NG boson, dilaton, in addition to the pions (and kaons). The explicit breaking due to the trace anomaly and the quark mass makes dilaton a pseudo-NG.

As far as I know, there is no lattice evidence for such an IR fixed point for small number – here 3 – of flavors whereas there is evidence for it at large $N_f \sim 8$ as discussed in this conference by lattice experts. When asked whether such an IR fixed point for $N_f \sim 3$ is plausible, the general answer from lattice experts is "not in matter-free space." Although there is no dynamical lattice data, there is, however, a support for an IR fixed point for $N_f = 2$ coming from numerical stochastic perturbation calculation at high fermion-loop order and Padé approximation[15]. While awaiting definitive confirmation/falsification of the existence of such an IR fixed point, I will simply assume it in the application to dense medium where such an IR fixed point could be envisioned to arise as an emergent phenomenon.

Now to implement this scale symmetry to (1) to obtain scale-symmetric HLS (sHLS for short), I will use the standard trick with "conformal compensator (or conformalon) field" with scale dimension 1,

$$\chi = f e^{\sigma/f} \tag{2}$$

where f is a spurion field for $f = \langle\chi\rangle$, the vev for a given vacuum (later defined in medium with density-modified ground state). The σ represents the pNG scalar in nonlinear realization. What plays an important role in the CT theory is the anomalous dimension of the operator G^2 given by $\beta' > 0$, derivative of the β function with respect to α_s near the IR fixed point. This corresponds to the slope with which the β function approaches the IR fixed point and this can be large with positive sign. Due to this, writing down a fully general effective sHLS Lagrangian is rather involved and is not straightforward.[d]

Confining to the simplied possible form, one can write

$$\mathcal{L}_{\rm sHLS} = (\frac{\chi}{f_\sigma})^2 \mathcal{L}^{\rm HLS}_{(2)} + \mathcal{L}^{\rm HLS}_{(4)} + \mathcal{L}^{\rm HLS}_{\rm 4anom} + (\frac{\chi}{f_\sigma})^3 \mathcal{L}_{\rm mass} + V(\chi) + \cdots . \tag{3}$$

Here $f_\sigma \equiv \langle 0|\chi|0\rangle$, vev in the matter-free vacuum (MFV for short) and $V(\chi)$ is the dilaton potential chosen so that together with the quark mass term, it is minimized at $\langle\chi\rangle = f_\sigma$. The $\mathcal{O}(p^4)$ terms (including the hWZ term) in (1) are of scale dimension 4 and hence not multplied by the conformalons. Again the terms that introduce the anomalous dimension β' are ignored in both of them.

In what follows, I apply the simple Lagrangian (3) to baryonic matter and point out a few qualitatively striking features that could arise at high density.

4. Scale-Chiral Symmetry in Medium

4.1. *General structure*

I start with several general features that are known of the Lagrangian (3).

- Vector manifestation (VM) fixed point

Let's first consider the chiral limit. Wilsonian-matched to QCD at a matching scale $\Lambda_M \lesssim \Lambda_\chi \approx 4\pi f_\pi$, the Lagrangian (1) has the VM fixed point at which the ρ mass goes to zero as[5]

$$m_\rho|_{\Sigma\to 0} \sim g \sim \Sigma \to 0 . \tag{4}$$

This is a feature that does not depend on how the VM fixed point is reached. We would like to see whether high density does it and if so, how it does. Let me denote the critical density as n_c. The appropriate value for n_c – thus far unknown in QCD – required for making predictions will be specified later.

---

[d] To illustrate what is involved, take the leading chiral order term in (1), $\mathcal{L}^{\rm HLS}_{(2)}$. It has scale dimension 2, so the standard procedure of implementing scale symmetry would be to multiply it with $\bar{\chi}^2$ (where $\bar{\chi} \equiv \chi/f$). That would give a scale-invariant Lagrangian $\bar{\chi}^2 \mathcal{L}^{\rm HLS}_{(2)}$. However due to the presence of β', one can have $c\bar{\chi}^2 \mathcal{L}^{\rm HLS}_{(2)} + (1-c)\bar{\chi}^{2+\beta'} \mathcal{L}^{\rm HLS}_{(2)}$ with an unknown constant c. Such a construction would make the resulting Lagrangian considerably more complicated with doubled number of parameters and hence unmanageable. To proceed however I will make the assumption that given that the dynamics of the matter fields that involve no scalar excitations is reliably given by (1), c could be taken to be near 1. How the anomalous dimension of G^2 affects the matter field dynamics in various different observables remains to be investigated. It will be seen that in the presence of nuclear interactions in medium, there is locking of scale symmetry to chiral symmetry as soon as the matter density is non-zero. This suggests that $c \neq 0$ may have to be taken into account for a more realistic treatment. I will leave this matter for future work.

Next we need to have baryons in the system. We first treat baryons as solitons, i.e., skyrmions, arising from the Lagrangian (3). This is a natural way in large N_c QCD. This endows the system with topology. That offers certain features of dense matter not manifestly visible in treatments with baryon fields explicitly incorporated into the theory. Dense bayonic system can be generated by putting skyrnions on crystal lattice and squeezing the crystal size to simulate density[16]. This approach has an advantage that both mesons and baryons, elementary as well as multi-body systems, are treated on the same footing. Unfortunately It has the disadvantage that rigorous formulation is mathematically involved and for that reason has not been fully developed to enable one to confront experiments quantitatively. Some of the highly intricate issues are discussed in recent reviews[17]. I am not going to attempt this approach here. However I will exploit some of the robust features that topology provides in setting up the tools for doing EFT calculations.

The most important aspect of the skyrmion description for dense-matter physics is that at a density that we denote as $n_{1/2}$ that lies somewhat above the normal nuclear matter density $n_0 \simeq 0.16\text{fm}^{-3}$, the skyrmions in dense matter fractionize to half-skyrmions. This is a topological transition involving no local order parameters. When this happens, the chiral condensate Σ which is related to the (bilinear) quark condensate in QCD language goes to zero *on average* within the unit cell of the crystal. The condensate however is not locally zero, so in fact chiral symmetry remains spontaneously broken. The order parameter for this is therefore not the bilinear condensate $\langle \bar{q}q \rangle$ but presumably of multiquark condensates $\langle (\bar{q}q)^n \rangle$ for $n > 1$ which are non-zero in the half-skyrmion phase. Thus the pion decay constant, which is usually associated with the quark condensate in continuum description, is not zero. The appearance of half-skyrmions with vanishing bilinear chiral condensate had been discussed before, but that chiral symmetry is not restored to Wigner phase was first observed in[18]. It turns out however that while chiral symmetry is not restored, parity-doubling takes place in the half-skyrmion phase[19] [e] with a *surprising consequence that some* $\gtrsim 70\%$ *of the proton mass (call it* M_0*), largely associated with gluon condensates, remain non-zero when the condensate* Σ *averages to zero.* Thus a large portion of the proton mass is not due to the standard Nambu-Goldstone mechanism of chiral symmetry spontaneous breaking as was generally thought[9]. One can *approximately* identify M_0 as a chiral invariant mass in parity-doubled baryon models. What is equally significant is that the appearance of this half-skyrmion phase, in this formulation, is found to play the key role in the structure of the EoS for compact stars.

- Topological demarcation of density regimes R-I and R-II

The impact of the topological change at $n_{1/2}$ is found in the change in the structure of the Lagrangian for the density regions below and above $n_{1/2}$. It will

[e] This is reminiscent of Georgi's vector symmetry where $\langle \bar{q}q \rangle = 0$ with $f_\pi \neq 0$[20].

be convenient to make the demarcation into the two density regions

$$\mathrm{R(egion)-I:}\quad 0 < n < n_{1/2}, \tag{5}$$

$$\mathrm{R(egion)-II:}\quad n_{1/2} \le n \le n_c. \tag{6}$$

At present, the demarcation density $n_{1/2}$ is difficult to pin down precisely. However on phenomenological considerations and its robustness to dilaton properties such as mass and coupling constants, one can make an estimate. It ranges $n_{1/2} \sim (1.5 - 2.0)n_0$. It is found in numerical applications that $n_{1/2} \sim 2.0n_0$ fits in well. Hopefully this region will soon be accessed experimentally in future heavy-ion accelerators.

We will see that one can infer from scale-chiral symmetry considerations how the "bare" parameters of the Lagrangian behave in the two regions.

- Cusp in symmetry energy

For EoS for compact-star matter, the important quantity is the energy per nucleon for matter with P protons and N neutrons, $E(n,x)$,

$$E(n,x) = E(n,0) + Sx^2 + \cdots \tag{7}$$

with $x = (N-P)/(N+P)$. Here S is what is referred to as "symmetry energy factor." The pure neutron matter is for $x = 1$ and the symmetric nuclear matter is for $x = 0$.

The most striking feature characterizing the topological property of skyrmion matter observed on crystal is a cusp structure in S at $n = n_{1/2}$[22]: S drops as density approaches $n_{1/2}$ and then turns up and increases at $n_{1/2}$. One might wonder whether this feature is not just an artifact of the crystal structure. It has however been shown that it is not an artifact but a rather robust feature of the topology change and can be justified by a microscopic argument.

The S can be derived in the skyrmion description by collective-quantization of the skyrmion neutron matter. It is a semi-classical quantity. Yet it contains certain features of quantum many-body features of standard many-body approaches. For instance, this sort of "classicalized" quantum effects are manifested in certain scattering processes computed in classical skyrmion processes[23]. In fact, implementing the VM (4) in the Lagrangian (3), it has been shown that the change in the structure of the tensor forces at $n_{1/2}$ in quantum many-body calculation reproduces the far-from-obvious cusp structure[22]. Let me stress that this represents one of my main themes in the talk, that is, combining HLS with topology leads to a novel prediction not found in nontopological approaches. In terms of the nuclear tensor forces, the density $n_{1/2}$ sets the demarcation of different tensor force structures, below $n_{1/2}$ an intricate interplay of pion and ρ, the two compensating each other, and above $n_{1/2}$ the pion field taking over the tensor forces[24].

4.2. *In-medium effective Lagrangian*

Let me focus on a Lagrangian of the form (3), the "bare" parameters of which are Wilsonian-matched via correlators to QCD[5]. The matching is made at a matching scale commensurate with the chiral scale $4\pi f_\pi$[f] but the Lagrangian can be brought down to a phenomenologically more relevant scale, say, just above the ρ mass scale in the matter-free space. The power of the matching is that it endows the parameters of the Lagrangian with nonperturbative quantities of QCD, typically of various condensates – such as gluon, quark etc. condensates. It turns out that the main condensates are the quark condensate $\langle\bar{q}q\rangle$, and the gluon condensate $\langle G^2\rangle$ but there are also various mixed condensates. The condensates, being the vacuum expectation values (vev) of local operators, reflect the vacuum structure. Therefore when the vacuum is modified by – in our case temperature, density etc.– the condensates will "slide" with the complex vacuum. This implies the "bare" parameters of the Lagrangian determined at the matching scale will depend on the vacuum structure, i.e., on density, in nuclear medium. We therefore implement this sliding vacuum structure in applying the Lagrangian to dense nuclear matter.

To take into account the vacuum property in the presence of baryonic matter, one then expands the scalar field

$$\chi = \langle 0^\star|\chi|0^\star\rangle + \chi' \tag{8}$$

where $|0^\star\rangle$ stands for the ground state with baryonic matter, say, $\star$-vacuum. Thus $f_\sigma^\star = \langle 0^\star|\chi|0^\star\rangle$, the in-medium decay constant ($f_\sigma = \langle 0|\chi|0\rangle$ is the decay constant in the matter-free vacuum). The expanded Lagrangian takes the form

$$\mathcal{L}_{\rm sHLS} = (\frac{f_\sigma^*}{f_\sigma})^2\mathcal{L}_{(2)}^{\rm HLS} + \mathcal{L}_{(4)}^{\rm HLS} + \mathcal{L}_{(4)\rm anom}^{\rm HLS} + (\frac{f_\sigma^*}{f_\sigma})^3\mathcal{L}_{\rm mass} + \cdots . \tag{9}$$

The dilaton terms including χ' coupling to pions, matter fields etc. are in the ellipsis and not shown explicitly.

In setting up the Lagrangian to be applied to dense matter, I will incorporate the nucleon fields explicitly into the Lagrangian (9) in chiral-scale symmetric way[25]. From now on, I will call it *bs*HLS Lagrangian.

4.2.1. *Scaling "bare" parameters of effective Lagrangian*

Frst I discuss the structure of the Lagrangian applicable in R-I.

In terms of the topological structure given by the sHLS Lagrangian (9), this region is populated by skyrmions. So it corresponds to normal skyrmion matter. This region is characterized by one density-scaling parameter

$$\Phi(n) = f_\sigma^\star/f_\sigma. \tag{10}$$

The first observation that follows naturally from the locking of chiral symmetry and scale symmetry is that[g]

$$f_\pi^*/f_\pi \approx \Phi \tag{11}$$

[f]Whether this matching can be reliably done for the problem is not clear. Although calculations are feasible in principle, I won't specifically rely on the specific form obtained in[5].
[g]I use the approximate relation instead of equality because as noted above, the Lagrangian (3) ignores possible effects of the anomalous dimension β'.

which leads to

$$m_\pi^*/m_\pi \approx \Phi^{\frac{1}{2}}. \tag{12}$$

Up to nuclear matter density $n_0 = 0.16\text{fm}^{-3}$, Φ can be obtained from experiments (by looking at deeply bound pionic nuclear systems) or perhaps from chiral perturbation theory. I will assume that one can extend this up to $n_{1/2} \approx 2n_0$.

It also follows straightforwardly that

$$m_N^*/m_N \approx m_\rho^*/m_\rho \approx m_\omega^*/m_\omega \approx m_\sigma^*/m_\sigma \approx \Phi\,, \tag{13}$$

$$g_{\sigma NN}^*/g_{\sigma NN} \approx g_{\omega NN}^*/g_{\omega NN} \approx g_{\rho NN}^*/g_{\rho NN} \approx 1 \tag{14}$$

and

$$g_A^*/g_A \approx g_{\pi NN}^*/g_{\pi NN} \approx \Phi\,. \tag{15}$$

Far away from the critical density n_c, one can parameterize

$$\Phi(n) \approx \frac{1}{1 + c_I n/n_0} \tag{16}$$

with $c_I < 1$.

In sum, one constant of the scaling parameter c_I – which can be connected to data on deeply bound pionic nuclear systems – completely specifies the "bare" Lagrangian in R-I with which one can do EFT calculations in medium up to density $n_{1/2}$. So the physics around – and in the vicinity above – n_0 is given almost entirely by the one parameter c_I which is controlled at least up to n_0 by available data.

The situation in R-II is much less clear. There is no direct help from QCD, i.e., no lattice QCD nor any reliable theoretical tools at high density. In skyrmon picture, due to the topology change at $n_{1/2}$ from skyrmions to half-skyrmions in R-II, when translated to the structure of the effective Lagrangian, *bs*HLS, the "bare" parameters will undergo drastic modification. I will first infer the structure of the parameters from what we know from the skyrmion-crystal simulation.

The first thing to note is that the nucleon mass in R-II goes to a constant M_0 when one writes the bare parameter for the nucleon mass as

$$m_N^* = M_0 + \Delta(\Sigma) \tag{17}$$

with $\Sigma \propto \langle \bar{q}q \rangle$ going to zero approaching n_c in R-II. As noted, although $\Sigma = 0$ in the half-skyrmion phase, the pion is still excited with a nonzero pion decay constant. This means that chiral symmetry is in the NG mode, with massive hadrons, that are parity-doubled. This is seen in the skyrmion crystal model with (3)[12,19]. This is reproduced also in a renormalization-group analysist of hidden local symmetric parity-doublet model with dilaton limit fixed point (DLFP)[21]. In both cases, the "nearly" chiral invariant mass[h] M_0 comes out to be $M_0 \approx (0.7 - 0.8)m_N$. I will therefore take $m_N^* \approx \kappa m_N$ with $\kappa \approx M_0/m_N$ in R-II.

[h]Note that in the skyrmion crystal formulation, Σ averages to zero in a unit cell but chiral symmetry is still broken. However chiral symmetry breaking is characterized by highly suppressed multi-quark condensates, with strongly reduced pion mass.

I should point out that QCD lattice calculations at high density for non-zero β in strong coupling expansion *do* provide a support for parity-doubling in the confined phase[26].

Both the skyrmion crystal and mean-field approaches[12,21] lead to the relation $f_\sigma^*/f_\sigma \approx f_\pi^*/f_\pi \approx m_N^*/m_N$. As stressed in[25], this relation together with $m_\sigma^*/m_\sigma \approx f_\sigma^*/f_\sigma$ follows from low-energy theorems for chiral-scale symmetry as applied in medium. Thus I take

$$f_\sigma^*/f_\sigma \approx f_\pi^*/f_\pi \approx m_N^*/m_N \approx m_\sigma^*/m_\sigma \approx \kappa. \tag{18}$$

I now turn to the vector mesons.

For the ρ meson, I will assume for numerics that the VM fixed point comes at $n_c \approx (6-7)n_0$. Considering that R-II is a density regime "closer" to the VM fixed point, I will simply take from Eq. (4) what could be applicable for the whole range $n_{1/2} \lesssim n < n_c$ of densities,

$$m_\rho^*/m_\rho \approx g_{\rho NN}^*/g_{\rho NN} \approx (1 - n/n_c). \tag{19}$$

This is undoubtedly a drastic simplification. For instance, near $n_{1/2}$ which is not very close to n_c, the density dependence could be more involved. Nonetheless I think it captures the essential feature of the process.

Now the ω meson. As noted, the $U(2)$ symmetry may be applicable in R-I, so

$$m_\omega^*/m_\omega \approx \Phi, \tag{20}$$

the same as (16). However the properties of the ω mass and ω-nucleon coupling are unknown in the whole region of R-II. These constitute the main uncertainty in the effective bsHLS Lagrangian. One can however make some progress from the present understanding of what's going on in massive compact stars.

What is most significant in the present framework is that once the symmetry energy factor S is suitably controlled by the pion and ρ properties, i.e., through the net tensor force that dominates the factor S with very little effect from ω and σ, it is the interplay between the dilaton and the ω meson in the symmetric ($x = 0$) part of the energy (7) that plays the most important role in the EoS of dense compact star medium. One can easily see what happens by looking at the potential energies associated with the dilaton and ω exchange which have the form $\{-(\frac{g_{\sigma NN}^*}{m_\sigma^*})^2 n_s + (\frac{g_{\omega NN}^*}{m_\omega^*})^2 n_B\}$ where $n_s(n_B)$ is the scalar (baryon number) density. Then it is the competition between the dilaton attraction and the ω repulsion that controls the game. The ratio $Y_\sigma \equiv (\frac{g_{\sigma NN}^*}{m_\sigma^*})^2$ is roughly constant without scaling whereas the ratio $Y_\omega \equiv (\frac{g_{\omega NN}^*}{m_\omega^*})^2$ is likely to scale in density. Taking into account that $n_B \gtrsim n_s$ for $m_N^* < \infty$, given that in equilibrium nuclear matter, the two roughly cancel giving rise to a small binding energy observed in nature, namely, a BPS structure in the language of skyrmion, one can see that the repulsion could overcome the attraction at some density above n_0 with a dropping ω mass. This repulsion is needed to resolve the strangeness (hyperons and kaon condensation) problem and to account for the massive ~ 2 solar mass neutron stars.

How the repulsion overcomes the attraction is constrained by two crucial factors: one is that the sound velocity of the matter not violate causality in the density regime relevant for the stars and the other is that the EoS not become too softened by the appearance of the hyperons or kaon condensations. The latter will obstruct the formation of the observed massive ($\gtrsim$ 2-solar mass) stars. Both can be avoided by tuning the repulsion due to the scaling ω mass. What is found necessary is that the ω-NN coupling $g_{\omega NN}$ and the ω mass m_ω scale slowly in density in the same way $\sim m_\omega^*/m_\omega \approx hg_{\omega NN}^*/g_{\omega NN}$ with $h < 1$.

The effective *bs*HLS Lagrangian so constructed sliding in density has been applied to compact stars with rather satisfactory results[27]. Given the space limit for this note, I will skip the details here and mention merely that with only 3 "bare parameters, 1 in R-I (c_I) and 2 in R-II (κ and h) with some fine-tuning – presumably associated with $1/N_c$ corrections – within the range of parameters obtained, both normal nuclear matter and massive compact stars come out in consistency with Nature.

5. Conclusions

In the description that combines hidden local symmetry of the ρ vector meson and scale invariance of the dilaton scalar σ, a large portion of the proton mass $M_0 \sim 0.7m_N$ is found to remain non-vanishing in the limit that the quark condensate, dialled by density, goes to zero in the chiral limit. The non-vanishing mass M_0 is chiral invariant, hence implies parity doubling. The source for M_0 then cannot be accounted for by the standard Nambu-Goldstone mechanism anchored on spontaneous breaking of chiral symmetry for mass generation. It is in a sense "mass without mass"[2], perhaps lying outside of the Higgs paradigm. This picture is arrived at when scale invariance associated with an as yet unproven IR fixed point in QCD gauge coupling is implemented to hidden local symmetry unified into a combined chiral-scale symmetry. A striking consequence of this picture is the structure of the EoS of compact star matter accommodating the massive neutron stars recently observed that follows from topological transition of matter from skyrmions to half-skyrmions predicted at large N_c of QCD. While QCD proper is incapable of addressing the high density regime involved, Nature seems to indicate rather unambiguously that the ω meson is *not* on the trajectory that leads to the VM fixed point of the ρ meson, signalling how $U(2)$ HLS for the light vector mesons could be breaking down. As far as I know, this is the first hint from Nature that the ω meson may not behave the same as the ρ nearing chiral restoration.

How the intricate locking of chiral symmetry and scale symmetry and the unbreaking of mended symmetries[28] involving pseudo-Nambu-Goldstone bosons – pions, kaons and dilaton (σ) – and vector mesons – ρ, ω and a_1 – remain an open issue in nuclear physics as well as in astrophysics.

Acknowledgments

The work described here is based on works done largely in collaboration with Masa Harada, Tom Kuo, Hyun Kyu Lee, Yong-Liang Ma and Won-Gi Paeng. I would like to thank them for discussions and help. I am particularly grateful for encouragements and support of Koichi Yamawaki and of course for his invitation to this meeting.

References

1. H. S. Fukano, M. Kurachi, S. Matsuzaki, K. Terashi and K. Yamawaki, "2 TeV walking technirho at LHC?," arXiv:1506.03751 [hep-ph].
2. F. Wilczek, "The origin of mass," Mod. Phys. Lett. A **21**, 701 (2006).
3. S. R. Beane and U. van Kolck, "The dilated chiral quark model," Phys. Lett. B **328**, 137 (1994) [hep-ph/9401218]; C. Sasaki, H. K. Lee, W. G. Paeng and M. Rho, "Conformal anomaly and the vector coupling in dense matter," Phys. Rev. D **84**, 034011 (2011) [arXiv:1103.0184 [hep-ph]].
4. R. J. Crewther and L. C. Tunstall, "$\Delta I = 1/2$ rule for kaon decays derived from QCD Infrared fixed point," Phys. Rev. D **91**, no. 3, 034016 (2015), arXiv:1312.3319 [hep-ph].
5. M. Harada and K. Yamawaki, "Hidden local symmetry at loop: A New perspective of composite gauge boson and chiral phase transition," Phys. Rept. **381**, 1 (2003) [hep-ph/0302103].
6. T. Sakai and S. Sugimoto, "Low energy hadron physics in holographic QCD," Prog. Theor. Phys. **113**, 843 (2005) [hep-th/0412141].
7. D. K. Hong, M. Rho, H. U. Yee and P. Yi, "Dynamics of baryons from string theory and vector dominance," JHEP **0709**, 063 (2007) [arXiv:0705.2632 [hep-th]].
8. G. E. Brown, M. Rho and W. Weise, "Phenomenological delineation of the quark - gluon structure from nucleon electromagnetic form-factors," Nucl. Phys. A **454**, 669 (1986).
9. M. Schumacher, "Mass generation via the Higgs boson and the quark condensate of the QCD vacuum," arXiv:1506.00410 [hep-ph].
10. C. Adami and G. E. Brown, "Order of the QCD transition and QCD sum rules," Phys. Rev. D **46**, 478 (1992).
11. M. Bando, T. Kugo, S. Uehara, K. Yamawaki and T. Yanagida, "Is ρ meson a dynamical gauge boson of hidden local symmetry?," Phys. Rev. Lett. **54**, 1215 (1985).
12. Y. L. Ma, et al., "Dense baryonic matter in conformally-compensated hidden local symmetry: Vector manifestation and chiral symmetry restoration," Phys. Rev. D **90**, no. 3, 034015 (2014) [arXiv:1308.6476 [hep-ph]].
13. G. E. Brown and M. Rho, "Scaling effective Lagrangians in a dense medium," Phys. Rev. Lett. **66**, 2720 (1991).

14. G. E. Brown and M. Rho, "Double decimation and sliding vacua in the nuclear many body system," Phys. Rept. **396**, 1 (2004) [nucl-th/0305089].
15. R. Horsley, H. Perlt, P. E. L. Rakow, G. Schierholz and A. Schiller, Phys. Lett. B **728**, 1 (2014) [arXiv:1309.4311 [hep-lat]].
16. B.Y. Park and V. Vento in *The multifaceted skyrmion* (World Scientific, Singapore, 2011) ed. G,E. Brown and M. Rho.
17. Reviews by P. Sutcliffe and by V. Kaplunovsky, D. Melnikov and J. Sonnenschein, in Special Issue on "Skyrmions in strongly correlated systems," MPLB **29**, No. 16 (29 June 2015).
18. H. J. Lee, B. Y. Park, D. P. Min, M. Rho and V. Vento, "A unified approach to high density: Pion fluctuations in skyrmion matter," Nucl. Phys. A **723**, 427 (2003) [hep-ph/0302019]; B.Y. Park and V. Vento [16]
19. D. Suenaga, B. R. He, Y. L. Ma and M. Harada, "Mass degeneracy of heavy-light mesons with chiral partner structure in the half-Skyrmion phase," Phys. Rev. D **91**, no. 3, 036001 (2015) [arXiv:1412.2462 [hep-ph]].
20. H. Georgi, "New realization of chiral symmetry," Phys. Rev. Lett. **63**, 1917 (1989).
21. W. G. Paeng, H. K. Lee, M. Rho and C. Sasaki, "Interplay between ω-nucleon interaction and nucleon mass in dense baryonic matter," Phys. Rev. D **88**, 105019 (2013) [arXiv:1303.2898 [nucl-th]].
22. H. K. Lee, B. Y. Park and M. Rho, "Half-skyrmions, tensor forces and symmetry energy in cold dense matter," Phys. Rev. C **83**, 025206 (2011) [Phys. Rev. C **84**, 059902 (2011)] [arXiv:1005.0255 [nucl-th]].
23. D. Foster and N.S. Manton, "Scattering of nucleons in the classical Skyrme model," arXiv: 1505.96843 [nucl-th].
24. H. K. Lee and M. Rho, "Topology change and tensor forces for the EoS of dense baryonic matter," Eur. Phys. J. A **50**, 14 (2014) [arXiv:1306.4584 [nucl-th]].
25. H. K. Lee, W. G. Paeng and M. Rho, "Scalar pseudo-Nambu-Goldstone boson in nuclei and dense nuclear matter," arXiv:1504.00908 [nucl-th].
26. E.T. Tomboulis, "Chiral symmetry in $SU(N_c)$ gauge theories at high density," IJMPA **29**, 14450043 (2014).
27. W.-G. Paeng, T.T.S. Kuo, H.K. Lee and M. Rho, "Scale-invariant hidden local symmetry, topology change and dense baryonic matter," in preparation.
28. S. Weinberg, "Mended symmetries," Phys. Rev. Lett. **65**, 1177 (1990); "Unbreaking symmetries," Salamfestschrift (World Scientific, Singapore 1994) ed. by Ali, Ellis, Randjbar-Daemi.

Holographic Estimate of Isospin splitting in Hadron Mass

Deog Ki Hong
Department of Physics, Pusan National University,
Busan 609-735, Korea
E-mail: dkhong@pusan.ac.kr

We present the calculations of isospin splitting mass of pions and nucleons in the Sakai-Sugimoto model of holographic QCD. We find that $m_{\pi^\pm} - m_{\pi^0} \simeq 1.8$ MeV and the electromagnetic mass difference of nucleons, $(M_p - M_n)_{\rm QED} = 0.48$ MeV, taking the Kaluza-Klein scale, $M_{\rm KK} = 949$ MeV.

Keywords: Isospin splitting mass, Hadron mass, Holographic QCD, Gauge-gravity duality.

1. Introduction

Quantum chromodynamics (QCD) is the well-established theory of strong interactions of elementary particles. However, it is notoriously difficult to solve, because the quarks and gluons are not the relevant degrees of freedom at low energy, though they are the fundamental degrees of freedom of QCD. The only available method at present to solve QCD from the first principle is the lattice calculation, which has produced quite impressive results in recent years. For instance, the nucleon mass, which is a totally nonperturbative quantity, has been calculated by the QCDSF collaboration among others to obtain the value very close to the physical one[1], as shown in Fig. 1.

Another interesting progress, made recently in lattice QCD, is to calculate the isospin splitting in the hadron masses. Accurately measuring the isospin-splitting mass of hadrons is vital to determine the current quark masses from the hadron masses. A current status on the lattice calculation of the isospin splitting in the nucleon mass has been reported by Portelli *et al.* (See Fig. 2.)

Though the lattice QCD is producing impressive results, it lacks the analytic understanding from QCD. In this talk we present a holographic estimate of the electromagnetic (EM) mass of hadrons, based on the gauge/gravity duality found in string theory[3]. Our results will be complementary to the lattice calculations in solving QCD. The gauge/gravity duality, sometimes called Anti-de Sitter/Conformal Field theory correspondence, that states the equivalence of gauge theory and a gravity theory in one or more higher dimension, is not proven rigorously but many checks have been made successfully at least for a large N and a large 't Hooft coupling, λ, limit.

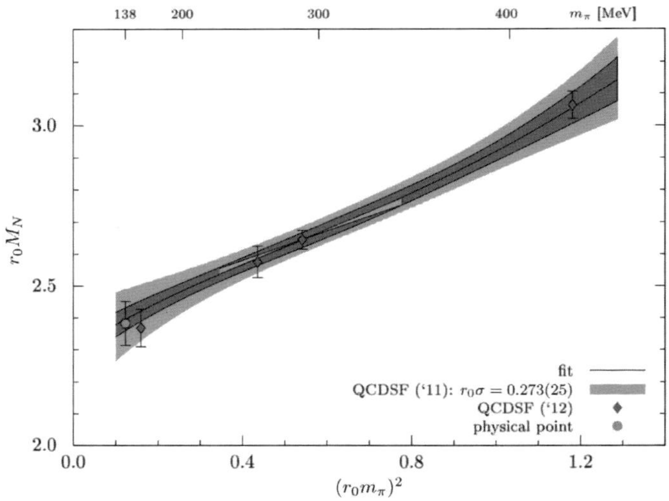

Fig. 1. The combined fit to the nucleon mass (red diamonds) of 2012 QCDSF collaboration. r_0 is the Sommer parameter and σ is the so-called pion-nucleon sigma term.

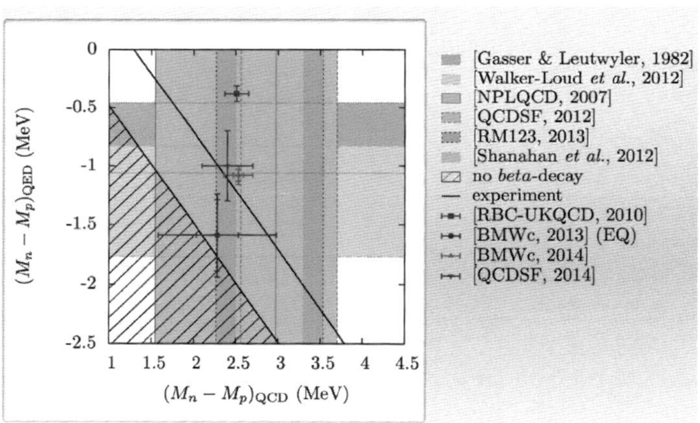

Fig. 2. Summary of the lattice calculations of isospin splitting in nucleon mass[2].

2. EM mass of pions - Vacuum alignment

QCD has a chiral symmetry, G, which rotates the left-handed quarks and the right-handed quark indepently when one neglects the current quark mass. The chiral symmetry is believed to be spontaneously broken to its (maximal) diagonal sub-group, H, which rotates the left and right-handed quarks simultaneosly. For the three-flavor case, $G = \mathrm{SU}(3)_L \times \mathrm{SU}(3)_R$ and $H = \mathrm{SU}(3)_V$. Pions and kaons are then identified as the Nambu-Goldstone bosons, associated with the spontaneous breaking of the chiral symmetry. In the chiral limit pions and kaons are massless, because they are fluctuations on the vacuum manifold, G/H, along the flat direction of broken generators[4–6]. When the chiral symmetry is approximate, however, the flat direction in the vacuum manifold is lifted and they become massive and called the pseudo-Nambu-Goldstone bosons[7].

The current quark mass lifts the vacuum degeneracy and gives pion mass. In the isospin limit one obtains by the PCAC hypothesis a so-called Gell-Mann-Oaks-Renner relation,

$$f_\pi^2 m_\pi^2 = 2\,m_q \langle q\bar{q}\rangle \,, \tag{1}$$

which relates the pion mass to the current quark mass and the quark condensation. The pion mass due to the current quark mass in the isospin limit has been calculated in holographic QCD or hQCD[8,9]. Since the action of holographic QCD becomes the vacuum energy for the static boundary fields, once the holographic action is obtained, the pion mass is simply evaluated by the formula, given as

$$m_\pi^2 = \frac{1}{TV}\,\frac{\delta^2}{\delta\pi^2} S_{\mathrm{hQCD}}\Big|_{\pi=0} \,. \tag{2}$$

The electromagnetic (EM) interaction breaks the isospin symmetry and contributes to the vacuum energy, lifting its degeneracy. At the leading order in the $\alpha_{\mathrm{em}} = e^2/4\pi$ expansion the vacuum energy is given as

$$\Delta\mathcal{E}_{\mathrm{vac}} = -\frac{e^2}{2}\int \mathrm{d}^4x\, D^{\mu\nu}(x)\,\langle 0|\,U^{-1} J_\mu^{Q_{\mathrm{em}}}(x) J_\nu^{Q_{\mathrm{em}}}(0)\,U\,|0\rangle \,, \tag{3}$$

where $U = e^{2i\pi/f_\pi}$ describes the pion fields, $J_\mu^{Q_{\mathrm{em}}}$ the electromagnetic currents of quarks, and $D^{\mu\nu}(x)$ denotes the photon propagator. The electric charge operator, Q_{em}, is defined as a sum of isospin and hypercharge, $Q_{\mathrm{em}} = I_3 + \frac{1}{2}Y$ up to a G_f-rotation. In Fig. 3 we have shown the leading EM contributions to the vacuum energy. We note that the vacuum energy is ultra-violet (UV) divergent and thus we

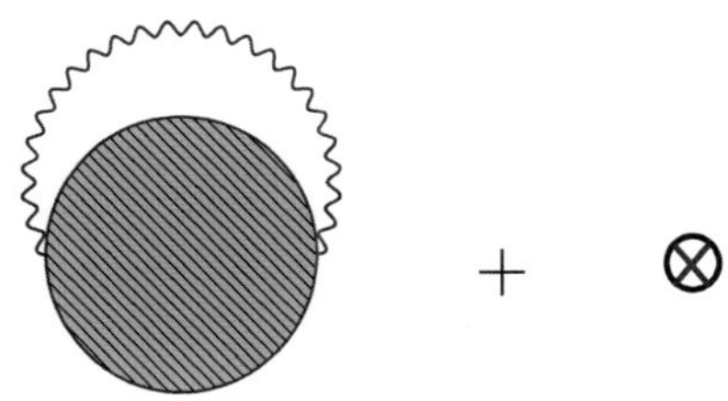

Fig. 3. The blob denotes the vacuum polarization and the wiggly line denotes the photon propagator. A counter term to the vacuum energy is denotes as ⊗.

need a counter term to remove the divergence. But, the counter term is independent of the pion fields, because it is the UV property of the theory. It therefore does not affect the mass of pions.

The EM mass of pions is now given as

$$m_{\pi^\pm}^2 - m_{\pi^0}^2 = \frac{\partial^2}{\partial\pi_+\partial\pi_-}\Delta\mathcal{E}_{\rm vac}[U]\bigg|_{U=1} = e^2 M^2\,, \tag{4}$$

where

$$M^2 = \frac{1}{f_\pi^2}\int \mathrm{d}^4x\, D^{\mu\nu}(x)\,\langle 0|\,T\left[V_\mu^3(x)V_\nu^3(0) - A_\mu^3(x)A_\nu^3(0)\right]|0\rangle \tag{5}$$

and $V_\mu^a(x)$ and $A_\mu^a(x)$ are the vector and axial vector flavor currents, respectively. The vacuum correlator of vector flavor currents is defined as

$$\Pi_{V\mu\nu}^{ab}(q) = \int \mathrm{d}^4x\, e^{iq\cdot x}\,\langle 0|\,V_\mu^a(x)V_\nu^b(0)\,|0\rangle = \delta^{ab}\left(q_\mu q_\nu - q^2 g_{\mu\nu}\right)\Pi_V(q) \tag{6}$$

and similarly the vacuum correlator of axial vector currents is given as

$$\Pi_{A\mu\nu}^{ab} = \int_x e^{iq\cdot x}\,\langle 0|\,A_\mu^a(x)A_\nu^b(0)\,|0\rangle = \delta^{ab}\left(q_\mu q_\nu - q^2 g_{\mu\nu}\right)\Pi_A(q) + \frac{q^\mu q^\nu}{q^2} f_\pi^2\delta^{ab}\,. \tag{7}$$

The vacuum correlators are nonperturbative and is very hard to calculate, but it is straightforward to calculate them in holographic QCD, since the holographic QCD action serves as the generating functional for the flavor-current correlators by the AdS/CFT correspondence.

The holographic QCD is described in general by a five-dimensional Chern-Simons-Yang-Mills theory for the flavor symmetry of QCD, the boundary theory,

$$S_{\rm hQCD} = \kappa\int \mathrm{d}^5x\,\mathrm{Tr}\left[-\frac{1}{2g^2(z)}F_{\mu\nu}^2 + f^2(z)F_{z\mu}^2\right] + S_{\rm CS}\,, \tag{8}$$

where $S_{\rm CS} = N_c/(24\pi^2)\int\omega_5(A)$ is the 5D Chern-Simons action that reproduces the QCD flavor anomalies. In the Sakai-Sugimoto model[10,11] the warp factors $g^3 = f = \sqrt{1+z^2}$ and $\kappa = N_c\lambda/(216\pi^3)$.

By the AdS/CFT prescription the (axial) vector current-current correlators in the momentum space are given by the bulk action evaluated on-shell as

$$\Pi_{V(A)}(q^2) = -\frac{2\kappa}{q^2}\left(1+z^2\right)\partial_z\,A(z,q^2)\bigg|_{z_m}\,, \tag{9}$$

where z_m is the UV boundary of the holographic coordinate that goes to infinity in the Sakai-Sugimoto model and $A(z,q^2)$ is the non-normalizable gauge field, up to the polarization vector, that satisfies the equation of motion with (anti-) symmetric UV boundary conditions for (axial) vector currents. The equation of motion of hQCD is a typical Sturm-Liouville equation. Therefore by the Sturm-Liouville theorem we may expand the non-normalizable mode in terms of the eigenfunctions of the Sturm-Liouville equation, which are the normalizable modes, forming a complete basis. The correlators are then given as

$$\Pi_V(q^2) = \sum_{n=1}^{\infty}\frac{g_{v^n}^2}{(q^2 - m_{v^n}^2)m_{v^n}^2}\,, \quad \Pi_A(q^2) = -\frac{f_\pi^2}{q^2} + \sum_{n=1}^{\infty}\frac{g_{a^n}^2}{(q^2 - m_{a^n}^2)m_{a^n}^2}\,, \tag{10}$$

where g_{v^n} (g_{a^n}) and m_{v^n} (m_{a^n}) are the decay constants and mass of n-th (axial) vector mesons, respectively. The EM mass of pions now becomes

$$e^2 M^2 = \frac{3e^2}{f_\pi^2}\int \frac{\mathrm{d}^4Q}{(2\pi)^4}\left[\Pi_V(-Q^2) - \Pi_A(-Q^2)\right] \tag{11}$$

$$= \frac{3e^2}{8\pi^2 f_\pi^2}\sum_n \left[g_{v^n}^2 \ln\left(\frac{\Lambda}{m_{v^n}}\right) - g_{a^n}^2 \ln\left(\frac{\Lambda}{m_{a^n}}\right)\right], \tag{12}$$

where Λ is the UV cutoff and we have used the Weinberg first sum rule in the second line. Since the Weinberg second sum rule on the spectral functions also holds in holographic QCD, the pion EM mass is finite, though each term in the sum in (12) is divergent. Indeed numerically one can check the correlators of flavor currents converge to each other rather quickly for large $Q^2 \gg M_{\mathrm{KK}}^2$, where M_{KK} is the Kaluza-Klein scale of Sakai-Sugimoto model, taken to be 949 MeV. By numerically evaluating Eq. (11), we find the EM mass of pions in the Sakai-Sugimoto model to be

$$m_{\pi^\pm} - m_{\pi^0} \simeq 1.8\ \mathrm{MeV}, \tag{13}$$

which is roughly half of the experimental value, 4.6 MeV and is much less than the hard-wall estimate, obtained by Da Rold and Pomarol[12].

3. Baryons in holographic QCD

Baryons in large N_c QCD is realized as solitons, since the baryon mass is proportional to the inverse of the expansion parameter of large N_c QCD[13]. Being a 5d gauge theory, holographic QCD admits a topologically conserved current

$$J^M = \frac{1}{32\pi^2}\epsilon^{MNLPQ}\,\mathrm{tr}\left(F_{NL}F_{PQ}\right) \tag{14}$$

which is automatically conserved by the Bianchi identity, $D_{[M,}F_{NL]} = 0$. Since this topological current couples to U(1) bulk (quark number) gauge fields through the Chern-Simons term, S_{CS}, we may define the baryon number current (of the boundary theory) as

$$B^\mu = \frac{1}{8\pi^2}\int \mathrm{d}z\,\epsilon^{\mu\nu\rho\sigma}\mathrm{tr}\left(F_{\nu\rho}F_{\sigma\,z}\right), \tag{15}$$

which is conserved, since $\partial_M J^M = 0$ and the field strength vanishes at $z = \pm\infty$. The baryon number is then nothing but the instanton number of the bulk gauge fields:

$$N_B = \int \mathrm{d}^3x B^0(x) = \frac{1}{32\pi^2}\int \mathrm{d}^3x\,\mathrm{d}z\,F^a_{IJ}\tilde{F}^{aIJ}, \tag{16}$$

where $I, J = 1, 2, 3, z$.

To find the soliton configuration we have to minimize its energy. The energy from the DBI action at the leading order in α' is found to be

$$\mathcal{E}_0 = \kappa\int \mathrm{d}^3x\,\mathrm{d}z\left(1+z^2\right)^{1/3}\vec{E}^a\cdot\vec{B}^a, \tag{17}$$

where we have used the self-duality condition for the gauge fields. For a small-size baryon located at $z = 0$, the space is almost flat and we may take the $O(4)$-invariant instanton solution to get

$$\frac{1}{8\pi^2}\,\vec{E}^a \cdot \vec{B}^a = \frac{6}{\pi^2} \cdot \frac{\rho^4}{\left(\rho^2 + \vec{x}^2 + z^2\right)^4}\,, \tag{18}$$

where ρ is the size of the instanton soliton. We then find the soliton energy from the DBI action to be, in powers of the ρ^2, $\mathcal{E}_0 = m_B^{(0)}\left(1 + \frac{1}{6}\rho^2 + \cdots\right)$, where $m_B^{(0)} = 8\pi^2\kappa = \lambda N_c\, M_{\rm KK}/(27\pi)$. We see that the warped geometry prefers a smaller soliton. On the other hand, since the soliton is charged under the bulk U(1), it has a Coulomb energy, which makes the soliton expand:

$$\mathcal{E}_{\rm C} = -\frac{\kappa}{2}\int \mathrm{d}^3x\mathrm{d}z\left[\left(1+z^2\right)^{-\frac{1}{2}}E_i^2 + \left(1+z^2\right)E_z^2\right] + \frac{N_c}{2}\int \mathrm{d}^3x\mathrm{d}z\, A_0\, J^0 \tag{19}$$

$$= \frac{1}{4}N_c\int \mathrm{d}^3x\mathrm{d}z\, A_0 J^0 \simeq \frac{N_c^2}{40\pi^2\kappa}\left[\frac{1}{\rho^2} + \mathcal{O}(1)\right], \tag{20}$$

where we used the equation of motion in the second line. By minimizing the soliton energy $\mathcal{E}_0 + \mathcal{E}_C$ we find the size of the soliton

$$\rho_B \simeq \frac{1}{\pi}\left(\frac{3}{40}\right)^{1/4}\sqrt{\frac{N_c}{2\kappa}} = \frac{9.6}{\sqrt{\lambda}}\,, \tag{21}$$

which is very small, compared to the generic scale of hQCD, which is $\mathcal{O}(1)$ in the unit of $M_{\rm KK}$, as the $\lambda \to 0$ limit is taken in hQCD. After quantizing the zero modes of the soliton, one can show its wave function transforms like an isospin doublet spinor for N_C odd. Since the size of soliton is quite small, we may then write down a low-energy effective (holographic) action for the baryons, consistent with bulk gauge symmetry and the instanton-soliton nature of holographic baryons[14], as

$$S_{\rm hQCD} = \int_{x,w}\left[-i\bar{\mathcal{B}}\gamma^m D_m\mathcal{B} - i m_b(w)\bar{\mathcal{B}}\mathcal{B} + \kappa(w)\bar{\mathcal{B}}\gamma^{mn}F_{mn}^{SU(2)_I}\mathcal{B} + \cdots\right], \tag{22}$$

where $\kappa(w) \simeq \frac{0.18N_c}{M_{KK}}$ and $m_b(w)$ is the position-dependent baryon mass[14,15]. This holographic action for baryons serves as the generating functional for the one-particle irreducible diagrams of baryons and mesons. Also by taking the non-normalizable modes for the (bulk) gauge fields in (22) one can get the various form-factors of baryons, including the EM form-factors[15].

The EM mass of nucleons at the leading order in $\alpha_{\rm em}$ comes from two diagrams in general, shown in Fig. 4. However, since the Pauli term in the effective action (22) does not contain any part of $Q_{\rm em}^2$ term, the diagram Fig. 4(b) vanishes at all orders in $1/N_c$. The blobs in Fig. 4(a) are nothing but the nucleon form factors, among which the Dirac EM form-factor is dominant. We therefore obtain the EM mass of nucleons to be

$$\delta M_{p\,(n)} = e^2\int\frac{\mathrm{d}^4Q}{(2\pi)^4}\left[F_1^{p\,(n)}(Q^2)\right]^2\frac{3M_N}{Q^2+M_N^2}\cdot\frac{1}{Q^2}\,, \tag{23}$$

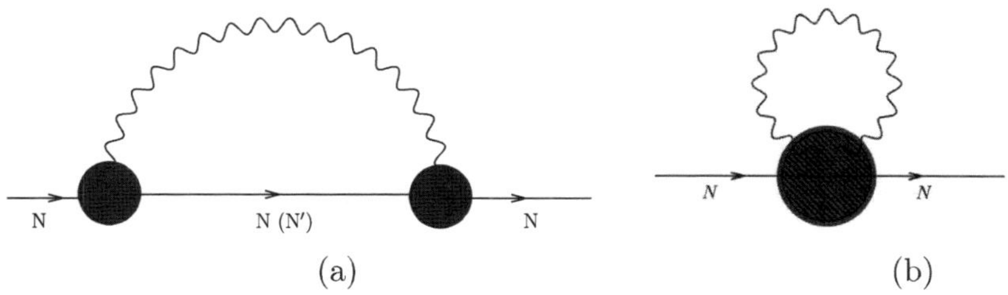

Fig. 4. Leading EM self-energy diagrams for nucleons

where $F_1(Q^2)$ is the Dirac EM form-factor and we used the Landau gauge for the photon propagator to be consistent with the vertex structure. Since the form-factor vanishes as $1/Q^2$ for large Q^2, the integration is finite. We find, keeping the first four vector mesons for the form-factor,

$$\delta M_p = 0.494 \text{ MeV}, \quad \delta M_n = 0.018 \text{ MeV}. \tag{24}$$

Acknowledgments

We would thank the organizers of SCGT15 for the invitation, especially Koichi Yamawaki for his support and encouragements. This research was supported by Basic Science Research Program through the National Research Foundation of Korea (NRF) funded by the Ministry of Education (NRF-2013R1A1A2011933).

References

1. G. S. Bali *et al.*, Nucl. Phys. B **866**, 1 (2013) [arXiv:1206.7034 [hep-lat]].
2. A. Portelli, PoS LATTICE **2014**, 013 (2015) [arXiv:1505.07057 [hep-lat]].
3. D. K. Hong, arXiv:1409.8139 [hep-ph].
4. Y. Nambu, Phys. Rev. **117**, 648 (1960).
5. J. Goldstone, Nuovo Cim. **19**, 154 (1961).
6. J. Goldstone, A. Salam and S. Weinberg, Phys. Rev. **127**, 965 (1962).
7. S. Weinberg, Phys. Rev. Lett. **29**, 1698 (1972).
8. O. Aharony and D. Kutasov, Phys. Rev. D **78**, 026005 (2008).
9. K. Hashimoto, T. Hirayama, F. L. Lin and H. U. Yee, JHEP **0807**, 089 (2008) [arXiv:0803.4192 [hep-th]].
10. T. Sakai and S. Sugimoto, Prog. Theor. Phys. **113**, 843 (2005) [hep-th/0412141].
11. T. Sakai and S. Sugimoto, Prog. Theor. Phys. **114**, 1083 (2005) [hep-th/0507073].
12. L. Da Rold and A. Pomarol, Nucl. Phys. B **721**, 79 (2005) [hep-ph/0501218].
13. E. Witten, Nucl. Phys. B **160**, 57 (1979).
14. D. K. Hong, M. Rho, H. U. Yee and P. Yi, Phys. Rev. D **76**, 061901 (2007) [hep-th/0701276 [HEP-TH]]:
15. D. K. Hong, M. Rho, H. U. Yee and P. Yi, JHEP **0709**, 063 (2007) [arXiv:0705.2632 [hep-th]].

Scalar Mesons in Lattice QCD

Chiho Nonaka[1,2], Teiji Kunihiro[3], Shin Muroya[4], Atsushi Nakamura[5], Motoo Sekiguchi,[6]
Hiroaki Wada[7] and Masayuki Wakayama[2]

[1]*Kobayashi Maskawa Institute, Nagoya University, Nagoya 464-8602, Japan*
[2]*Department of Physics, Nagoya University, Nagoya 464-8602, Japan*
[3]*Department of Physics, Kyoto University, Kyoto 606-8502, Japan*
[4]*Matsumoto University, Matsumoto 390-1295, Japan*
[5]*RCNP, Osaka University, Osaka 567-0047, Japan*
[6]*School of Science and Engineering, Kokushikan University, Tokyo 154-8515, Japan*
[7]*Faculty of Political Science and Economics, Kokushikan University, Tokyo 154-8515, Japan*

1. Introduction

The σ meson ($I = 0$, $J^{PC} = 0^{++}$) plays an important role in the low energy hadron physics from point of view of chiral symmetry in QCD. From re-analyses of the $\pi - \pi$ scattering phase shift the existence of the pole of the σ pole is established[1–3]. Furthermore the existence of κ meson ($I = 1/2$)[4,5] supports the nonet scalar states of chiral SU(3)$\otimes$ SU(3) symmetry together with σ meson. However there are still long standing controversies on nature of the low lying scalar mesons. For example, proposed possible states of the σ meson are a usual $q\bar{q}$ meson, four-quark states ($qq\bar{q}\bar{q}$), π-π molecular state, or their mixing state. In addition, the glueball has the same quantum number as the σ meson. Here we show our recent investigation for the σ meson in lattice QCD[6,7].

2. Sigma meson as two-quark state

First we explore whether the σ meson exists as a pole in QCD, by a full lattice QCD simulation using the plaquette action and Wilson fermions. Inspiring the conventional quark model, we employ the following interpolation operator for the σ meson as two-quark state,

$$\sigma(x) = \sum_{c=1}^{3}\sum_{\alpha=1}^{4} \frac{\bar{u}_\alpha^c(x)u_\alpha^c(x) + \bar{d}_\alpha^c(x)d_\alpha^c(x)}{\sqrt{2}}, \tag{1}$$

where $u(d)$ stands for the up-quark (down-quark) operator with c and α being the color and Dirac-spinor indices, respectively. The σ meson propagator is obtained by $G(y,x) = \langle \sigma(y)\sigma(x)^\dagger \rangle$,

$$G(y,x) = -\langle \mathrm{Tr}W^{-1}(x,y)W^{-1}(y,x)\rangle + 2\langle(\sigma(y) - \langle\sigma(y)\rangle\rangle\langle(\sigma(x) - \langle\sigma(x)\rangle\rangle, \tag{2}$$

where $\sigma(x) = \mathrm{Tr}W^{-1}(x,x)$ with the "Tr" for operation of the color and Dirac indices. It has the connected and disconnected diagrams.

We perform the full-QCD calculation with the same parameters of CP-PACS[8] except lattice size, *i.e.*, $\beta = 4.8$ and the hopping parameter, $\kappa = 0.1846, 0.1874, 0.1891$ and our lattice size, $8^3 \times 16^6$. Figure 1 shows the connected and disconnected parts of the σ meson propagators, respectively. The propagator of the disconnected part is dominant of the σ propagator and has a flatter slope compared with the connected part. In Fig.1 (b), propagators of the π, ρ and σ for $\kappa = 0.1874$ are shown. Figures 1 (a) and (b) indicate that the sigma meson as a light meson comes from the existent of the disconnected diagram in σ propagators. Figure 2 shows the m_π^2, m_ρ, m_σ and $2m_\pi$ as a function of the inverse hopping parameter. As the chiral limit ($\kappa_c = 0.195(3)$) is approached, the σ meson mass decreases rapidly. Then, in the chiral limit the mass of σ meson is smaller than that of ρ meson. For the light scalar meson, the disconnected diagram in the σ meson propagators is important, which indicates the possible physical contents to the light σ such as the glueball mixing and four-quark states, other than two-quark states.

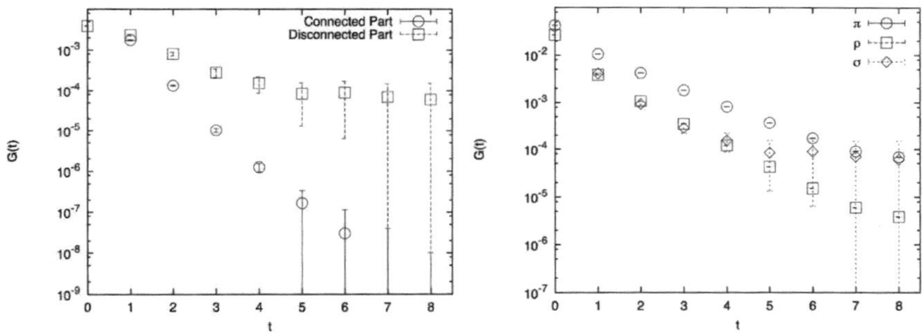

(a) (b)

Fig. 1. (a) The propagators of the connected and disconnected diagrams of the σ meson at $\kappa = 0.1874$. (b) Propagators of the π, ρ and σ at $\kappa = 0.1874$.

3. Sigma meson as four-quark state

Next we explore the possible significance of the four-quark states in the iso-singlet scalar mesons. Here we take into account of the molecular state $(\bar{q}q)(\bar{q}q)$ and/or the tetra-quark state $(\bar{q}\bar{q}qq)$ for the four-quark state and prepare interpolation operators for them, respectively.

3.1. *Four quark operators*

The molecular interpolation operators are given by

$$\mathcal{O}^{\rm molec}(t) = \frac{1}{\sqrt{3}}\left[\mathcal{O}^{\pi^+}(t)\mathcal{O}^{\pi^-}(t) - \mathcal{O}^{\pi^0}(t)\mathcal{O}^{\pi^0}(t) + \mathcal{O}^{\pi^-}(t)\mathcal{O}^{\pi^+}(t)\right], \quad (3)$$

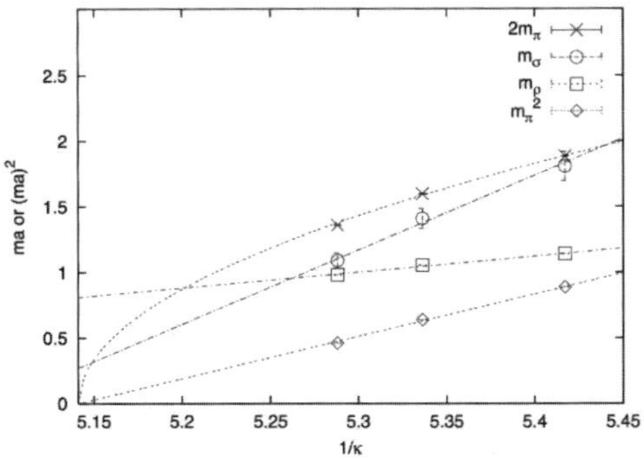

Fig. 2. The quark mass dependence of $2m_\pi$, m_π, m_σ,m_ρ and m_π^2.

where $\mathcal{O}^{\pi^+}(t)$, $\mathcal{O}^{\pi^-}(t)$, and $\mathcal{O}^{\pi^0}(t)$ are the π meson operators made up of two quarks. They are written by

$$\begin{aligned}
\mathcal{O}^{\pi^+}(t) &= -\sum_{\mathbf{x}\,a} \bar{d}^a(t,\mathbf{x})\gamma_5 u^a(t,\mathbf{x}) \ , \\
\mathcal{O}^{\pi^-}(t) &= \sum_{\mathbf{x}\,a} \bar{u}^a(t,\mathbf{x})\gamma_5 d^a(t,\mathbf{x}) \ , \\
\mathcal{O}^{\pi^0}(t) &= \frac{1}{\sqrt{2}}\sum_{\mathbf{x}\,a} \left[\bar{u}^a(t,\mathbf{x})\gamma_5 u^a(t,\mathbf{x}) - \bar{d}^a(t,\mathbf{x})\gamma_5 d^a(t,\mathbf{x})\right] \ ,
\end{aligned} \tag{4}$$

where a is the index of the color.

The tetra-quark interpolation operators are given by

$$\mathcal{O}^{\rm tetra}(t) = \sum_a [ud]^a(t)[\bar{u}\bar{d}]^a(t) \ , \tag{5}$$

where $[ud]^a(t)$ and $[\bar{u}\bar{d}]^a(t)$ are diquark and antidiquark operators, respectively, written as

$$\begin{aligned}
[ud]^a(t) &= \frac{1}{2}\sum_{\mathbf{x}\,b,c} \epsilon^{abc}\left[u^{Tb}(t,\mathbf{x})\ C\gamma_5\ d^c(t,\mathbf{x}) - d^{Tb}(t,\mathbf{x})\ C\gamma_5\ u^c(t,\mathbf{x})\right] \ , \\
[\bar{u}\bar{d}]^a(t) &= \frac{1}{2}\sum_{\mathbf{x}\,b,c} \epsilon^{abc}\left[\bar{u}^b(t,\mathbf{x})\ C\gamma_5\ \bar{d}^{Tc}(t,\mathbf{x}) - \bar{d}^b(t,\mathbf{x})\ C\gamma_5\ \bar{u}^{Tc}(t,\mathbf{x})\right] \ ,
\end{aligned} \tag{6}$$

with the charge conjugation matrix C.

The propagator $G^i(t)$ for the four-quark operators is written as

$$G^i(t) = \left\langle \mathcal{O}^i(t)\mathcal{O}^{i\,\dagger}(0)\right\rangle \ , \ i = \text{molec or tetra,} \tag{7}$$

where $\mathcal{O}^i$ is the molecular or tetra-quark interpolation operator.

We show the diagrams for the elements of the propagator $G^i(t)$, the molecule $G^{\rm molec}(t)$ and the tetra-quark $G^{\rm tetra}(t)$. Through the functional integral of Eq. (7)

with the quark fields, the propagator of the molecular operator $G^{\rm molec}(t)$ is written as

$$G^{\rm molec}(t) = 2\left[D(t) + \frac{1}{2}C(t) - 3A(t) + \frac{3}{2}V(t)\right], \quad (8)$$

where $D(t)$, $C(t)$, $A(t)$, and $V(t)$ correspond to direct, crossed, single annihilation (singly disconnected), and vacuum (doubly disconnected) diagrams which are shown in Fig.3. On the other hand, the tetra propagator is given by

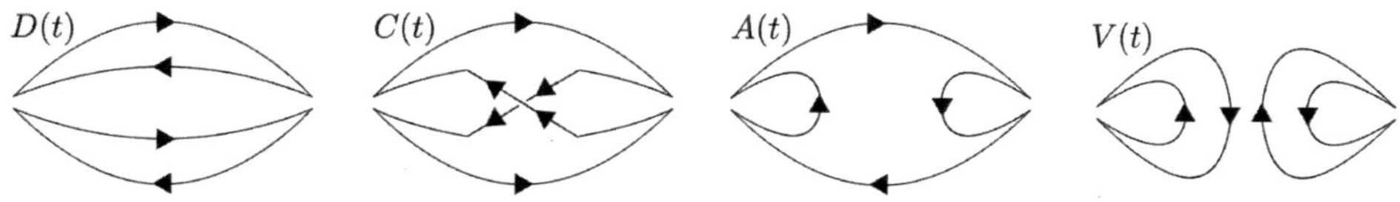

Fig. 3. The diagrams for the propagator of the molecular operator $G^{\rm molec}(t)$.

$$\begin{aligned} G^{\rm tetra}(t) &= 2\Big[2\left(D_1'(t) + D_2'(t)\right) - 2\left(A_1'(t) + A_2'(t) + A_3'(t) + A_4'(t)\right) \\ &\qquad + \left(V_1'(t) + V_2'(t) + V_3'(t) + V_4'(t)\right)\Big], \quad (9) \end{aligned}$$

where $D'(t)$, $A'(t)$, and $V'(t)$ are shown in Fig.4. One can find the difference be-

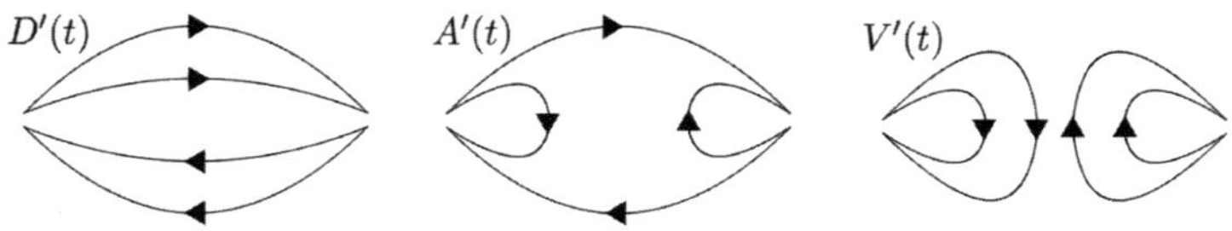

Fig. 4. The diagrams for the propagator of the tetra-quark operator $G^{\rm tetra}(t)$.

tween Figs. 3 and 4 in the directions of the arrows on the quark lines. In the propagators of the molecular and tetra-quark operators,we neglect the doubly disconnected diagrams $V(t)$ and $V'(t)$, assuming that the N_c counting scheme also works for $N_c = 3$. For numerical calculation of the singly disconnected diagram we carry out the Z_2-noise method with the truncated eigenmode approach.

3.2. *Calculated results*

We generate two-flavor full QCD configurations using the same simulation parameters in Ref.[9] (clover coefficient $C_{SW} = 1.68$ and coupling $\beta = 1.7$) on the $8^3 \times 16$ lattice. The parameters for the numerical simulation are listed in Tab.1.

In Fig.5 (a) the propagators of the molecular operators at $\kappa = 0.146$, together with the propagators of diagrams $D(t)$, $C(t)$, and $A(t)$ are shown. We emphasize

Table 1. Masses of π and ρ and number of configurations.

κ	$m_\pi a$	m_π MeV	$m_\rho a$	m_ρ MeV	configurations
0.146	1.018(2)	747(27)	1.431(4)	1050(39)	16496
0.147	0.930(2)	682(25)	1.358(6)	996(38)	14344
0.148	0.827(4)	607(23)	1.304(10)	956(39)	11720

Note: Number of configurations separated by 10 trajectories to each other.

that the contribution from the singly disconnected diagram $A(t)$ is the same order of magnitude as that from the connected diagram $D(t)$, which suggests that the singly disconnected diagram should not be neglected in the propagator of the molecule. In Fig. 5 (b) we plot the propagator of the tetra-quark operator and the elements of the tetra-quark diagrams $D'(t)$ and $A'(t)$. We can see that the main component of the propagator of the tetra-quark operator originates from the singly disconnected diagrams $A'(t)$. The absolute values of the propagator of the connected diagrams $D'(t)$ are much smaller than those of the singly disconnected diagrams $A'(t)$. We thus cannot neglect the singly disconnected diagram in the investigation of the tetra-quark. From the comparison between Figs. 5(a) and 5(b), the propagators of the molecule have smaller errors than those of the tetra-quark. The propagators of the molecule have only small errors because the interpolation operators of it are composed of two pseudo-scalar mesons whose propagators have small errors. The propagators of molecular and tetra-quark operators at $\kappa = 0.147$ and 0.148 are shown in Ref.[7]. In Fig.6 (a) and (b) effective masses as a function

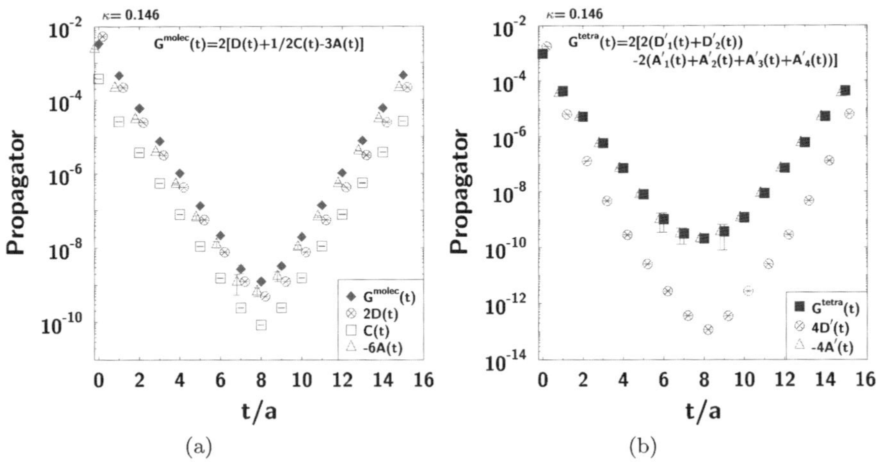

Fig. 5. (a) The propagators of the molecule and its components at $\kappa = 0.146$. (b) The propagators of the tetra-quark operator and its components at $\kappa = 0.146$.

of time for the molecule and tetra-quark with or without singly disconnected diagrams at $\kappa = 0.146$ are shown. The behavior of the effective masses of the molecule with the singly disconnected diagrams is almost the same as that without the singly disconnected diagram, in which we can observe the clear plateau around $2m_\pi$. On the other hand, the behavior of the effective masses of tetra-quark with the singly disconnected diagrams is different from that without the singly disconnected diagrams. In the case of existence of the singly disconnected diagrams, a plateau-like structure appears at small t which suggests that the tetra-quark has a small overlap to the lowest state in the molecule. Finally, we display the m_π^2, $2m_\pi$, m_ρ, $m^{\rm molec}$,

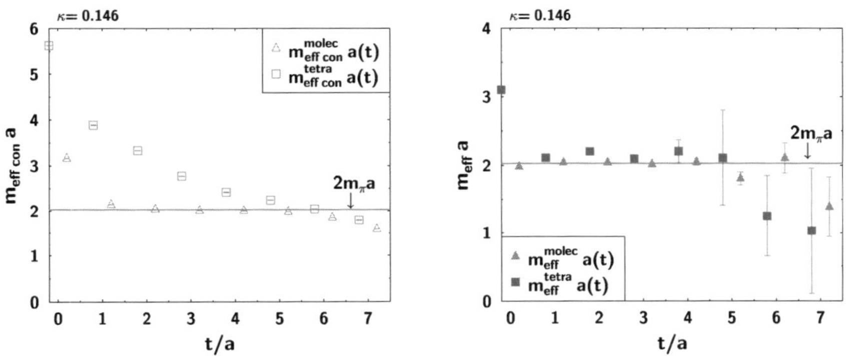

(a) (b)

Fig. 6. (a) The effective masses of the molecule (open triangles) and tetra-quark (open squares) without the singly disconnected diagrams at $\kappa = 0.146$. (b)The effective masses of the molecule (solid triangles) and tetra-quark (solid squares) with the singly disconnected diagrams at $\kappa = 0.146$.

$m_{\rm con}^{\rm molec}$, $m^{\rm tetra}$ and $m_{\rm con}^{\rm tetra}$ as a function of the inverse of the hopping parameter in Fig. 7. The difference between the masses of the molecular operator with the singly disconnected diagram and those without the singly disconnected diagram is small at all κ. In both cases, the extracted masses are identical to $2m_\pi$ which would suggest that the molecular operators have a large overlap to the 2-particle $\pi - \pi$ scattering state. Assuming that the plateau-like structure of the effective masses of the tetra-quark with the singly disconnected diagrams, we evaluate the mass of the tetra-quark. The mass from the tetra-quark operator is larger than that of the molecular operator and the difference between the masses becomes larger at smaller quark mass. It indicates that the tetra-quark operators have smaller overlap to the lowest state in the molecular operators and the mass of them can be a mixture of the excited state. In the calculation, we do not observe any bound four-quark states in the molecular and tetra-quark operators.

4. Summary

We investigate the structure of the σ meson using two-quark and four-quark interpolation operators in lattice QCD. First using the two-quark operators we show that there exists a σ pole, whose mass is similar to that to the ρ meson. The dominance

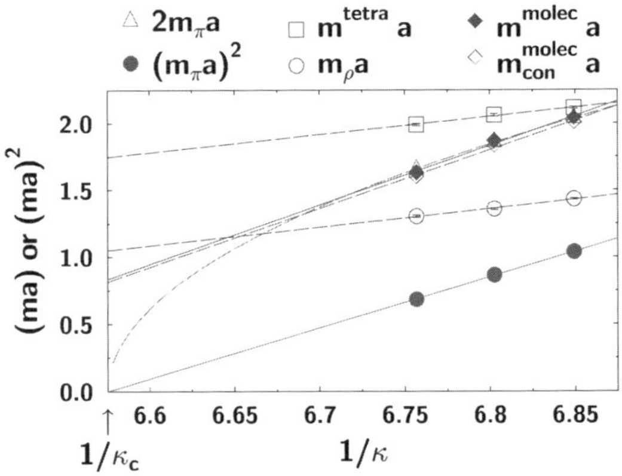

Fig. 7. The quark mass dependence of m_π^2 (solid circles), $2m_\pi$ (open triangles), m_ρ (open circles), m^{molec} (solid diamonds), $m_{\text{con}}^{\text{molec}}$ (open diamonds), m^{tetra} (solid squares and $m_{\text{con}}^{\text{tetra}}$ (open squares).

of the disconnected diagram in the σ propagator is important for the existence of σ as a light meson. For the four-quark state, unfortunately, we do not observe any bound four-quark states in molecular and tetra-quark operators. However, for the conclusive results, further improvement of the computation is indispensable such as calculation with lighter quark masses on a larger lattice, estimation of the doubly disconnected diagrams in the molecule and tetra-quark operators, inclusion of other possible interpolation operators for the four-quark states, and the use of the variational method with possible interpolation operators. Furthermore for the comprehensive understanding of the iso-singlet scalar mesons, we need to take into account of not only two-quark and four-quark operators themselves but also possible other operators glueballs (gg), hybrid operators ($\bar{q}qg$) and so on.

References

1. K. Igi and K.-I. Hikasa, Phys. Rev. **D**59, 034005 (1999).
2. J. A. Oller, E. Oset and A. Ramos, Prog. Part. Nucl. Phys. 45, 157 (2000), and the references therein.
3. G. Colangelo, J. Gasser and H. Leutwyler, Nucl. Phys. **B** 603, 125 (2001).
4. E791 collaboration, Phys. Rev. Lett. 89, 121801 (2002).
5. J.Z. Bai, et al., BES collaboration, hep-ex/0304001.
6. T. Kunihiro *et al.* [SCALAR Collaboration], Phys. Rev. D **70**, 034504 (2004).
7. M.Wakayama *et al.* [SCALAR Collaboration], Phys. Rev. D **91**, no. 9, 094508 (2015).
8. S. Aoki, et al. [CP-PACS Collaboration], Phys. Rev. **D** 60, 114508 (1999).
9. A. Ali Khan et al. [CP-PACS Collaboration], PhysRev. **D** 63, 034502 (2001).

Magnetic Monopole Versus Vortex as Gauge-Invariant Topological Objects for Quark Confinement

Kei-Ichi Kondo*, Takaaki Sasago and Toru Shinohara
Department of Physics, Graduate School of Science, Chiba University, Chiba 263-8522, Japan

Akihiro Shibata
High Energy Accelerator Research Organization (KEK), Tsukuba 305-0801, Japan

Seikou Kato
Fukui National College of Technology, Sabae 916-8507, Japan

First, we give a gauge-independent definition of chromomagnetic monopoles in $SU(N)$ Yang-Mills theory which is derived through a non-Abelian Stokes theorem for the Wilson loop operator. Then we discuss how such magnetic monopoles can give a nontrivial contribution to the Wilson loop operator for understanding the area law of the Wilson loop average. Next, we discuss how the magnetic monopole condensation picture are compatible with the vortex condensation picture as another promising scenario for quark confinement. We analyze the profile function of the magnetic flux tube as the non-Abelian vortex solution of $U(N)$ gauge-Higgs model, which is to be compared with numerical simulations of the $SU(N)$ Yang-Mills theory on a lattice. This analysis gives an estimate of the string tension based on the vortex condensation picture, and possible interactions between two non-Abelian vortices.

Keywords: Quark confinement, Dual superconductivity, Non-Abelian magnetic monopole.

1. Introduction

To begin with, let me summarize the current status of numerical simulations of the **static quark–antiquark potential**. Fig. 1 shows that the static quark–antiquark potential $V_{q\bar{q}}(r)$ as a function of the quark–antiquark distance r is well fitted by the form of the **Cornell type**: Coulomb+Linear (See Fig. 1)

$$V_{q\bar{q}}(r) = -\frac{\alpha}{r} + \sigma r + c,$$

where the parameters have the following dimensions, the Coulomb coefficient $\alpha = [\text{mass}^0]$, the string tension $\sigma = [\text{mass}^2]$, and a constant $c = [\text{mass}^1]$. Therefore, the potential goes to infinity $V_{q\bar{q}}(r) \to \infty$ as the distance increases $r \to \infty$, leading to **quark confinement**.

*Speaker at the workshop SCGT15.

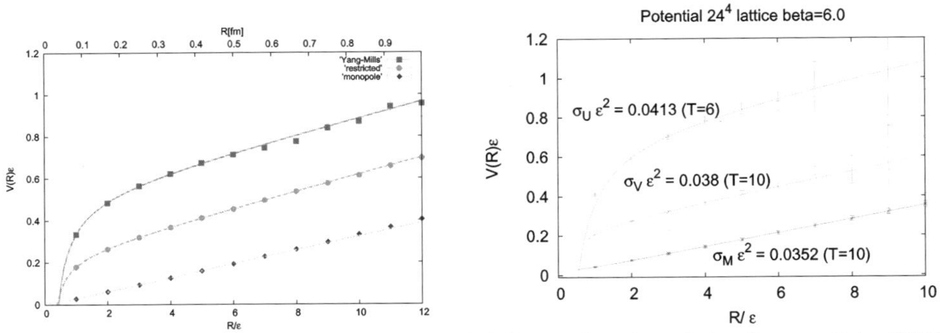

Fig. 1. The static quark–antiquark potential $V(R)$ as a function of the distance R in $SU(N)$ Yang-Mills theory. (from above to below): The full potential $V_{\text{full}}(R)$, restricted (or "Abelian") part $V_{\text{rest}}(R)$ and magnetic–monopole part $V_{\text{mono}}(R)$. (Left) $SU(2)$ at $\beta = 2.5$ on 24^4 lattice [8], (Right) $SU(3)$ at $\beta = 6.0$ on 24^4 lattice [9].

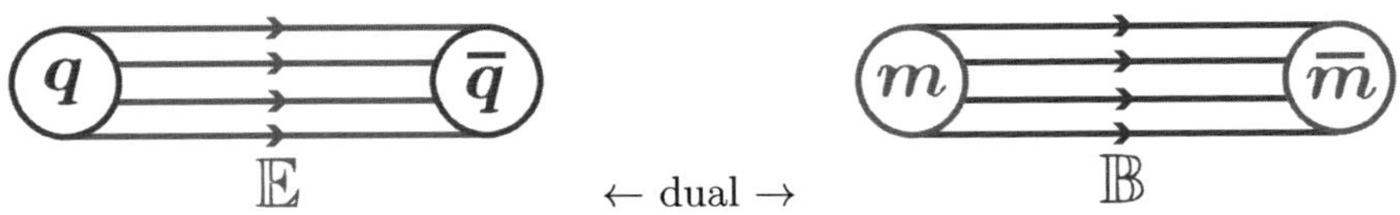

Fig. 2. The electro-magnetic duality: electric charge is replaced by the magnetic charge, and the electric field is replaced by the magnetic field, and vice versa.

A promising scenario for understanding quark confinement is the **dual superconductor hypothesis for quark confinement** based on the electro-magnetic duality (See Fig. 2) proposed by Nambu (1974), 't Hooft (1975), Mandelstam (1976) [1] and Polyakov (1975,1977) [2]. The key ingredients for the dual superconductivity are as follows. For reviews, see reviews [3–5].

- **dual Meissner effect**
  In the dual superconductor, the chromoelectric flux must be squeezed into tubes. [← In the ordinary superconductor (of the type II), the magnetic flux is squeezed into tubes.]
- **condensation of chromomagnetic monopoles**
  The dual superconductivity will be caused by condensation of magnetic monopoles (called the chromomagnetic monopoles). [← The ordinary superconductivity is cased by condensation of electric charges into Cooper pairs.]

In order to establish the dual superconductivity, we must answer the following questions:

- How to introduce **chromomagnetic monopoles** in the Yang-Mills theory without scalar fields? [This should be compared with the 't Hooft-Polyakov magnetic monopole.]

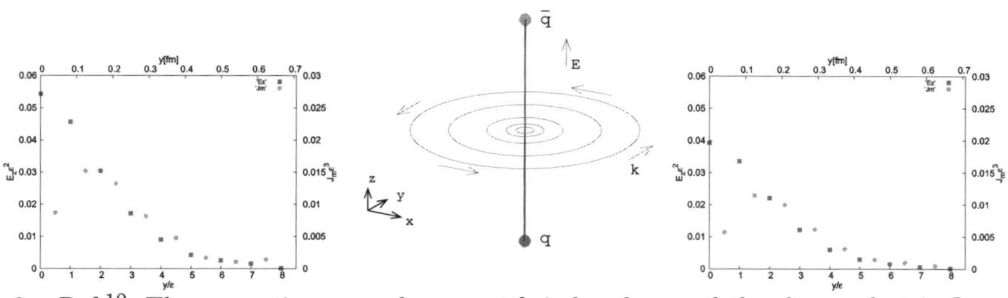

Fig. 3. Ref.[10]: The magnetic-monopole current **k** induced around the chromoelectric flux along the z axis connecting a pair of quark and antiquark. (Center panel) The positional relationship between the chromoelectric field E_z and the magnetic current **k**. (Left panel) The magnitude of the chromoelectric field E_z and the magnetic current $J_m = |\mathbf{k}|$ as functions of the distance y from the z axis calculated from the original full variables. (Right panel) The counterparts of the left graph calculated from the restricted variables.

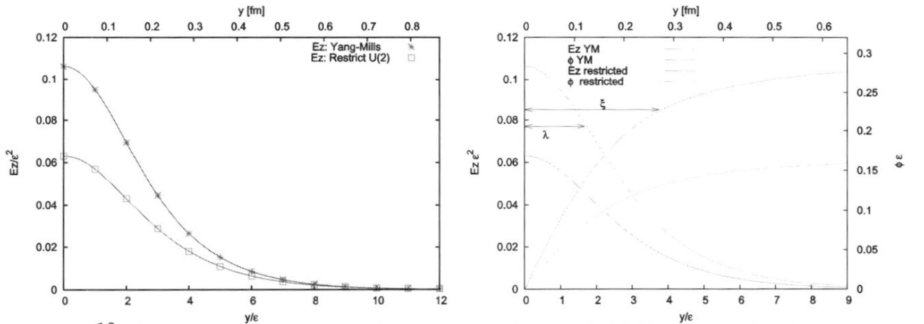

Fig. 4. Ref. [10]: (Left panel) The plot of the chromoelectric field E_z versus the distance y in units of the lattice spacing ϵ and the fitting as a function $E_z(y)$ of y. The red cross for the original $SU(3)$ field and the green square symbol for the restricted field. (Right panel) The order parameter ϕ reproduced as a function $\phi(y)$ of y.

- How to define the **electric-magnetic non-Abelian duality** in the non-Abelian gauge theory?
- How to extract the **infrared dominant mode** for confinement?

In this talk,

- We give a **gauge-invariant definition of (chromo)magnetic monopoles** in the $SU(N)$ Yang-Mills theory (in the absence of the scalar fields) from the non-Abelian Wilson loop operator. This is achieved by using a **non-Abelian Stokes theorem** for the Wilson loop operator [6]. This leads to the **non-Abelian magnetic monopoles** for the Wilson loop operator in the fundamental representation. This definition is independent of the gauge fixing. One does not need to use the conventional prescription called the **Abelian projection** proposed by 't Hooft (1981) [7] which realizes magnetic monopoles by a partial gauge fixing as **gauge-fixing defects**.

In fact, the validity of this definition has been confirmed by numerical simulations on a lattice as follows.

- The magnetic monopole reproduces the linear potential with almost the same string tension $\sigma_{\rm mono}$ as the original one $\sigma_{\rm full}$: 85% for $SU(2)$ [8] 80% for $SU(3)$ [9] This is called the **magnetic monopole dominance** in the string tension.
- The **dual Meissner effect** occur in $SU(N)$ Yang–Mills theory as signaled by the simultaneous formation of the **chromoelectric flux tube** and the **associated magnetic-monopole current** induced around it. Only the component E_z of the chromoelectric field $(E_x, E_y, E_z) = (F_{14}, F_{24}, F_{34})$ connecting q and $\bar{q}$ has non-zero value. The other components are zero consistently within the numerical errors. Therefore, the chromomagnetic field $(B_x, B_y, B_z) = (F_{23}, F_{31}, F_{12})$ connecting q and $\bar{q}$ does not exist. The magnitude of the chromoelectric field E_z decreases quickly as the distance y in the direction perpendicular to the line increases. Therefore, we have confirmed the formation of the chromoelectric flux in Yang–Mills theory on a lattice [8,10].
- We have also shown that the restricted field V reproduces the dual Meissner effect in the $SU(N)$ Yang–Mills theory on a lattice [8,10].

The superconductor is characterized by the penetration depth δ, the coherence length ξ, and the Ginzburg-Landau (GL) parameter κ defined by

$$\kappa := \frac{\delta}{\xi} \begin{cases} < \frac{1}{\sqrt{2}} & \text{(type I)} \\ = \frac{1}{\sqrt{2}} & \text{(BPS)} \\ > \frac{1}{\sqrt{2}} & \text{(type II)} \end{cases} . \tag{1}$$

In the type-I superconductor, the attractive force acts between two flux tubes, while the repulsive force in the type-II superconductor. There is no interaction at $\kappa = \frac{1}{\sqrt{2}} \simeq 0.707$.

In the Abelian-Higgs model, two vortices are attractive in the type I, repulsive for the type II, and there are no force between two vortices in the BPS limit. In the Abelian case, therefore, the ordinary **superconductivity** must be type II or BPS. Otherwise, the initial configuration of the vortices will be eventually collapsed by the attractive force among the vortices. In fact, this is consistent with a fact that the vortices of magnetic flux tube form the **Abrikosov lattice** (hexagonal lattice rather than forming a square lattice) in the superconductor as stable configuration, which is verified experimentally.

In the non-Abelian case, recent numerical simulations show that the dual superconductivity of the Yang-Mills vacuum is **type I**, in contrast to the preceding results which claim the border between type I and type II, i.e., BPS limit.
For $SU(2)$, Kato et al.(2014) [8] reports the GL parameter, the penetration depth, the coherence length:

$$\kappa = 0.48 \pm 0.07,\ \delta = 0.12\text{fm},\ \xi = 0.25\text{fm}, (m_A = 1.64\text{GeV},\ m_\phi = 1.1\text{GeV}) \tag{2}$$

For $SU(3)$, Shibata et al.(2013) [10]. reports

$$\kappa = 0.45 \pm 0.01,\ \delta = 0.12\text{fm},\ \xi = 0.27\text{fm}, (m_A = 1.64\text{GeV},\ m_\phi = 1.0\text{GeV}) \quad (3)$$

These results are obtained using the novel reformulation of the Yang-Mills theory, which is Cho-Duan-Ge decomposition[11–17] for $SU(2)$, and Cho-Faddeev-Niemi decomposition[18–20] for $SU(3)$.

2. Non-Abelian vortex

The **type I dual superconductivity** for the Yang-Mills vacuum yields the attractive force between two non-Abelian vortices. How this result is consistent with the above considerations? The crucial point is that the **non-Abelian vortices** have internal degrees of freedom, i.e., **orientational moduli**, in addition to the degrees of freedom related to the positions in space. This has a possibility for preventing the vortices from collapse due to the attractive force. Whereas the Abelian vortices have only the positions and the collapse of the lattice structure for the Abelian vortices in the type I superconductor will not be avoidable. We must examine the interaction among vortices depending on the orientational moduli in more detail. This analysis gives an estimate of the string tension based on the vortex condensation picture, and possible interactions between two non-Abelian vortices.

We discuss how the **magnetic monopole condensation** picture are compatible with the **vortex condensation** picture as another promising scenario for quark confinement. As a dual or complementary point of view, we can define a gauge-equivalent thin vortex (non-orientable) which has the magnetic monopole and the anti-magnetic monopole as the boundaries [21].

We want to construct the non-Abelian vortex which ends at the non-Abelian magnetic monopole, that is to say, the non-Abelian orientational zero modes of the vortex endow the endpoint non-Abelian magnetic monopole and antimonopole with same CP^{N-1} zero modes.

Such a vortex will be obtained in the Higgs phase of a $U(N)$ gauge theory with $SU(N)$ flavor symmetry where the vortices with non-Abelian CP^{N-1} orientational zero modes exist. This is a candidate of the dual gauge theory for describing the magnetic flux tube as a vortex solution. This is different from the dual Abelian-Higgs model with the magnetic gauge symmetry $U(1)^{N-1}$ suggested from the Abelian projection.

3. Vortex picture towards the area law

Then we discuss how such magnetic monopoles can give a nontrivial contribution to the Wilson loop operator for understanding the area law of the Wilson loop average.

The vortex condensation picture gives an easy way to understand the area law. Let us assume that the vacuum is filled with percolating thin vortices. Suppose that N random vortices intersect a plane of area L^2. For simplicity, we consider the $SU(2)$ gauge group in what follows. Each intersection multiplicatively contributes a factor $(-1)^{2J}$ to the Wilson loop average and n intersect within the loop take the value $(-1)^{2Jn}$. Then the probability that n of the intersections occur within an area S spanned by a Wilson loop is given by $(-1)^{2Jn}\left(\frac{S}{L^2}\right)^n(+1)^{N-n}\left(1-\frac{S}{L^2}\right)^{N-n}$.

Summing over all possibilities with the proper binomial weight yields

$$\begin{aligned}
W_C &= \sum_{n=0}^{N}\binom{N}{n}(-1)^{2Jn}\left(\frac{S}{L^2}\right)^n(+1)^{N-n}\left(1-\frac{S}{L^2}\right)^{N-n} \\
&= \left(1-\frac{S}{L^2}+(-1)^{2J}\frac{S}{L^2}\right)^N \\
&= \left(1-\frac{[1-(-1)^{2J}]\rho S}{N}\right)^N \to \exp\{-\sigma_J S\} \quad (N\to\infty), \quad \rho := \frac{N}{L^2},
\end{aligned} \tag{4}$$

where

$$\sigma_J = [1-(-1)^{2J}]\rho = \begin{cases}\sigma_F = 2\rho & (J=\frac{1}{2},\frac{3}{2},\cdots) \\ 0 & (J=1,2,\cdots)\end{cases}, \tag{5}$$

where L has been eliminated in favor of the planar vortex density $\rho := N/L^2$. The limit of a large $N\to\infty$ is taken with a constant ρ. Thus one obtains an area law for the Wilson loop average with the string tension σ_J determined by the vortex density ρ.

The crucial assumption in this argument is the independence of the intersection points. The asymptotic string tensions are zero for all integer-J representations (with N-ality or "biality" being equal to 0), while they are nonzero and equal for all half-integer J representations (with N-ality or "biality" being equal to 1). For the details, see Ref. [22].

4. Conclusion and discussion

(1) [Magnetic monopole picture] We have given a gauge-invariant definition of the magnetic monopole in the $SU(N)$ pure Yang-Mills theory in the absence of the scalar field. This is achieved through a non-Abelian Stokes theorem for the Wilson loop operator. This enables one to estimate the magnetic monopole contribution to the Wilson loop average, which confirms the magnetic monopole dominance in the string tension. See Shibata's talk for the numerical simulations on a lattice.

(2) [Vortex picture] We have discussed the vortex solution in the Higgs phase of a gauge-Higgs model with $U(N)$ gauge fields and N Higgs fields. The model fulfills the requirement that a non-Abelian vortex has the same $\boldsymbol{CP}^{N-1}$ moduli space as those of the non-Abelian magnetic monopole. The non-Abelian magnetic monopole is regarded as a kink making a junction with N non-Abelian vortices with different internal orientation moduli. The internal moduli will be important to understand the type I dual superconductivity.

Acknowledgement

This work is supported by Grant-in-Aid for Scientific Research (C) 24540252 from the Japan Society for the Promotion Science (JSPS), and also in part by JSPS Grant-in-Aid for Scientific Research (S) 22224003. This work is in part supported by the Large Scale Simulation Program No.09-15 (FY2009), No.T11-15 (FY2011), No.12/13-20 (FY2012-2013) and No.13/14-23 (FY2013-2014) of High Energy Accelerator Research Organization (KEK).

References

1. Y. Nambu, Phys. Rev. D**10**, 4262–4268 (1974).
   G. 't Hooft, in: High Energy Physics, edited by A. Zichichi (Editorice Compositori, Bologna, 1975).
   S. Mandelstam, Phys. Report **23**, 245–249 (1976).
2. A.M. Polyakov, Phys. Lett. B**59**, 82–84 (1975). A.M. Polyakov, Nucl. Phys. B**120**, 429–458 (1977).
3. M.N. Chernodub, M.I. Polikarpov, [hep-th/9710205]
4. J. Greensite, Prog. Part. Nucl. Phys. **51**, 1–83 (2003). [hep-lat/0301023]
5. K.-I. Kondo, S. Kato, A. Shibata and T. Shinohara, Physics Report **579**, 1–226 (2015). arXiv:1409.1599 [hep-th]
6. K.-I. Kondo, Phys. Rev. D **77**, 085029 (2008). arXiv:0801.1274 [hep-th]
7. G. 't Hooft, Nucl.Phys. B**190** [FS3], 455–478 (1981).
8. S. Kato, K.-I. Kondo, and A. Shibata, Phys. Rev. D**91**, 034506 (2015). arXiv:1407.2808 [hep-lat]
9. K.-I. Kondo, A. Shibata, T. Shinohara, and S. Kato, Phys. Rev. D**83**, 114016 (2011). arXiv:1007.2696 [hep-th]
10. A. Shibata, K.-I. Kondo, S. Kato and T. Shinohara, Phys. Rev. D**87**, 054011 (2013). arXiv:1212.6512 [hep-lat]
11. Y.S. Duan and M.L. Ge, Sinica Sci., **11**, 1072–1081 (1979).
12. Y.M. Cho, Phys. Rev. D**21**, 1080–1088 (1980). Phys. Rev. D**23**, 2415–2426 (1981).
13. L. Faddeev and A.J. Niemi, Phys. Rev. Lett. **82**, 1624–1627 (1999). [hep-th/9807069]
    L.D. Faddeev and A.J. Niemi, Nucl. Phys. B**776**, 38-65 (2007). [hep-th/0608111]
14. S.V. Shabanov, Phys. Lett. B**458**, 322–330 (1999). [hep-th/0608111]
    S.V. Shabanov, Phys. Lett. B**463**, 263–272 (1999). [hep-th/9907182]
15. K.-I. Kondo, T. Murakami and T. Shinohara, Prog. Theor. Phys. **115**, 201–216 (2006). [hep-th/0504107]
16. K.-I. Kondo, T. Murakami and T. Shinohara, Eur. Phys. J. C**42**, 475–481 (2005). [hep-th/0504198]
17. K.-I. Kondo, Phys. Rev. D**74**, 125003 (2006). [hep-th/0609166]
18. K.-I. Kondo, T. Shinohara and T. Murakami, Prog. Theor. Phys. **120**, 1–50 (2008). arXiv:0803.0176 [hep-th]
19. Y.M. Cho, Unpublished preprint, MPI-PAE/PTh 14/80 (1980).
    Y.M. Cho, Phys. Rev. Lett. **44**, 1115–1118 (1980).
20. L. Faddeev and A.J. Niemi, Phys. Lett. B**449**, 214–218 (1999). [hep-th/9812090]
    L. Faddeev and A.J. Niemi, Phys. Lett. B**464**, 90–93 (1999). [hep-th/9907180],
    T.A. Bolokhov and L.D. Faddeev, Theoretical and Mathematical Physics, **139**, 679–692 (2004).
21. K.-I. Kondo, J. Phys. G: Nucl. Part. Phys. **35**, 085001 (2008). arXiv:0802.3829 [hep-th]
22. K.-I. Kondo, T. Sasago and T. Shinohara, in preparation.

Confinement/Deconfinement Phase Transition in SU(3) Yang-Mills Theory in View of Dual Superconductivity

Akihiro Shibata[a], Kei-Ichi Kondo[b], Seikou Kato[c] and Toru Shinohara[b]

[a] *Computing Research Center, High Energy Accelerator Research Organization (KEK), Tsukuba 305-0801, Japan*
[b] *Department of Physics, Graduate School of Science, Chiba University, Chiba 263-8522, Japan*
[c] *Oyama National College of Technology, Oyama, Tochigi 323-0806, Japan*

In the preceeding works, we have given a non-Abelian dual superconductivity picture for quark confinement, and demonstrated the numerical evidences on the lattice. In this talk, we discuss the confinement and deconfinement phase transition at finite temperature in view of the dual superconductivity. We investigate chromomagnetic monopole currents induced by chromoelectric flux in both confinement and deconfinement phase by the numerical simulations on a lattice at finite temperature, and discuss the role of the chromomagnetic monopole in the confinement/deconfinement phase transition.

Keywords: Quark confinement, Dual superconductivity, Dual Meissner effect, Phase transition.

1. Introduction

The dual superconductivity is a promising mechanism for quark confinement.[2] In order to establish this picture, we have to show evidences of the dual version of the superconductivity. For this purpose, we have presented a new formulation of the Yang-Mills theory and shown the numerical evidences on a lattice: the non-Abelian magnetic monopole dominantly reproduces the string tension in the linear potential in SU(3)Yang-Mills theory, and the SU(3) Yang-Mills vacuum is the type I dual superconductor profiled by the chromoelectric flux tube and the magnetic monopole current induced around it, which is a novel feature obtained by our simulations.[1 4 5]

In this talk, we further study the confinement and deconfinement phase transition at finite temperature in view of the dual superconductivity. We introduce a new formulation of the Yang-Mills theory on a lattice, and investigate confinement/deconfinement phase transition at finite temperature by using the new variable which extracts the dominant mode of the quark as well as original Yang-Mills fields. We first measure the space-averaged Polyakov-loop for each configuration and the Polyakov-loop average to investigate the role of the new variable at finite temperature. We then measure chromo fluxes and induced magnetic-monopole currents induced by a pair of quark and antiquark source to investigate the dual Meissner effect. We will demonstrate confinement/deconfinement phase transition in view of the non-Abelian dual superconductivity picture.

2. Gauge-link decomposition

We introduce a new formulation of the lattice Yang-Mills theory in the minimal option, which extracts the dominant mode of the quark confinement for $SU(3)$ Yang-Mills theory[14], since we consider the quark confinement in the fundamental representation. Let $U_{x,\mu} = X_{x,\mu}V_{x,\mu}$ be a decomposition of the Yang-Mills link variable $U_{x,\mu}$, where $V_{x,\mu}$ could be the dominant mode for quark confinement, and $X_{x,\mu}$ the remainder part. The Yang-Mills field and the decomposed new variables are transformed by full $SU(3)$ gauge transformation Ω_x such that $V_{x,\mu}$ is transformed as the gauge link variable and $X_{x,\mu}$ as the site variable:

$$U_{x,\mu} \longrightarrow U'_{x,\nu} = \Omega_x U_{x,\mu} \Omega^{\dagger}_{x+\mu}, \tag{1a}$$

$$V_{x,\mu} \longrightarrow V'_{x,\nu} = \Omega_x V_{x,\mu} \Omega^{\dagger}_{x+\mu}, \quad X_{x,\mu} \longrightarrow X'_{x,\nu} = \Omega_x X_{x,\mu} \Omega^{\dagger}_{x}. \tag{1b}$$

The decomposition is given by solving the defining equation:

$$D^{\epsilon}_{\mu}[V]\mathbf{h}_x := \frac{1}{\epsilon}\left[V_{x,\mu}\mathbf{h}_{x+\mu} - \mathbf{h}_x V_{x,\mu}\right] = 0, \tag{2a}$$

$$g_x := e^{i2\pi q/3} \exp(-ia^0_x \mathbf{h}_x - i\sum\nolimits_{j=1}^{3} a^{(j)}_x \mathbf{u}^{(j)}_x) = 1, \tag{2b}$$

where $\mathbf{h}_x$ is an introduced color field $\mathbf{h}_x = \xi_x(\lambda^8/2)\xi^{\dagger}_x \in [SU(3)/U(2)]$ with λ^8 being the Gell-Mann matrix and ξ_x an $SU(3)$ group element. The variable g_x is an undetermined parameter from Eq.(2a), $\mathbf{u}^{(j)}_x$'s are $su(2)$-Lie algebra valued, and q_x has an integer value $0, 1, 2$. These defining equations can be solved exactly[3], and the solution is given by

$$X_{x,\mu} = \widehat{L}^{\dagger}_{x,\mu} \det(\widehat{L}_{x,\mu})^{1/3} g_x^{-1}, \qquad V_{x,\mu} = X^{\dagger}_{x,\mu} U_{x,\mu} = g_x \widehat{L}_{x,\mu} U_{x,\mu}, \tag{3a}$$

$$\widehat{L}_{x,\mu} = \left(L_{x,\mu} L^{\dagger}_{x,\mu}\right)^{-1/2} L_{x,\mu}, \tag{3b}$$

$$L_{x,\mu} = \frac{5}{3}\mathbf{1} + \frac{2}{\sqrt{3}}(\mathbf{h}_x + U_{x,\mu}\mathbf{h}_{x+\mu}U^{\dagger}_{x,\mu}) + 8\mathbf{h}_x U_{x,\mu}\mathbf{h}_{x+\mu}U^{\dagger}_{x,\mu} \ . \tag{3c}$$

Note that the above defining equations correspond to the continuum version

The decomposition is uniquely obtained as the solution (3) of Eqs.(2), if color fields$\{\mathbf{h}_x\}$ are obtained. To determine the configuration of color fields, we use the reduction condition to formulate the new theory written by new variables $(X_{x,\mu}, V_{x,\mu})$ which is equipollent to the original Yang-Mills theory. Here, we use the reduction functional:

$$F_{\text{red}}[\mathbf{h}_x] = \sum_{x,\mu} \text{tr}\left\{(D^{\epsilon}_{\mu}[U_{x,\mu}]\mathbf{h}_x)^{\dagger}(D^{\epsilon}_{\mu}[U_{x,\mu}]\mathbf{h}_x)\right\}, \tag{4}$$

and then color fields $\{\mathbf{h}_x\}$ are obtained by minimizing the functional (4).

3. Lattice result

We generate the Yang-Mills gauge field configurations (link variables) $\{U_{x,\mu}\}$ for the standard Wilson action. We prepare data sets for finite temperature on the lattice

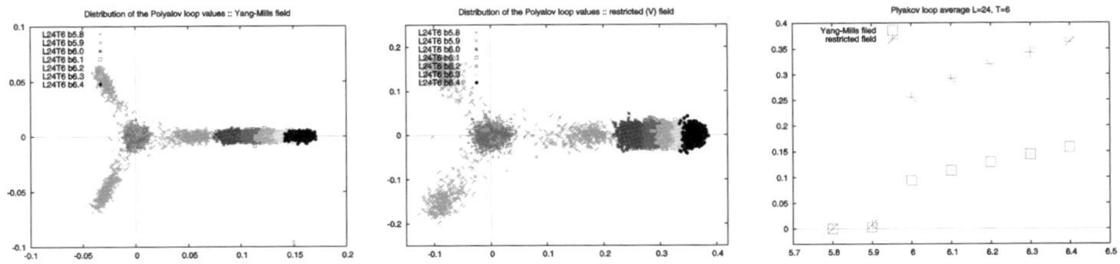

Fig. 1. The distribution of the space-averaged Polyakov loop on the complex plane: (Left) P_U obtained from a set of the original gauge field configurations, (Center) P_V from the restricted field. (Right) The Polyakov-loop average versus temperature. The red and green represent the Polyakov-loop average for the original field U and the restricted field V, respectively.

of size $L^3 \times N_T$ at finite temperature by using the pseudo heat bath algorithm. For the fixed spatial size L and the temporal size N_T: $L = 24$, $N_T = 6$: the temperature varies by changing the inverse gauge coupling constant $\beta = 2N_c/g^2$ ($N_c = 3$): $\beta = 5.8, 5.9, 6.0, 6.1, 6.2, 6.3$. In our simulations, we use the cold start to obtain the real-valued Polyakov loop average $\langle P \rangle$ at high temperature, see Fig. 1. We thermalized 8000 sweeps, and we used 500 configurations for measurements.

We perform the decomposition of the gauge link variable $U_{x,\mu} = X_{x,\mu}V_{x,\mu}$ by using the formula (3) given in the previous section, after the color-field configuration $\{\mathbf{h}_x\}$ is obtained by solving the reduction condition of minimizing the functional (4) for each set of the gauge field configurations $\{U_{x,\mu}\}$. In the measurement of the Polyakov loop average and the Wilson loop average defined below, we apply the APE smearing technique to reduce noises.

3.1. *Polyakov-loop average in the confinement/deconfinement transition*

First, we measure the Polyakov-loop average which is a conventional order parameter for detecting the confinement and deconfinement phase transition in the pure Yang-Mills theory. We define the space-averaged Polyakov loop (i.e., the value of the Polyakov loop which is averaged over the space volume) for a set of the original gauge field configurations $\{U_{x,\mu}\}$ and the restricted gauge field (V-field) configurations $\{V_{x,\mu}\}$:

$$P_U := L^{-3} \sum_{\vec{x}} \mathrm{tr}\left(\prod_{t=1}^{N_T} U_{(\vec{x},t),4}\right), \quad P_V := L^{-3} \sum_{\vec{x}} \mathrm{tr}\left(\prod_{t=1}^{N_T} V_{(\vec{x},t),4}\right). \tag{5}$$

Left and center panel of Fig. 1 show the plots of P_U (left panel) and P_V (center panel) on the complex plane measured from the original gauge field configurations and the restricted gauge field configurations respectively. Notice that the Polyakov loop average is in general complex-valued for the $SU(3)$ group.

Then, we measure the Polyakov-loop average $\langle P_U \rangle$ and $\langle P_V \rangle$ obtained by averaging the space-averaged Polyakov loop over the total sets of the original gauge field configurations and the restricted gauge field configurations respectively. Note that

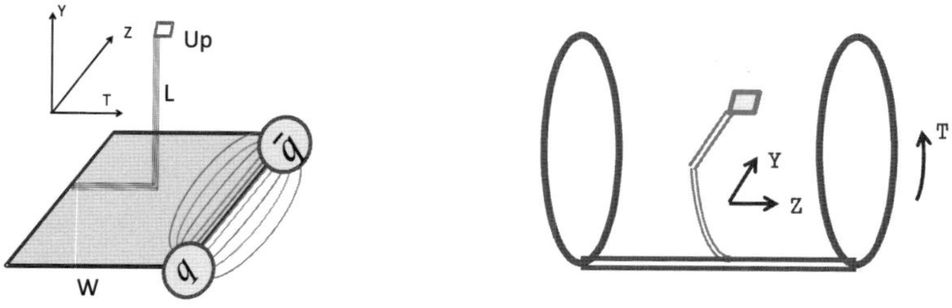

Fig. 2. The setup of the measurement of the chromo-field strength distribution. (Left) The gauge-invariant operator $\mathrm{tr}(WLU_pL^\dagger)$ between a plaquette U_p and the Wilson loop W. (Right) Measurement of the chromo-flux at finite temperature.

the symbol $\langle \mathcal{O} \rangle$ denotes the average of the operator $\mathcal{O}$ over the space and the ensemble of the configurations. Right panel of Fig. 1 shows the result for the Polyakov loop average for various temperature (β). We find that the behavior of $\langle P_U \rangle$ and $\langle P_V \rangle$ give the same critical temperature for the phase transition separating the low-temperature confined phase characterized by the vanishing Polyakov loop average $\langle P_U \rangle = \langle P_V \rangle = 0$ from the high-temperature deconfined phase characterized by the non-vanishing Polyakov loop average $\langle P_U \rangle \neq 0$ and $\langle P_V \rangle \neq 0$.

3.2. *Chromo-flux tube at finite temperature*

We proceed to investigate the non-Abelian dual Meissner effect at finite temperature. For this purpose, we measure the chromo-flux at finite temperature created by a quark-antiquark pair which is represented by the maximally extended Wilson loop W defined in Fig. 2. The chromo-field strength, i.e., the field strength of the chromo-flux at the position P is measured by using a plaquette variable U_p as the probe operator for the field strength. See Fig. 2. We use the same gauge-invariant correlation function as that used at zero temperature[6]:

$$\rho_{U_P} := \frac{\langle \mathrm{tr}\left(WLU_pL^\dagger\right)\rangle}{\langle \mathrm{tr}\,(W)\rangle} - \frac{1}{N_c}\frac{\langle \mathrm{tr}\,(U_p)\,\mathrm{tr}\,(W)\rangle}{\langle \mathrm{tr}\,(W)\rangle}, \tag{6}$$

where L is the Wilson line connecting the source W and the probe U_p needed to obtain the gauge-invariant result. Note that ρ_{U_P} is sensitive to the field strength rather than the disconnected one. Notice that the setup in the right figure of Fig. 2 is different from the correlation function for a pair of the Polyakov loop $P(z = 0)$ and the anti-Polyakov loop $\bar{P}(z = R)$ where each Polyakov loop is defined by the corresponding closed loop which is obtained by identifying the end points at $\tau = 0$ and $\tau = 1/T$ by the periodic boundary condition, and the probe U_P is attached to one the Polyakov loop or both the Polyakov and anti-Polyakov loops. Such operator was recently used to measure the chromo-flux in the work[9]. It should be remarked that the Polyakov loop correlation function, $\langle P_U(\vec{x})P_U^*(\vec{y})\rangle \simeq e^{-F_{q\bar{q}}/T}$, proportional to the partition function in the presence of a quark at $\vec{x}$ and an anti-quark at $\vec{y}$ is decomposed into the singlet and the adjoint combinations in the color space.

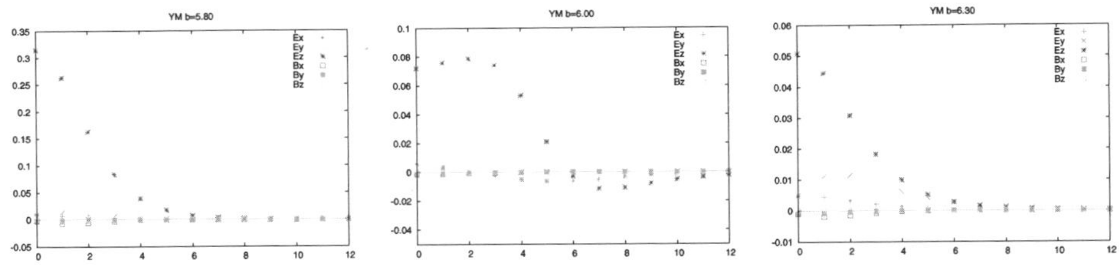

Fig. 3. Chromo-field strength of the original Yang-Mills field created by the quark-antiquark pair at finite temperature: (left) $\beta = 5.80$ (center) $\beta = 6.00$ (right) $\beta = 6.30$ on the lattice $24^3 \times 6$.

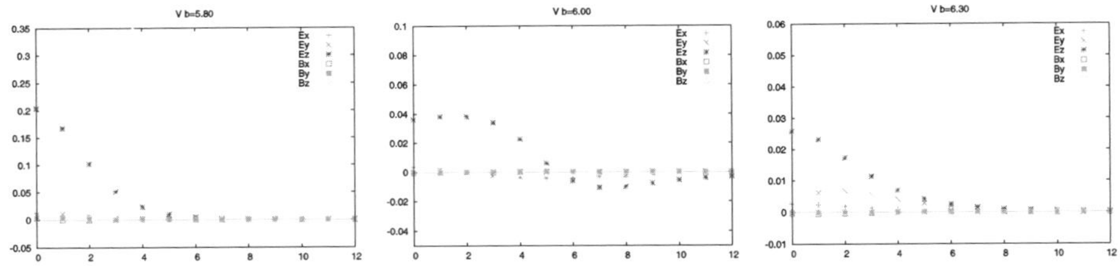

Fig. 4. Chromo-field strength of the restricted field created by the quark-antiquark pair at finite temperature: (left) $\beta = 5.80$ (center) $\beta = 6.00$ (right) $\beta = 6.30$ on the lattice $24^3 \times 6$.

Furthermore, the decomposed component is gauge-dependent and thus should be taken with care.[7,8] In sharp contrast to this fact, the potential obtained from the Wilson loop is the color singlet in the gauge-independent way.

Figures 3 and 4 show the results of the measurement of chromo-field strength at different temperatures obtained from the data set for the original Yang-Mills field and the restricted field (V-field), respectively. In the low temperature confined phase $T < T_c$ (see left panels of Fig.3 and Fig.4), we observe that only the E_z component of the chromoelectric flux tube, i.e. the flux in the direction connecting a quark and antiquark pair is non-vanishing, and that the other components take vanishing values. This is consistent with the result obtained by Cea et al.[9], although they use the different operator for measuring the flux. In the high temperature deconfined phase $T > T_c$ (see right panels of Fig. 3 and Fig. 4), we observe the non-vanishing E_y component in the chromoelectric flux, which means no more squeezing of the chromoelectric flux tube. This shows the disappearance of the dual Meissner effect in the high temperature deconfined phase.

3.3. *Magnetic–monopole current and dual Meissner effect at finite temperature*

Finally, we investigate the dual Meissner effect by measuring the magnetic–monopole current k induced around the chromo-flux tube created by the quark-

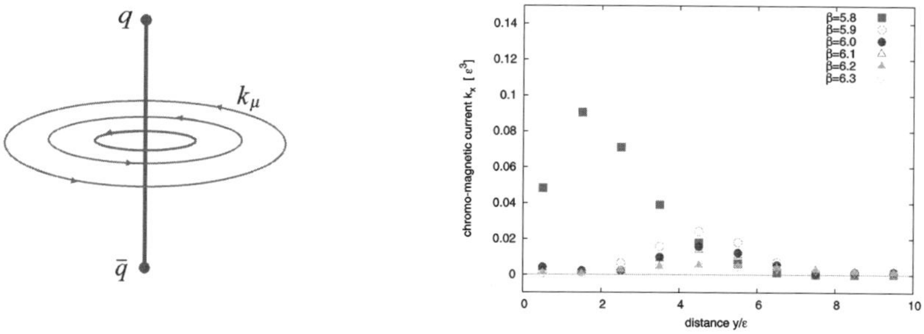

Fig. 5. (Left) Sketch of the induced magnetic-monopole current by a pair of quark and antiquark. (Right) The magnitude $\sqrt{k_\mu k_\mu}$ of the induced magnetic current k_μ around the flux tube connecting the quark-antiquark pair as a function of the distance y from the axis z for various values of β, i.e., temperature.

antiquark pair. We use the the magnetic-monopole current k defined by

$$k_\mu(x) = \frac{1}{2}\epsilon_{\mu\nu\alpha\beta}\left(F[V]_{\alpha\beta}(x+\hat{\nu}) - F[V]_{\alpha\beta}(x)\right). \tag{7}$$

Note that the magnetic–monopole current (7) must vanish due to the Bianchi identity, if there exist no singularity in the gauge potential. We show that the magnetic–monopole current defined in this way can be the order parameter for the confinement/deconfinement phase transition, as suggested from the dual superconductivity hypothesis. Figure 5 shows the result of the measurements of the magnitude $\sqrt{k_\mu k_\mu}$ of the induced magnetic current k_μ obtained according to (7). We observe the appearance and disappearance of the magnetic monopole current in the low temperature phase and high temperature phase, respectively.

4. Summary and outlook

Using a new formulation of the Yang-Mills theory on a lattice, we have examined the confinement/deconfinement phase transition and the (non-Abelian) dual superconductivity in the $SU(3)$ Yang-Mills theory at finite temperature. The reformulation enables one to extract the dominant mode for quark confinement. Indeed, we have extracted the restricted field (V-field) from the original Yang-Mills field which plays a dominant role in confining quark in the fundamental representation at finite temperature.

First, we have given the numerical evidences for the restricted field dominance in the Polyakov loop average P in the sense that the Polyakov loop average P_V written in terms of the restricted field V gives the same critical temperature T_c as that detected by the Polyakov loop average P_U written in terms of the original gauge field U: $P=0$ for $T<T_c$ and $P\neq 0$ for $T>T_c$.

However, the Polyakov loop average cannot be the direct signal of the dual Meissner effect or magnetic monopole condensation. Therefore, it is important to find an order parameter which enables one to detect the dual Meissner effect directly.

In view of these, we have measured the chromoelectric and chromomagnetic flux for both the original field and the restricted field. In the low–temperature confined phase $T < T_c$, we have obtained the numerical evidences of the dual Meissner effect in the $SU(3)$ Yang-Mills theory, i.e., the squeezing of the chromoelectric flux tube created by a quark-antiquark pair and the associated magnetic–monopole current induced around the flux tube. In the high–temperature deconfined phase $T > T_c$, on the other hand, we have observed the disappearance of the dual Meissner effect, no more squeezing of the chromoelectric flux tube detected by non-vanishing E_y component in the chromoelectric flux and the vanishing of the magnetic-monopole current associated with the chromo-flux tube. These results are also obtained by the restricted field alone. Therefore, we have confirmed the restricted field dominance in the dual Meissner effect even at finite temperature. Thus, we have given the evidences that the confinement/deconfinement phase transition is caused by appearance/disappearance of the non-Abelian dual superconductivity.

Acknowledgement

This work is supported by Grant-in-Aid for Scientific Research (C) 24540252 and 15K05042 from Japan Society for the Promotion Science (JSPS), and in part by JSPS Grant-in-Aid for Scientific Research (S) 22224003. The numerical calculations are supported by the Large Scale Simulation Program No.13/14-23 (2013-2014) and No.14/15-24 (2014-2015) of High Energy Accelerator Research Organization (KEK).

References

1. Kei-Ichi Kondo, Seikou Kato, Akihiro Shibata and Toru Shinohara, Phys.Rept. 579 (2015) 1-226
2. Y. Nambu, Phys. Rev. D10, 4262(1974); G. 't Hooft, in High Energy Physics, edited by A. Zichichi (Editorice Compositori, Bologna, 1975); S. Mandelstam, Phys. Report 23, 245(1976); A.M. Polyakov, Nucl. Phys. B120, 429(1977).
3. A. Shibata, K.-I. Kondo and T. Shinohara, Phys.Lett.B691:91-98 (2010)
4. K.-I. Kondo, A. Shibata, T. Shinohara, S. Kato, Phys.Rev. D83 (2011) 114016
5. A. Shibata, K.-I. Kondo, S. Kato and T. Shinohara, Phys.Rev. D87 (2013) 5, 05401
6. A. Di Giacomo, M. Maggiore, and S. Olejnik, Phys. Lett. B236, 199 (1990); Nucl. Phys. B347, 441 (1990).
7. K. Yagi, T. Hatsuda and Y. Miake, *Quark-Gluon Plasma* (Cambridge Univ. press, Cambridge, 2005).
8. L.S. Brown and W.I. Weisberger, Phys.Rev. D20 (1979) 3239
9. P. Cea, L. Cosmai, F. Cuteri, and A.Papa, Phys. Rev. D89, 094505 (2014)

Linking $U(2) \times U(2)$ to $O(4)$ via Decoupling

Norikazu Yamada

KEK Theory Center, Institute of Particle and Nuclear Studies, High Energy Accelerator Research Organization (KEK), Tsukuba 305-0801, Japan

The nature of chiral phase transition of massless two flavor QCD depends on the fate of flavor singlet axial symmetry $U_A(1)$ at the critical temperature (T_c). Assuming that a finite $U_A(1)$ breaking remains at T_c, the corresponding three dimensional effective theory is composed of four massless and four massive scalar fields. We study the renormalization group flow of the effective theory in the ϵ-expansion, using a mass dependent renormalization scheme, and determine the region of the attractive basin flowing into the $O(4)$ fixed point with a focus on its dependence on the size of the $U_A(1)$ breaking. The result is discussed from a perspective of the decoupling of massive fields. It is pointed out that, although the effective theory inside the attractive basin eventually reaches the $O(4)$ fixed point, the approaching rate, one of the universal exponents, is different from that of the standard $O(4)$ model. We present the reason for this peculiarity, and propose a novel possibility for chiral phase transition in two-flavor QCD.

Walking Dynamics from Gauge-Gravity Duality

Daniel Elander

Department of Physics, Purdue University, 525 Northwestern Avenue, West Lafayette, IN 47907-2036, USA

We will discuss solutions in Type IIB supergravity corresponding to either a system of wrapped D5-branes or a system of D3- and D5-branes. These can be thought of as arising as deformations of the Maldacena-Nunez or Klebanov-Strassler backgrounds. We will argue that the corresponding dual field theories exhibit walking dynamics. Computing the spectrum of scalar glueballs, we find a light state, the mass of which is suppressed by the length of the walking region, thus suggesting it is a dilaton.

Holography and the Conformal Window in the Veneziano Limit

M. Järvinen

Laboratoire de Physique Théorique, École Normale Supérieure &
Institut de Physique Théorique Philippe Meyer,
24 rue Lhomond, 75231 Paris Cedex 05, France
E-mail: jarvinen@lpt.ens.fr

We discuss holographic QCD in the Veneziano limit (the V-QCD models), concentrating on phenomena near the "conformal" phase transition taking place at a critical value of the ratio $x \equiv N_f/N_c$. In particular, we review the results for the S-parameter, the technidilaton, and the masses of the mesons.

Keywords: Holography, QCD, Veneziano limit, Conformal window, Walking.

1. Introduction and motivation

For QCD in the Veneziano limit [1],

$$N_c \to \infty\ , \quad N_f \to \infty\ , \quad x \equiv \frac{N_f}{N_c}\ \text{fixed}\ , \quad g^2 N_c\ \text{fixed}\ , \tag{1}$$

the number of quarks ($\sim N_f N_c$) and the number of gluons ($\sim N_c^2$) are comparable. Consequently, the flavor degrees of freedom are fully dynamical and there is a nontrivial interplay between the glue and flavor sectors of the theory. This backreaction of the quarks to the dynamics of the gluons would be absent in the 't Hooft limit, which is the same as (1) except that N_f is kept fixed instead of x such that $N_f \ll N_c$. Therefore studying QCD in the Veneziano limit is more complicated than in the 't Hooft limit, but it is well motivated because there are interesting phenomena which cannot be captured if $N_f \ll N_c$. In particular, the phase diagram of QCD is nontrivial as a function of the variable x which is continuous in the Veneziano limit.

The "standard" expectation for the phase diagram as a function of x (at zero temperature and quark mass) is shown in Fig. 1. Within the interval $0 < x < 11/2 \equiv x_{\rm BZ}$, where the theory is asymptotically free, the following regimes can be identified:

- *The QCD regime* with $0 < x < x_c$, where the infrared (IR) dynamics is similar to ordinary QCD (i.e., a theory with $N_c = 3$ and a few light quarks). That is, the theory is confining and breaks chiral symmetry.
- *The walking regime* with $0 < x < x_c$ and $x_c - x \ll 1$, where the coupling constant of the theory varies very slowly, i.e., "walks", over a large range of energies.

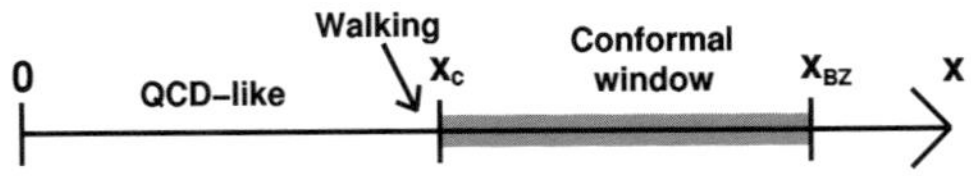

Fig. 1. The conjectured phase diagram of QCD in the Veneziano limit (at zero temperature and quark mass) as a function of $x = N_f/N_c$.

- *The conformal window* with $x_c < x < x_{\rm BZ}$, where the theory runs to an IR fixed point (IRFP).

The existence of the conformal window is guaranteed in the Banks-Zaks (BZ) limit $x \to x_{\rm BZ}$, because the value of the coupling is parametrically small for the entire renormalization group (RG) flow and perturbation theory is trustable. It is also credible that going to large N_c and N_f at small $x \lesssim 1$ does not change the dynamics much, and it is therefore similar to that of ordinary QCD. But the nature of the "conformal transition" at $x = x_c$, and the behavior of the theories near the transition, is much more difficult to study and remains essentially an open question.

The phase diagram of Fig. 1 with a walking regime is obtained in the Dyson-Schwinger approach[2]. The transition is then of the Berezinsky-Kosterlitz-Thouless (BKT) type, associated with the so-called Miransky scaling law. The existence of the walking regime is important, since theories in this region may have properties, which are desirable for technicolor candidates. However, it has also been suggested that the transition is discontinuous[3]. There is an ongoing effort to study the theories near the transition by using first-principles lattice simulations[4–7], but as it turns out, obtaining reliable results is much more difficult than for ordinary QCD.

Here we discuss the conformal transition and related by using holography and working in the Veneziano limit. We concentrate on a specific bottom-up approach, which tries to follow principles from string theory as closely as possible. That is, the model is derived by using five dimensional noncritical string theory with a certain brane configuration. However, as it turns out, some things do not work – in particular the ultra violet (UV) physics is not that of QCD. This can be fixed by modifying the various potentials of the model by hand, therefore switching from a top-down to a bottom-up approach. As the potentials are thereafter not those derived from string theory, the model should be understood as an effective framework which could in principle be used to model more general field theories than just QCD, depending on the choice of potentials. It makes sense to fit the remaining degrees of freedom in the potentials to QCD data, picking a specific model which has physics similar to QCD. When working in the Veneziano limit, this kind of approach leads to the V-QCD models[8] (where "V" stands for Veneziano). The V-QCD models also have the phase diagram of Fig. 1, including a BKT transition and walking.

In more detail, V-QCD is based on two building blocks, the improved holographic QCD (IHQCD)[9,10] for the gluon sector, and a method for adding flavor by inserting a pair of space filling $D4-\overline{D4}$ branes[11,12]. The glue and flavor sectors are fully backreacted in the Veneziano limit. The structure of V-QCD at finite temperature and chemical potential[13–16] as well as the two-point correlators and bound state masses[17–19] have also been analyzed.

In this talk we will first discuss IHQCD and V-QCD in more detail. Then we will give an overview of the results for V-QCD. We will concentrate on the physics related to the conformal transition and the conformal window.

2. Building blocks of V-QCD

2.1. *Improved holographic QCD*

Improved holographic QCD (IHQCD)[9,10] is a model for the Yang-Mills theory, which is loosely based on five dimensional noncritical string theory. The most relevant terms of the IHQCD action, which appear at leading order in the large N_c ('t Hooft) limit, are

$$S_{\mathrm{IHQCD}} = M^3 N_c^2 \int d^5x \sqrt{-\det g}\left(R - \frac{4}{3}(\partial\phi)^2 + V_g(\phi)\right), \tag{2}$$

where the first term defines the five dimensional Einstein gravity and ϕ is the dilaton.

The dilaton is linked to the 't Hooft coupling $g_s^2 N_c$. That is, we denote $\lambda = e^{\phi}$ and identify the vacuum solution $\lambda(r)$ (a function of the bulk coordinate r), which extremizes of the action (2), as the (running) 't Hooft coupling. The Ansatz for the metric is

$$ds^2 = e^{2A(r)}(dx_{1,3}^2 + dr^2)\,. \tag{3}$$

Here A is identified as the logarithm of the energy scale in field theory and can therefore be used to define the holographic RG flow. In our conventions the UV boundary lies at $r = 0$, and the bulk coordinate runs from zero to infinity. The metric will be close to the AdS near the UV ($A \sim -\log(r/\ell)$, where ℓ is the UV AdS radius). Therefore, r corresponds roughly to the inverse of the energy scale of the dual field theory.

As we already pointed out, the model deviates from the "top-down" derivation as the dilaton potential V_g is fixed by hand rather than derived from string theory. Its asymptotic behavior is determined by matching with general properties Yang-Mills theory as follows.

In the UV (i.e, as $\lambda \to 0$), V_g must asymptote to a constant, in order for the geometry to be asymptotically AdS. Moreover, if V_g has an analytic expansion at $\lambda = 0$, the coupling λ turns out to have logarithmic dependence on r in the UV, reminiscent of the RG flow of the coupling in Yang-Mills. Indeed, the holographic beta function defined by the above dictionary

$$\beta = \frac{\lambda'(r)}{A'(r)} \tag{4}$$

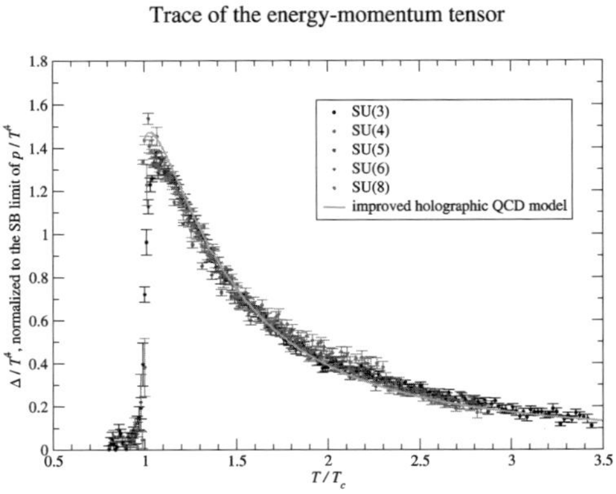

Fig. 2. The interaction measure $(p - 3\epsilon)/T^4$ in IHQCD compared to Yang-Mills lattice data at various number of colors[21].

admits a Taylor expansion at $\lambda = 0$ (with $\lambda = \lambda(r)$ evaluated on the vacuum solution). In order to fix V_g in the UV, the coefficients in its Taylor expansion are then chosen such that the holographic beta function (4) agrees with the perturbative beta function of Yang-Mills up to two-loop order. It may sound weird that a holographic model is matched with perturbative QCD, since holography typically cannot be used easily to weak coupling. This is however the best we can do to include the UV physics of QCD, and should provide as correct boundary conditions as possible for the more interesting IR dynamics in the model.

In the IR (i.e, at large λ), we require the potential to diverges as

$$V_g \sim \lambda^{4/3}\sqrt{\log\lambda}\,, \qquad (\lambda \to \infty)\,. \tag{5}$$

This ensures confinement, a mass gap, discrete spectrum and asymptotically linear glueball trajectories (for radial excitations)[9,10,20]. Interestingly, this asymptotic behavior matches with the behavior predicted by string theory $V_g \sim \lambda^{4/3}$ up to the logarithmic correction, which is important to ensure that the glueballs have exactly linear trajectories.

The behavior of V_g at intermediate couplings, $\lambda = \mathcal{O}(1)$, must be fitted to Yang-Mills data. Most precise constraints are obtained by studying the model at finite temperature[22,23]. An example of a two parameter fit[20] to lattice data for the interaction measure in $SU(3)$ Yang-Mills is given in Fig. 2, compared to lattice data at higher N_c which was computed afterwards[21].

2.2. *Adding flavor*

A framework for adding flavor in holographic models has been introduced in[11,12]. It is realized through adding a pair of $D4 - \overline{D4}$ branes which fill the whole space in a five dimensional holographic model. The string stretching between the branes gives rise to a tachyon field, which is dual to the $\bar{q}q$ operator on the field theory side, and therefore controls chiral symmetry breaking. The gauge fields on the branes are dual to the left and right handed currents $\bar{q}(1 \pm \gamma_5)q$ in QCD.

The dynamics of the flavor is governed by a tachyon-Dirac-Born-Infeld (TDBI) action, and taking a Sen-like exponential potential, $V_f(\tau) = \exp(-\tau^2)$, for the tachyon. This framework has been tested in the probe approximation ($x \to 0$ limit) for IHQCD[10] and in more detail[24] for the Kuperstein-Sonnenschein background[25] with encouraging results. Confining background was seen to trigger chiral symmetry breaking, the pion modes were identified and their mass were shown to obey the Gell-Mann-Oakes-Renner relation, and a good description of the masses of the light mesons in QCD was obtained.

3. V-QCD

The V-QCD models arise as the fusion of IHQCD with the framework for adding flavor, in the Veneziano limit where flavor fully backreacts to the glue[8]. The main degrees of freedom are therefore the two scalars: the "dilaton" $\lambda = e^{\phi}$ and the tachyon τ. The dictionary is as stated above, in particular (the background value of) λ is the 't Hooft coupling, and the tachyon is dual to the $\bar{q}q$ operator.

The terms in the V-QCD action which determine the vacuum solution (for which the gauge fields on the branes vanish) are

$$\begin{aligned} S_{\mathrm{VQCD}} = {} & M^3 N_c^2 \int d^5x \sqrt{-\det g}\left(R - \frac{4}{3}\frac{(\partial\lambda)^2}{\lambda^2} + V_g(\lambda)\right) \\ & - M^3 N_f N_c \int d^5x\, V_f(\lambda,\tau)\sqrt{-\det\left(g_{\mu\nu} + \kappa(\lambda)\partial_\mu\tau\partial_\nu\tau\right)}\,. \end{aligned} \tag{6}$$

The first term is simply the IHQCD action (2) and describes the dynamics of the glue in the model. The second term is a generalized TDBI action, which describes the dynamics of the flavor. From the explicit dependence on N_c and N_f we see that both terms are of the same order in the Veneziano limit and there is indeed full backreaction.

Notice that the form of the TDBI action is only known precisely in the probe limit. Therefore we have introduced a more general Ansatz in (6), which is consistent with the probe limit action. For the tachyon potential we assume the form

$$V_f(\lambda,\tau) = V_{f0}(\lambda)e^{-a(\lambda)\tau^2} \tag{7}$$

where the dependence on the tachyon is the same as in the probe limit approach[12]. The functions $V_{f0}(\lambda)$, $\kappa(\lambda)$, and $a(\lambda)$ would be power laws in the probe limit, but we have taken them to be (at this point) arbitrary functions of λ. This generalizes the approach of IHQCD, where the dilaton potential V_g was taken to be a more complicated function of λ than the power law predicted by string theory analysis, to the case with backreacted flavors.

In order to pin down the model, we need to specify the functions $V_g(\lambda)$, $V_{f0}(\lambda)$, $\kappa(\lambda)$, and $a(\lambda)$. It turns out that the dilaton potential $V_g(\lambda)$ must obey the same asymptotic constraints which were found in IHQCD. For the other functions similar constraints are found, both in the UV and in the IR.

In the UV ($\lambda \to 0$) the potentials must approach constant values in order for the dimension of the $\bar{q}q$ operator to be correctly reproduced, among other things. In analogy to the approach taken in IHQCD, we take the potentials to be analytic at $\lambda = 0$ and match the coefficients of their Taylor expansions with the perturbative QCD. As it turns out, the coefficients of $V_{f0}(\lambda)$ can be mapped to the (x dependent) coefficients of the QCD beta function, and the coefficients of the ratio $a(\lambda)/\kappa(\lambda)$ are mapped to the coefficients of the anomalous dimension of the quark mass $\gamma = -d \log m_q / d \log \mu$. The result from this mapping is that the tachyon has the following asymptotics in the UV ($r \to 0$):

$$\tau(r) \simeq m_q \; r \; (-\log r\Lambda)^{-\gamma_0/\beta_0} + \langle \bar{q}q \rangle \; r^3 \; (-\log r\Lambda)^{\gamma_0/\beta_0} \tag{8}$$

where β_0 and γ_0 are the leading coefficients of the beta and gamma functions, respectively, and the scale of the holographic RG flow Λ corresponds to Λ_{QCD} on the field theory side. The implemented RG flow (the logarithmic terms) is the same as the leading perturbative RG flow of the quark mass and the condensate in QCD. Therefore the UV matching is done at slightly higher level than in the traditional bottom-up models. As we stressed above for IHQCD, it should be understood that this matching is done in order to guarantee correct boundary conditions for the IR dynamics, which can be more reliably modeled by holography.

In the IR ($\lambda \to \infty$), the functions are constrained, among other things, by the asymptotics of the spectra at large excitation numbers, and by requiring that the IR geometry has a "good" kind of IR singularity that uniquely fixes all IR boundary conditions. We find the following asymptotics[18]:

$$\begin{aligned} a(\lambda) &\sim \text{const.} \\ \kappa(\lambda) &\sim \lambda^{-4/3} \sqrt{\log \lambda} \end{aligned} \qquad (\lambda \to \infty) \; . \tag{9}$$

Interestingly, these asymptotics agree with the string theory prediction, up to the logarithmic correction in $\kappa(\lambda)$, in analogy to what was found for $V_g(\lambda)$ in IHQCD. The function $V_{f0}(\lambda)$ is less constrained, but must diverge asymptotically in the IR.

The remaining freedom is the dependence of the potentials on λ at intermediate values, which could be fitted to lattice and experimental QCD data (such as the equation of state and meson masses) in order to build a concrete model for QCD. An overall fit is the topic of ongoing work. The results presented below were computed with potentials obeying the asymptotics found above but were not fitted to data.

4. Properties of V-QCD

4.1. *Background solutions*

In order to uncover the phase structure of V-QCD, we take an Ansatz where the fields $\lambda(r)$ and $\tau(r)$ depend on the bulk coordinate r, and the metric has the form (3).

The Ansatz is then inserted to the equations of motion derived from the action (6), which are solved numerically.

This procedure results in a zero temperature phase diagram that is essentially universal and shown schematically in Fig. 1. By universality, we mean that the phase diagram is largely independent on the exact choice of the potentials in the V-QCD action, and qualitatively similar to all potentials which smoothly interpolate between the UV and IR asymptotics discussed in the previous section. In addition, many results near the conformal transition hold for a little more general class of models than just V-QCD [19] as we will explain below.

As a function of x and at zero quark mass there are two phases (restricting to the range $0 < x < x_{\mathrm{BZ}}$) separated by a phase transition at some $x = x_c \simeq 4$:

- In the region $0 < x < x_c$, the theory has chiral symmetry breaking, and IR dynamics is similar to that of ordinary QCD. The corresponding background solution has nontrivial $\lambda(r)$, $A(r)$ and $\tau(r)$, with the tachyon diverging at the "good" IR singularity of the geometry.
- In the conformal window, i.e., when $x_c < x < x_{\mathrm{BZ}}$, the theory flows to a nontrivial IR fixed point and there is no chiral symmetry breaking. The background solution has zero tachyon $\tau(r) = 0$ (so that indeed the quark mass and condensate in (8) vanish) and nontrivial $\lambda(r)$ and $A(r)$, giving rise to a geometry which is asymptotically AdS both in the UV and in the IR [26].

As $x \to x_c$ from below, "walking" RG flow of the coupling constant is found: the background flows very close to the IRFP (i.e., the geometry is close to AdS) but eventually the nonzero value of the tachyon drives the system away from the fixed point very deep in the IR.

4.2. *BF bound and the conformal transition*

The location of the Banks-Zaks region as $x \to x_{\mathrm{BZ}}$ from below, and the existence of the perturbative fixed point in this region is a consequence of matching the potentials to the perturbative RG flow of QCD in the UV. The conformal transition and the walking regime, on the contrary, are generated by the dynamics of the model in a nontrivial way. Let us discuss the holographic mechanism leading to this structure.

The conformal transition is related to an instability of the tachyon field in the conformal window [27], which arises when the five dimensional bulk mass of the tachyon, evaluated at the IR fixed point, violates the so-called Breitenlohner-Freedman (BF) bound. That is, an infinitesimal tachyon field near an AdS fixed point obeys the flow

$$\tau(r) \simeq C_1 r^{\Delta_*} + C_2 r^{4-\Delta_*} \ . \tag{10}$$

Here the exponents are related to the squared bulk mass of the tachyon m_τ and the

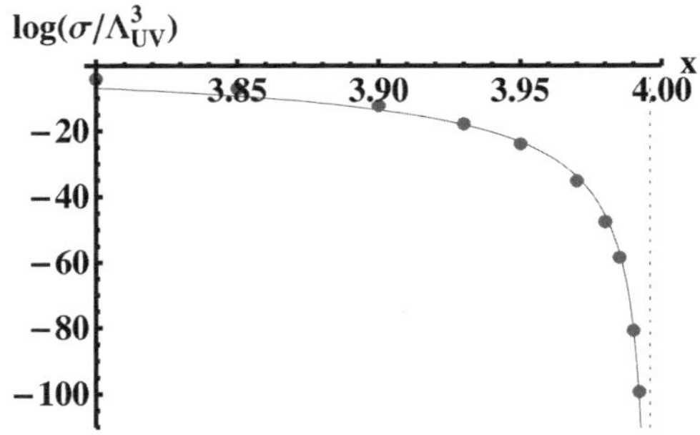

Fig. 3. The dependence of the chiral condensate on x near the conformal transition. The blue dots are data from V-QCD, and the red curve is a fit by the Miransky scaling law (12).

AdS radius ℓ_* as

$$-m_\tau^2 \ell_*^2 = \Delta_*(4-\Delta_*) \leq 4 \tag{11}$$

where the inequality, the BF bound, holds for real Δ_*. When the bound is violated, Δ_* becomes complex, leading to an oscillating tachyon solution which signals an instability towards forming a tachyon condensate in the bulk, and therefore a chiral condensate in field theory through (8). Therefore the point where the bound is saturated marks the location of the conformal transition $x = x_c$. At the transition, the anomalous dimension of the quark mass $\gamma_* = \Delta_* - 1 = 1$ which agrees with the expectation from Dyson-Schwinger analysis.

Transition triggered by the violation of the bound quite in general leads to walking and a BKT/Miransky phase transition of infinite order. This transition involves a characteristic exponential scaling law, called the Miransky scaling. The chiral condensate, for example, vanishes as

$$\langle \bar{q}q \rangle \sim \exp\left(-\frac{2K}{\sqrt{x_c - x}}\right) , \tag{12}$$

as $x \to x_c$ from below, see Fig. 3. The coefficient K is given by

$$K = \frac{\pi}{\sqrt{\frac{d}{dx}\left[m_\tau^2 \ell_*^2\right]_{x=x_c}}} . \tag{13}$$

The presence of the Miransky scaling also means that the theory involves two energy scales and a large scale hierarchy in the walking regime. We denote by $\Lambda = \Lambda_{\mathrm{UV}}$ the UV scale which already appeared in (8) and maps to Λ_{QCD} on the field theory side. The other scale is denoted by Λ_{IR}. It gives the scale of soft nonperturbative physics and for example the rho mass in the walking regime. The relation between the scales is given by

$$\frac{\Lambda_{\mathrm{UV}}}{\Lambda_{\mathrm{IR}}} \sim \exp\left(\frac{K}{\sqrt{x_c - x}}\right) , \qquad (x_c - x \ll 1) . \tag{14}$$

Notice that these arguments do not use directly the action of V-QCD. Indeed, as we already commented above, the physics near the transition is largely independent

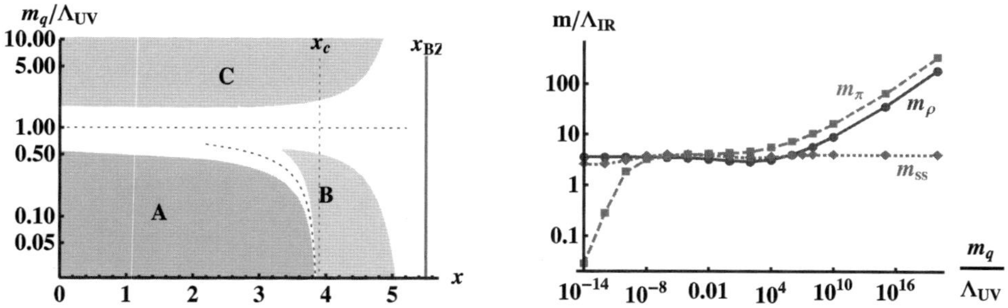

Fig. 4. Left: The regimes A, B, and C, where observables have qualitatively different dependence on the quark mass. The white regimes represent crossovers between the various scaling regimes. Right: The masses of the lowest vector (rho meson), pseudoscalar (pion), and singlet scalar states as a function of m_q and in units of $\Lambda_{\rm IR}$.

of the details of the holographic model. In particular, conformal transition (of the BKT type) has been found and analyzed earlier in other holographic models and the behavior near the transition is similar to that of V-QCD in these models[28–32].

4.3. *Physics at finite quark mass and at finite x*

So far we concentrated on the phase diagram of V-QCD at zero quark mass, but there is also interesting behavior at finite quark mass m_q. When the quark mass is turned on, chiral symmetry is always broken, and the conformal transition changes to a crossover. Therefore there are no phase transitions on the (x, m_q)-plane for $0 < x < x_{\rm BZ}$. Instead, one can identify different regimes where observables (such as the chiral condensate and meson masses) have qualitatively different mass dependencies, see Fig. 4 (left)[19]. In this diagram, the white regions represent crossovers between the different regimes, which are

- A The regime where quark mass acts as a small (linear) perturbation to the theory at $m_q = 0$.
- B The regime where the amount of walking in the model is controlled by the quark mass, leading to the so called "hyperscaling"[33].
- C The regime of large quark mass.

The results in regimes A and B are not sensitive to the details of the V-QCD action, but in regime C there is quite a lot of dependence on the model. Interestingly, requiring match with QCD in regime C is consistent with (9) which was obtained from considerations at zero m_q. In particular, both requiring linear meson trajectories at large excitation numbers, and a mass gap $\propto m_q$ at large m_q, independently imply that $a(\lambda)$ is constant.

The structure of Fig. 4 (left) can be verified, e.g., by stuying the messon masses which are found by identifying the normalizable fluctuation modes of V-QCD around the vacuum solution. Full analysis requires including flavor structure and gauge fields in the TDBI action. There are both flavor singlet (corresponding to quark

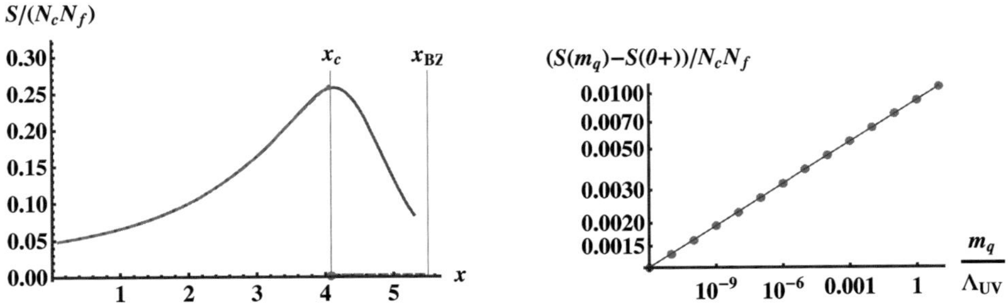

Fig. 5. Left: The S-parameter as a function of x in V-QCD. The dashed red curve shows the value at zero quark mass, whereas the solid blue curve has $m_q = 10^{-6}$. Right: The mass dependence of the S-parameter for $x = 4.5$ in log-log scale. The red dots are the data and the blue line is a power-law fit.

bilinear operators with the unit matrix $\mathbf{1}_{N_f\times N_f}$ in flavor space) and nonsinglet (corresponding to operators with the traceless Hermitean generators t^a) mesons.

In regime B, the mass gap of the mesons and the condensate obey the standard hyperscaling relations

$$m_{\text{gap}} \sim m_q^{\frac{1}{\Delta_*}} = m_q^{\frac{1}{1+\gamma_*}} \ , \quad \langle \bar{q}q \rangle \sim m_q^{\frac{4-\Delta_*}{\Delta_*}} = m_q^{\frac{3-\gamma_*}{1+\gamma_*}} \tag{15}$$

as $m_q \to 0$, which have also been studied in the dynamic AdS/QCD models[32]. Notice that regime B also extends to $x < x_c$, where one should set $\gamma_* = \Delta_* - 1 = 1$ in these formulae.

As an example of the dependence of the meson masses on m_q, we plot the pion (lowest nonsinglet pseudoscalar) mass m_π, the rho (lowest nonsinglet vector) mass m_ρ, and the mass of the lowest singlet scalar state m_{ss} as a function of m_q in Fig. 4 (right). We fixed the value of x right below the critical value x_c such that the plot intersects all three regimes. In regime A (lowest m_q) the pion mass is suppressed as required by the Gell-Mann-Oakes-Renner relation. In regime B (intermediate m_q), all masses have the same dependence on m_q. The masses obey (15) in units of Λ_{UV}, but we have chosen to plot the masses in units of Λ_{IR} so that the hyperscaling behavior is divided out. In regime C (largest m_q) the scalar singlet state is a glueball and therefore its mass is suppressed with respect to the $\bar{q}q$ states by a factor of m_q.

4.4. *The S-parameter*

Another important quantity which can be determined from the fluctuations is the S-parameter, which can be defined in terms of the difference of the vector-vector and axial-axial correlators. For a strongly interacting theory to be a viable candidate for the realization of technicolor, i.e., in order for the theory to pass the precision tests at LEP, this dimensionless parameter must be much smaller than one. It has been predicted that this parameter is suppressed in the walking regime[34–36], suggesting that walking technicolor theories are viable.

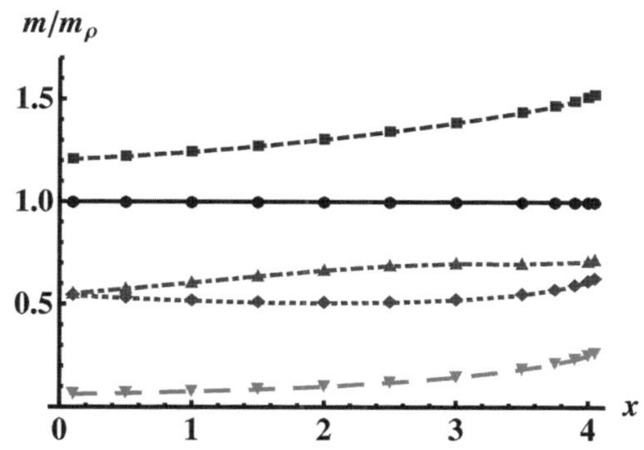

Fig. 6. The mass gaps of the different meson sectors normalized to the rho mass as a function of x. The black line is the rho mass, i.e., the vector mass gap. The blue dashed, red dotted, and magenta dot-dashed, curves are the axial, nonsinglet scalar, and singlet scalar mass gaps, respectively, while the green long-dashed curve is the pion decay constant.

The (normalized) S-parameter in V-QCD is shown in Fig. 5 (left). The blue solid (red dashed) curves have $m_q = 10^{-6}$ ($m_q = 0$), respectively. We notice that the normalized S-parameter is enhanced rather than suppressed in the walking region (as $x \to x_c$ from below). For finite quark mass, the S-parameter is also nontrivial inside the conformal window[19], whereas at zero quark mass the S-parameter vanishes there because chiral symmetry is intact. Our results show that as a tiny quark mass is turned on in the conformal window, the S-parameter immediately jumps to a finite $\mathcal{O}(N_f N_c)$ value. Such a discontinuity is consistent with recent field theory analysis[37].

At small but finite values of m_q inside the conformal window, the dependence of S on m_q is a power law, see Fig. 5 (right)[19]. The power turns out to be connected to the dimension Δ_{FF} of the $\mathbb{T}rF^2$ operator:

$$S(m_q) - S(0^+) \sim m_q^{\frac{\Delta_{FF}-4}{\Delta_*}} , \qquad (m_q \to 0) . \tag{16}$$

4.5. *(Non)existence of a technidilaton mode*

It has been conjectured that the scalar singlet sector contains a "dilaton" as $x \to x_c$, an anomalously light Goldstone mode due to the breaking of conformal symmetry[38]. We plot the masses of the lowest states for the various meson sectors in V-QCD, normalized to the rho mass and as a function of x, in Fig. 6. The scalar singlet mass gap is the magenta dot-dashed curve, which approaches a constant for $x \to x_c$, so there is no anomalously light technidilaton in V-QCD.

The example of V-QCD therefore shows that the mechanism with BF bound violation does not automatically yield a light technidilaton state (while it does yield Miransky scaling and hyperscaling relations). Interestingly, a detailed analysis[18,39] shows that the scalar singlet sector is special. The fluctuation equation for the scalar singlets is "critical" for values of r where the coupling walks, which suggests

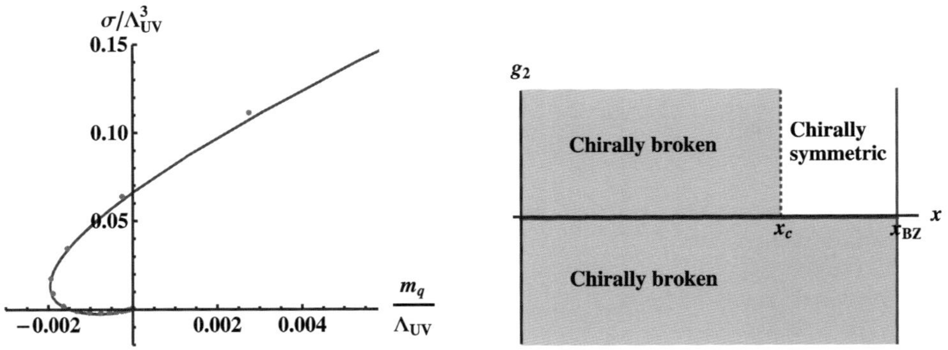

Fig. 7. Left: All regular V-QCD solutions on the (m_q, σ)-plane in the QCD regime. The red dots are data from V-QCD and the blue curve is an analytic fit. Right: The phase diagram of V-QCD on the (x, g_2)-plane, where g_2 is the four-fermion coupling, at zero quark mass.

the presence of a light state. In the end it turns out, however, that the mass of the light state depends on the definition of the model in the deep IR, so that there is also room for model dependence. In V-QCD a relatively light, but not parametrically light, scalar is found as $x \to x_c$. In dynamic AdS/QCD models with a BKT transition, one does find a parametrically light scalar mode[40], but in the top-down $D3 - D7$ defect model[28] the results are qualitatively similar to V-QCD: there is a light, but no parametrically light scalar state.

4.6. *Efimov spirals and four fermion operators*

Finally we comment briefly on the dependence of the chiral condensate on the quark mass and how this is used to include four-fermion operators in the model. In the QCD regime, plotting the values of m_q and the chiral condensate $\propto \sigma$ on the plane reveal a spiral[41] structure, see Fig. 7 (left). (The spiral winds around the origin infinitely many times but this is not visible due to limited resolution.) We call this structure the Efimov spiral, and the additional solutions at fixed quark mass (which are found at intersection points of the spiral with a vertical line placed at the value of the quark mass) Efimov vacua. Such vacua have been shown to be subdominant[8,19]: the solution furthest away from the origin is the "standard" stable solution. In the conformal window there is no spiral, but $\sigma(m_q)$ is a simple curve.

The curve $\sigma(m_q)$ can be used to add four-fermion operators in the model by modifying the UV boundary conditions for the tachyon as suggested by Witten[42]. The operator we want to add is

$$\propto g_2 \int d^4x \, (\bar{q}q)^2 \ . \tag{17}$$

The recipe is, that we first replace the coefficient m_q by α in the UV expansion (8) for the tachyon, and then set the new boundary condition

$$\alpha = m_q + g_2 \sigma \ . \tag{18}$$

This means that all solutions at fixed m_q are found by overlapping a tilted line instead of a vertical line with the Efimov spiral of Fig. 7 and searching for the intersection points. Computing the free energy of each solution, we find for $m_q = 0$ the phase diagram of Fig. 7 (right). For $g_2 > 0$, there is no qualitative change with respect to the solutions at $g_2 = 0$, and there is still a BKT transition at $x = x_c$. But turning on any negative g_2 leads to an instability, a new vacuum appears which is energetically favorable with respect to the standard vacuum. Therefore there is a chirally broken phase for $g_2 < 0$ which is disconnected from the dominant vacuum at $g_2 = 0$.

5. Conclusions and outlook

In this talk we have discussed results and predictions from the holographic V-QCD models, mostly near the conformal transition and in the conformal window. Many of these results appear universal, and may apply even slightly more in general than only in V-QCD. The outcome of the studies is encouraging: the model is consistent, at qualitative level, well with expectations based on field theory analyses.

Several additional extensions and applications of V-QCD are also under study. These include a study of the CP-odd sector (physics of the theta angle and axial anomaly of QCD) and finally fitting the model to QCD data, for example Yang-Mills thermodynamics from the lattice, experimental values of the QCD meson masses, and the equation of state at finite x from the lattice.

Acknowledgments

The author wishes to thank K. Yamawaki for the invitation to SCGT15 and the participants of the conference for several interesting discussions. This work was partially supported by the "ARISTEIA II" Action of the "Operational Programme Education and Lifelong Learning" and was co-funded by the European Social Fund (ESF) and National Resources.

References

1. G. Veneziano, U(1) Without Instantons, *Nucl.Phys.* **B159**, 213 (1979).
2. T. Appelquist, K. D. Lane and U. Mahanta, On the Ladder Approximation for Spontaneous Chiral Symmetry Breaking, *Phys.Rev.Lett.* **61**, p. 1553 (1988).
3. F. Sannino, Jumping Dynamics, *Mod. Phys. Lett.* **A28**, p. 1350127 (2013).
4. E. T. Neil, Exploring Models for New Physics on the Lattice, *PoS* **LATTICE2011**, p. 009 (2011).
5. J. Giedt, Lattice gauge theory and physics beyond the standard model, *PoS* **LATTICE2012**, p. 006 (2012).
6. J. Kuti, The Higgs particle and the lattice, *PoS* **LATTICE2013**, p. 004 (2014).
7. M. P. Lombardo, K. Miura, T. J. Nunes da Silva and E. Pallante, One, two, zero: Scales of strong interactions, *Int. J. Mod. Phys.* **A29**, p. 1445007 (2014).

8. M. Järvinen and E. Kiritsis, Holographic Models for QCD in the Veneziano Limit, *JHEP* **1203**, p. 002 (2012).
9. U. Gursoy and E. Kiritsis, Exploring improved holographic theories for QCD: Part I, *JHEP* **0802**, p. 032 (2008).
10. U. Gursoy, E. Kiritsis and F. Nitti, Exploring improved holographic theories for QCD: Part II, *JHEP* **0802**, p. 019 (2008).
11. F. Bigazzi, R. Casero, A. Cotrone, E. Kiritsis and A. Paredes, Non-critical holography and four-dimensional CFT's with fundamentals, *JHEP* **0510**, p. 012 (2005).
12. R. Casero, E. Kiritsis and A. Paredes, Chiral symmetry breaking as open string tachyon condensation, *Nucl.Phys.* **B787**, 98 (2007).
13. T. Alho, M. Järvinen, K. Kajantie, E. Kiritsis and K. Tuominen, On finite-temperature holographic QCD in the Veneziano limit, *JHEP* **1301**, p. 093 (2013).
14. T. Alho, M. Järvinen, K. Kajantie, E. Kiritsis, C. Rosen and K. Tuominen, A holographic model for QCD in the Veneziano limit at finite temperature and density, *JHEP* **04**, p. 124 (2014), [Erratum: JHEP02,033(2015)].
15. I. Iatrakis and I. Zahed, Spectral Functions in V-QCD with Matter: Masses, Susceptibilities, Diffusion and Conductivity, *JHEP* **04**, p. 080 (2015).
16. T. Alho, M. Järvinen, K. Kajantie, E. Kiritsis and K. Tuominen, Quantum and stringy corrections to the equation of state of holographic QCD matter and the nature of the chiral transition, *Phys. Rev.* **D91**, p. 055017 (2015).
17. D. Arean, I. Iatrakis, M. Järvinen and E. Kiritsis, V-QCD: Spectra, the dilaton and the S-parameter, *Phys.Lett.* **B720**, 219 (2013).
18. D. Arean, I. Iatrakis, M. Järvinen and E. Kiritsis, The discontinuities of conformal transitions and mass spectra of V-QCD, *JHEP* **11**, p. 068 (2013).
19. M. Järvinen, Massive holographic QCD in the Veneziano limit, *JHEP* **07**, p. 033 (2015).
20. U. Gursoy, E. Kiritsis, L. Mazzanti and F. Nitti, Improved Holographic Yang-Mills at Finite Temperature: Comparison with Data, *Nucl.Phys.* **B820**, 148 (2009).
21. M. Panero, Thermodynamics of the QCD plasma and the large-N limit, *Phys.Rev.Lett.* **103**, p. 232001 (2009).
22. U. Gursoy, E. Kiritsis, L. Mazzanti and F. Nitti, Deconfinement and Gluon Plasma Dynamics in Improved Holographic QCD, *Phys.Rev.Lett.* **101**, p. 181601 (2008).
23. U. Gursoy, E. Kiritsis, L. Mazzanti and F. Nitti, Holography and Thermodynamics of 5D Dilaton-gravity, *JHEP* **0905**, p. 033 (2009).
24. I. Iatrakis, E. Kiritsis and A. Paredes, An AdS/QCD model from tachyon condensation: II, *JHEP* **1011**, p. 123 (2010).
25. S. Kuperstein and J. Sonnenschein, Non-critical, near extremal AdS(6) background as a holographic laboratory of four dimensional YM theory, *JHEP* **11**,

p. 026 (2004).
26. M. Järvinen and F. Sannino, Holographic Conformal Window - A Bottom Up Approach, *JHEP* **1005**, p. 041 (2010).
27. D. B. Kaplan, J.-W. Lee, D. T. Son and M. A. Stephanov, Conformality Lost, *Phys.Rev.* **D80**, p. 125005 (2009).
28. D. Kutasov, J. Lin and A. Parnachev, Conformal Phase Transitions at Weak and Strong Coupling, *Nucl.Phys.* **B858**, 155 (2012).
29. D. Kutasov, J. Lin and A. Parnachev, Holographic Walking from Tachyon DBI, *Nucl.Phys.* **B863**, 361 (2012).
30. M. Goykhman and A. Parnachev, S-parameter, Technimesons, and Phase Transitions in Holographic Tachyon DBI Models, *Phys.Rev.* **D87**, p. 026007 (2013).
31. R. Alvares, N. Evans and K.-Y. Kim, Holography of the Conformal Window, *Phys.Rev.* **D86**, p. 026008 (2012).
32. N. Evans and M. Scott, Hyper-Scaling Relations in the Conformal Window from Dynamic AdS/QCD, *Phys. Rev.* **D90**, p. 065025 (2014).
33. L. Del Debbio and R. Zwicky, Hyperscaling relations in mass-deformed conformal gauge theories, *Phys. Rev.* **D82**, p. 014502 (2010).
34. T. Appelquist and F. Sannino, The Physical spectrum of conformal SU(N) gauge theories, *Phys.Rev.* **D59**, p. 067702 (1999).
35. M. Harada, M. Kurachi and K. Yamawaki, The pi+ - pi0 mass difference and the S parameter in large N(f) QCD, *Prog.Theor.Phys.* **115**, 765 (2006).
36. M. Kurachi and R. Shrock, Behavior of the S Parameter in the Crossover Region Between Walking and QCD-Like Regimes of an SU(N) Gauge Theory, *Phys.Rev.* **D74**, p. 056003 (2006).
37. F. Sannino, Mass Deformed Exact S-parameter in Conformal Theories, *Phys.Rev.* **D82**, p. 081701 (2010).
38. K. Yamawaki, M. Bando and K.-i. Matumoto, Scale Invariant Technicolor Model and a Technidilaton, *Phys.Rev.Lett.* **56**, p. 1335 (1986).
39. J. Alanen, T. Alho, K. Kajantie and K. Tuominen, Mass spectrum and thermodynamics of quasi-conformal gauge theories from gauge/gravity duality, *Phys.Rev.* **D84**, p. 086007 (2011).
40. T. Alho, N. Evans and K. Tuominen, Dynamic AdS/QCD and the Spectrum of Walking Gauge Theories, *Phys.Rev.* **D88**, p. 105016 (2013).
41. N. Iqbal, H. Liu and M. Mezei, Quantum phase transitions in semilocal quantum liquids, *Phys. Rev.* **D91**, p. 025024 (2015).
42. E. Witten, Multitrace operators, boundary conditions, and AdS / CFT correspondence (2001).

Gauge/Gravity Duality
– From Lattice Gauge Theory to Black Hole

Daisuke Kadoh

KEK Theory Center, High Energy Accelerator Research Organization (KEK), Tsukuba, Ibaraki 305-0801, Japan
E-mail: kadoh@post.kek.jp

The duality conjecture states that $p+1$-dimensional maximally supersymmetric Yang-Mills theory at finite temperature is expected to be dual to the non extremal black p-brane at large N. We perform the lattice simulations of SYM for $p = 0, 1$ to investigate the validity of the conjecture. We show that the conjecture is qualitatively valid by comparing lattice results of the black p-branes mass with analytic expectations in the gravity side.

1. Introduction

The duality between gauge theory and the theory of gravity is an important subject in theoretical physics, which was originally proposed as the AdS/CFT correspondence[1]. Once we accept it, there are many interesting applications. It gives us deep insights into some long standing problems, such as the black hole information loss paradox. Moreover, there are many attempts to solve strongly coupled theories, like QCD, analytically in the gravity side. To ensure the validity of the applications, it is important to give strong evidence of the conjecture. In this study, we aim to verify it in maximally supersymmetric Yang-Mills theories in 0+1 and 1+1 dimensions by using lattice simulations.

Lattice theory provides a framework to study strongly coupled gauge theories. In the context of the duality, gauge theory is typically a strongly coupled one and thus is a good target for lattice theory. We can test it by comparing lattice results in the gauge theory side with theoretical predictions in the gravity side. Although lattice theory achieves great success for Yang-Mills theories, the nature of supersymmetry algebra makes it hard to put supersymmetric Yang-Mills theories on the lattice.

In the last decade, some developments of lattice supersymmetry changed the situation completely. Several lattice formulations of SYM theories were proposed[2,3,4,5,6,7,8,9]. In one dimensional SYM with sixteen supercharges, some evidences of the correspondence between the gauge theory and the dual black hole[10] have been obtained from numerical simulations[12,13]. In two-dimensional $\mathcal{N} = (8, 8)$ SYM, lattice simulations have been already performed and the phase structure of the theory was investigated[14]. However, these subjects deserve more studies because some physical quantities, like the internal energy of the black 1-branes[15], have not been estimated yet.

In this paper, we perform the simulations of maximally supersymmetric Yang-Mills theories in 0+1 and 1+1 dimensions. We employ Sugino lattice action[5] with two exact supercharges. Our results show that the duality between the gauge theories and the black p-branes is qualitatively valid in the phase quenched approximation for $p = 0, 1$.

2. Maximally supersymmetric Yang-Mills theory

Maximally supersymmetric Yang-Mills theory in $p+1$ dimensions has $p+1$ gauge fields $A_\mu(\mu = 0, 1, \cdots, p)$, and $9-p$ scalar fields $X_i(i = p+1, \cdots, 8, 9)$, and sixteen fermions $\psi_\alpha(\alpha = 1, 2, \cdots, 16)$ as the field variables. The scalars and fermions, as well as the gauge fields, are in adjoint representation of $SU(N)$ and can be represented as matrix-valued fields. [a]

The Euclidean action is given by

$$S = \frac{N}{\lambda}\int d^{p+1}x \ \mathrm{tr}\left\{\frac{1}{4}F_{\mu\nu}^2 + \frac{1}{2}(D_\mu X_i)^2 - \frac{1}{4}[X_i, X_j]^2 - \frac{i}{2}\psi\cdot\gamma_\mu D_\mu\psi + \frac{1}{2}\psi\cdot\gamma_i[X_i, \psi]\right\}, \tag{1}$$

where the field strength is $F_{\mu\nu} = \partial_\mu A_\nu - \partial_\nu A_\mu + i[A_\mu, A_\nu]$ and the covariant derivatives are defined by $D_\mu\varphi = \partial_\mu\varphi + i[A_\mu, \varphi]$. In addition, $\gamma_0 = i$ for simplicity, and the real symmetric matrices $\gamma_a(a = 1, \cdots, 9)$ satisfy the nine-dimensional Euclidean Clifford algebra, $\{\gamma_a, \gamma_b\} = 2\delta_{ab}$.

The action (1) is invariant under the sixteen supersymmetry transformations,

$$Q_\alpha A_\mu = -i(\gamma_\mu)_{\alpha\beta}\Psi_\beta, \tag{2}$$

$$Q_\alpha X_i = -i(\gamma_i)_{\alpha\beta}\Psi_\beta, \tag{3}$$

$$Q_\alpha\Psi_\beta = \frac{1}{2}(\gamma_\mu\gamma_\nu)_{\alpha\beta}F_{\mu\nu} + (\gamma_\mu\gamma_i)_{\alpha\beta}D_\mu X_i + \frac{i}{2}(\gamma_i\gamma_j)_{\alpha\beta}[X_i, X_j]. \tag{4}$$

Furthermore, the $p+1$-dimensional Poincaré group and the internal $O(9-p)$ rotations are symmetries of this theory.

In four dimensions ($p = 3$), the action (1) can be written in the $Q_\pm$-exact form with the four-dimensional topological field theory notation,

$$S = Q_+Q_-\frac{N}{2\lambda}\int d^2x \ \mathrm{tr}\left\{-2iB_i\left(F_{i3} + \frac{1}{2}\epsilon_{ijk}F_{jk}\right) - \frac{2}{3}\epsilon_{ijk}B_iB_jB_k - \psi_{+\mu}\psi_{-\mu} - \chi_{+i}\chi_{-i} - \frac{1}{4}\eta_+\eta_-\right\}, \tag{5}$$

where μ runs from 0 to 3, and i, j, k from 0 to 2, and ϵ_{ijk} is a totally antisymmetric tensor that satisfy $\epsilon_{012} = 1$. The new boson fields $B_i, C, \phi_\pm$ and the new fermion fields $\psi_{\pm\mu}, \chi_{\pm i}, \eta_\pm$ are some linear combinations of the original boson and fermion

[a] In this paper, any fields φ can be expanded as $\varphi = \sum_a \varphi^a T^a$ where T^a are group generators satisfying $\mathrm{tr}(T^aT^b) = \delta_{ab}$,

fields, respectively.[16] For $p \leq 3$, $\partial_\mu = 0$ for $\mu = p+1, \cdots, 3$. The associated $Q_\pm$-transformations[5] satisfy

$$Q_\pm^2 = i\delta_{\phi_\pm}, \qquad \{Q_+, Q_-\} = -i\delta_C \tag{6}$$

with δ_ω being a gauge transformation with the gauge parameter ω.

Sugino introduced the lattice action with two exact supercharges in $p \leq 3$ from the twisted form (5) as

$$S = Q_+ Q_- \frac{N}{2\lambda_0} \sum_{x \in \Gamma_p} \mathrm{tr}\left\{ -2iB_i \left(F_{i3} + \frac{1}{2}\epsilon_{ijk}F_{jk} \right) - \frac{1}{3}\epsilon_{ijk}B_i[B_j, B_k] \right.$$
$$\left. -\psi_{+\mu}\psi_{-\mu} - \chi_{+i}\chi_{-i} - \frac{1}{4}\eta_+\eta_- \right\}, \tag{7}$$

where Γ_p is a set of $p+1$-dimensional lattice sites. The lattice version of $F_{\mu\nu}(x)$ and $Q_\pm$-transformations are given in Refs.[16,17]. The lattice action (7) has two exact supercharges since the lattice $Q_\pm$-transformations are nilpotent up to the gauge transformations even on the finite lattice.

The Sugino lattice action breaks the remaining fourteen supersymmetries on the lattice. In Refs.[16,17], we measured the supersymmetry Ward-Takahashi identity at finite temperature for $SU(2)$ theory in 1+0 and 1+1 dimensions and check that the all of the supersymmetry are restored in the continuum limit

3. Thermodynamics of black branes

The free energy of the black p-branes is a fundamental quantity to test the duality conjecture. It is, however, difficult to measure the free energy in the lattice simulations. Instead, in this section, we see the lattice results of the internal energy in 1+0 dimensions and the internal energy minus pressure in 1+1 dimensions, which can be easily estimated from the expectation value of the SYM action itself.

The numerical simulations were done using the Hybrid Monte Carlo method. The fermion contribution was included using the pseudo fermion method with the rational approximation. The phase of the fermion pfaffian was quenched for the results presented here.

3.1. *The case of black hole*

The internal energy of the non-extremal black hole, which is the gravity dual of the target gauge theory, behaves

$$\frac{1}{N^2}\left(\frac{E}{\lambda^{1/3}}\right) = c_1 \left(\frac{T}{\lambda^{1/3}}\right)^{2.8} + c_2 \left(\frac{T}{\lambda^{1/3}}\right)^{4.6} + \cdots, \tag{8}$$

in the large N limit. The leading order coefficient c_1 can be calculated by hand[15]: $c_1 = 9(4^{13}15^2(\pi/7)^{14})^{\frac{1}{5}}/14 = 7.407\cdots$. In contrast, the next-to-leading order coefficient c_2 is unknown.

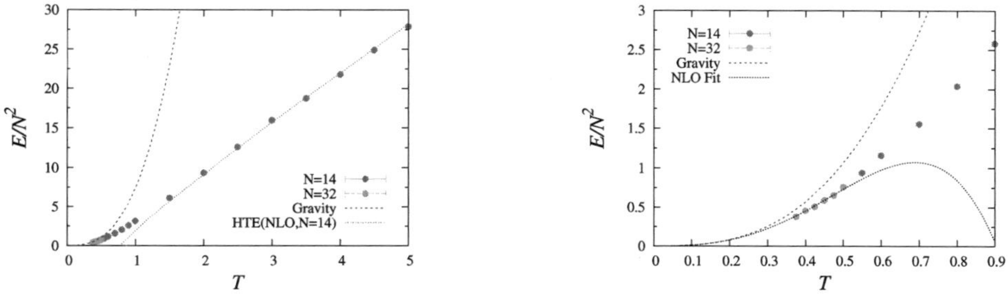

Fig. 1. Internal energy against temperature. The right figure shows low temperature of the left one. The results coincide with the result of the high temperature expansion (dashed orange curve) at high temperature and approach the theoretical prediction (dashed blue curve) as the temperature decreases. The dotted curve in the right figure denotes the NLO-fit result.

In the gauge theory side, one can estimate the internal energy from the expectation value of the action[13] as

$$E = -\frac{3}{\beta}\langle S\rangle, \tag{9}$$

where S is given by (5). On the lattice, we measure the internal energy from (9) with (7) instead of (5).

Figure 1 shows the results of the internal energy, where the red and green points denote the results for $N = 14$ and ones for $N = 32$, respectively.[17] The lattice size is $L = 16$ for $T < 1$ and $L = 8$ for $1 \leq T$ with the lattice spacing $aT_{\text{eff}} = 1/L$. The x-axis denotes dimensionless temperature $T_{\text{eff}} = T/\lambda^{1/3}$. The dashed blue curve represents the theoretical prediction of the gravity side at the leading order, (8) with $c_2 = 0$. The slightly curved orange line is the result of the high temperature expansion[11,18]. As can be seen in the left figure, the lattice data reproduces the high temperature expansion at high temperature, while as the temperature decreases, they smoothly approach the theoretical prediction of the gravity side.

One can study the contribution of the next-to-leading order term by fitting the lattice results using the following formula,

$$f(x) = 7.41x^{2.8} + Cx^p, \tag{10}$$

where C and p are the fitting parameters. From (8), if the duality conjecture is really true, the obtained p should be 4.6. We performed the fit using the 5 points within $0.375 \leq T \leq 0.475$ and obtained

$$C = 9.0 \pm 2.6, \qquad p = 4.74 \pm 0.35, \tag{11}$$

which is consistent with the theoretical prediction of the gravity side within about seven percent statistical error. This is the first lattice result of the NLO term, which quantitatively shows the validity of the duality conjecture in this system.[17]

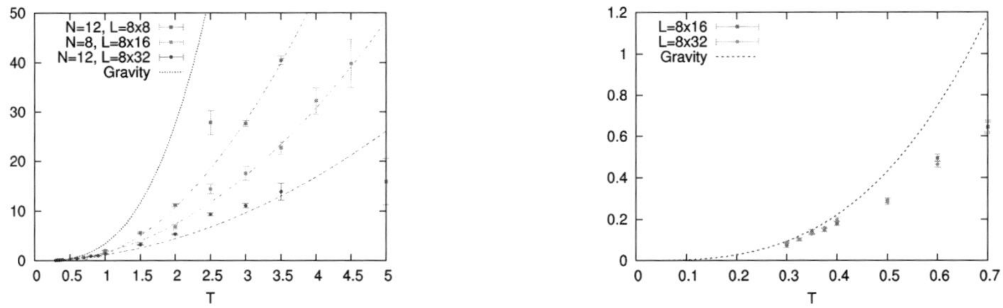

Fig. 2. Left figure shows $(\epsilon - p)/N^2$ against temperature, and the right figure shows low temperature of the left one with $N = 12$. The line named gravity indicates the theoretical prediction in the gravity side. The dashed lines are the results of the high temperature expansion up to the next-to-leading order.

3.2. *The case of black string*

In 1+1 dimensions, the expectation value of the action is related to the internal energy density ϵ and pressure p as

$$\epsilon - p = -\frac{2}{\beta L}\langle S\rangle. \tag{12}$$

On the other hand, in the gravity side,

$$\epsilon - p = cN^2T^3, \tag{13}$$

where $c = 2^4\pi^{5/2}/3^4 (= 3.12...)$.

Figure 2 shows the preliminary results of $\epsilon - p$ in the $\lambda = 1$ unit.[20] The dotted line correspond to the theoretical prediction (13) and the colored curves in the left figure are the results of the high temperature expansion[19]. Roughly speaking, the left figure shows that the lattice results are consistent with high temperature expansion within large statistical errors since the deviations which are seen at several temperatures are caused by insufficient statistics.

The right show low temperature of the left one. The dotted line again represents the theoretical prediction of the gravity side It can be clearly seen that our simulation results tends toward the leading order prediction of the gravity side. Therefore the duality conjecture is qualitatively valid. However, further studies are needed to obtained quantitative results. Further simulations to obtain sufficient statistics are currently running.

Acknowledgments

The numerical simulations were performed by using RICC, Hokusai and K computer at RIKEN, KEK supercomputer and SR16000 at YITP in Kyoto University.

References

1. J. M. Maldacena, *The Large N limit of superconformal field theories and supergravity,* Int. J. Theor. Phys. **38** (1999) 1113 [Adv. Theor. Math. Phys. **2** (1998) 231] [hep-th/9711200].
2. D. B. Kaplan, E. Katz and M. Unsal, *Supersymmetry on a spatial lattice,* JHEP **0305** (2003) 037 [hep-lat/0206019].
3. F. Sugino, *A Lattice formulation of superYang-Mills theories with exact supersymmetry,* JHEP **0401** (2004) 015 [hep-lat/0311021].
4. F. Sugino, *SuperYang-Mills theories on the two-dimensional lattice with exact supersymmetry,* JHEP **0403** (2004) 067 [hep-lat/0401017].
5. F. Sugino, *Various super Yang-Mills theories with exact supersymmetry on the lattice,* JHEP **0501** (2005) 016 [hep-lat/0410035].
6. F. Sugino, *Two-dimensional compact N=(2,2) lattice super Yang-Mills theory with exact supersymmetry,* Phys. Lett. B **635** (2006) 218 [hep-lat/0601024]. *Lattice Formulation of Two-Dimensional N=(2,2) SQCD with Exact Supersymmetry,* Nucl. Phys. B **808** (2009) 292 [arXiv:0807.2683 [hep-lat]]. Y. Kikukawa and F. Sugino, *Ginsparg-Wilson Formulation of 2D N = (2,2) SQCD with Exact Lattice Supersymmetry,* Nucl. Phys. B **819** (2009) 76 [arXiv:0811.0916 [hep-lat]]. D. Kadoh, F. Sugino and H. Suzuki, *Lattice formulation of 2D N = (2,2) SQCD based on the B model twist,* Nucl. Phys. B **820** (2009) 99 [arXiv:0903.5398 [hep-lat]].
7. S. Matsuura and F. Sugino, *Lattice formulation for 2d = (2, 2), (4, 4) super Yang-Mills theories without admissibility conditions,* JHEP **1404** (2014) 088 [arXiv:1402.0952 [hep-lat]].
8. A. D'Adda, I. Kanamori, N. Kawamoto and K. Nagata, *Exact extended supersymmetry on a lattice: Twisted N=2 super Yang-Mills in two dimensions,* Phys. Lett. B **633** (2006) 645 [hep-lat/0507029]. *Exact Extended Supersymmetry on a Lattice: Twisted N=4 Super Yang-Mills in Three Dimensions,* Nucl. Phys. B **798** (2008) 168 [arXiv:0707.3533 [hep-lat]].
9. S. Matsuura, *Two-dimensional N=(2,2) Supersymmetric Lattice Gauge Theory with Matter Fields in the Fundamental Representation,* JHEP **0807** (2008) 127 [arXiv:0805.4491 [hep-th]].
10. N. Itzhaki, J. M. Maldacena, J. Sonnenschein and S. Yankielowicz, *Supergravity and the large N limit of theories with sixteen supercharges,* Phys. Rev. D **58** (1998) 046004 [hep-th/9802042].
11. N. Kawahara, J. Nishimura and S. Takeuchi, *High temperature expansion in supersymmetric matrix quantum mechanics,* JHEP **0712** (2007) 103 [arXiv:0710.2188 [hep-th]].
12. K. N. Anagnostopoulos, M. Hanada, J. Nishimura and S. Takeuchi, *Monte Carlo studies of supersymmetric matrix quantum mechanics with sixteen supercharges at finite temperature,* Phys. Rev. Lett. **100** (2008) 021601 [arXiv:0707.4454 [hep-th]]. M. Hanada, A. Miwa, J. Nishimura and S. Takeuchi,

Schwarzschild radius from Monte Carlo calculation of the Wilson loop in supersymmetric matrix quantum mechanics, Phys. Rev. Lett. **102** (2009) 181602 [arXiv:0811.2081 [hep-th]]. M. Hanada, Y. Hyakutake, J. Nishimura and S. Takeuchi, *Higher derivative corrections to black hole thermodynamics from supersymmetric matrix quantum mechanics,* Phys. Rev. Lett. **102** (2009) 191602 [arXiv:0811.3102 [hep-th]]. M. Hanada, J. Nishimura, Y. Sekino and T. Yoneya, *Monte Carlo studies of Matrix theory correlation functions,* Phys. Rev. Lett. **104** (2010) 151601 [arXiv:0911.1623 [hep-th]]. *Direct test of the gauge-gravity correspondence for Matrix theory correlation functions,* JHEP **1112** (2011) 020 [arXiv:1108.5153 [hep-th]]. M. Hanada, Y. Hyakutake, G. Ishiki and J. Nishimura, *Holographic description of quantum black hole on a computer,* Science **344** (2014) 882 [arXiv:1311.5607 [hep-th]].

13. S. Catterall and T. Wiseman, *Towards lattice simulation of the gauge theory duals to black holes and hot strings,* JHEP **0712** (2007) 104 [arXiv:0706.3518 [hep-lat]]. *Black hole thermodynamics from simulations of lattice Yang-Mills theory,* Phys. Rev. D **78** (2008) 041502 [arXiv:0803.4273 [hep-th]]. *Extracting black hole physics from the lattice,* JHEP **1004** (2010) 077 [arXiv:0909.4947 [hep-th]].
14. S. Catterall, A. Joseph and T. Wiseman, *Gauge theory duals of black hole - black string transitions of gravitational theories on a circle,* J. Phys. Conf. Ser. **462** (2013) 1, 012022 [arXiv:1009.0529 [hep-th]]. *Thermal phases of D1-branes on a circle from lattice super Yang-Mills,* JHEP **1012** (2010) 022 [arXiv:1008.4964 [hep-th]].
15. I. R. Klebanov and A. A. Tseytlin, *Entropy of near extremal black p-branes,* Nucl. Phys. B **475** (1996) 164 [hep-th/9604089].
16. E. Giguère and D. Kadoh, *Restoration of supersymmetry in two-dimensional SYM with sixteen supercharges on the lattice,* JHEP **1505** (2015) 082 [arXiv:1503.04416 [hep-lat]].
17. D. Kadoh and S. Kamata, *Gauge/gravity duality and lattice simulations of one dimensional SYM with sixteen supercharges,* arXiv:1503.08499 [hep-lat].
18. D. Kadoh and S .Kamata, *High temperature expansion of one-dimensional supersymmetric Yang-Mills theory,* in preparation.
19. E. Giguère and D. Kadoh, *High temperature expansion of two-dimensional supersymmetric Yang-Mills theory,* in preparation.
20. E. Giguère and D. Kadoh, *Thermodynamics of black string from the lattice simulations,* in preparation.

Maximally Supersymmetric Yang–Mills on the Lattice

David Schaich* and Simon Catterall

Department of Physics, Syracuse University, Syracuse, New York 13244, USA
**E-mail: dschaich@syr.edu*

We summarize recent progress in lattice studies of four-dimensional $\mathcal{N} = 4$ supersymmetric Yang–Mills theory and present preliminary results from ongoing investigations. Our work is based on a construction that exactly preserves a single supersymmetry at non-zero lattice spacing, and we review a new procedure to regulate flat directions by modifying the moduli equations in a manner compatible with this supersymmetry. This procedure defines an improved lattice action that we have begun to use in numerical calculations. We discuss some highlights of these investigations, including the static potential and an update on the question of a possible sign problem in the lattice theory.

In recent years there has been tremendous progress in lattice studies of four-dimensional $\mathcal{N} = 4$ supersymmetric Yang–Mills (SYM) theory, employing a lattice formulation that exactly preserves a single supercharge $\mathcal{Q}$.[1–5] Advances both in the scale of computations and in the algorithms employed[6] have provided the first ab initio numerical results to be confronted with perturbative and holographic predictions for quantities such as the static potential.[7,8] At the same time this work has also led to improvements of the lattice construction, in particular the development of a new procedure to regulate flat directions in a manner compatible with the $\mathcal{Q}$ supersymmetry.[9] Given the central role of $\mathcal{N} = 4$ SYM in the AdS/CFT correspondence that relates it to quantum gravity, it is important to continue large-scale lattice investigations of the theory away from the regime of weak coupling and for arbitrary numbers of colors N. The recent progress, although significant, is only the beginning of this effort.

In this proceedings we briefly review the latest developments and present some new preliminary results from ongoing lattice $\mathcal{N} = 4$ SYM studies. We begin in the next section by summarizing the new procedure to regulate flat directions without breaking the exact $\mathcal{Q}$ supersymmetry. Numerical calculations using the resulting improved lattice action exhibit dramatically reduced violations of supersymmetric Ward identities and much more rapid approach to the continuum limit. We are now carrying out large-scale studies using this improved action. In Section 2 we revisit the static potential, checking that the improved action reproduces previous results. The improved action has also allowed us to gain new insight into the possible sign problem of lattice $\mathcal{N} = 4$ SYM, which we present in Section 3. We conclude with some discussion of other work currently underway and some of the next steps in our wide-ranging investigations.

1. Supersymmetric deformation of the moduli space

A convenient starting point for lattice $\mathcal{N} = 4$ SYM is the direct discretization of the continuum action produced by topological twisting,[10,11]

$$S_{\rm formal} = S_{\rm exact} + S_{\rm closed} \tag{1}$$

$$S_{\rm exact} = \frac{N}{2\lambda_{\rm lat}} \sum_n \text{Tr}\left[\mathcal{Q}\left(\chi_{ab}(n)\mathcal{D}_a^{(+)}\mathcal{U}_b(n) + \eta(n)\overline{\mathcal{D}}_a^{(-)}\mathcal{U}_a(n) - \frac{1}{2}\eta(n)d(n)\right)\right]$$

$$S_{\rm closed} = -\frac{N}{8\lambda_{\rm lat}} \sum_n \text{Tr}\left[\epsilon_{abcde}\ \chi_{de}(n + \widehat{\mu}_a + \widehat{\mu}_b + \widehat{\mu}_c)\overline{\mathcal{D}}_c^{(-)}\chi_{ab}(n)\right],$$

with repeated indices summed. All indices run from 1 through 5, requiring that the theory be formulated on the A_4^* lattice with five linearly dependent basis vectors symmetrically spanning four space-time dimensions.[2,8] The five complexified gauge links $\mathcal{U}_a$ contain both the gauge and scalar fields, and result in an enlarged $\text{U}(N) = \text{SU}(N) \otimes \text{U}(1)$ gauge invariance, where N is the number of colors.

Both the SU(N) and U(1) sectors necessarily possess flat directions that destabilize the vacuum of the lattice theory. We recently introduced a general method that can be applied to regulate these flat directions in a manner compatible with the $\mathcal{Q}$ supersymmetry.[9] The procedure is to deform the $\mathcal{Q}\left(\eta\overline{\mathcal{D}}_a^{(-)}\mathcal{U}_a\right)$ term in the action, Eq. 1, as

$$\mathcal{Q}\ \text{Tr}\left[\eta(n)\left(\overline{\mathcal{D}}_a^{(-)}\mathcal{U}_a(n)\right)\right] \to \mathcal{Q}\ \text{Tr}\left[\eta(n)\left(\overline{\mathcal{D}}_a^{(-)}\mathcal{U}_a(n) + G\mathcal{O}(n)\mathbb{I}_N\right)\right] \tag{2}$$

where G is a new tunable coupling and $\mathcal{O}(n)$ is an appropriate gauge-invariant bosonic operator. This $\mathcal{Q}$-exact deformation modifies the auxiliary field equations of motion, $d(n) = \overline{\mathcal{D}}_a^{(-)}\mathcal{U}_a(n) + G\mathcal{O}(n)\mathbb{I}_N$, which determine the moduli space of the system. Since $\mathcal{Q}\,\eta = d$, this also modifies the $\mathcal{Q}$ Ward identity

$$\left\langle \sum_n \text{Tr}\left[\mathcal{Q}\,\eta(n)\right]\right\rangle = NG\left\langle\sum_n \mathcal{O}(n)\right\rangle = 0, \tag{3}$$

where the $\sum_n \text{Tr}\left[\overline{\mathcal{D}}_a\mathcal{U}_a(n)\right]$ term vanishes due to the sum over all lattice sites and the structure of the finite-difference operator $\overline{\mathcal{D}}_a$.[3,4]

Because the flat directions associated with the U(1) sector produce especially severe lattice artifacts,[7,9] we take

$$\mathcal{O}(n) = \sum_{a\neq b}\left(\det\mathcal{P}_{ab}(n) - 1\right) = 2\text{Re}\sum_{a<b}\left(\det\mathcal{P}_{ab}(n) - 1\right), \tag{4}$$

where $\mathcal{P}_{ab} = \mathcal{P}_{ba}^*$ is the oriented plaquette in the a–b plane. The corresponding Ward identity is $\sum_n\sum_{a\neq b}\langle\det\mathcal{P}_{ab}(n) - 1\rangle = 0$, implying $\langle\text{Re}\ \det\mathcal{P}_{ab}\rangle = 1$. However, $\langle\text{Im}\ \det\mathcal{P}_{ab}\rangle$ is still unconstrained, and the SU(N) flat directions still must be regulated as well. We address both of these issues by adding to the action a soft $\mathcal{Q}$ breaking scalar potential,

$$S_{\rm soft} = \frac{N}{2\lambda_{\rm lat}}\mu^2\sum_n\sum_a\left(\frac{1}{N}\text{Tr}\left[\mathcal{U}_a(n)\overline{\mathcal{U}}_a(n)\right] - 1\right)^2. \tag{5}$$

Previous numerical studies used both this scalar potential as well as an additional soft $\mathcal{Q}$ breaking potential term that constrained the plaquette determinant.[6–8] Fig. 1 illustrates the improvement that results from our new method of lifting the U(1) flat directions. Both plots in this figure quantify lattice artifacts by considering violations of $\mathcal{Q}$ Ward identities, which must vanish in the $a/L \to 0$ continuum limit. The left plot addresses the Ward identity that fixes the value of the bosonic action per lattice site, $s_B = 9N^2/2$. The right plot shows $1 - \langle \text{Re } \det \mathcal{P} \rangle$ as discussed above; while only the improved action constrains this through a Ward identity it must still vanish for the unimproved action as the U(1) sector decouples in the continuum limit. In both cases the improved action produces much smaller lattice artifacts and approaches the continuum limit much more rapidly, roughly $\propto (a/L)^2$. This scaling suggests that the improved action preserves the $\mathcal{Q}$ supersymmetry well enough to eliminate any significant effects of dimension-5 operators, all of which are forbidden by $\mathcal{Q}$.[9]

2. Revisiting the static potential

The main physical quantity considered by previous studies using the unimproved action was the static potential. In our earlier work[7,8] we found the static potential to be coulombic at both weak and strong coupling, $V(r) = A + C/r$ with vanishing string tension. For gauge group U(2) our results for the Coulomb coefficient $C(\lambda)$ agreed with perturbation theory[12–14] for all $\lambda \lesssim 2$ that we could investigate, where $\lambda \equiv \lambda_{\text{lat}}/\sqrt{5}$ is the continuum-normalized bare 't Hooft coupling. For $N = 3$, however, we observed a significant departure from perturbation theory for $\lambda \gtrsim 1$, which could conceivably be the first sign that our results were approaching the famous large-N prediction $C \propto \sqrt{\lambda}$ at strong coupling $1 \ll \lambda \ll N$.[15,16]

We are currently repeating and extending these studies using the improved action that will reduce any effects of lattice artifacts. As shown in Fig. 2, our preliminary results from this ongoing work are consistent with the earlier observations

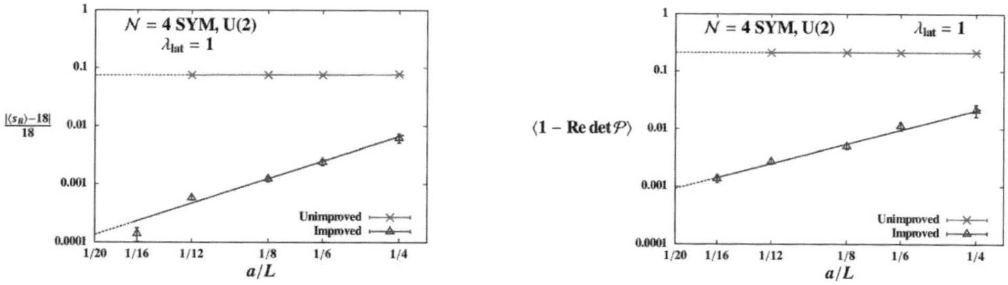

Fig. 1. Continuum extrapolations of $\mathcal{Q}$ Ward identity violations on log–log axes with power-law fits, comparing the new improved lattice action[9] with the unimproved action used in previous works.[6–8] Both deviations of the bosonic action from its exact supersymmetric value (left) and $\langle 1 - \text{Re } \det \mathcal{P} \rangle$ (right) are much smaller for the improved action and approach the $a/L \to 0$ continuum limit much more rapidly, roughly $\propto (a/L)^2$. For both actions we fix $\lambda_{\text{lat}} = 1$ and the plaquette determinant coupling (G or κ), while scaling $\mu \propto 1/L$.

summarized above. Since the improved action allows us to consider significantly stronger couplings, we are in the process of exploring larger λ, as well as investigating $N = 4$. Comparing results for all three gauge groups U(2), U(3) and U(4) will be extremely useful to identify systematic trends, but is numerically challenging[6] due to the rapid growth of computational costs $\propto N^5$.

3. New insight into the pfaffian phase

The improved action allows access to two new sources of information on the possible sign problem of lattice $\mathcal{N} = 4$ SYM. A sign problem is possible because gaussian integration over the fermion fields in the lattice path integral produces a pfaffian that is potentially complex, $\int [d\Psi]\, e^{-\Psi^T \mathcal{D} \Psi} \propto \text{pf}\,\mathcal{D} = |\text{pf}\,\mathcal{D}| e^{i\alpha}$ where $\mathcal{D}$ is the fermion operator. All our numerical calculations "quench" the phase $e^{i\alpha} \to 1$ in order to take advantage of efficient importance sampling algorithms.[6] For the unimproved action subsequent measurements of the pfaffian displayed no indication of a sign problem: the phase $e^{i\alpha}$ exhibited very small and volume-independent fluctuations around unity.[7]

However, most of our unimproved investigations of the pfaffian phase focused on relatively weak couplings, and for volumes larger than $3^3 \times 4$ we considered only $\lambda_{\text{lat}} = 1$. As mentioned in the previous section, the improved action allows us to study significantly stronger couplings while still controlling lattice artifacts, and we are taking advantage of this to begin measuring $\langle e^{i\alpha} \rangle$ for larger $\lambda_{\text{lat}} \leq 8$. Preliminary results from 4^4 lattices are shown in Fig. 3, and reveal that fluctuations in the pfaffian phase increase significantly at stronger couplings $\lambda_{\text{lat}} \gtrsim 4$, apparently indicating a sign problem. Part of this increase may result from the larger values of the auxiliary coupling G (Eq. 2) that have to be used as λ_{lat} increases;[9] larger G appears to be correlated with larger pfaffian phase fluctuations at fixed λ_{lat}.

The other new development is related to the fact that the $\mathcal{Q}$-invariant deformation in Eq. 2 breaks an $\eta \to \eta + c\mathbb{I}_N$ shift symmetry and lifts the corresponding

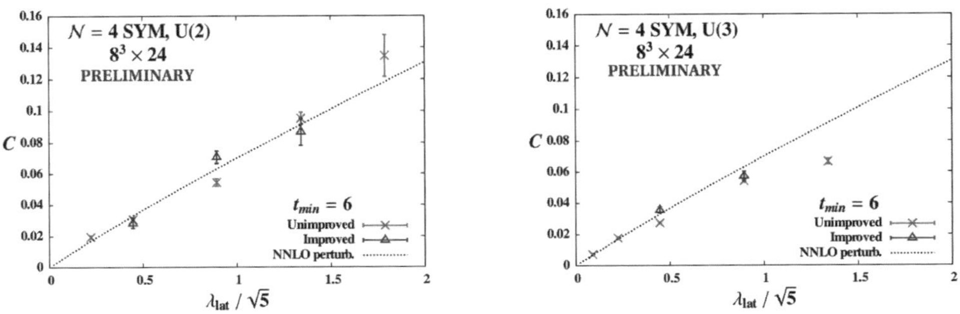

Fig. 2. Static potential Coulomb coefficients C vs. the bare 't Hooft coupling $\lambda_{\text{lat}}/\sqrt{5}$, comparing preliminary results obtained using the improved action with previously published unimproved results.[7,8] For gauge group U(2) (left) we observe agreement with perturbation theory. The $N = 3$ results (right) fall significantly below the NNLO perturbative prediction[12–14] for $\lambda \gtrsim 1$.

U(1) fermion zero mode.[9] The improved action therefore allows us to compute the pfaffian of $\mathcal{D}$ with fully periodic boundary conditions (BCs). This is not possible for the unimproved lattice action, which relies on the fermions' anti-periodic (thermal) temporal BCs to lift this zero mode. In Fig. 4 we compare pfaffian phase measurements with either periodic or anti-periodic temporal BCs for the fermion fields and everything else fixed. The contrast is dramatic: anti-periodic BCs produce $e^{i\alpha} \approx 1$ while fully periodic BCs lead to uncontrolled fluctuations and $\langle e^{i\alpha} \rangle$ consistent with zero. It is not yet clear to us why the pfaffian is so sensitive to these BCs. Even more mysteriously, all other observables change very little between these two ensembles, despite the apparent presence of a sign problem in one case.

4. Work in progress and next steps

Significant work remains necessary to complete both of the ongoing studies discussed above. We are still accumulating statistics for several points in each of Figs. 2 and 3. For the static potential we are in the process of studying stronger couplings and gauge group U(4) to confirm the departure from perturbation theory and explore whether this behavior is consistent with holographic expectations.[15,16] We are also improving our static potential data analysis, to account for short-distance discretization effects when fitting $V(r)$ and to compute renormalized couplings for more consistent comparisons with continuum predictions.

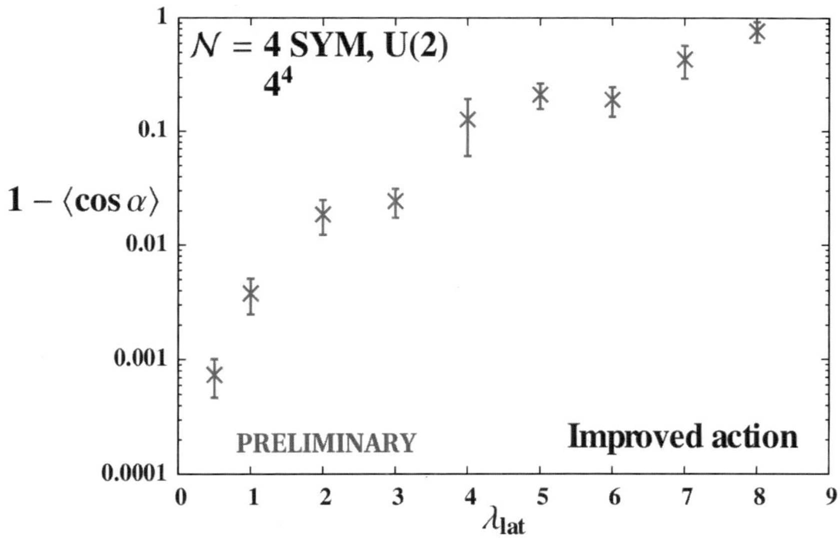

Fig. 3. Dependence of the pfaffian phase fluctuations on the bare 't Hooft coupling λ_{lat}, for 4^4 lattices and gauge group U(2). $1 - \langle \cos\alpha \rangle$ vanishes when the pfaffian is real and positive, and its large value at strong coupling indicates significant fluctuations in the phase, i.e., an apparent sign problem. For each λ_{lat} we use the smallest acceptable μ and G (Eqs. 5 and 2), both of which must increase with the coupling. Larger λ_{lat} and larger G both appear to produce larger fluctuations in $e^{i\alpha}$, potentially compounding the sign problem at strong coupling.

For the pfaffian phase, our new results at stronger 't Hooft couplings need to be supplemented by measurements on larger volumes V, to see whether the volume-independence observed for the unimproved action persists with the improved action. We also want to investigate $e^{i\alpha}$ for $\mathrm{U}(N)$ gauge groups with larger $N = 3$ and 4. Both of these goals are challenging;[6] the computational costs of our pfaffian measurements increase $\propto N^6$ and $\propto V^3 \sim L^{12}$. At the same time we continue to search for improved qualitative understanding of the system's insensitivity to apparent sign problems at strong coupling or with fully periodic BCs.

Of course there are many other interesting aspects of lattice $\mathcal{N} = 4$ SYM that we lack the space to discuss in this proceedings. One example is the restoration of the other fifteen supersymmetries $\mathcal{Q}_a$ and $\mathcal{Q}_{ab}$ in the continuum limit. This can be investigated by measuring violations of discrete R symmetries,[17] as we previously explored with the unimproved action.[7,8] We are currently revisiting this study using the improved action, and will also address the related tuning of a single marginal operator that may be required to obtain the correct continuum limit.[8,18]

Our highest priority, however, is to compute the anomalous dimension of the Konishi operator $\mathcal{O}_K = \sum_I \mathrm{Tr}\left[\Phi^I \Phi^I\right]$, where Φ^I are the six scalar fields of $\mathcal{N} = 4$ SYM. Lattice predictions for this quantity will be complementary to those obtainable in perturbation theory,[19] from holography,[20,21] or via the conformal bootstrap program.[22] This work is advancing rapidly, exploiting both finite-size scaling and Monte Carlo renormalization group techniques,[18] and we hope to report initial results in the near future.

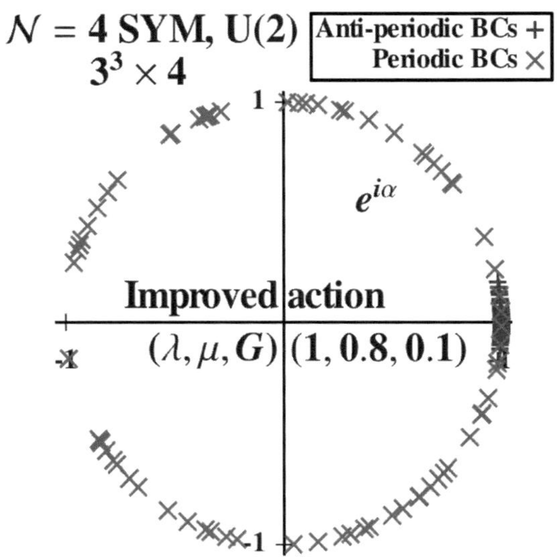

Fig. 4. Scatter plot of pfaffian phase measurements from independent $3^3 \times 4$ ensembles using either periodic or anti-periodic temporal BCs for the fermion fields (and periodic BCs for all other fields and all other directions), with all couplings fixed. While anti-periodic BCs produce $e^{i\alpha} \approx 1$, periodic BCs lead to uncontrolled fluctuations and $\langle e^{i\alpha} \rangle$ consistent with zero.

Acknowledgments

We thank Joel Giedt, Poul Damgaard, Tom DeGrand and Aarti Veernala for helpful conversations and continuing collaboration on lattice $\mathcal{N} = 4$ SYM, as well as Julius Kuti, Valentina Forini and Rainer Sommer for useful discussions about the static potential. We are grateful for the hospitality of the Aspen Center for Physics, supported by the U.S. National Science Foundation under Grant No. PHYS-1066293. This work was supported by the U.S. Department of Energy (DOE), Office of Science, Office of High Energy Physics, under Award Numbers DE-SC0008669 and DE-SC0009998. Numerical calculations were carried out on the HEP-TH cluster at the University of Colorado and on the DOE-funded USQCD facilities at Fermilab.

References

1. D. B. Kaplan and M. Ünsal, *JHEP* **0509**, 042 (2005).
2. M. Ünsal, *JHEP* **0610**, 089 (2006).
3. S. Catterall, *JHEP* **0801**, 048 (2008).
4. P. H. Damgaard and S. Matsuura, *Phys. Lett.* **B661**, 52 (2008).
5. S. Catterall, D. B. Kaplan and M. Unsal, *Phys. Rept.* **484**, 71 (2009).
6. D. Schaich and T. DeGrand, *Comput. Phys. Commun.* **190**, 200 (2015).
7. S. Catterall, D. Schaich, P. H. Damgaard, T. DeGrand and J. Giedt, *Phys. Rev.* **D90**, 065013 (2014).
8. S. Catterall, J. Giedt, D. Schaich, P. H. Damgaard and T. DeGrand, *PoS* **LATTICE2014**, 267 (2014).
9. S. Catterall and D. Schaich, *JHEP* **07**, 057 (2015).
10. N. Marcus, *Nucl. Phys.* **B452**, 331 (1995).
11. A. Kapustin and E. Witten, *Commun. Num. Theor. Phys.* **1**, 1 (2007).
12. A. Pineda, *Phys. Rev.* **D77**, 021701 (2008).
13. M. Stahlhofen, *JHEP* **1211**, 155 (2012).
14. M. Prausa and M. Steinhauser, *Phys. Rev.* **D88**, 025029 (2013).
15. S.-J. Rey and J.-T. Yee, *Eur. Phys. J.* **C22**, 379 (2001).
16. J. M. Maldacena, *Phys. Rev. Lett.* **80**, 4859 (1998).
17. S. Catterall, J. Giedt and A. Joseph, *JHEP* **1310**, 166 (2013).
18. S. Catterall and J. Giedt, *JHEP* **1411**, 050 (2014).
19. C. Beem, L. Rastelli, A. Sen and B. C. van Rees, *JHEP* **1404**, 122 (2014).
20. S. S. Gubser, I. R. Klebanov and A. M. Polyakov, *Phys. Lett.* **B428**, 105 (1998).
21. N. Gromov, V. Kazakov and P. Vieira, *Phys. Rev. Lett.* **104**, 211601 (2010).
22. C. Beem, L. Rastelli and B. C. van Rees, *Phys. Rev. Lett.* **111**, 071601 (2013).

Higgs Mass in D-Term Triggered Dynamical SUSY Breaking

Nobuhito Maru

Department of Mathematics and Physics, Osaka City University,
Osaka, 558-8585, Japan

We discuss a new mechanism of D-term dynamical supersymmetry breaking in the context of Dirac gaugino scenario. The existence of a nontrivial solution of the gap equation for D-term is shown. It is also shown that an observed 126 GeV Higgs mass is realized by tree level D-term effects in a broad range of parameters.

Keywords: Dynamical supersymmetry breaking, Gap equation, Higgs mass.

1. Introduction

Supersymmetry (SUSY) is one of the leading candidates of physics beyond the Standard Model (SM) solving the hierarchy problem. Unfortunately, we have no indication of SUSY at present. Furthermore, an observed Higgs boson mass of 126 GeV discovered at LHC puts severe constraints on the minimal supersymmetric standard model (MSSM) parameter space, namely the MSSM with light sparticles. In this situation, we have two approaches to proceed. One is the MSSM with heavy sparticles, the so-called high scale SUSY breaking scenario which gives up to solve the naturalness problem. The other is extensions of the MSSM. We take the latter approach and focus on the Dirac gaugino scenario[1] as one of the many interesting extensions. The setup to realize Dirac gaugino scenario is as follows. The gauge sector in the MSSM is extended to $\mathcal{N}=2$ matter content by introducing chiral superfields in the adjoint representations of the SM gauge group, $\Phi_a = (\phi_a, \psi_a, F_a)$ $(a = SU(3), SU(2), U(1))$. The matter sector is $\mathcal{N}=1$. Dirac gaugino masses are generated from the dimension five operator

$$\mathcal{L} = \int d^2\theta \sqrt{2} \frac{\mathcal{W}^0_\alpha \mathcal{W}^\alpha_a \Phi_a}{\Lambda} = \frac{\langle D^0 \rangle}{\Lambda} \psi_a \lambda_a + \cdots , \tag{1}$$

if a $U(1)$ D-term vacuum expectation value (VEV) $\langle D^0 \rangle \subset \mathcal{W}^0_\alpha = \theta_\alpha D^0$ in the hidden sector is nonzero. $\mathcal{W}^0_\alpha, \mathcal{W}^a_\alpha$ represents the field strength tensor superfield for the hidden $U(1)$ and the SM gauge groups, respectively. Λ is a cutoff scale.

Sfermion masses are generated at 1-loop,

$$M^2_{\tilde{f}} \simeq \frac{C_a(f)\alpha_a}{\pi} M^2_{\lambda_a} \log \left(\frac{m^2_{\phi_a}}{M^2_{\lambda_a}} \right) \tag{2}$$

where $C_a(f)$ is a group theoretical factor. α_a denotes fine structure constant for the SM gauge coupling constant. $M_{\lambda_a}, m_{\phi_a}$ are gaugino and adjoint scalar masses.

These masses are flavor-blind since it is induced by the SM gauge interactions. Therefore, we have no SUSY flavor and CP problems.

In Dirac gaugino scenario, gaugino masses are typically heavier than the sfermion masses by a factor $4\pi/g$. This fact relaxes the bounds from LHC experiment for spectrum since the gluino and squark production are suppressed comparing to the Majorana gaugino case.

2. A New Mechanism of D-term Dynamical SUSY Breaking

In this section, we discuss a new mechanism of D-term dynamical SUSY breaking generating mixed Majorana-Dirac gaugino masses, which we proposed several years ago[2].

Consider a SUSY $U(N)$ gauge theory with adjoint chiral superfield Φ^a. The Lagrangian is given by

$$\mathcal{L} = \int d^4\theta K(\Phi^a, \overline{\Phi}^a, V) + \int d^2\theta \mathrm{Im}\frac{1}{2}\mathcal{F}_{ab}(\Phi^a)\mathcal{W}^{a\alpha}\mathcal{W}^a_\alpha + \left[\int d^2\theta W(\Phi^a) + \mathrm{h.c.}\right] \quad (3)$$

where $K, \mathcal{F}$ and W are Kähler potential, the gauge kinetic function which is holomorphic in Φ^a, and W is a superpotential whose F-term is assumed to be vanished. The subscripts a, b of $\mathcal{F}$ denote the derivative with respect to $\Phi^{a,b}$. We regard an overall $U(1)$ in $U(N)$ gauge group as the hidden gauge group and denotes the components of the hidden $U(1)$ fields with superscript or subscript "0".

Focusing on the fermion mass terms derived from eq.(3), we find

$$-\frac{1}{2}(\lambda^a, \psi^a)\begin{pmatrix} 0 & -\frac{\sqrt{2}}{4}\mathcal{F}_{ab0}D^0 \\ -\frac{\sqrt{2}}{4}\mathcal{F}_{ab0}D^0 & \partial_a\partial_b W \end{pmatrix}\begin{pmatrix} \lambda^b \\ \psi^b \end{pmatrix}. \quad (4)$$

The mass eigenvalues of (seesaw-type) mass matrix are easily obtained

$$m_\pm = \frac{1}{2}\langle \partial_a\partial_a W\rangle \left[1 \pm \sqrt{1 + \left(\frac{2\langle D\rangle}{\langle \partial_a\partial_a W\rangle}\right)^2}\right], \quad D \equiv -\frac{\sqrt{2}}{4}\mathcal{F}_{0aa}D^0 \quad (5)$$

if $\langle D\rangle \neq 0$ and $\langle \partial_a\partial_a W\rangle \neq 0$. Gaugino masses ($m_-$) becomes massive by nonzero D^0 and SUSY is spontaneously broken.

D-term equation of motion tells us that the D-term VEV is given by Dirac bilinear condensate between the gaugino and the adjoint fermion,

$$\langle D^0\rangle \sim \langle \mathcal{F}_{0ab}\psi^a\lambda^b + \bar{\mathcal{F}}_{0ab}\bar{\psi}^a\bar{\lambda}^b\rangle. \quad (6)$$

As in the BCS and the NJL models, the D-term VEV is determined by solving the gap equation for D-term.

Our strategy of the potential analysis is as follows. Note that we have three constant background fields ϕ^0, D^0, F^0. We simply work in a region where $\langle D^0\rangle \gg \langle F^0\rangle$, namely F-term is perturbative comparing to D-term. Some sample points satisfying $\langle D^0 \rangle \gg \langle F^0 \rangle$ were numerically found in[3]. We first solve the stationary conditions

$$\frac{\partial V(D,\phi,\bar{\phi},F=\bar{F}=0)}{\partial D} = 0, \tag{7}$$

$$\frac{\partial V(D,\phi,\bar{\phi},F=\bar{F}=0)}{\partial \phi} = \frac{\partial V(D,\phi,\bar{\phi},F=\bar{F}=0)}{\partial \bar{\phi}} = 0. \tag{8}$$

Then, we find the stationary values $(D_*, \phi_*, \bar{\phi}_*)$. Using these stationary values, the stationary conditions for F-term is solved perturbatively,

$$\frac{\partial V(D_*,\phi_*,\bar{\phi}_*,F,\bar{F})}{\partial F} = \frac{\partial V(D_*,\phi_*,\bar{\phi}_*,F,\bar{F})}{\partial \bar{F}} = 0 \tag{9}$$

and the stationary values of F-term $(F_*, \bar{F}_*)$ are found. Note that the stationary condition for D-term (7) is nothing but the gap equation which we wish to solve.

The 1-loop effective potential is calculated

$$V = N^2|m_\phi|^4 \left[c_1(\phi,\bar{\phi})\Delta_0^2 + \frac{1}{32\pi^2}\left(c_2\Delta_0^4 - |\lambda^{(+)}|^4 \log|\lambda^{(+)}|^2 - |\lambda^{(-)}|^4 \log|\lambda^{(-)}|^2 \right) \right] \tag{10}$$

where $\lambda^{(\pm)} \equiv m_\pm/\langle \partial_a \partial_a W \rangle = \frac{1}{2}(1 \pm \sqrt{1+\Delta_0^2}), \Delta_0 \equiv 2\langle D \rangle / \langle \partial_a \partial_a W \rangle$. c_1 is some function of ϕ and $\bar{\phi}$. c_2 is a constant[4].

We can immediately obtain the gap equation for D-term from the stationary condition of D-term.

$$0 = \frac{\partial V}{\partial D} = \Delta_0 \left[c_1 + \frac{1}{64\pi^2} \left\{ 4c_2\Delta_0^2 - \frac{1}{\sqrt{1+\Delta_0^2}} \left(\lambda^{(+)3}(2\log\lambda^{(+)2}+1) - \lambda^{(-)3}(2\log\lambda^{(-)2}+1) \right) \right\} \right]. \tag{11}$$

Note that the gap equation has a trivial solution $\Delta_0 = 0$, which represents a SUSY vacuum. In Fig.1 (a), the plot of the quantity $\partial V/(\Delta_0 \partial D)$ as a function of Δ_0 is shown. From this plot, we can find a nontrivial solution representing SUSY breaking vacuum. The stationary value of the adjoint scalar fields is determined by the intersection point of the two curves representing the gap equation and the stationary condition for ϕ in the $(\Delta_0, \phi = \bar{\phi})$ plane. The schematic picture is shown in Fig.1 (b).

The SUSY breaking vacuum cannot be a global minimum of the potential since the energy of SUSY theory is positive semi-definite and a SUSY minimum with the vanishing energy is a solution of the gap equation as mentioned above. We have to check whether our SUSY breaking is sufficiently long-lived by taking into account a tunneling effect from our false SUSY breaking vacuum to the true SUSY one. Its tunneling rate can be roughly estimated and should be small

$$\exp\left[-\frac{\langle \Delta\phi \rangle^4}{\langle \Delta V \rangle} \right] \sim \exp\left[-\frac{(\Delta_0 \Lambda)^2}{m_\phi^2} \right] \ll 1 \tag{12}$$

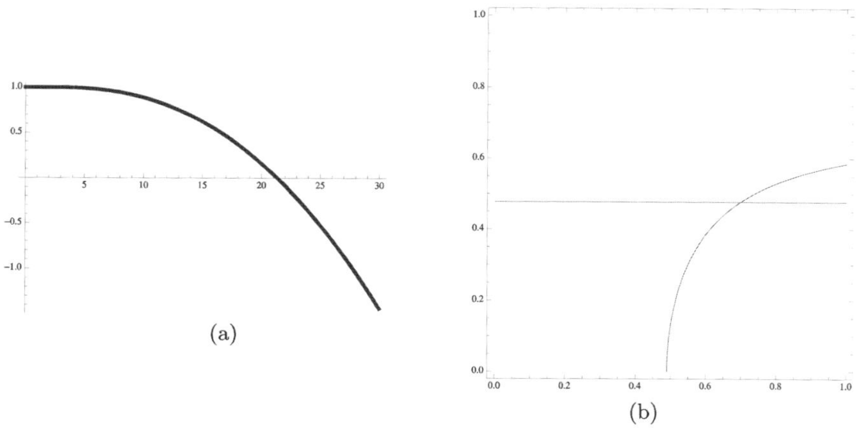

Fig. 1. (a) The plot of the quantity $\partial V/(\Delta_0 \partial D)$ as a function of Δ_0. (b) The schematic picture of the intersection of the two curves representing the gap equation (the straight line) and the ϕ flat condition (the curved one). The horizontal (vertical) axis is $\phi/\Lambda(\Delta_0)$.

where $\Delta\phi$ is a field distance between the two vacua, which is roughly given by $\Delta_0\Lambda$. ΔV is a potential height between them, which is given by D-term squared $(m_\phi \Delta_0 \Lambda)^2$. In order for our false vacuum to be sufficiently long-lived, $m_\phi \ll \Delta_0\Lambda$ is required, which can be always satisfied since the adjoint scalar mass is a free superpotential mass parameter.

3. Higgs Mass via D-term Effects

In this section, we discuss whether an observed 126 GeV Higgs mass can be realized in our framework[4]. The Lagrangian of Higgs sector is

$$\begin{aligned}\mathcal{L}_{\text{Higgs}} &= \int d^4\theta \left[H_u^\dagger e^{-g_Y V_1 - g_2 V_2 - 2q_u g V_0} H_u + H_d^\dagger e^{g_Y V_1 - g_2 V_2 - 2q_d g V_0} H_d \right] \\ &\quad + \left[\left(\int d^2\theta \mu H_u \cdot H_d \right) - B\mu H_u \cdot H_d + \text{h.c.} \right]. \end{aligned} \tag{13}$$

We have adopted notation $X \cdot Y \equiv \epsilon_{AB} X^A Y^B = X^A Y_A = -Y \cdot X$, $\epsilon_{12} = -\epsilon_{21} = \epsilon^{21} = -\epsilon^{12} = 1$. $V_{1,2,0}$ are vector superfields of the SM gauge group and that of the overall $U(1)$ respectively and the corresponding gauge couplings are denoted by $g_{Y,2}$ and g, respectively. Unlike the MSSM case, the soft scalar Higgs masses $m_{H_u}^2 |H_u|^2, m_{H_d}^2 |H_d|^2$ are not introduced since they are induced by D-term contributions in our framework.

The Higgs potential is derived from (13)

$$\begin{aligned}
V_{\rm Higgs} &= \frac{g_2^2}{2(1+{\rm Im}\mathcal{F}'''\langle\Phi_0\rangle)}\left(H_u^\dagger\frac{\sigma^a}{2}H_u + H_d^\dagger\frac{\sigma^a}{2}H_d\right)^2 \\
&+\frac{g_Y^2}{8(1+{\rm Im}\mathcal{F}'''\langle\Phi_0\rangle)}\left(|H_u|^2-|H_d|^2\right)^2 \\
&+\frac{1}{2(1+{\rm Im}\mathcal{F}'''\langle\Phi_0\rangle)}\left(q_u g|H_u|^2+q_d g|H_d|^2-\left.\frac{\partial\Gamma^{1-{\rm loop}}(D^0)}{\partial D^0}\right|_{D^0=D^{0*}}\right)^2 \\
&+|\mu|^2(|H_u|^2+|H_d|^2)+(B\mu H_u\cdot H_d+{\rm h.c.}). \qquad (14)
\end{aligned}$$

Here we have denoted by D^{0*} the solution to the improved gap equation,

$$0=(1+{\rm Im}\mathcal{F}'''\langle\Phi_0\rangle)D^0-q_u g|H_u|^2-q_d g|H_d|^2+\frac{\partial\Gamma^{1-{\rm loop}}(D^0)}{\partial D^0}. \qquad (15)$$

The deviation δD^{0*} of the value from D^{0*} in[4] is in fact small by the ratio of electroweak scale and SUSY breaking scale. Therefore, we approximate the solution to the improved gap equation by the value of D^{0*} in[4] denoted as $\langle D^0\rangle$. Taking into account the fact that ${\rm Im}\mathcal{F}'''\langle\Phi_0\rangle\sim\langle\Phi_0\rangle/\Lambda\ll 1$, we neglect the term ${\rm Im}\mathcal{F}'''\langle S\rangle$ at the leading order. The resulting Higgs potential at the leading order is given by

$$\begin{aligned}
V_{\rm Higgs} &\simeq \frac{g_2^2+g_Y^2}{8}\left[|H_u^0|^2-|H_d^0|^2\right]^2+\frac{1}{2}\left(q_u g|H_u^0|^2+q_d g|H_d^0|^2-\langle D^0\rangle\right)^2 \\
&+|\mu|^2\left(|H_u^0|^2+|H_d^0|^2\right)-\left(B\mu H_u^0 H_d^0+{\rm h.c.}\right) \\
&= \frac{g_2^2+g_Y^2}{32}v^4c_{2\beta}^2+\frac{v^2}{2}\left[\mu^2-B\mu s_{2\beta}\right]+\frac{1}{8}\left((q_u g s_\beta^2+q_d g c_\beta^2)v^2-2\langle D^0\rangle\right)^2 \qquad (16)
\end{aligned}$$

where we have restricted the potential to the CP-even neutral sector of Higgs doublets $H_u=(H_u^+,H_u^0)^T$, $H_d=(H_d^0,H_d^-)^T$. In the last line, the neutral components of Higgs fields are parametrized as

$$H_u^0=\frac{1}{\sqrt{2}}\left[s_\beta(v+h)+c_\beta H+i(c_\beta A-s_\beta G^0)\right], \qquad (17)$$

$$H_d^0=\frac{1}{\sqrt{2}}\left[c_\beta(v+h)-s_\beta H+i(s_\beta A+c_\beta G^0)\right] \qquad (18)$$

where $s_\beta\equiv\sin\beta, c_\beta\equiv\cos\beta$. h,H,A are the SM Higgs, the CP even heavy Higgs and CP odd Higgs, respectively. G^0 is the would-be Nambu-Goldstone boson eaten as the longitudinal component of Z-boson. The VEV of Higgs field is $v\simeq 246$ GeV and $\frac{g_Y^2+g_2^2}{4}v^2=M_Z^2$ in this convention.

We are now ready to calculate Higgs mass. As in the MSSM, the minimization of the scalar potential $\partial V_{\rm Higgs}/\partial v^2=\partial V_{\rm Higgs}/\partial\beta=0$ allows us to express μ and

$B\mu$ in terms of other parameters.

$$\mu^2 + \frac{M_Z^2}{2} = \frac{g}{2c_{2\beta}} \left((q_u s_\beta^2 + q_d c_\beta^2) g v^2 - 2\langle D^0 \rangle\right) \left(q_u s_\beta^2 - q_d c_\beta^2\right), \tag{19}$$

$$\begin{aligned} M_A^2 \equiv \frac{2B\mu}{s_{2\beta}} &= 2\mu^2 + \frac{q_u + q_d}{2} g \left((q_u s_\beta^2 + q_d c_\beta^2) g v^2 - 2\langle D^0 \rangle\right) \\ &= -M_Z^2 + \frac{q_u - q_d}{2c_{2\beta}} g \left((q_u s_\beta^2 + q_d c_\beta^2) g v^2 - 2\langle D^0 \rangle\right). \end{aligned} \tag{20}$$

The Higgs mass is obtained as

$$m_{\text{Higgs}}^2 = \frac{1}{2}\left[\tilde{M}_Z^2 + M_A^2 - \sqrt{\left(\tilde{M}_Z^2 + M_A^2\right)^2 - 4\tilde{M}_Z^2 M_A^2 c_{2\beta}^2}\right] \tag{21}$$

where $\tilde{M}_Z^2 \equiv M_Z^2 + q_u^2 g^2 v^2$. It is interesting to see the correspondence between our expression of Higgs mass (21) and that in the MSSM,

$$m_{\text{MSSM Higgs}}^2 = \frac{1}{2}\left[M_Z^2 + M_A^2 - \sqrt{(M_Z^2 + M_A^2)^2 - 4M_Z^2 M_A^2 c_{2\beta}^2}\right]. \tag{22}$$

As in the case of MSSM, the upper bound of Higgs mass can be obtained by taking a decoupling limit $M_A^2 \to \infty$,

$$m_{\text{Higgs}}^2 \to \tilde{M}_Z^2 c_{2\beta}^2. \tag{23}$$

$\tilde{M}_Z$ can be large enough by taking $\mathcal{O}(1)$ charge and coupling $q_u g$

$$\tilde{M}_Z \sim \sqrt{(90 \text{ GeV})^2 + (246 \text{ GeV})^2} \sim 262 \text{ GeV}. \tag{24}$$

Note that the minimization conditions of Higgs potential with $q_u + q_d = 0$ to allow μ-term in the superpotential leads to

$$M_Z^2 + M_A^2 = -\frac{q_u g}{c_{2\beta}} \left(c_{2\beta} q_u g v^2 + 2\langle D^0 \rangle\right). \tag{25}$$

In order to satisfy this condition, the dominant part in the right-hand side of (25) $q_u g \langle D^0 \rangle / c_{2\beta}$ is required to be negative.

Using this condition, we can eliminate M_A^2 in Higgs mass (21).

$$m_{\text{Higgs}}^2 = \frac{1}{2}\left[-\frac{2q_u g}{c_{2\beta}}\langle D^0 \rangle - \sqrt{\left(-\frac{2q_u g}{c_{2\beta}}\langle D^0 \rangle\right)^2 + 8c_{2\beta} q_u g \tilde{M}_Z^2 \langle D^0 \rangle + 4c_{2\beta}^2 \tilde{M}_Z^4}\right] \tag{26}$$

A plot for 126 GeV Higgs mass as a function of $\tan\beta$ and $q_u g$ is shown below. Here we have taken $q_u g > 0$ and $\cos 2\beta < 0$ to satisfy the condition $q_u g \langle D^0 \rangle / c_{2\beta} < 0$. We can immediately see that 126 GeV Higgs mass is realized in a broad range of parameters. Also, we found that the result is insensitive to the values of D-term VEV. This fact is naturally expected from the non-decoupling nature of Higgs mass.

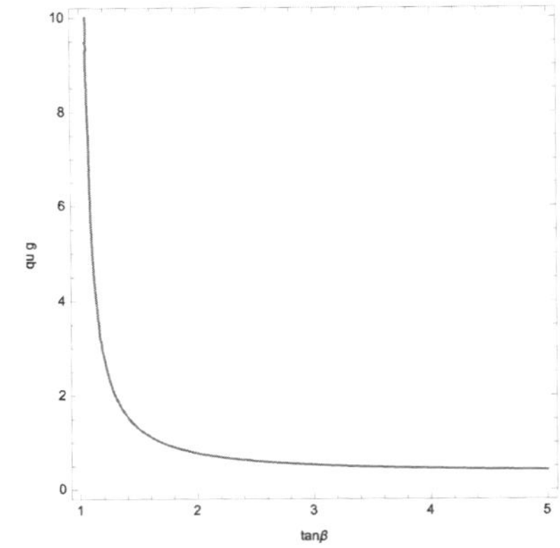

Fig. 2. A plot for the 126 GeV Higgs mass as a function of $\tan\beta$ and $q_u g$.

4. Summary

Dirac gaugino scenario is an interesting extension of the MSSM, which relaxes LHC constraints for the SUSY breaking parameters by gluino and squark production suppression comparing to Majorana gaugino case. Dirac gaugino mass is generated by the nonvanishing D-term VEV. We have proposed a new mechanism of D-term dynamical SUSY breaking in the context of Dirac gaugino scenario, where the gap equation of D-term has a nontrivial solution[2,3]. The SUSY breaking vacuum is a necessarily local minimum since we have a trivial solution of the gap equation, i.e. SUSY vacuum, and the vacuum energy in SUSY theories is positive semi-definite. It was shown that the lifetime of our SUSY breaking vacuum can be sufficiently long-lived by adjusting the superpotential mass parameter. As a phenomenological application, we have shown that an observed 126 GeV Higgs mass can be realized by tree level D-term effects in a broad range of parameters.

Acknowledgments

The work of the author is supported in part by the Grant-in-Aid for Scientific Research from the Ministry of Education, Science and Culture, Japan No. 24540283.

References

1. P. Fayet, Phys. Lett. B **78**, 417 (1978); J. Polchinski and L. Susskind, Phys. Rev. D **26**, 3661 (1982); L. J. Hall and L. Randall, Nucl. Phys. B **352**, 289 (1991); P. J. Fox, A. E. Nelson and N. Weiner, JHEP **0208**, 035 (2002).
2. H. Itoyama and N. Maru, Int. J. Mod. Phys. A **27**, 1250159 (2012).
3. H. Itoyama and N. Maru, Phys. Rev. D **88**, no. 2, 025012 (2013).
4. H. Itoyama and N. Maru, Symmetry **2015** 7(1) 193.

Large-Scale Magnetic Fields, Non-Gaussianity, and Gravitational Waves from Inflation

Kazuharu Bamba[1,2,3,4]

[1] *Leading Graduate School Promotion Center, Ochanomizu University, 2-1-1 Ohtsuka, Bunkyo-ku, Tokyo 112-8610, Japan*
[2] *Department of Physics, Graduate School of Humanities and Sciences, Ochanomizu University, Tokyo 112-8610, Japan*
[3] *Kobayashi-Maskawa Institute for the Origin of Particles and the Universe, Nagoya University, Nagoya 464-8602, Japan*
[4*] *Division of Human Support System, Faculty of Symbiotic Systems Science, Fukushima University, Fukushima 960-1296, Japan*
E-mail: bamba@sss.fukushima-u.ac.jp

We explore the generation of large-scale magnetic fields in the so-called moduli inflation. The hypercharge electromagnetic fields couple to not only a scalar field but also a pseudoscalar one, so that the conformal invariance of the hypercharge electromagnetic fields can be broken. We explicitly analyze the strength of the magnetic fields on the Hubble horizon scale at the present time, the local non-Gaussianity of the curvature perturbations originating from the massive gauge fields, and the tensor-to-scalar ratio of the density perturbations. As a consequence, we find that the local non-Gaussianity and the tensor-to-scalar ratio are compatible with the recent Planck results.

Keywords: Cosmology, Particle-theory and field-theory models of the early Universe, Axions and other Nambu-Goldstone bosons, String and brane phenomenology.

1. Introduction

Magnetic fields with the current strength $\sim 10^{-6}$G on 1–10kpc scale and those with 10^{-7}–10^{-6}G on 10 kpc–1Mpc scale have been detected in galaxies and clusters of galaxies, respectively. The generation mechanism of the large-scale magnetic fields in clusters of galaxies have not been understood well.[1–7]

Electromagnetic quantum fluctuations generated during inflation are considered to be the most natural origin of such large-scale magnetic fields. This is because inflation extends the coherent scale of the magnetic fields to be larger than the Hubble horizon.[8] The homogeneous and isotropic Friedmann-Lemaître-Robertson-Walker (FLRW) universe is conformally flat. In addition, the electromagnetic fields are conformally invariant. Accordingly, the conformal invariance of the electromagnetic fields has to be broken at the inflationary stage. As a result, the quantum fluctuations of the electromagnetic fields can be generated.[8,9] As representative mechanisms to break the conformal invariance of the electromagnetic fields, there have

*The author's current affiliation

been proposed the non-minimal coupling of the electromagnetic fields to the scalar curvature,[8,10–13] its coupling to the scalar fields,[14–18] and the trace anomaly.[19]

In this paper, we review our main results in Ref. 20. We explore a toy model of moduli inflation inspired by racetrack inflation[21] in the framework of the Type IIB string theory. The new point of our model is to take into account the coupling of the hypercharge electromagnetic fields to a scalar field as well as to a pseudoscalar one, which plays a role of the inflaton field. Only the latter coupling has been studied in the past works.[22–25] We explicitly estimate the values of current magnetic fields on the Hubble horizon scale, local non-Gaussianity of the curvature perturbations, and tensor-to-scalar ratio of density perturbations. This is the most significant consequence of the present work. We use the units $k_{\mathrm{B}} = c = \hbar = 1$ and describe the Newton's constant by $G = 1/M_{\mathrm{P}}^2$, where $M_{\mathrm{P}} = 2.43 \times 10^{18}$ GeV is the reduced Planck mass. In terms of electromagnetism, we adopt Heaviside-Lorentz units.

The paper is organized as follows. In Sec. 2, the model Lagrangian is explained, and the field equations are derived. In Sec. 3, we investigate the present magnetic field strength on the Hubble horizon scale. Furthermore, we study the non-Gaussianity of the curvature perturbations in Sec. 4, and we explore the tensor-to-scalar ratio of the density perturbations in Sec. 5. Finally, summary is given in Sec. 6.

2. Model Lagrangian

The Lagrangian is represented as

$$\begin{aligned}\mathcal{L} = & \frac{M_{\mathrm{P}}^2}{2} R - \frac{1}{4} X F_{\mu\nu} F^{\mu\nu} - \frac{1}{4} g_{\mathrm{ps}} \frac{Y}{M} F_{\mu\nu} \tilde{F}^{\mu\nu} \\ & - \frac{1}{2} g^{\mu\nu} \partial_\mu \Phi \partial_\nu \Phi - U(\Phi) - \frac{1}{2} g^{\mu\nu} \partial_\mu Y \partial_\nu Y - V(Y) \,. \end{aligned} \tag{1}$$

Here, R is the scalar curvature, Φ is the canonical scalar field, $X \equiv \exp(-\lambda\Phi/M_P)$ with a constant λ, and $V(Y) \approx \bar{V} - (1/2)\, m^2 Y^2$, where $\bar{V}$ is a constant and m is the mass of Y. Moreover, $U(\Phi)$ and $V(Y)$ are the potentials of Φ and Y, respectively, g_{ps} is a dimensionless coupling constant, and M is a constant with the mass dimension, which corresponds to the decay constant of Y. In addition, the field strength of the $\mathrm{U}(1)_Y$ hypercharge gauge field F_μ is given by $F_{\mu\nu} = \nabla_\mu F_\nu - \nabla_\nu F_\mu$ with ∇_μ the covariant derivative, and the dual field strength of F_μ is $\tilde{F}^{\mu\nu}$. The pseudoscalar field Y corresponds to the inflaton field.

We assume the flat FLRW universe with the metric $ds^2 = -dt^2 + a^2(t) d\boldsymbol{x}^2$, where a is the scale factor. In this background, the field equations of Φ (i.e., X) and Y read $\ddot{\Phi} + 3H\dot{\Phi} + dU(\Phi)/d\Phi = 0$ and $\ddot{Y} + 3H\dot{Y} + dV(Y)/dY = 0$, respectively, where $H \equiv \dot{a}/a$ is the Hubble parameter and the dot means the derivative with respect to the cosmic time t. With the Coulomb gauge $F_0(t, \boldsymbol{x}) = 0$ and $\partial_j F^j(t, \boldsymbol{x}) = 0$, we also obtain the field equation of F_μ.

3. Magnetic field strength at the present time

For the slow-roll inflation driven by the potential of Y, the scale factor $a(t)$ can be described by $a(t) = a_k \exp\left[H_{\rm inf}\left(t - t_k\right)\right]$, where $a_k = a(t_k)$, t_k is the time when a comoving wavelength $2\pi/k$ of the $\mathrm{U}(1)_Y$ gauge field first crosses the horizon during inflation ($k/(a_k H_{\rm inf}) = 1$), and $H_{\rm inf}$ is the Hubble parameter during inflation. A solution of equation of motion for Y is derived as[15] $Y(t) = Y_k \exp\left[(3/2)\,\mathcal{D} H_{\rm inf}\left(t - t_k\right)\right]$, where $\mathcal{D} \equiv -1 \pm \sqrt{1 + \left[2m/\left(3H_{\rm inf}\right)\right]}$ and $Y_k \equiv Y(t = t_k)$.

We execute the cannonical quantization of the $\mathrm{U}(1)_Y$ gauge field $F_\mu(t, \boldsymbol{x})$. We express the comoving wave number by $\boldsymbol{k}$ ($k = |\boldsymbol{k}|$). We set the x^3 axis to lie along the direction of the spatial momentum $\boldsymbol{k}$ and express the transverse directions as x^1 and x^2. By using the circular polarizations $F_\pm(k,t) \equiv F_1(k,t) \pm iF_2(k,t)$ with the Fourier modes $F_1(k,t)$ and $F_2(k,t)$ of the $\mathrm{U}(1)_Y$ gauge field, we have

$$\ddot{F}_\pm(k,t) + \left(H_{\rm inf} + \frac{\dot{X}}{X}\right)\dot{F}_\pm(k,t) + \left[1 \pm \frac{g_{\rm ps}}{M}\frac{\dot{Y}}{X}\left(\frac{k}{a}\right)^{-1}\right]\left(\frac{k}{a}\right)^2 F_\pm(k,t) = 0\,. \quad (2)$$

We numerically calculate the solution of this equation at the inflationary stage by defining the ratio $C_+(k,t) \equiv F_+(k,t)/F_+(k,t_k)$, where t_k is taken as the initial time of the numerical calculation. Here, in the short-wavelength limit of $k \to \infty$, we set $F_+(k,t) = 1/\sqrt{2kX(t)}$. Namely, we take the so-called Bunch-Davies vacuum so that in this limit the vacuum can be the one in the Minkowski space-time. As a result, without sensitive dependence of the model parameters, $C_+(k,t)$ becomes a constant after ~ 10 Hubble expansion time. after the horizon crossing during inflation. The proper hypermagnetic field is expressed as[14] ${B_Y}_i^{\rm proper}(t,\boldsymbol{x}) = \left(1/a^2\right) {B_Y}_i(t,\boldsymbol{x}) = \left(1/a^2\right)\epsilon_{ijk}\partial_j F_k(t,\boldsymbol{x})$, where ${B_Y}_i(t,\boldsymbol{x})$ is the comoving hypermagnetic field, and ϵ_{ijk} is the totally antisymmetric tensor ($\epsilon_{123} = 1$). We suppose that after inflation, the instantaneous reheating happens $t = t_{\rm R}$ (much before the electroweak phase transition (EWPT), where $T_{\rm EW} \sim 100$GeV). Carged particles are produced at the reheating stage, so that the cosmic conductivity of $\sigma_{\rm c}$ can be much larger than H. Thus, after reheating, B_Y behaves as $B_Y \propto a^{-2}$. On the other hand, the hyperelectric fields accelerats the charged particles and eventually vanish. The energy density of the magnetic fields $\rho_B(L,t)$ in the position space reads[26]

$$\rho_B(L,t) \simeq \frac{1}{8\pi^2}\frac{1}{X(t_k)}\frac{1}{\sqrt{2\xi_k}}\exp\left[2\left(\pi\xi_k - 2\sqrt{2\xi_k}\right)\right]\left(\frac{k}{a}\right)^4 \left|C_+(k,t_{\rm R})\right|^2, \quad (3)$$

with $\xi_k = \xi(t = t_k)$, where $\xi \equiv g_{\rm ps}\dot{Y}/\left(2MXH_{\rm inf}\right)$. Here, we have imposed $X(t_{\rm R}) = 1$ in order or the Maxwell theory is recovered after the instantaneous reheating ($t \geq T_{\rm R}$). We have also neglected the difference between the coefficient of the $\mathrm{U}(1)_Y$ magnetic field and that of the $\mathrm{U}(1)_{\rm em}$ one, which is $\mathcal{O}(1)$.

In Table 1, we show the cases in which $B(H_0^{-1}, t_0) = \mathcal{O}(10^{-64})$ G on the Hubble horizon scale at the present time t_0 is generated for $X(t_k) = \exp\left(\chi_k\right)$ with $\chi_k = -0.940$, $g_{\rm ps} = 1.0$, $\xi_k = 2.5590616$, and $k = 2\pi/\left(2997.9h^{-1}\right)$ Mpc^{-1} with $h = 0.673$ (which is consistent with the recent Planck result[27]). In the cases (j) (j = a, b, c,

Table 1. Magnetic field strength on the Hubble horizon scale at the present time.

	$B(H_0^{-1}, t_0)$ [G]	$B(1\text{Mpc}, t_0)$ [G]	H_{inf} [GeV]	m [GeV]	Y_k/M_{P}
(a)	7.15×10^{-64}	1.42×10^{-56}	1.0×10^{11}	2.44×10^{10}	7.70×10^{-2}
(b)	7.15×10^{-64}	1.42×10^{-56}	1.0×10^{10}	2.44×10^{9}	7.70×10^{-2}
(c)	2.33×10^{-64}	4.62×10^{-57}	1.0×10^{8}	1.0×10^{7}	1.62×10^{1}
(d)	2.33×10^{-64}	4.62×10^{-57}	1.0×10^{6}	1.0×10^{5}	1.62×10^{1}
(e)	2.85×10^{-64}	5.66×10^{-57}	1.0×10^{4}	8.0×10^{2}	2.23×10^{1}
(f)	2.85×10^{-64}	5.66×10^{-57}	1.0×10^{2}	8.0	2.23×10^{1}

d, e, f), we obtain $C_+(k, t_{\text{R}}) = (0.528, 0.528, 0.172, 0.172, 0.211, 0.211,)$, T_{R} [GeV] $= (1.02\times10^{14},\ 3.22\times10^{13},\ 3.22\times10^{12},\ 3.22\times10^{11},\ 3.22\times10^{10},\ 3.22\times10^{9})$, and $\bar{V}/M_{\text{P}}^4 = (5.07\times10^{-15},\ 5.07\times10^{-17},\ 5.07\times10^{-21},\ 5.07\times10^{-25},\ 5.07\times10^{-29},\ 5.07\times10^{-33})$. It is confirmed that $C_+(k, t_{\text{R}})$ is $\mathcal{O}(0.1)$ for the wide range of H_{inf} and m. All the results in Table 1 satisfy the constraints on the non-Gaussianity [28] and the tensor-to-scalar ratio [29] from the Planck analysis. The magnetic field strength of $\mathcal{O}(10^{-64})$ G is also compatible with both the theoretical backreaction problem [30] and the recent observational constraints from the Planck satellite. [31]

4. Non-Gaussianity of the curvature perturbations

In the context of string theories, the gauge symmetry is spontaneously broken, so that the gauge field can have the mass. Therefore, we study such a spontaneous symmetry breaking by considering the case that the $\text{U}(1)_Y$ gauge field has its coupling to other Higgs-like field φ, which can evolve to the vacuum expectation value. The kinetic term of φ reads $|D\varphi|^2$, where the covariant derivative for φ is defined by $D_\mu \equiv \partial_\mu + ig' F_\mu$ with g' the gauge coupling[24]. The gauge field acquires the mass thanks to the Higgs mechanism in terms of φ. The quantum fluctuations of φ lead to the quantum fluctuations on the mass of the gauge field. Hence, the amount of the quantum fluctuations is equal to that of quanta of the generated gauge field. Consequently, the gauge field generation yields the perturbations of number of e-folds during inflation δN. The local type non-Gaussinanity of anisotropy of the CMB radiation originates from these perturbations. The non-Gaussianity can be analyzed by exploring the curvature perturbations coming from the quantum fluctuations of φ. For the model of inflation in Ref. 25, with the COBE normalization of power spectrum of the curvature perturbations [32] $\Delta_{\mathcal{R}}^2(k) = 2.4 \times 10^{-9}$ at $k = k_* = 0.002\,\text{Mpc}^{-1}$, the local type non-Gaussianity $f_{\text{NL}}^{\text{local}}$ is given by [24] $f_{\text{NL}}^{\text{local}} \approx 1.0 \times 10^{14} \left(g'^4/\xi^6\right) \left(m^2/H_{\text{inf}}^2\right)$. In Table 2, we list the local non-Gaussianity $f_{\text{NL}}^{\text{local}}$ of the curvature perturbations for $M = 1.0 \times 10^{-1} M_{\text{P}} = 2.43 \times 10^{17}\,\text{GeV}$, $g_{\text{ps}} = 1.0$, and $k = 2\pi/\left(2997.9h^{-1}\right)\,\text{Mpc}^{-1}$ with $h = 0.673$. Through the relation $\bar{V} = 3H_{\text{inf}}^2 M_{\text{P}}^2$, we determine the value of $\bar{V}$ as $\bar{V} = 5.07\times10^{-11} M_{\text{P}}^4$ for the case (A) and $\bar{V} = 5.07 \times 10^{-13} M_{\text{P}}^4$ for the case (B). The value of $C_+(k, t_{\text{R}})$ is (the case (A), the case (B)) $= (0.528, 0.528)$. The magnetic field strength on the Hubble horizon scale at the present time $B(H_0^{-1}, t_0)$ [G] is estimated as (the case (A), the case

Table 2. Local type non-Gaussianity.

	$f_{\rm NL}^{\rm local}$	g'^2	$H_{\rm inf}$ [GeV]	m [GeV]	$Y_k/M_{\rm P}$
(A)	2.70	1.13×10^{-5}	1.0×10^{13}	2.44×10^{12}	7.70×10^{-2}
(B)	2.12×10^{8}	1.0×10^{-1}	1.0×10^{12}	2.44×10^{11}	7.70×10^{-2}

Table 3. Tensor-to-scalar ratio.

	r	$H_{\rm inf}$ [GeV]	m [GeV]	$Y_k/M_{\rm P}$
(A)	1.87×10^{-5}	1.0×10^{13}	2.44×10^{12}	7.70×10^{-2}
(B)	1.87×10^{-5}	1.0×10^{12}	2.44×10^{11}	7.70×10^{-2}

(B)) $= (7.15 \times 10^{-64}, 7.15 \times 10^{-64})$, where we have used Eq. (3) with the absolute value of $C_+(k, t_{\rm R})$. The constraint on $f_{\rm NL}^{\rm local}$ acquired from the Planck satellite[28] is $f_{\rm NL}^{\rm local} = 2.7 \pm 5.8\,(68\%\,{\rm CL})$. It follows from Table 2 that the value of $f_{\rm NL}^{\rm local}$ in the case (A) can satisfy the constraints of the Planck results, but that of $f_{\rm NL}^{\rm local}$ in the case (B) cannot. The parameter sets whose values are close to those of the case (A) in Table 2 can also meet the constraints on $f_{\rm NL}^{\rm local}$, although the parameter space to meet the constraints of $f_{\rm NL}^{\rm local}$ is small. The relation of the primordial magnetic fields to the non-Gaussianity of the curvature perturbations has also been examined[33].

5. Tensor-to-scalar ratio of the density perturbations

The definition of the tensor-to-scalar ratio r is the amplitude of tensor modes (i.e., the primordial gravitational waves) of the density perturbations divided by that of their scalar modes. With a slow-roll parameter $\epsilon \equiv \left(M_{\rm P}^2/2\right)\left[\left(dV(Y)/dY\right)/V(Y)\right]^2$, r is represented as[23] $r = 16\epsilon(t_k)$, where $\epsilon(t_k) = \epsilon(t = t_k) = 2M_{\rm P}^2 m^4 Y_k^2 / \left(2\bar{V} - m^2 Y_k\right)^2$. In Table 3, we display the values of the tensor-to-scalar ratio r in the cases (A) and (B) (which are the same in Table 2). The upper bound estimated by the Planck analysis is $r < 0.11(95\%\,{\rm CL})$.[29] It is clearly seen that this upper limit can be met in these cases. Thus, if the magnetic fields on the Hubble horizon scale with the current field strength $\mathcal{O}(10^{-64})$ G are generated, not only the local non-Gaussianity but also the tensor-to-scalar ratio in terms of the CMB radiation, which are compatible with the recent Planck data, can be produced for a certain parameter space.

6. Summary

In the present paper, we have investigated the generation of the large-scale magnetic fields from a kind of moduli inflation. We have first analyzed the values of three cosmological observables: Current strength of the magnetic fields on the Hubble horizon scale, local non-Gaussianity, and the tensor-to-scalar ratio. We have found that the local non-Gaussianity and tensor-to-scalar ratio obtained in this model can satisfy the recent constraints acquired from the Planck satellite.

Acknowledgments

The author would like to sincerely appreciate the significant discussions with Professor Tatsuo Kobayashi and Professor Osamu Seto and their kind meaningful suggestions and comments. Furthermore, he is really grateful to Professor Akio Sugamoto and Professor Masahiro Morikawa for their very warm advice and discussions. This work was partially supported by the JSPS Grant-in-Aid for Young Scientists (B) # 25800136 and the research-funds supplied from Fukushima University.

References

1. P. P. Kronberg, Extragalactic magnetic fields, *Rept. Prog. Phys.* **57**, 325 (1994).
2. D. Grasso and H. R. Rubinstein, Magnetic fields in the early universe, *Phys. Rept.* **348**, 163 (2001).
3. L. M. Widrow, Origin of Galactic and Extragalactic Magnetic Fields, *Rev. Mod. Phys.* **74**, 775 (2002).
4. M. Giovannini, The magnetized universe, *Int. J. Mod. Phys. D* **13**, 391 (2004).
5. A. Kandus, K. E. Kunze and C. G. Tsagas, Primordial magnetogenesis, *Phys. Rept.* **505**, 1 (2011).
6. D. G. Yamazaki, T. Kajino, G. J. Mathew and K. Ichiki, The Search for a Primordial Magnetic Field, *Phys. Rept.* **517**, 141 (2012).
7. R. Durrer and A. Neronov, Cosmological Magnetic Fields: Their Generation, Evolution and Observation, *Astron. Astrophys. Rev.* **21**, 62 (2013).
8. M. S. Turner and L. M. Widrow, Inflation Produced, Large Scale Magnetic Fields, *Phys. Rev. D* **37**, 2743 (1988).
9. L. Parker, Particle creation in expanding universes, *Phys. Rev. Lett.* **21**, 562 (1968).
10. I. T. Drummond and S. J. Hathrell, QED Vacuum Polarization In A Background Gravitational Field And Its Effect On The Velocity Of Photons, *Phys. Rev. D* **22**, 343 (1980).
11. K. Bamba and M. Sasaki, Large-scale magnetic fields in the inflationary universe, *JCAP* **0702**, 030 (2007).
12. K. Bamba and S. D. Odintsov, Inflation and late-time cosmic acceleration in non-minimal Maxwell-$F(R)$ gravity and the generation of large-scale magnetic fields, *JCAP* **0804**, 024 (2008).
13. K. Bamba, S. Nojiri and S. D. Odintsov, Inflationary cosmology and the late-time accelerated expansion of the universe in non-minimal Yang-Mills-F(R) gravity and non-minimal vector-F(R) gravity, *Phys. Rev. D* **77**, 123532 (2008).
14. B. Ratra, Cosmological 'seed' magnetic field from inflation, *Astrophys. J.* **391**, L1 (1992).
15. W. D. Garretson, G. B. Field and S. M. Carroll, Primordial magnetic fields from pseudoGoldstone bosons, *Phys. Rev. D* **46**, 5346 (1992).
16. K. Bamba and J. Yokoyama, Large scale magnetic fields from inflation in dilaton electromagnetism, *Phys. Rev. D* **69**, 043507 (2004).

17. K. Bamba and J. Yokoyama, Large-scale magnetic fields from dilaton inflation in noncommutative spacetime, *Phys. Rev. D* **70**, 083508 (2004).
18. J. Martin and J. Yokoyama, Generation of Large-Scale Magnetic Fields in Single-Field Inflation, *JCAP* **0801**, 025 (2008).
19. A. D. Dolgov, Breaking of conformal invariance and electromagnetic field generation in the universe, *Phys. Rev. D* **48**, 2499 (1993).
20. K. Bamba, Generation of large-scale magnetic fields, non-Gaussianity, and primordial gravitational waves in inflationary cosmology, *Phys. Rev. D* **91**, 043509 (2015).
21. J. J. Blanco-Pillado, C. P. Burgess, J. M. Cline, C. Escoda, M. Gomez-Reino, R. Kallosh, A. D. Linde and F. Quevedo, Racetrack inflation, *JHEP* **0411**, 063 (2004).
22. N. Barnaby and M. Peloso, Large Nongaussianity in Axion Inflation, *Phys. Rev. Lett.* **106**, 181301 (2011).
23. N. Barnaby, R. Namba and M. Peloso, Phenomenology of a Pseudo-Scalar Inflaton: Naturally Large Nongaussianity, *JCAP* **1104**, 009 (2011).
24. P. D. Meerburg and E. Pajer, Observational Constraints on Gauge Field Production in Axion Inflation, *JCAP* **1302**, 017 (2013).
25. A. Linde, S. Mooij and E. Pajer, Gauge field production in SUGRA inflation: local non-Gaussianity and primordial black holes, *Phys. Rev. D* **87**, 103506 (2013).
26. K. Bamba, Baryon asymmetry from hypermagnetic helicity in dilaton hypercharge electromagnetism, *Phys. Rev. D* **74**, 123504 (2006).
27. P. A. R. Ade *et al.* [Planck Collaboration], Planck 2015 results. XIII. Cosmological parameters, arXiv:1502.01589 [astro-ph.CO].
28. P. A. R. Ade *et al.* [Planck Collaboration], Planck 2015 results. XVII. Constraints on primordial non-Gaussianity, arXiv:1502.01592 [astro-ph.CO].
29. P. A. R. Ade *et al.* [Planck Collaboration], Planck 2015 results. XX. Constraints on inflation, arXiv:1502.02114 [astro-ph.CO].
30. V. Demozzi, V. Mukhanov and H. Rubinstein, Magnetic fields from inflation?, *JCAP* **0908**, 025 (2009).
31. P. A. R. Ade *et al.* [Planck Collaboration], Planck 2015 results. XIX. Constraints on primordial magnetic fields, arXiv:1502.01594 [astro-ph.CO].
32. G. F. Smoot, C. L. Bennett, A. Kogut, E. L. Wright, J. Aymon, N. W. Boggess, E. S. Cheng and G. De Amici *et al.*, Structure in the COBE differential microwave radiometer first year maps, *Astrophys. J.* **396**, L1 (1992).
33. R. G. Cai, B. Hu and H. B. Zhang, Acoustic signatures in the Cosmic Microwave Background bispectrum from primordial magnetic fields, *JCAP* **1008**, 025 (2010).

The Jet Energy Profile: A BSM Analysis Tool*

R. S. Chivukula#†, E. H. Simmons and N. Vignaroli

Department of Physics and Astronomy, Michigan State University, East Lansing, MI, 48825, USA
 # E-mail: sekhar@msu.edu

A new heavy di-jet resonance could be discovered at the 14 TeV LHC. In this talk we present a strategy to reveal the nature of such a particle; in particular to discern whether it is a quark-antiquark ($q\bar{q}$), quark-gluon (qg), or gluon-gluon (gg) resonance. The strategy is based on the study of the energy profiles of the two leading jets in the di-jet channel. Including statistical uncertainties in the signal and the QCD backgrounds, we show that one can distinguish between gg, qg, and $q\bar{q}$ resonances; an evaluation of systematic uncertainties in the measurement of the jet energy profile will require a detailed detector study once sufficient 14 TeV di-jet data is in hand.

Keywords: Extended strong interactions, Experimental tests, BSM physics.

1. Introduction

Searches for heavy resonances produced in the s-channel and decaying into a pair of jets offer a simple and powerful probe of many different scenarios of new physics at the Large Hadron Collider. ATLAS[2] and CMS[3,4] have recently presented the results of the searches for narrow di-jet resonances at the LHC with $\sqrt{s} = 8$ TeV. Lower limits on the masses of new hypothetical particles in a variety of beyond the standard model theories have been obtained. The upcoming LHC run at $\sqrt{s} = 14$ TeV will have the capability to greatly extend the discovery reach in the di-jet channel[5]. If a hadronic resonance is discovered in the di-jet channel a major challenge will be the identification of the nature and of the properties of the newly discovered particle. In this talk we present a strategy to reveal if such a particle is a quark-antiquark ($\bar{q}q$), a quark-gluon (qg), or a gluon-gluon (gg) resonance. This strategy[1] analyzes the energy profiles of the two final jets[6].

We consider three compelling benchmark scenarios to describe different di-jet resonances: the flavor universal coloron model[7,8] for $\bar{q}q$ resonances, the excited quark model of Ref.[9,10] for qg resonances and the general parameterization in[11] of color-octet scalar interactions for gg resonances. All of the results are shown in the relevant mass-coupling parameter space that is both not excluded by the 8 TeV LHC analyses[2–4] and conducive to a 5σ discovery of the resonance in the di-jet

*This contribution is an abbreviated version of ref. [1].
†Dr. Chivukula presented this talk at SCGT15.

channel at the 14 TeV LHC. The LHC-8 excluded regions are extracted from the ATLAS[2] and CMS[3,4] searches; the LHC-14 discovery reach is evaluated based on Monte Carlo simulations.

Information on the partonic origin of a di-jet resonance is provided by the analysis of the jets' substructure, in particular by the study of the energy profiles[6] of the two final jets. Quark-initiated jets have more quickly rising profiles compared to gluon jets, so that discrimination among the different $\bar{q}q$, qg, and gg di-jet resonances is possible from analyzing the di-jet energy profiles. We evaluate the (mean) jet energy profiles by applying the theoretical calculations in perturbative QCD, at next-to-leading logarithmic accuracy, which have been developed in Ref.[12,13]. Statistical fluctuations on the jet energy profiles are generated through Monte Carlo simulations and, consequently, the statistical efficiency of our discriminating tool based on the di-jet energy profiles is evaluated. Including statistical uncertainties in the signal and the background, we show that one can distinguish between gg, qg, and $q\bar{q}$ resonances; an evaluation of systematic uncertainties in the measurement of the jet energy profile will require a detailed detector study once sufficient 14 TeV di-jet data is in hand and is beyond the scope of this paper. [a]

2. Benchmark models for di-jet resonances

We consider three benchmark models for $\bar{q}q$, qg and gg resonances. For gg and qg resonances, we will refer to the same models as were considered in the recent CMS[3,4] and ATLAS[2] analyses. For the $\bar{q}q$ resonance, we will consider the flavor universal coloron model considered in the CMS analysis[3,4].

2.1. *Flavor universal colorons* (C)

Quark-antiquark resonances are present in many different kinds of new physics scenarios. We focus here on the coloron model presented in Ref.[7,8]. This model belongs to the class of theories predicting an extended strongly interacting sector $SU(3)_1 \times SU(3)_2$ that spontaneously breaks to $SU(3)_{QCD}$[19–21]. The model can be flavor universal, which is the case we will consider here.

At high energies, the model features an enlarged color gauge structure $SU(3)_1 \times SU(3)_2$. This extended color symmetry is broken down to $SU(3)_C$ by the (diagonal) expectation value of a scalar field, which transforms as a $(\mathbf{3}, \bar{\mathbf{3}})$ under $SU(3)_1 \times SU(3)_2$. It is assumed that each standard model (SM) quark transforms as a $(\mathbf{1}, \mathbf{3})$

[a] Additional techniques based on the study of jet substructure and aimed at identifying di-jet resonances and/or improving the signal-to-background ratio have been extensively considered in the literature. Recent examples are the study of the color flow in Ref.[14], the analysis of the charge track multiplicity and the p_T-weighted linear radial moment (girth) in Refs.[15,16], and the study of generalized angularities in[17] aimed at distinguishing quark and gluon jets on an event-by-event basis. The current status of jet substructure techniques, covering both experimental and theoretical efforts, is reviewed in[18]. More sophisticated analyses employing these methods could yield even more discrimination power than the simpler method considered here.

under the extended strong gauge group. The color symmetry breaking induces a mixing between the original $SU(3)_1$ and $SU(3)_2$ gauge fields, which is diagonalized by a field rotation determined by

$$\tan\theta = \frac{g_2}{g_1} \qquad g_S = g_1 \sin\theta = g_2 \cos\theta \ , \tag{1}$$

where g_S is the QCD strong coupling and g_1, g_2 are the $SU(3)_1$ and $SU(3)_2$ gauge couplings, respectively. The diagonalization reveals two classes of color-octet vector boson mass eigenstates – the massless SM gluons and the new colorons C^a, which are massive,

$$m_C = \frac{g_S u}{\sin\theta\cos\theta} \ , \tag{2}$$

where u is the breaking scale for the extended color symmetry. The coloron's interactions with quarks are determined by a new QCD-like coupling

$$-g_S \tan\theta \sum_f \bar{q}_f \gamma^\mu \frac{\lambda^a}{2} q_f C^a_\mu \ . \tag{3}$$

A coloron that decays to all six quark flavors ($m_c > 2m_{top}$) has a decay width:

$$\Gamma(C) = \alpha_S m_C \tan^2\theta \tag{4}$$

Colorons can be produced at the LHC by quark-antiquark fusion at a rate determined by the C coupling to light quarks, $g_s \tan\theta$. Gluon-gluon fusion production, on the other hand, is forbidden at tree level by $SU(3)_C$ gauge invariance[22–24], and has been found to be insignificant at the one-loop level[25]. The CMS search for di-jet resonances[3,4] has considered the hypothesis of a flavor universal coloron, taking this model as a benchmark and fixing $\tan\theta = 1$.

2.2. *Excited quarks* (q*)

Quark-gluon resonances are a general prediction of composite models with excited quarks[9,10]. In this work we will take as our exemplar the phenomenological model of[9], which describes an electroweak doublet of excited color-triplet vector-like quarks $q^* = (u^*, d^*)$ coupled to first-generation ordinary quarks. In this model, right-handed excited quarks interact with gauge bosons and ordinary (left-handed) quarks through magnetic moment interactions described by the effective Lagrangian:

$$\mathcal{L}_{int} = \frac{1}{2\Lambda}\bar{q}^*_R \sigma^{\mu\nu}\left[g_S f_S \frac{\lambda^a}{2} G^a_{\mu\nu} + g f \frac{\tau}{2}\cdot \mathbf{W}_{\mu\nu} + g' f' \frac{Y}{2} B_{\mu\nu}\right] q_L + \text{H.c.} \tag{5}$$

The excited quarks can decay into qg or into a quark plus a gauge boson. The corresponding decay rates are:

$$\begin{aligned}\Gamma(q^* \to qg) &= \frac{1}{3}\alpha_S f_S^2 \frac{m_{q*}^3}{\Lambda^2} \\ \Gamma(q^* \to q\gamma) &= \frac{1}{4}\alpha f_\gamma^2 \frac{m_{q*}^3}{\Lambda^2} \\ \Gamma(q^* \to qV)) &= \frac{1}{8}\frac{g_V^2}{4\pi} f_V^2 \frac{m_{q*}^3}{\Lambda^2}\left[1 - \frac{m_V^2}{m_{q*}^2}\right]^2 \left[2 + \frac{m_V^2}{m_{q*}^2}\right]\end{aligned} \tag{6}$$

with $V = W, Z$ and with the definitions

$$\begin{aligned}f_\gamma &= fT_3 + f'\frac{Y}{2} \\ f_Z &= fT_3 \cos^2\theta_W - f'\frac{Y}{2}\sin^2\theta_W \\ f_W &= \frac{f}{\sqrt{2}} \ .\end{aligned} \tag{7}$$

The $q^* \to qg$ branching ratio is about 0.8 for $f_S = f = f'$.

Excited quarks are singly produced at the LHC through quark-gluon annihilation and, as just noted, they dominantly decay into qg. For our analysis, we choose the benchmark parameters $\Lambda = m_{q^*}$ and $f_S = f = f'$, while allowing the overall coupling strength to vary. By way of comparison, recent LHC searches, CMS[3,4] and ATLAS[2] have used the same value of Λ with $f_S = f = f' = 1$.

2.3. *Color-octet scalars* ($\mathbf{S_8}$)

A gluon-gluon final state can generally arise from decay of colored scalars in models with extended color gauge structures[19–21,26,27]. In this work we adopt the general effective interaction for a color octet scalar, S_8, introduced in[11]:

$$\mathcal{L}_{S_8} = g_S d^{ABC} \frac{k_S}{\Lambda_S} S_8^A G_{\mu\nu}^B G^{C,\mu\nu} \ , \tag{8}$$

where d is the QCD totally symmetric tensor.

A colored scalar of this kind is singly-produced at the LHC through gluon-gluon annihilation. We consider the case in which it decays entirely (or almost entirely) into gluons. The corresponding decay rate reads:

$$\Gamma(S_8) = \frac{5}{3}\alpha_S \frac{k_S^2}{\Lambda_S^2} m_{S_8}^3 \ . \tag{9}$$

We set $\Lambda_S = m_{S_8}$ and we present results for different couplings k_S. Similarly, CMS[3,4] and ATLAS[2] present searches for $\Lambda_S = m_{S_8}$ and $k_S = 1$.

3. LHC discovery reach

For each type of dijet resonance, we begin by deriving the relevant mass and coupling parameter space for our analysis, namely the region that is not yet excluded by LHC-8 analyses and in which a 5σ discovery will be possible at the 14 TeV LHC.

We derive the excluded parameter region for colorons from the CMS analysis in[3]; for excited quarks and scalar-octets, we obtain constraints by considering the strongest limits within the CMS[3] and ATLAS analyses[2]. Note that CMS and ATLAS searches have a poor sensitivity to resonances whose width is large compared to the detector di-jet mass resolution, *i.e.* with a width-over-mass value of greater than $\sim$0.15[28]. In what follows, we assume that the new resonances are sufficiently narrow to be discovered in di-jets and that they decay only (or at least predominantly) to pairs of jets: $\bar{q}q$, qg, or gg.

The 5σ discovery reach at the 14 TeV LHC is estimated by evaluating $S/\sqrt{S+B}$, where S and B are, respectively, the total number of signal and background di-jet events passing the CMS kinematic selection criteria in[3]:

$$p_T(j_{1,2}) > 30 \text{ GeV}, \qquad |\eta(j_{1,2})| < 2.5 \;, \qquad |\Delta\eta(j_1 j_2)| < 1.3 \;. \tag{10}$$

For a given potential resonance mass M we also require the invariant mass of the two leading jets to be within a range of $\pm 0.15M$ from the di-jet invariant mass peak. The standard model di-jet background is taken from Ref.[29], where it has been carefully estimated by applying the same CMS cuts to matched samples of two- and three-jet final states using MADGRAPH[30] and PYTHIA[31]. The simulated di-jet signals at the 14 TeV LHC for the different resonances are generated at parton level with MADGRAPHv5 and the CT10[32] set of parton distribution functions, after implementing the benchmark models with Feynrules[33]. We find an acceptance rate for the CMS kinematic selection criteria[3] of about 0.58 for S_8 or q^* and of about 0.5 for C, for the mass ranges of interest.

Fig. 4 shows our estimates of the 5σ reach at the 14 TeV LHC in the mass-vs.-coupling plane for colorons, excited quarks, and scalar-octets, for integrated luminosities of 30 – 3000 fb^{-1}. The discovery reach we find for the coloron is very similar to those already derived in[29,34] and in[35].

Within each pane of Fig. 4, we may identify a "region of interest" where a resonance of a given mass and coupling is not excluded by LHC (i.e., is not in the blue region at left), is relatively narrow (lies below the horizontal dashed curve) and would be detectable at LHC-14 at the indicated luminosity (is within the central light-grey region).

4. Jet energy profiles

In this section we will examine the use of jet energy profiles (JEPs)[6] to statistically distinguish $\bar{q}q$, qg and gg di-jet resonances. For a jet of size R, the (integrated) JEP, $\psi(r)$, is defined as the fraction of jet transverse momentum that lies inside a

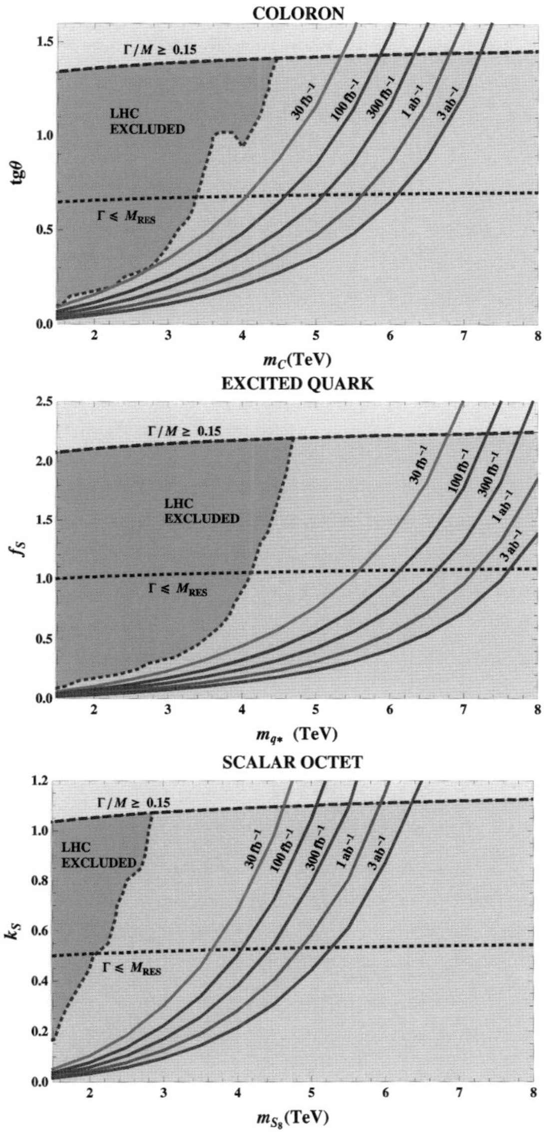

Fig. 1. In each pane from left to right: regions of coupling-mass parameter space excluded by LHC-8 (blue), regions accessible to LHC-14 (pale grey), region inaccessible at LHC-14 (pink). Thick colored curves show the 5σ reach at luminosities from 30 – 3000 fb^{-1}. Above the upper dashed line, the resonance is too broad to detect ($\Gamma = 0.15M$); below the lower dotted line it is narrower than the experimental resolution ($\Gamma \leq M_{res}$), where $M_{res} = 0.035M^{3}$. Resonance widths are calculated as shown in section 2.

sub-cone of size r ($< R$),

$$\psi(r) = \frac{\sum\limits_{r'<r} p_T(r')}{\sum\limits_{r'<R} p_T(r')}. \tag{11}$$

Gluon-initiated jets radiate more and produce a slowly rising JEP. Quark initiated jets, on the other hand, radiate less and have a quickly rising JEP.

We will begin by considering the statistical limitations of measuring the JEP of a simulated sample of *pure* signal events – di-jet events arising solely from a coloron, excited quark, or color-octet scalar. We will subsequently consider the effect of QCD background on the statistical significance of measuring the jet energy profile of the signal. We will show that the measurement is not statistically limited and, if systematic errors can be controlled, will clearly distinguish between the types of di-jet resonances we consider. An evaluation of systematic uncertainties in the measurement of the jet energy profile will require a detailed detector study once sufficient 14 TeV di-jet data is in hand, and is beyond the scope of this paper.

4.1. *JEP measurement based on signal events*

We consider first the measurement of the jet energy profile of a sample of di-jet events arising solely from the production of a coloron, excited quark, or color-octet scalar. As explained above, due to the differing pattern of soft gluon radiation from quarks and from gluons, in principle a measurement of the JEP could distinguish among these different types of resonances since they decay into different final states. Experimentally measured JEPs, of course, include not just the effects of the initial high-Q^2 radiation arising from the quarks and gluons produced in the hard event, but the subsequent low-Q^2 showering and hadronization of these objects – a description of which depends on tune-dependent Monte Carlo event generators such as PYTHIA[31,36]. JEPs have been recently measured at ATLAS[37] and at CMS[38] and, indeed, the results of the JEP measurements[37,38] show that the data can be reproduced only after a careful calibration of the shower/hadronization parameters.

The copious di-jet data available from the 14 TeV LHC will allow for the necessary calibration – and, as we will show, we expect to find clear differences between the di-jet JEP measured in the resonance region and that measured from the purely SM background events at off-resonance di-jet invariant masses. However, since 14 TeV LHC data and tuned event generators are not yet available, we will rely on a theoretical calculation to estimate the *average* shape of the JEPs for colorons, excited quarks, and color-octet scalars, and we will use MC simulations – MADGRAPH interfaced with PYTHIA v6.4 (default tune) – to evaluate the statistical uncertainties on the measurement of these profiles. Specifically, we calculate the mean values of the jet energy profiles in perturbative QCD (pQCD) by using the jet functions derived in[12,13], which apply a next-to-leading-logarithm (NLL) resummation[b]. Indeed, we find this procedure yields very good agreement with the experimental data from CMS at 7 TeV[38] and the Tevatron[39]. The pQCD calculation depends on two phenomenological parameters that take into account the effect

[b]Terms of the form $\alpha_S^n(\log(R/r))^{2n}$, $\alpha_S^n(\log(R/r))^{2n-2}$ are resummed to all orders in α_S. The studies in[12,13] show that NLL resummation is necessary for a correct description of the data; fixed NLO calculations overestimate the JEPs and fail to describe the data.

of uncalculated sub-sub-leading logarithmic contributions. We will fix these parameters at the values that reproduce the Tevatron data[39]. Once calibration becomes possible, these parameters, too, will need to be fixed at the values that reproduce the 14 TeV LHC data[c]. Since we are not using calculations tuned to LHC energies, our *absolute* results for the jet energy profiles will not precisely match those to be expected at the LHC — however, we expect the *relative differences* in the JEPs we find between the various kinds of resonances to be representative of what would be seen there.

We consider first the signal of a 4 TeV di-jet resonance, coming from an S_8, C or q^*, which can be discovered with approximately 30 fb^{-1} at the 14 TeV LHC and which has not been excluded by the present LHC-8 searches. In particular, we consider an S_8 resonance with a coupling $k_S = 0.65$, a coloron with $\tan\theta = 0.6$ and a q^* with $f_S = 0.4$. After the CMS selection cuts (10), all of these three types of resonances give, approximately, the same signal cross section around the resonance peak. We will analyze the jet-energy profiles for di-jet resonance events passing the CMS kinematic cuts (10) and in a region $|M_{jj} - M| < \Gamma/2$; we take conservatively $\Gamma = \Gamma(S_8)$, corresponding to the largest possible width among those of the three types of resonances. The choice of focusing our analysis in a narrow region around the resonance peak is intended to minimize the SM di-jet background which, as we will see in the next subsection, will affect the uncertainty of our discriminating tool. After selection we obtain a di-jet resonance signal cross-section of 22 fb.

The predicted JEPs for a quark or gluon jet, $\psi(r)$, are obtained as in Ref.[40] by fitting a functional form to the results of a full perturbative QCD calculation done at several values of r.[d] Since the resonances we are studying each decay to two jets, we then calculate a predicted di-jet profile ψ_{jj} for the resonance decay as

$$\psi_{jj}(r) = \psi_1(r) + \psi_2(r) \tag{12}$$

where $\psi_1(r)$, $\psi_2(r)$, respectively, denote the JEPs of the leading and next-to-leading jet.

In order to quantify the power of JEPs to discriminate between different types of resonances, we will apply a one-parameter fit, so that we can unequivocally assign a specific value of the fit parameter to each signal di-jet profile. Specifically, we can parameterize the generic di-jet profile of the signal as

$$\psi_S(r) = f\psi_{\bar{q}q}(r) + (1-f)\psi_{gg}(r)\ . \tag{13}$$

Here, f is our fit parameter that indicates the fraction of quark-jets in a generic di-jet resonance: $f = 0, 0.5, 1$ for a gg, qg, or $\bar{q}q$ resonance respectively. The mean

[c] *E.g.*, Z +jets events with jets in a kinematic region similar to that of di-jet resonances could be used as calibration samples.

[d] To be more specific, the predicted $\psi(r)$ JEPs for either quark jets or gluon jets are obtained by fitting an exponential function of the type $(1 - be^{-ar})/(1 - be^{-aR})$[40] to the discrete $\psi(r)$ values obtained from the full pQCD calculation at several fixed r points. We calculate $\psi(r)$ at $\Delta r = 0.1$ steps, starting from $r = 0.1$ up to $r = R = 0.5$.

values of the different jet-energy profiles determined by pQCD are shown as the central values of the curves in Fig. 3 – note the difference between the JEPs arising from $\bar{q}q$, qg, and gg dijet events.

We estimate the statistical uncertainty on the mean values of the JEPs for a sample of pure signal events by running pseudo-experiments through MC simulations. We evaluate the statistical errors in the JEPs at $\Delta r = 0.1$ steps. Signal sample events are generated with MADGRAPH v.5[30] and interfaced with PYTHIA v.6 for shower and hadronization. The jets are clustered through FASTJET[41] by an anti-kT algorithm with cone size $R = 0.5$. JEPs are then obtained by analyzing the jet substructure, according to the formula in (11). We find, as expected, that the statistical fluctuations in ψ, and hence f, follow Gaussian distributions and that the errors scale as the square root of the number of events. In particular, we find that the uncertainty in the value of $\psi(r)$ at $r = 0.1$ (which yields the largest error) scales as

$$(\delta\psi_S(0.1))^2 \approx \frac{\sigma^2(0.1)}{S}\,, \tag{14}$$

where $\sigma(0.1) \approx 0.4$ and S is the total number of signal events.

4.2. *Including QCD background*

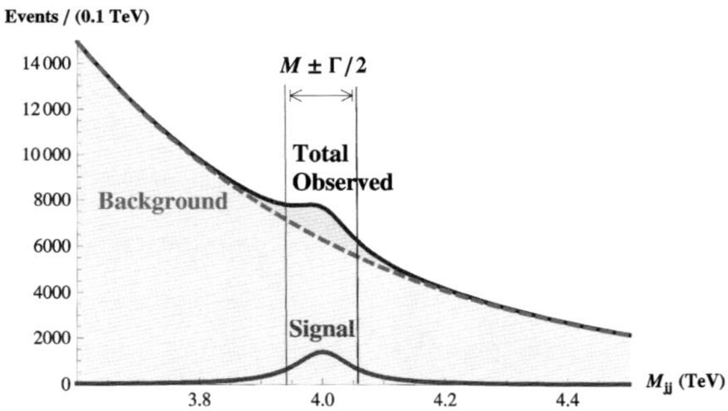

Fig. 2. Sketch of how the number of signal (blue), background (red) and total observed events (black) could depend on dijet invariant mass (M_{jj}) if a dijet resonance is discovered at the LHC. This figure illustrates issues raised in the discussion of JEP measurements and uncertainties in Section 4.2.

Next, we consider the impact of QCD background on our analysis. The resonance will appear as a "bump" in a plot of the di-jet invariant mass distribution, as sketched in Fig. 2. In the signal region ($|M_{jj} - M| < \Gamma/2$) there will be S signal events and B QCD background events. As mentioned above, for the benchmark 4 TeV di-jet resonance we find a signal cross section of 22 fb, and extracting the background from Ref.[29], we find a signal-to-background ratio of 1/23.

It is not possible to measure the jet energy profile of the signal alone; measurements of the JEP in the signal region, $\psi_{OBS}(r)$, will include both signal and background. One can also measure the jet energy profiles in "side-bands", regions of di-jet invariant mass immediately adjacent to but outside the resonance region; this yields an experimentally determined measurement of the JEP of the QCD background, $\psi_B(r)$. We expect that the experimental uncertainties on these individual measurements will scale analogously to what is shown in eq. (14):

$$(\delta\psi_{OBS}(r))^2 \approx \frac{\sigma^2(r)}{S+B} \qquad (\delta\psi_B(r))^2 \approx \frac{\sigma^2(r)}{B} \; . \tag{15}$$

The desired quantity $\psi_S(r)$ is now related to the measurable JEPs by

$$\psi_{OBS}(r) = \frac{S}{S+B}\psi_S(r) + \frac{B}{S+B}\psi_B(r) \; , \tag{16}$$

and hence

$$\psi_S(r) = \psi_{OBS}(r) + \frac{B}{S}(\psi_{OBS}(r) - \psi_B(r)) \; . \tag{17}$$

The statistical uncertainties in the quantities ψ_{OBS} and ψ_B are given by eq. (15). Since we are working in a regime in which $B \gg S$, the uncertainty in B/S is dominated by fluctuations in the number of signal events and is roughly $B/S^{3/2}$. From eq. (17), we find the mean-square error on ψ_S to be

$$(\delta\psi_S)^2 \approx \frac{\sigma^2}{S}\left[1 + 2\frac{B}{S}\right] + \frac{(\psi_S - \psi_B)^2}{S} \tag{18}$$

where we have neglected terms suppressed by S/B. The first term in eq. (18) represents the "dilution" in the measurement of ψ_S due to QCD background, relative to the sample-only error of eq. (14), and the second term is due to the uncertainty in the number of signal events. From Fig. 3, we see that the difference in JEPs (which is maximal for the difference between pure qq and gg states at $r = 0.1$) is bounded from above by about 0.5; in the regions in which the di-jet resonance can be observed, the second term in eq. (18) is negligible. Fig. 3 shows the resulting di-jet energy profiles, with uncertainty bands including the effect of the background subtraction, for the $\bar{q}q$ (coloron), qg (excited quark) and gg (scalar octet) 4 TeV di-jet resonance.

We can translate the statistical error on $\psi(r)_S$ into a statistical uncertainty on the f parameter [e]. Results predicted for the 14 TeV LHC with 100 fb^{-1} of data are $f = 1.00 \pm 0.06$ for a $q\bar{q}$ resonance, 0.50 ± 0.07 for a qg resonance, and 0.00 ± 0.08 for a gg resonance.

[e]This is obtained via the following procedure. Using a step size $\Delta r = 0.1$, we generate a large number of $\psi(r)$ values according to the Gaussian fluctuations which we have calculated by running pseudo-experiments. The generated $\psi(r)$ points are fitted by the function $(1 - be^{-ar})/(1 - be^{-aR})$ and the resulting profiles are translated into f values according to eq. (13). We thus obtain the statistical fluctuation on f and we are able to calculate the corresponding standard deviation.

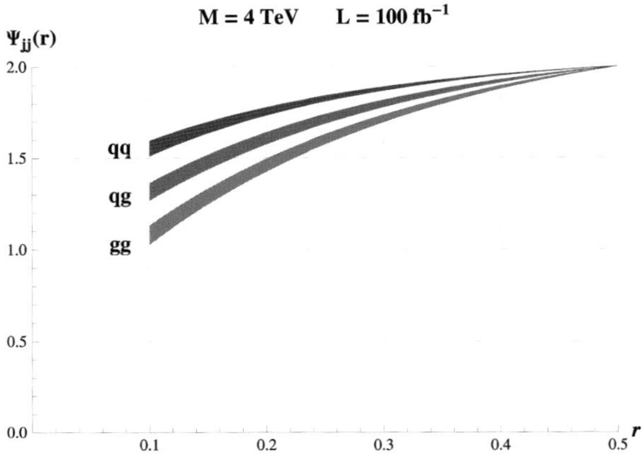

Fig. 3. Di-jet energy profiles for $\bar{q}q$ (coloron), qg (excited quark) and gg (scalar octet) 4 TeV di-jet resonances (the respective resonance couplings are fixed to $\tan\theta = 0.6$, $f_S = 0.4$, and $k_S = 0.65$). Each band shows a $\pm 1\sigma$ statistical variation from the mean curve. The effect of background subtraction, eq. (18), is included.

We can also evaluate the statistical efficiency of our discriminating tool according to a t-test. We find that even the most challenging discrimination, that between qg and gg resonances, can be performed at a high statistical level, of $\sim 5\sigma$, with 100 fb^{-1} at the 14 TeV LHC. A $\bar{q}q$ resonance can be well distinguished from a gg resonance: a $\sim 5\sigma$ level of distinction can be achieved with only ~ 30 fb^{-1}.

We can repeat this analysis for several different di-jet resonance mass values and consequently estimate the statistical uncertainty on the quark-jet fraction parameter, Δf, for different resonance couplings and LHC luminosities by considering that Δf scales as

$$\Delta f \sim \frac{\sqrt{1 + 2\frac{B}{S}}}{\sqrt{S}} \tag{19}$$

with the total number of signal (S) and background (B) events. $S = \sigma_S L$, $B = \sigma_B L$, where L is the integrated luminosity, σ_S is the di-jet signal cross section (which depends on the resonance mass and coupling), and σ_B is the background cross section (which depends on the di-jet invariant mass cut). As in the previous analysis at $M = 4$ TeV, we apply the CMS selection cuts in (10) and we restrict to a di-jet invariant mass region $|M_{jj} - M| < \Gamma/2$, where, conservatively, we take $\Gamma = 0.15M$.

Through this analysis we can establish the region of masses and couplings where the quark jet-fraction parameter f can be measured sufficiently well to distinguish between colorons, excited quarks, and color-octet scalars. In particular, we obtain contours of constant statistical uncertainty in the signal quark-jet fraction, Δf, in the parameter space for the three di-jet resonances at different 14 TeV LHC integrated luminosities, as shown in Fig. 4. Together with the Δf contours we show (in grey) the regions illustrated in Fig. 4, that are still allowed by LHC-8 data and where a 5σ discovery of the specific di-jet resonance is achievable with the given luminosity. In the case $L = 100$ fb^{-1} we also indicate with a red dot the mass-coupling values considered in the analysis at $M = 4$ TeV.

The results show that if a 5σ discovery of a di-jet resonance occurs at the 14 TeV LHC, the statistical uncertainty on the corresponding f parameter will be small; we have $\Delta f \leq 0.1$ for all of the three types of resonances in essentially the entire relevant parameter space where we can reach a 5σ discovery at the 14 TeV LHC. Thus, it should be possible to use the analysis of JEPs to distinguish among gg, qg, and $\bar{q}q$ dijet resonances.

We must reiterate, however, that our study only examines the statistical significance of the di-jet resonance discrimination through JEPs. We make no attempt to estimate the effects of possible systematic uncertainties on the JEPs, as this will require a detailed detector study and is only likely to be possible with data in hand – and is therefore beyond the scope of this paper.

5. Conclusions

We have presented a strategy of distinguishing different possible dijet resonances through the study of the energy profiles of the decay jets. Fig. 4 summarizes our results for the analysis of the di-jet energy profiles of $\bar{q}q$, qg and gg resonances, including the statistical uncertainties and the effect of background subtraction. We find that the analysis of JEPs can distinguish gg, qg, and $\bar{q}q$ resonances even after accounting for statistical uncertainties in the signal and the background.

We look forward to exciting results from the upcoming run of the LHC, and the possible discovery of a heavy di-jet resonance.

Acknowledgments

This material is based upon work supported by the National Science Foundation under Grant No. PHY-0854889. We wish to acknowledge the support of the Michigan State University High Performance Computing Center and the Institute for Cyber Enabled Research. PI is supported by Development and Promotion of Science and Technology Talents Project (DPST), Thailand. RSC and EHS thank the Kobayashi Maskawa Institute for the Origins of Particles and the Universe for hospitality during the SGT15, and also thank the Aspen Center for Physics and the NSF Grant #1066293 for hospitality during the writing of this Proceedings.

References

1. R. Sekhar Chivukula, E. H. Simmons and N. Vignaroli, Distinguishing dijet resonances at the LHC, *Phys. Rev.* **D91**, p. 055019 (2015).
2. G. Aad *et al.*, Search for new phenomena in the dijet mass distribution using $p-p$ collision data at $\sqrt{s} = 8$ TeV with the ATLAS detector, *Phys. Rev.* **D91**, p. 052007 (2015).

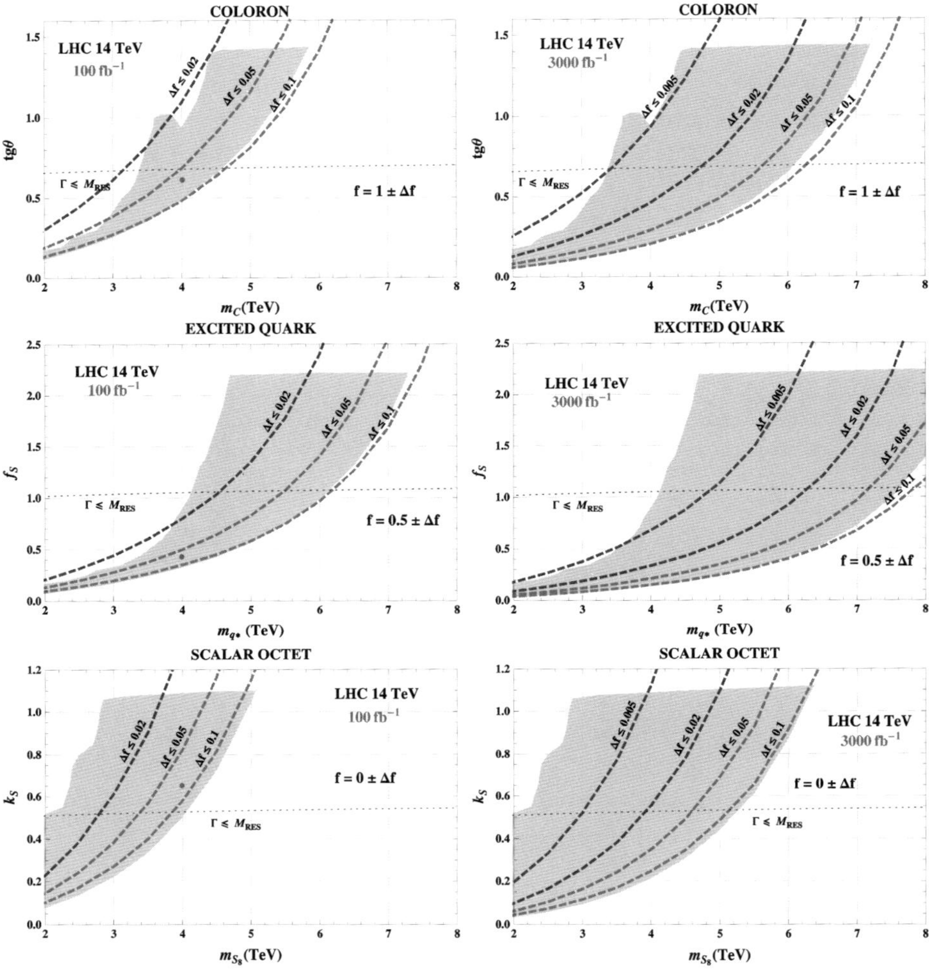

Fig. 4. Contours of constant statistical uncertainty in the quark-jet signal fraction Δf (dashed lines) in the mass-coupling parameter space for the three di-jet resonances at different 14 TeV LHC integrated luminosities (Left Plots: 100 fb^{-1}, Right Plots: 3000 fb^{-1}). The shaded regions show the areas that are allowed by the LHC-8 data, and where we can reach a 5-σ discovery of the specific di-jet resonance with the given luminosity. The dotted black line indicates the narrow-width limit; below this line, the color discriminant analysis is not possible but analysis of the JEPs is still valuable. In the case $L = 100$ fb^{-1} we also indicate with a red dot the mass-coupling values considered in the benchmark analyses of a $M = 4$ TeV di-jet resonance, discussed in the text. Note that the use of JEPs can yield a statistically significant measurement distinguishing between the different types of resonances over the entire parameter space accessible to the LHC.

3. *Search for Narrow Resonances using the Dijet Mass Spectrum with 19.6fb-1 of pp Collisions at sqrts=8 TeV*, Tech. Rep. CMS-PAS-EXO-12-059, CERN (Geneva, 2013).

4. S. Chatrchyan *et al.*, Search for narrow resonances using the dijet mass spectrum in pp collisions at sqrt(s) = 8 TeV (2013).
5. Y. Gershtein *et al.*, Working Group Report: New Particles, Forces, and Dimensions, in *Community Summer Study 2013: Snowmass on the Mississippi (CSS2013) Minneapolis, MN, USA, July 29-August 6, 2013*, 2013.
6. S. D. Ellis, Z. Kunszt and D. E. Soper, Jets at hadron colliders at order $\alpha - s^{3:}$ A Look inside, *Phys. Rev. Lett.* **69**, 3615 (1992).
7. E. H. Simmons, Coloron phenomenology, *Phys.Rev.* **D55**, 1678 (1997).
8. R. Chivukula, A. G. Cohen and E. H. Simmons, New strong interactions at the Tevatron?, *Phys.Lett.* **B380**, 92 (1996).
9. U. Baur, M. Spira and P. M. Zerwas, Excited Quark and Lepton Production at Hadron Colliders, *Phys. Rev.* **D42**, 815 (1990).
10. U. Baur, I. Hinchliffe and D. Zeppenfeld, Excited Quark Production at Hadron Colliders, *Int. J. Mod. Phys.* **A2**, p. 1285 (1987).
11. T. Han, I. Lewis and Z. Liu, Colored Resonant Signals at the LHC: Largest Rate and Simplest Topology, *JHEP* **1012**, p. 085 (2010).
12. H.-n. Li, Z. Li and C. P. Yuan, QCD resummation for jet substructures, *Phys. Rev. Lett.* **107**, p. 152001 (2011).
13. H.-n. Li, Z. Li and C. P. Yuan, QCD resummation for light-particle jets, *Phys. Rev.* **D87**, p. 074025 (2013).
14. J. Gallicchio and M. D. Schwartz, Seeing in Color: Jet Superstructure, *Phys. Rev. Lett.* **105**, p. 022001 (2010).
15. J. Gallicchio and M. D. Schwartz, Pure Samples of Quark and Gluon Jets at the LHC, *JHEP* **10**, p. 103 (2011).
16. J. Gallicchio and M. D. Schwartz, Quark and Gluon Tagging at the LHC, *Phys. Rev. Lett.* **107**, p. 172001 (2011).
17. A. J. Larkoski, J. Thaler and W. J. Waalewijn, Gaining (Mutual) Information about Quark/Gluon Discrimination, *JHEP* **11**, p. 129 (2014).
18. A. Altheimer *et al.*, Jet Substructure at the Tevatron and LHC: New results, new tools, new benchmarks, *J. Phys.* **G39**, p. 063001 (2012).
19. C. T. Hill, Topcolor: Top quark condensation in a gauge extension of the standard model, *Phys.Lett.* **B266**, 419 (1991).
20. P. H. Frampton and S. L. Glashow, Chiral Color: An Alternative to the Standard Model, *Phys.Lett.* **B190**, p. 157 (1987).
21. M. V. Martynov and A. D. Smirnov, Chiral color symmetry and possible G-prime-boson effects at the Tevatron and LHC, *Mod. Phys. Lett.* **A24**, 1897 (2009).
22. P. L. Cho and A. K. Leibovich, Color octet quarkonia production, *Phys. Rev.* **D53**, 150 (1996).
23. A. R. Zerwekh and R. Rosenfeld, Gauge invariance, color octet vector resonances and double technieta production at the Tevatron, *Phys. Lett.* **B503**, 325 (2001).

24. R. S. Chivukula, A. Grant and E. H. Simmons, Two gluon coupling and collider phenomenology of color octet technirho mesons, *Phys. Lett.* **B521**, 239 (2001).
25. R. S. Chivukula, A. Farzinnia, J. Ren and E. H. Simmons, Hadron Collider Production of Massive Color-Octet Vector Bosons at Next-to-Leading Order, *Phys.Rev.* **D87**, p. 094011 (2013).
26. R. S. Chivukula, E. H. Simmons and N. Vignaroli, Same-Sign Dileptons from Colored Scalars in the Flavorful Top-Coloron Model, *Phys.Rev.* **D88**, p. 034006 (2013).
27. Y. Bai and B. A. Dobrescu, Heavy octets and Tevatron signals with three or four b jets, *JHEP* **07**, p. 100 (2011).
28. R. M. Harris and K. Kousouris, Searches for Dijet Resonances at Hadron Colliders, *Int.J.Mod.Phys.* **A26**, 5005 (2011).
29. F. Yu, Di-jet resonances at future hadron colliders: A Snowmass whitepaper (2013).
30. J. Alwall, M. Herquet, F. Maltoni, O. Mattelaer and T. Stelzer, MadGraph 5 : Going Beyond, *JHEP* **1106**, p. 128 (2011).
31. T. Sjostrand, S. Mrenna and P. Z. Skands, A Brief Introduction to PYTHIA 8.1, *Comput. Phys. Commun.* **178**, 852 (2008).
32. H.-L. Lai, M. Guzzi, J. Huston, Z. Li, P. M. Nadolsky *et al.*, New parton distributions for collider physics, *Phys.Rev.* **D82**, p. 074024 (2010).
33. N. D. Christensen and C. Duhr, FeynRules - Feynman rules made easy, *Comput. Phys. Commun.* **180**, 1614 (2009).
34. B. A. Dobrescu and F. Yu, Coupling-mass mapping of dijet peak searches, *Phys.Rev.* **D88**, p. 035021 (2013).
35. A. Atre, R. S. Chivukula, P. Ittisamai and E. H. Simmons, Distinguishing Color-Octet and Color-Singlet Resonances at the Large Hadron Collider, *Phys.Rev.* **D88**, p. 055021 (2013).
36. Summary of ATLAS Pythia 8 tunes (2012).
37. G. Aad *et al.*, Study of Jet Shapes in Inclusive Jet Production in pp Collisions at $\sqrt{s} = 7$ TeV using the ATLAS Detector, *Phys. Rev.* **D83**, p. 052003 (2011).
38. S. Chatrchyan *et al.*, Shape, Transverse Size, and Charged Hadron Multiplicity of Jets in pp Collisions at 7 TeV, *JHEP* **06**, p. 160 (2012).
39. D. Acosta *et al.*, Study of jet shapes in inclusive jet production in $p\bar{p}$ collisions at $\sqrt{s} = 1.96$ TeV, *Phys. Rev.* **D71**, p. 112002 (2005).
40. V. Rentala, N. Vignaroli, H.-n. Li, Z. Li and C.-P. Yuan, Discriminating Higgs production mechanisms using jet energy profiles, *Phys.Rev.* **D88**, p. 073007 (2013).
41. M. Cacciari, G. P. Salam and G. Soyez, FastJet User Manual, *Eur. Phys. J.* **C72**, p. 1896 (2012).

Separating Dijet Resonances Using the Color Discriminant Variable

E. H. Simmons[#*], R. S. Chivukula, P. Ittisamai and N. Vignaroli

Department of Physics and Astronomy, Michigan State University, East Lansing, MI, 48825, USA
[#]E-mail: esimmons@msu.edu

Color-singlet and color-octet vector bosons predicted in theories beyond the Standard Model have the potential to be discovered as dijet resonances at the LHC. A color-singlet resonance that has leptophobic couplings needs further investigation to be distinguished from a color-octet one. In previous work, we introduced a method for discriminating between the two kinds of resonances when their couplings are flavor-universal, using measurements of the dijet resonance mass, total decay width and production cross-section. Here, we describe two extensions of that work. First, we broaden the method to the case where the vector resonances have flavor non-universal couplings, by incorporating measurements of the heavy-flavor decays of the resonance. Second, we apply the method to separating vector bosons from color-octet scalars and excited quarks.

Keywords: Extended strong interactions, Experimental tests, BSM physics.

1. Introduction

There have been many collider searches for Beyond-the-Standard-Model resonances decaying to dijet final states. The current LHC run will be able to detect such resonances out to larger masses. The question that we are pursuing in this talk is: once a new dijet resonance has been discovered, what can we deduce about it using information readily available after the discovery?

In previous work[1], we introduced a way to distinguish whether a vector resonance is a leptophobic Z' or a coloron, using a construct that we called a "color discriminant variable", $D_{\rm col}$. The variable is constructed from the dijet cross-section for the resonance (σ_{jj}), its mass (M), and its total decay width (Γ), observables that will be available from the dijet channel measurements of the resonance:

$$D_{\rm col} \equiv \frac{M^3}{\Gamma}\sigma_{jj}, \tag{1}$$

For a narrow-width resonance, the color discriminant variable is independent of the resonance's overall coupling strength. We also illustrated applications of the color discriminant variable technique for two simple cases[1]. The first was a flavor universal model with identical couplings to all quarks. In the second, the overall strength of couplings to quarks in the third generation was allowed to be different

*Dr. Simmons presented this talk at SCGT15.

from those in the first two (couplings to top and bottom were kept equal). While these two scenarios clearly illustrate the application of the method, they did not encompass the features of a typical Z', whose up- and down-type couplings are usually also different from one another.

In this talk we extend the D_{col} method to more general scenarios where couplings to quarks within the same generation (e.g., up vs. down, left-handed vs right-handed) are different[2]. We also show that the method may be used to distinguish among resonances decaying to $q\bar{q}$ (colored vector boson), qg (excited quark) or gg (color-octet scalar) final states[3].

2. Vector Resonances: General Parameterization and Assumptions

A coloron (C) or a Z' that manifests as a dijet resonance is produced at hadron colliders via quark-antiquark annihilation. The interaction of a C with the SM quarks q_i is described by

$$\mathcal{L}_C = i g_{QCD} C^a_\mu \sum_{i=u,d,c,s,t,b} \bar{q}_i \gamma^\mu t^a \left(g^i_{C_L} P_L + g^i_{C_R} P_R \right) q_i, \tag{2}$$

where t^a is an SU(3) generator, $g^i_{C_L}$ and $g^i_{C_R}$ denote left and right chiral coupling strengths, relative to the strong coupling g_{QCD}, of the color-octet to the SM quarks. The projection operators have the form $P_{L,R} = (1 \mp \gamma_5)/2$ and the quark flavor index i runs over $i = u, d, c, s, t, b$. Similarly, the interactions of a leptophobic Z' with the SM quarks are given by

$$\mathcal{L}_{Z'} = i g_w {Z'}_\mu \sum_{i=u,d,c,s,t,b} \bar{q}_i \gamma^\mu \left(g^i_{Z'_L} P_L + g^i_{Z'_R} P_R \right) q_i, \tag{3}$$

where $g^i_{Z'_L}$ and $g^i_{Z'_R}$ denote left and right chiral coupling strengths of the leptophobic Z' to the SM quarks relative to the weak coupling $g_w = e/\sin\theta_W$.

Couplings between the vector boson and the left- and right-handed forms of the up- and down-type fermions need not be equal in general – and the couplings can generally be different among the three generations of quarks. However, the observed suppressions of flavor-changing neutral currents disfavor a TeV-scale resonance with non-universal couplings to the first two generations[4]; we therefore study only scenarios in which the couplings for the first two generations are identical. With this in mind, a coloron would have six free parameters describing its couplings to quarks of different flavors, generations, and chiralities, while a leptophobic Z' boson would have eight. However, the observables of interest to us (width, mass, dijet cross-section) are insensitive to the chiral structure of the couplings. So we may denote

$$g^{q2} \equiv g^{q2}_L + g^{q2}_R \tag{4}$$

and conclude that there are only four relevant quark couplings for a coloron or Z':

$$g^{u\,2}_C = g^{c\,2}_C, \quad g^{d\,2}_C = g^{s\,2}_C, \quad g^{t\,2}_C, \quad g^{b\,2}_C \tag{5}$$

$$g^{u\,2}_{Z'} = g^{c\,2}_{Z'}, \quad g^{d\,2}_{Z'} = g^{s\,2}_{Z'}, \quad g^{t\,2}_{Z'}, \quad g^{b\,2}_{Z'}\,. \tag{6}$$

Finally, in discussing a dijet cross section ($\sigma(pp \to V \to jj)$), we classify quarks from the first two generations as those yielding final-state jets: $j = u, d, c, s$. The cross sections involving decays to $t\bar{t}$ and $b\bar{b}$ final states (respectively, $\sigma_{tt} \equiv \sigma(pp \to V \to t\bar{t})$ and $\sigma_{bb} \equiv \sigma(pp \to V \to b\bar{b})$) will provide separate and valuable information.

3. The Color Discriminant Variable

3.1. *Review of the flavor universal scenario*

We first review the color discriminant variable in a flavor-universal scenario. We work in the limit ($\Gamma/M \ll 1$) where the narrow-width approximation applies

$$\sigma^V_{jj} \equiv \sigma(pp \to V \to jj) \simeq \sigma(pp \to V) Br(V \to jj), \tag{7}$$

where $\sigma(pp \to V)$ is the cross section for producing the resonance. Note that $Br(V \to jj)$ is the boson's dijet branching fraction, which equals 4/6 for a flavor universal vector resonance that is heavy enough to decay to top quarks. In this limit, the total decay width for a heavy coloron or a Z' is

$$\Gamma_C = \frac{\alpha_s}{2} M_C g_C^2, \qquad \Gamma_{Z'} = 3\alpha_w M_{Z'} g_{Z'}^2\,, \tag{8}$$

where $g^2_{C/Z'} = \left(g^2_{C/Z'_L} + g^2_{C/Z'_R}\right)$ denotes the flavor-universal coupling of the resonance to quarks. These, respectively, lead to the dijet cross sections:

$$\sigma^C_{jj} = \frac{8}{9}\frac{\Gamma_C}{M_C^3} \sum_q W_q(M_C) Br(C \to jj), \qquad \sigma^{Z'}_{jj} = \frac{1}{9}\frac{\Gamma_{Z'}}{M^3_{Z'}} \sum_q W_q(M_{Z'}) Br(Z' \to jj)\,. \tag{9}$$

Here the function W_q, which is constructed from the parton luminosity for the production of the vector resonance with mass M_V via $q\bar{q}$ annihilation at the center-of-mass energy squared s, is defined by

$$W_q(M_V) = 2\pi^2 \frac{M_V^2}{s} \int_{M_V^2/s}^1 \frac{dx}{x} \left[f_q\left(x, \mu_F^2\right) f_{\bar{q}}\left(\frac{M_V^2}{sx}, \mu_F^2\right) + f_{\bar{q}}\left(x, \mu_F^2\right) f_q\left(\frac{M_V^2}{sx}, \mu_F^2\right)\right], \tag{10}$$

where $f_q\left(x, \mu_F^2\right)$ is the parton distribution function at the factorization scale μ_F^2. Throughout this article, we set $\mu_F^2 = M_V^2$.

The color discriminant variables are then

$$D^C_{\text{col}} = \frac{M_C^3}{\Gamma_C}\sigma^C_{jj} = \frac{8}{9}\left[\sum_q W_q(M_C) Br(C \to jj)\right] \tag{11}$$

$$D^{Z'}_{\text{col}} = \frac{M^3_{Z'}}{\Gamma_{Z'}}\sigma^{Z'}_{jj} = \frac{1}{9}\left[\sum_q W_q(M_{Z'}) Br(Z' \to jj)\right] \tag{12}$$

for the coloron and Z', respectively. The factors in the square brackets in (11) and (12) are the same for flavor-universal resonances having a particular mass; only the initial numerical factors differ. In other words, the different values of the color discriminant variables for the the two types of flavor-universal resonance

$$D_{\text{col}}^{C} = 8D_{\text{col}}^{Z'} \tag{13}$$

will help pinpoint the nature of the color structure of the discovered particle.

3.2. *Flavor non-universal scenario*

In a flavor non-universal scenario, we still follow the parameterization and assumptions introduced above. The production cross section and decay width for the coloron and Z' are discussed in detail in Ref.[2]. Combining them with the mass yields the color discriminant variables for the coloron,

$$\begin{aligned} D_{\text{col}}^{C} = \frac{16}{3}(W_u + W_c)&\left[\frac{g_C^{u\,2}}{g_C^{u\,2}+g_C^{d\,2}} + \left(1-\frac{g_C^{u\,2}}{g_C^{u\,2}+g_C^{d\,2}}\right)\left(\frac{W_d+W_s}{W_u+W_c}\right)\right. \\ &\left.+\frac{g_C^{b\,2}}{g_C^{u\,2}+g_C^{d\,2}}\left(\frac{W_b}{W_u+W_c}\right)\right] \times \left\{\frac{2}{\left(2+\frac{g_C^{t\,2}}{g_C^{u\,2}+g_C^{d\,2}}+\frac{g_C^{b\,2}}{g_C^{u\,2}+g_C^{d\,2}}\right)^2}\right\} \end{aligned} \tag{14}$$

and for the Z',

$$\begin{aligned} D_{\text{col}}^{Z'} = \frac{2}{3}(W_u + W_c)&\left[\frac{g_{Z'}^{u\,2}}{g_{Z'}^{u\,2}+g_{Z'}^{d\,2}} + \left(1-\frac{g_{Z'}^{u\,2}}{g_{Z'}^{u\,2}+g_{Z'}^{d\,2}}\right)\left(\frac{W_d+W_s}{W_u+W_c}\right)\right. \\ &\left.+\frac{g_{Z'}^{b\,2}}{g_{Z'}^{u\,2}+g_{Z'}^{d\,2}}\left(\frac{W_b}{W_u+W_c}\right)\right] \times \left\{\frac{2}{\left(2+\frac{g_{Z'}^{t\,2}}{g_{Z'}^{u\,2}+g_{Z'}^{d\,2}}+\frac{g_{Z'}^{b\,2}}{g_{Z'}^{u\,2}+g_{Z'}^{d\,2}}\right)^2}\right\} \end{aligned} \tag{15}$$

where parts related to resonance production are grouped within the square brackets, while those related to decay are grouped within curly braces.

The relative strength with which the vector boson couples to the u- and d-type quarks of the light SM generations, $g_u^2/\left(g_u^2+g_d^2\right)$, which we call the "up ratio", is not accessible by experiments available in the dijet channel. However, equivalent information for quarks in the third generation can be measured by comparing the dijet and heavy flavor cross sections. As detailed in Ref.[2], we have

$$\frac{g_t^2}{g_u^2+g_d^2} = 2\frac{\sigma_{t\bar{t}}^V}{\sigma_{jj}^V} \equiv \text{“top ratio''} \qquad \frac{g_b^2}{g_u^2+g_d^2} = 2\frac{\sigma_{b\bar{b}}^V}{\sigma_{jj}^V}\,, \equiv \text{“bottom ratio''}\,. \tag{16}$$

Supplementary measurements of these ratios of cross sections will help pinpoint the structure of couplings of the resonance.

Two items are worth noting at this point: First, as detailed in Ref.[2], unless the resonance has a *much* stronger coupling to the b than to quarks in first two generations, the precise strength of the couplings to third-generation quarks becomes relevant to D_{col} only through the decay part of the expressions (14) and (15), the part in curly braces. Second, as we will demonstrate below, the experimentally inaccessible parameter that we call the up ratio appears unlikely to leave us confused as to whether a new dijet resonance is a coloron or a leptophobic Z'.

4. Accessible Vector Boson Parameter Space at the 14 TeV LHC

We now find the region of parameter space in which $D_{\rm col}$ can distinguish vector bosons of different color representations. In this region the resonance has not already been excluded by the current searches, is within the reach of future searches, and has a total width that is measurable and consistent with the designation "narrow".

The limits on the production cross section times branching ratio ($\sigma \times Br(jj)$) from the (null) ATLAS and CMS[5–7] at $\sqrt{s} = 8\,\text{TeV}$ searches for narrow resonances establish bounds on dijet resonances. We use the most stringent constraint, from CMS[7]. As the exclusion limit is provided as $\sigma \times Br(jj) \times$ (Acceptance), we estimate the acceptance for each value of the resonance mass by comparing, within the same theoretical model, $\sigma \times Br(jj)$ as calculated by us vs. $\sigma \times Br(jj) \times$ (Acceptance) provided by CMS. The acceptance is a characteristic of properties of the detector and kinematics, the latter being the same for coloron and Z' to leading order; thus we use throughout our analysis the acceptance deduced from such a comparison made within a sequential Z' model. The excluded region of parameter space is displayed in gray in Fig. 1.

Sensitivity to a dijet resonance in future LHC experiments with $\sqrt{s} = 14\,\text{TeV}$ depends on the knowledge of QCD backgrounds, the measurements of dijet mass distributions, and statistical and systematic uncertainties. CMS [8] has estimated the limits on $\sigma \times Br(jj) \times$ (Acceptance) that will be required in order to attain a 5σ discovery at CMS with integrated luminosities up to $10\,\text{fb}^{-1}$, including both statistical and systematic uncertainties. We obtain the acceptance for CMS at $\sqrt{s} = 14\,\text{TeV}$ as described above. The sensitivity for the dijet discovery from $10\,\text{fb}^{-1}$ is then scaled to the integrated luminosities $\mathcal{L} = 30, 100, 300, 1000\,\text{fb}^{-1}$. The predicted discovery reaches for these luminosities are shown in varying shades of blues for coloron and greens for Z' in Fig. 1.

The total decay width also constrains the absolute values of the coupling constants. On the one hand, experimental searches are designed for narrow-width dijet resonances: $\Gamma/M < 0.15$ [9–11]. On the other, widths smaller than the experimental dijet mass resolution, $M_{\rm res}$, cannot be measured. The region of parameter space that meets both constraints and is relevant to our analysis is shown in Fig. 1 as the region between the two dashed horizontal curves labeled $\Gamma \geq 0.15M$ and $\Gamma \leq M_{\rm res}$. Regions where the width is too broad or too narrow are shown with a cloudy overlay to indicate that they are not accessible via our analysis.

5. Applying $D_{\rm col}$ to Flavor-non-universal Vector Bosons

We now illustrate how the color discriminant variable $D_{\rm col}$ can distinguish whether a dijet resonance is a coloron or a leptophobic Z' even if it is flavor non-universal.

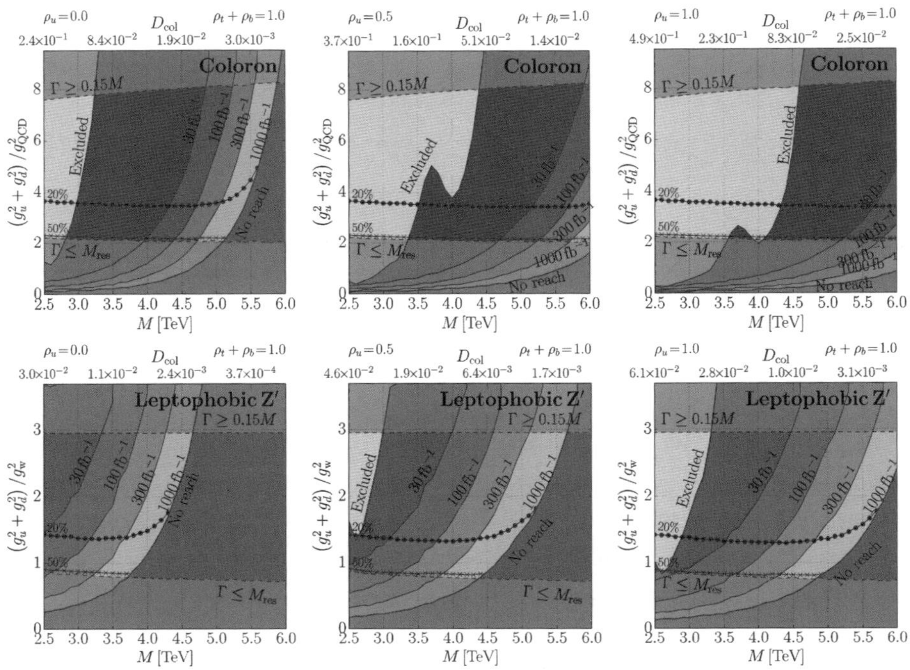

Fig. 1. Parameter space for colorons (upper row) and Z' bosons (lower row) where the color discriminant variable analysis applies at the LHC-14, for different sets of coupling ratios, $\rho_q \equiv g_q^2/[g_u^2 + g_d^2]$; for a wider array of coupling ratios, see Ref. [2]. The 5σ discovery reach, with statistical and systematic uncertainties included, is shown in varying shades of blue (green) for luminosities ranging from 30 fb^{-1} to 1000 fb^{-1}. The red area marked "no reach" lies beyond the discovery reach at 1000 fb^{-1} The gray area marked "excluded" has been excluded[7] by LHC-8. Above the dashed line marked $\Gamma \geq 0.15M$, the narrow-width approximation is not valid. Below the horizontal dashed line marked $\Gamma \leq M_{\text{res}}$, the experimental mass resolution is larger than the intrinsic width. In each figure, for fixed ρ_q, D_{col} is a function of resonance mass only, with values shown along the upper horizontal axis. The contours marked 20% and 50% indicate the region above which the uncertainty in measuring D_{col}, as estimated in Ref. [2], is lower than 20% and 50%, respectively.

5.1. *C and Z′ lie in different regions of coupling ratio space*

The value of D_{col} at a fixed mass and dijet cross section may correspond to a variety of combinations of values of the three ratios of couplings, the up ratio $(\frac{g_u^2}{g_u^2+g_d^2})$, the top ratio $(\frac{g_t^2}{g_u^2+g_d^2})$, and the bottom ratio $(\frac{g_b^2}{g_u^2+g_d^2})$. The last two are directly determined from the measurements of $\sigma_{t\bar{t}}$ and $\sigma_{b\bar{b}}$ while D_{col} is relatively insensitive to the first ratio. We now show that measuring the mass, width, dijet cross-section, $\sigma_{t\bar{t}}$ and $\sigma_{b\bar{b}}$ can definitively identify the color charge of a newly discovered resonance. We focus on resonances of mass 3 TeV and 4 TeV, as we have seen in Fig. 1 that most colorons with lower masses are excluded by the current experiments and most Z' bosons with higher masses are not within reach of 1000 fb^{-1} of LHC data.

In Fig. 2, we show the region of parameter space of the three coupling ratios (using the observables $\frac{\sigma_{t\bar{t}}}{\sigma_{jj}}$ and $\frac{\sigma_{b\bar{b}}}{\sigma_{jj}}$ in place of the top and bottom ratios, respectively) in which coloron or Z' models (each displayed as a point) with the same mass lead to a certain range of dijet cross-section and $D_{\rm col}$. We choose the range for the dijet cross-section to be within 1 standard deviation of the value that allows a 5σ discovery at luminosity $1000\,\mathrm{fb}^{-1}$. We selected $D_{\rm col}$ to be within 50% of the value 3×10^{-3} for this illustration, as it permits the required measurements to be made for either a coloron or Z'. Points in the accessible area of parameter space are highlighted in blue if the discoverable resonance is a C and in green if it is a Z'. These points lie in the blue (green) regions of Fig. 1 for a coloron (Z').

The 3-dimensional plots in the left panel of Fig. 2 show that colorons and leptophobic Z' which correspond to measurable observables appear in different regions of $\frac{\sigma_{t\bar{t}}}{\sigma_{jj}}$ vs $\frac{\sigma_{b\bar{b}}}{\sigma_{jj}}$ vs $\frac{g_u^2}{g_u^2+g_d^2}$ three-dimensional parameter space. The symmetry between the $\frac{\sigma_{t\bar{t}}}{\sigma_{jj}}$ and $\frac{\sigma_{b\bar{b}}}{\sigma_{jj}}$ axes illustrates the rarity of having a heavy resonance produced via $b\bar{b}$ annihilation - the only process in which b quarks could contribute to $D_{\rm col}$ without a corresponding contribution from t quarks. In addition, we see that while the up ratio is experimentally inaccessible, the top-view figures displayed in the right panels show that this does not typically lead to confusion between a coloron and a Z'. Even having projected the 3D data for all up ratio values onto the bottom-ratio vs. top-ratio plane, the blue coloron and green Z' points still lie in distinct regions.

5.2. *Heavy flavor measurements can separate C from Z'*

Fig. 2 not only shows that the coloron and Z' lie in different regions of parameter space, but also implies that measurements of the top and bottom decays of the new resonance will almost always enable us to determine its color structure. Let us assume that a new resonance has been found and that its mass, dijet cross-section, and $D_{\rm col}$ have been measured. For illustration, take the values of these observables to be those used in Fig. 2 (for the same LHC energy, luminosity and estimated uncertainties). Because there is a gap between the Z' and C regions of that figure when the up ratio is 0, and because the boundaries of the regions are angled rather than vertical, we can see that a Z' with the minimum up ratio value of 0 would not be mistaken for a coloron. But a Z' with the maximum up ratio (equal to 1) lies as close as possible to the coloron region of parameter space. So, to see how close the two regions can get, we should compare a Z' with an up ratio of 1 (one that does not couple to d or s quarks) to a coloron with varying values of the up ratio.

This very comparison is presented in Fig. 3, which is plotted in the top-ratio vs bottom-ratio plane. The up ratio for the Z' is fixed to be 1; the up ratio for the C is varied from 0 (left panels) to 1 (right panels). As the up ratio for the coloron increases, the blue coloron band of parameter space moves out from the origin, away from the green Z' band of parameter space. Conversely, if we decreased the up ratio of the Z' boson, the green Z' band would shift closer to the origin, away from the blue coloron band. In general, the coloron and Z' regions do not overlap.

Given the shape and orientation of the regions corresponding to color-singlet and color-octet resonances in the plots, measuring both the top ratio and bottom ratio

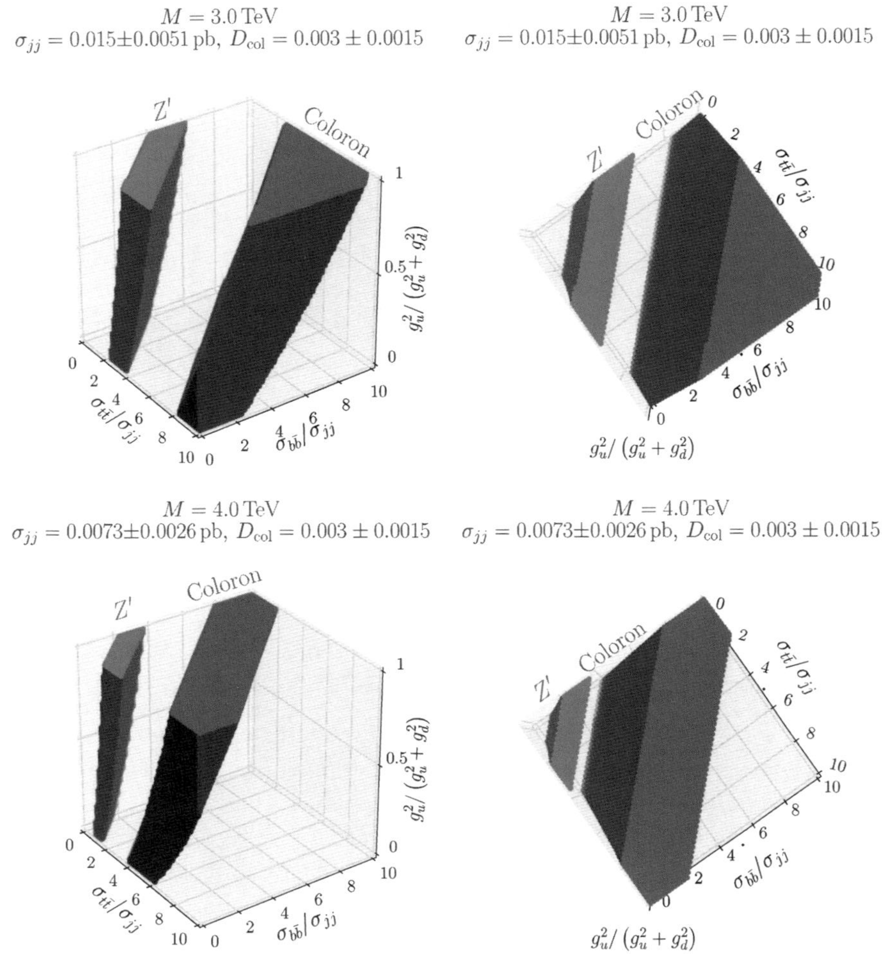

Fig. 2. Illustration that the proposed measurements suffice to identify the color structure of a new dijet resonance. These plots show regions of the 3-d parameter space $\frac{g_u^2}{g_u^2+g_d^2}$ vs. $\frac{\sigma_{t\bar{t}}}{\sigma_{jj}}$ vs. $\frac{\sigma_{b\bar{b}}}{\sigma_{jj}}$, at fixed values of mass (3 TeV on the top panels and 4 TeV on the bottom panels) where dijet cross section and D_{col} fall within a certain range. The cross section lies within about 35% of the value required for a 5σ discovery at the LHC with $\sqrt{s} = 14\,\text{TeV}$ at $\mathcal{L} = 1000\,\text{fb}^{-1}$. Similarly, D_{col} is chosen to lie within $\pm 50\%$ of 3×10^{-3} for illustration. Points in parameter space that are accessible to the LHC are circles highlighted in blue for C and triangles highlighted in green for Z'. Each plot on the left panel is shown again, as viewed from above, on the right. The two views make clear that colorons and Z' bosons lie in distinguishably separate regions of parameter space.

would clearly allow us to distinguish the new resonance's color structure. Moreover, we see that if either the top ratio or bottom ratio were measured to be sufficiently large, we would know that the resonance must be a coloron (because the Z' region is already at its maximum distance from the origin). For example, a measurement of $\sigma_{t\bar{t}}/\sigma_{jj} \gtrsim 6$ for a 3 TeV resonance or $\sigma_{t\bar{t}}/\sigma_{jj} \gtrsim 3$ for a 4 TeV resonance, for the values of σ_{jj} used in Fig. 3, would identify it as a color-octet.

There could still be a rare situation where our inability to measure the up ratio would prevent us from determining the color structure of a new resonance. The regions of parameter space corresponding to the extreme cases of a coloron with only down-type light quark couplings and a Z' with only up-type light quark couplings could potentially overlap. As discussed in Ref.[2], this is more likely to happen for heavier resonances. For example, given our estimates of uncertainties, such an overlap could potentially occur for a 4 TeV resonance as illustrated by the close approach of the Z' and coloron bands in the lower left panel of Fig. 3.

Determining the color structure of a resonance generally requires measurements of both $\sigma_{t\bar{t}}$ and $\sigma_{b\bar{b}}$. As with the dijet cross sections, systematic uncertainties for these measurements will be obtained after the experiment (at 14 TeV) has started. Our results illustrate[2] that measuring the $t\bar{t}$ and $b\bar{b}$ cross sections to an uncertainty of $\mathcal{O}(1)$ would still provide significant information.

6. Using D_{col} to Separate Vectors, Scalars, and Fermions

We now report on work[3] using the color discriminant variable to compare resonances that decay to $\bar{q}q$, qg, and gg final states. We consider three compelling benchmark scenarios to describe the different di-jet resonances: the flavor universal coloron model[12,13] for $\bar{q}q$ resonances, the excited quark model of Ref.[14,15] for qg resonances and the general parameterization in[16] of color-octet scalar interactions for gg resonances. All of the results are shown in the relevant mass-coupling parameter space that is both not excluded by the 8 TeV LHC analyses[6,7,17] and conducive to a 5σ discovery of the resonance in the di-jet channel at the 14 TeV LHC. The LHC-8 excluded regions are extracted from the ATLAS[17] and CMS[6,7] searches; the LHC-14 discovery reach is evaluated based on Monte Carlo simulations.

6.1. *Benchmark models*

Let us review the essential features of our three benchmark models. The flavor universal coloron model was described above. The other two will be sketched here.

6.1.1. *Excited quarks* ($\mathbf{q}^*$)

Quark-gluon resonances are a general prediction of composite models with excited quarks[14,15]. We will focus on the phenomenological model of[14], which describes an electroweak doublet of excited color-triplet vector-like quarks $q^* = (u^*, d^*)$ coupled to first-generation ordinary quarks. In this model, right-handed excited quarks interact with gauge bosons and ordinary (left-handed) quarks through magnetic moment interactions described by the effective Lagrangian:

$$\mathcal{L}_{int} = \frac{1}{2\Lambda}\bar{q}^*_R\sigma^{\mu\nu}\left[g_S f_S \frac{\lambda^a}{2} G^a_{\mu\nu} + g f \frac{\tau}{2}\cdot \mathbf{W}_{\mu\nu} + g' f' \frac{Y}{2} B_{\mu\nu}\right] q_L + \text{H.c.} \qquad (17)$$

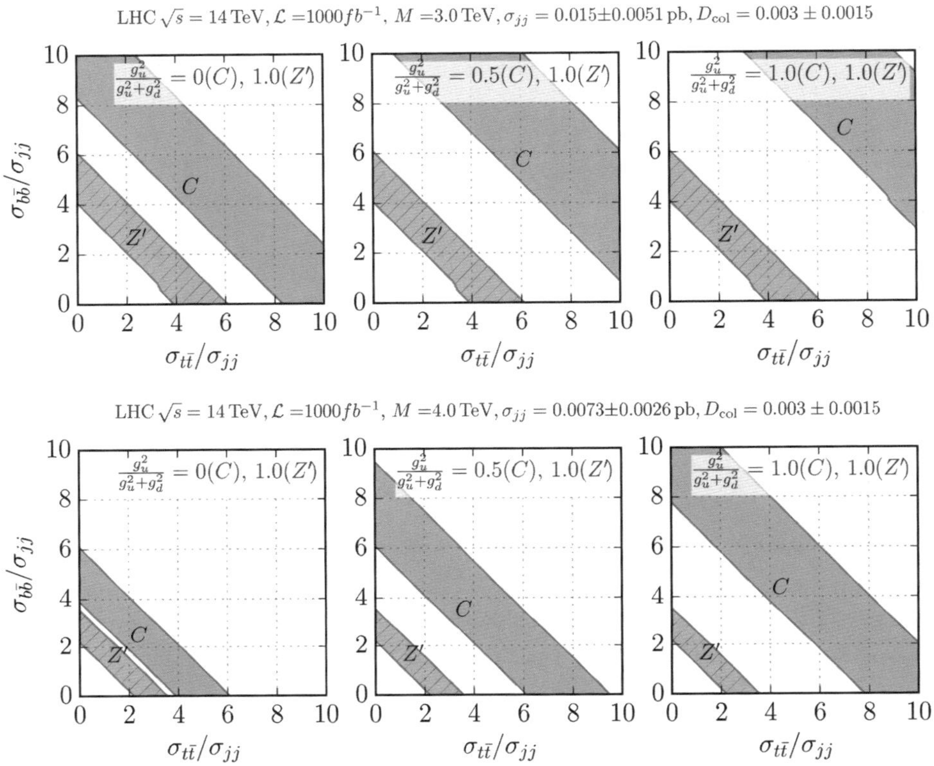

Fig. 3. Measuring $\sigma_{t\bar{t}}$ and $\sigma_{b\bar{b}}$ will show whether a resonance of a given mass, dijet cross-section, and D_{col} is a coloron or Z'. We display regions of parameter space corresponding to a dijet cross section within 1σ uncertainty of the value required for a discovery at the LHC with $\sqrt{s} = 14\,\text{TeV}$ at $\mathcal{L} = 1000\,\text{fb}^{-1}$, and that have the color discriminant variable within a range of 50% around the value of $D_{\text{col}} = 3 \times 10^{-3}$, for resonances with masses 3 TeV (top panel) and 4 TeV (bottom panel). Coloron and Z' models within reach are, respectively, displayed in a blue region and a diagonally-hatched green region. The colorons and Z' bosons lie in different regions of parameter space. However, at the bottom left panel for a 4 TeV resonance, the bands for a C coupling only to down-type light quarks and a Z' coupling only to up-type light quarks approach closely.

The excited quarks can decay into qg or into a quark plus a gauge boson. The corresponding decay rates are:

$$\begin{aligned} \Gamma(q^* \to qg) &= \frac{1}{3}\alpha_S f_S^2 \frac{m_{q*}^3}{\Lambda^2} \qquad\qquad \Gamma(q^* \to q\gamma) = \frac{1}{4}\alpha f_\gamma^2 \frac{m_{q*}^3}{\Lambda^2} \\ \Gamma(q^* \to qV)) &= \frac{1}{8}\frac{g_V^2}{4\pi} f_V^2 \frac{m_{q*}^3}{\Lambda^2}\left[1 - \frac{m_V^2}{m_{q*}^2}\right]^2 \left[2 + \frac{m_V^2}{m_{q*}^2}\right] \end{aligned} \tag{18}$$

with $V = W, Z$ and with the definitions

$$f_\gamma = fT_3 + f'\frac{Y}{2} \qquad f_Z = fT_3\cos^2\theta_W - f'\frac{Y}{2}\sin^2\theta_W \qquad f_W = \frac{f}{\sqrt{2}}\,. \tag{19}$$

The $q^* \to qg$ branching ratio is about 0.8 for $f_S = f = f'$.

Excited quarks are singly produced at the LHC through quark-gluon annihilation and, as just noted, they dominantly decay into qg. For our analysis, we choose the benchmark parameters $\Lambda = m_{q^*}$ and $f_S = f = f'$, while allowing the overall coupling strength to vary. By way of comparison, recent LHC searches, CMS[6,7] and ATLAS[17] have used the same value of Λ with $f_S = f = f' = 1$.

6.1.2. *Color-octet scalars* ($\mathbf{S_8}$)

A gluon-gluon final state can generally arise from decay of colored scalars in models with extended color gauge structures[18–22]. In this work we adopt the general effective interaction for a color octet scalar, S_8, introduced in[16]:

$$\mathcal{L}_{S_8} = g_S d^{ABC} \frac{k_S}{\Lambda_S} S_8^A G_{\mu\nu}^B G^{C,\mu\nu} \ , \tag{20}$$

where d is the QCD totally symmetric tensor.

A colored scalar of this kind is singly-produced at the LHC through gluon-gluon annihilation. We consider the case in which it decays entirely (or almost entirely) into gluons. The corresponding decay rate reads:

$$\Gamma(S_8) = \frac{5}{3} \alpha_S \frac{k_S^2}{\Lambda_S^2} m_{S_8}^3 \ . \tag{21}$$

We set $\Lambda_S = m_{S_8}$ and we present results for different couplings k_S. Similarly, CMS[6,7] and ATLAS[17] present searches for $\Lambda_S = m_{S_8}$ and $k_S = 1$.

6.2. *LHC discovery reach*

For each type of dijet resonance, we have derived[3] the relevant mass and coupling parameter space for our analysis, namely the region that is not yet excluded by LHC-8 analyses and in which a 5σ discovery will be possible at the 14 TeV LHC. Fig. 4 summarizes our estimates of the 5σ reach at the 14 TeV LHC in the mass-vs.-coupling plane for colorons, excited quarks, and scalar-octets, for integrated luminosities of 30 – 3000 fb^{-1}. The discovery reach we find for the coloron is very similar to those already derived in[23,24] and in[1].

Within each pane of Fig. 4, we may identify a "region of interest" where a resonance of a given mass and coupling is not excluded by LHC (i.e., is not in the blue region at left), is relatively narrow (lies below the horizontal dashed curve) and would be detectable at LHC-14 at the indicated luminosity (is within the central light-grey region). The portion of this region of interest that lies above the horizontal dotted curve is accessible to coloron discriminant variable analysis, while the area below the dotted curve region is also accessible to jet energy profile analysis[3].

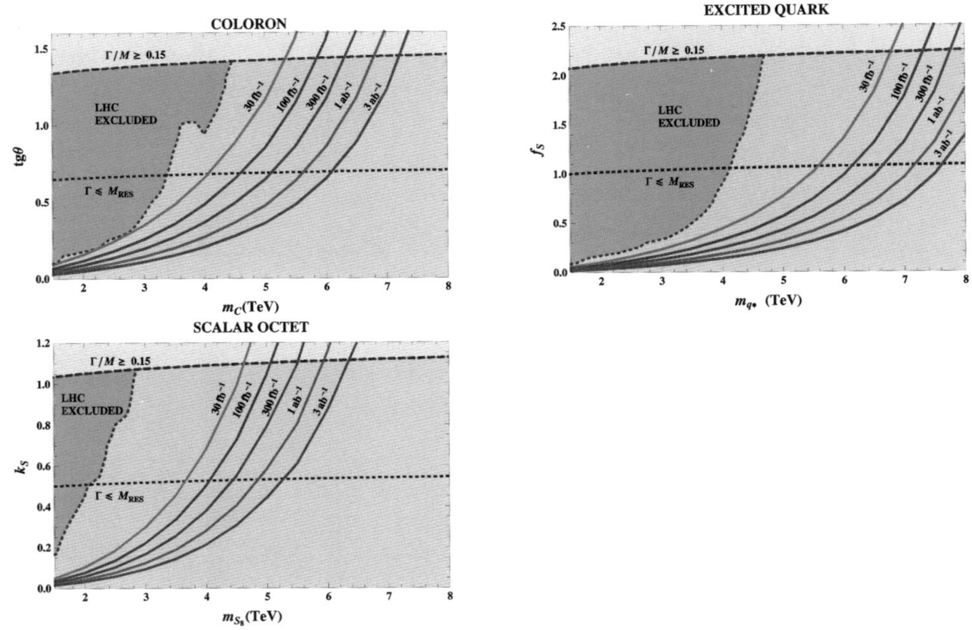

Fig. 4. In each pane from left to right: regions of parameter space excluded by LHC-8 (blue), accessible to LHC-14 (pale grey), or inaccessible at LHC-14 (pink). Thick colored curves show the 5σ reach at luminosities from 30 – 3000 fb^{-1}. Above the upper dashed line, the resonance is too broad to detect ($\Gamma = 0.15M$); below the lower dotted line it is narrower than the experimental resolution ($\Gamma \leq M_{res}$), where $M_{res} = 0.035M$[7].

6.3. *The color discriminant variable*

We have evaluated the value of D_{col} for our benchmark di-jet resonances – flavor universal colorons, excited quarks and scalar octets, in the allowed and accessible range of resonance masses. The dependence of D_{col} on the di-jet mass is controlled by the parton distribution functions (PDFs), since the quark and gluon parton content vary with the energy scale of the di-jet process. We calculate the di-jet resonance production cross section by using the CT10[25] next-to-leading-order PDF set with factorization and renormalization scales fixed at the resonance mass value. The measurement of D_{col} at the LHC is affected by the statistical and systematic uncertainties on measurements of the di-jet cross section, the resonance mass and the resonance width. Furthermore, D_{col} is only experimentally accessible if the mass resolution of the detector is less than the intrinsic width of the resonance. Our analysis[3] has fully included these uncertainties.

Fig. 5 shows the $\log_{10} D_{col}$ values, including the statistical and systematic uncertainties, for the three types of di-jet resonances q^*, C, S_8, as a function of the di-jet resonance mass at the 14 TeV LHC for different integrated luminosities. We observe that an excited quark resonance can be efficiently distinguished from either a coloron or a scalar octet resonance by the color discriminant variable at the 14

TeV LHC. Discriminating between colorons and scalar octets using the color discriminant variable is more challenging, but we find it should be possible to establish a separation which ranges from $\sim 2\sigma$ at $M \simeq 4$ TeV to $\sim 3\sigma$ at $M \simeq 6$ TeV.

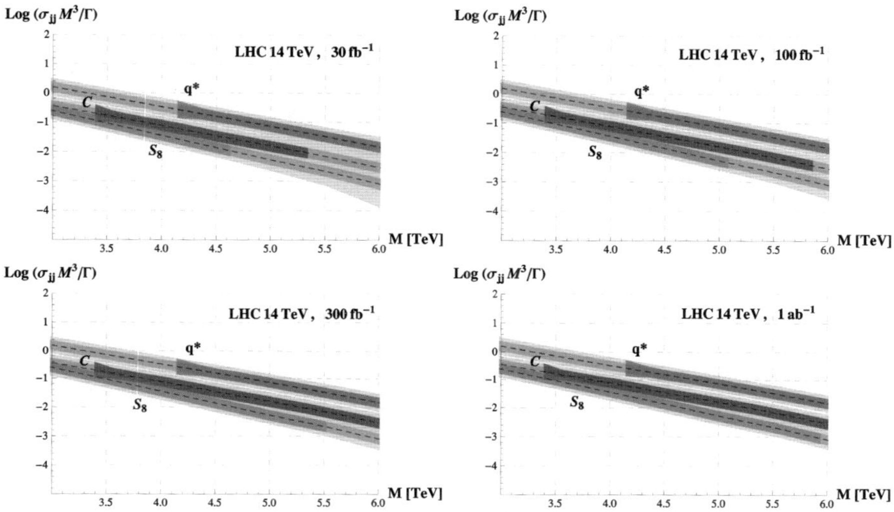

Fig. 5. Each pane shows $\log_{10} D_{col}$ for flavor universal colorons (C), excited quarks (q^*) and scalar octets (S_8) as a function of the dijet resonance mass at the 14 TeV LHC with a particular integrated luminosity. The outer (inner) grey (light grey) bands represent the $\pm$1-sigma statistical plus systematic uncertainty on $\log_{10} D_{col}$ when the resonance width is narrow (broad): $\Gamma = M_{res}$ ($\Gamma = 0.15M$). The colored bands [red for $q*$, blue for (C), green for (S_8)] show the $\log_{10} D_{col} \pm 1\sigma$ values obtained in the region of parameter space where the resonance is allowed by LHC-8 analyses, is neither too broad nor too narrow, and is amenable to discovery at LHC-14.

7. Discussion

If the LHC discovers a new dijet resonance, the color discriminant variable can help identify what has been found. The variable $D_{\rm col}$ is constructed from measurements available directly after discovery: namely, the resonance's mass, its total decay width, and its dijet cross section. This talk has reviewed recent results that apply the color discriminant variable to two new situations: (a) distinguishing between color-singlet and color-octet vector bosons whose couplings to quarks are not flavor-universal and (b) telling apart vector bosons, colored scalars, and excited quarks.

The first analysis assumes the new resonance couples identically to quarks of the first two generations, in which case $D_{\rm col}$ depends on three model-specific ratios of coupling constants: the up ratio ($g_u^2/[g_u^2 + g_d^2]$), the top ratio ($g_t^2/[g_u^2 + g_d^2]$), and the bottom ratio ($g_b^2/[g_u^2 + g_d^2]$). We find the method is generally not dependent on knowing the up ratio, a quantity which is not presently accessible to experiment.

Since $D_{\rm col}$ is insensitive to chiral structure, discriminating between color-singlet and color-octet resonances with flavor non-universal couplings requires only measurements of the $t\bar{t}$ and $b\bar{b}$ resonance cross sections.

The second analysis shows that an excited quark can be cleanly distinguished from either a coloron or a color-octet scalar by the color discriminant variable at the 14 TeV LHC. Establishing the distinction between colorons and color-octet scalars using the color discriminant variable is more challenging, but we still find the possibility of a $\sim 2(3)\sigma$ separation for resonance masses of order 4(6) TeV.

To summarize: we have generalized the color discriminant variable for use in determining the color structure of new bosons that may have flavor non-universal couplings to quarks or for comparing $q\bar{q}$, qg and gg dijet resonances. . We focused on resonances having masses $2.5 - 6.0\,\text{TeV}$ for the LHC with center-of-mass energy $\sqrt{s} = 14\,\text{TeV}$ and integrated luminosities up to 1000, fb^{-1}. After taking into account the relevant uncertainties and exclusion limits from current experiment and sensitivity for future experiments, we find that the future runs of the LHC can reliably determine the color structure of a resonance decaying to the dijets.

Acknowledgments

This material is based upon work supported by the National Science Foundation under Grant No. PHY-0854889. We wish to acknowledge the support of the Michigan State University High Performance Computing Center and the Institute for Cyber Enabled Research. PI is supported by Development and Promotion of Science and Technology Talents Project (DPST), Thailand. EHS and RSC thank the Kobayashi Maskawa Institute for the Origins of Particles and the Universe for hospitality during the SGT15, and also thank the Aspen Center for Physics and the NSF Grant #1066293 for hospitality during the writing of this Proceedings.

References

1. A. Atre, R. S. Chivukula, P. Ittisamai and E. H. Simmons, Distinguishing Color-Octet and Color-Singlet Resonances at the Large Hadron Collider, *Phys.Rev.* **D88**, p. 055021 (2013).
2. R. Sekhar Chivukula, P. Ittisamai and E. H. Simmons, Distinguishing flavor nonuniversal colorons from Z bosons at the LHC, *Phys. Rev.* **D91**, p. 055021 (2015).
3. R. Sekhar Chivukula, E. H. Simmons and N. Vignaroli, Distinguishing dijet resonances at the LHC, *Phys. Rev.* **D91**, p. 055019 (2015).
4. M. Bona *et al.*, Model-independent constraints on $\Delta F = 2$ operators and the scale of new physics, *JHEP* **0803**, p. 049 (2008).
5. G. Aad *et al.*, ATLAS search for new phenomena in dijet mass and angular distributions using pp collisions at $\sqrt{s} = 7$ TeV, *JHEP* **1301**, p. 029 (2013).
6. S. Chatrchyan *et al.*, Search for narrow resonances using the dijet mass spectrum in pp collisions at sqrt(s) = 8 TeV (2013).

7. *Search for Narrow Resonances using the Dijet Mass Spectrum with 19.6fb-1 of pp Collisions at sqrts=8 TeV*, Tech. Rep. CMS-PAS-EXO-12-059, CERN (Geneva, 2013).
8. K. Gumus, N. Akchurin, S. Esen and R. M. Harris, *CMS Sensitivity to Dijet Resonances*, Tech. Rep. CMS-NOTE-2006-070, CERN (Geneva, 2006).
9. Y. Bai, J. L. Hewett, J. Kaplan and T. G. Rizzo, LHC Predictions from a Tevatron Anomaly in the Top Quark Forward-Backward Asymmetry, *JHEP* **1103**, p. 003 (2011).
10. U. Haisch and S. Westhoff, Massive Color-Octet Bosons: Bounds on Effects in Top-Quark Pair Production, *JHEP* **1108**, p. 088 (2011).
11. R. M. Harris and K. Kousouris, Searches for Dijet Resonances at Hadron Colliders, *Int.J.Mod.Phys.* **A26**, 5005 (2011).
12. E. H. Simmons, Coloron phenomenology, *Phys.Rev.* **D55**, 1678 (1997).
13. R. Chivukula, A. G. Cohen and E. H. Simmons, New strong interactions at the Tevatron?, *Phys.Lett.* **B380**, 92 (1996).
14. U. Baur, M. Spira and P. M. Zerwas, Excited Quark and Lepton Production at Hadron Colliders, *Phys. Rev.* **D42**, 815 (1990).
15. U. Baur, I. Hinchliffe and D. Zeppenfeld, Excited Quark Production at Hadron Colliders, *Int. J. Mod. Phys.* **A2**, p. 1285 (1987).
16. T. Han, I. Lewis and Z. Liu, Colored Resonant Signals at the LHC: Largest Rate and Simplest Topology, *JHEP* **1012**, p. 085 (2010).
17. G. Aad *et al.*, Search for new phenomena in the dijet mass distribution using $p-p$ collision data at $\sqrt{s} = 8$ TeV with the ATLAS detector, *Phys. Rev.* **D91**, p. 052007 (2015).
18. C. T. Hill, Topcolor: Top quark condensation in a gauge extension of the standard model, *Phys.Lett.* **B266**, 419 (1991).
19. P. H. Frampton and S. L. Glashow, Chiral Color: An Alternative to the Standard Model, *Phys.Lett.* **B190**, p. 157 (1987).
20. M. V. Martynov and A. D. Smirnov, Chiral color symmetry and possible G-prime-boson effects at the Tevatron and LHC, *Mod. Phys. Lett.* **A24**, 1897 (2009).
21. R. S. Chivukula, E. H. Simmons and N. Vignaroli, Same-Sign Dileptons from Colored Scalars in the Flavorful Top-Coloron Model, *Phys.Rev.* **D88**, p. 034006 (2013).
22. Y. Bai and B. A. Dobrescu, Heavy octets and Tevatron signals with three or four b jets, *JHEP* **07**, p. 100 (2011).
23. F. Yu, Di-jet resonances at future hadron colliders: A Snowmass whitepaper (2013).
24. B. A. Dobrescu and F. Yu, Coupling-mass mapping of dijet peak searches, *Phys.Rev.* **D88**, p. 035021 (2013).
25. H.-L. Lai, M. Guzzi, J. Huston, Z. Li, P. M. Nadolsky *et al.*, New parton distributions for collider physics, *Phys.Rev.* **D82**, p. 074024 (2010).

Hidden Local Symmetry as Magnetic Gauge Theory

Ryuichiro Kitano

Theory Center, High Energy Accelerator Research Organization (KEK), Tsukuba 305-0801, Japan

The color-flavor locking phenomenon in the magnetic picture can be the microscopic description of the quark confinement in QCD. We demonstrate it in an $N = 2$ supersymmetric $SU(N_c) \times SU(N_c)$ quiver gauge theory coupled to N_f flavors of quarks ($N_f < N_c$). This model reduces to $SU(N_c)$ gauge theory with N_f flavors when the vacuum expectations value of the link field is much larger than the dynamical scales, and thus provides a continuous deformation of the $N = 2$ supersymmetric QCD. We study a vacuum which survives upon adding a superpotential term to reduce to $N = 1$ while preserving the vectorial $SU(N_f)$ flavor symmetry. We find a region of the parameter space where the confinement is described by the Higgsing of a weakly coupled magnetic $SU(N_f) \times U(1)$ gauge theory. The Higgsing locks the quantum numbers of $SU(N_f)$ magnetic color to those of $SU(N_f)$ flavor symmetry, and thus the massive magnetic gauge bosons become the singlet and adjoint representations of the flavor group, i.e, the vector mesons. If the qualitative picture remains valid in non-supersymmetric QCD, one can understand the Hidden Local Symmetry as the magnetic dual description of QCD, and the confining string is identified as the vortex of vector meson fields.

A Symmetry Breaking Mechanism by Parity Assignment in the Noncommutative Higgs Model

Masaki J. S. Yang

Institute of Particle and Nuclear Studies,
High Energy Accelerator Research Organization (KEK),
Tsukuba 305-0801, Japan
E-mail: yang@post.kek.jp

We apply the orbifold grand unified theory (GUT) mechanism to the noncommutative Higgs model. An assignment of Z_2 parity to the "constituent fields" induces parity assignments of both the gauge and Higgs bosons, because these bosons are treated as some kind of composite fields in this formalism.

Keywords: Noncommutative geometry, Induced gauge theory, Orbifold GUT.

1. Introduction

The discovery of the Higgs boson [1,2], a great triumph of science, has completed the particle contents of the Standard Model (SM). However, the theoretical origin of the Higgs boson is still unclear. Innumerable theories and models have been considered so far, in order to explain the existence of the boson and clarify the underlying physics beyond the SM.

Among these attempts, one interesting possibility is the Yang–Mills–Higgs model inspired noncommutative geometry, founded by A. Connes [3,4]. As a shortened name, we call this model noncommutative Higgs model (NHM). In this picture, the underlying spacetime is assumed to be multi-sheet of ordinary Minkowski space $M^4 \times Z_n$. The Higgs boson is interpreted as a gauge boson along a discrete extra dimension, that has noncommutative differential algebra. For example, in the simplest model $M^4 \times Z_2$, the coordinates are represented as $(x^\mu, y = \pm 1)$. Thus $y^2 = 1$ holds and it leads to the anti-commutative relation $y\, dy = -dy\, y$. This noncommutative differential algebra generates nonzero Higgs mass and the spontaneous symmetry breaking (SSB) mechanism. This concept is widely followed by its successors [5–8]. It is also applied to supersymmetry [9], and the grand unified theories (GUT), e.g. $SU(5)$ [10–13] or $SO(10)$ models [14,15]. Meanwhile, when this kind of model is regarded as a theory with an extra dimension, several mechanisms and structures in the normal extra dimension can be recast to the noncommutative extra dimension.

In this study [16], as a typical example, we implement the orbifold GUT mechanism [17] to the $SU(5)$ NHM. In the previous standard studies of NHM, the symmetry breaking pattern is determined by selecting the *distance* matrices M_{nm}. By contrast, we imposed an assignment of Z_2 parity to the "constituent fields". This induces parity assignments of both gauge and Higgs bosons, because these bosons are treated as some kind of composite fields in this formalism. This is the main difference from the original orbifold GUT model that treats the assignments of bosons as independent conditions [16].

This proceedings is organized as follows. In the next section, we review the generalized gauge theory in NHM. In Section 3, we present the symmetry breaking mechanism by parity assignment, and discussions on a new field and the proton decay. Section 4 is devoted to conclusion.

2. Generalized gauge theory on $M^4 \times Z_n$

Here, we define several definitions and the main frame of formulation in the generalized gauge theory on $M^4 \times Z_n$. Since the formulation is the same as those in Ref. [12], a detailed description is omitted. The essential difference in the formulations between the original papers [10,11] and Ref. [12] is the flavor dependence of the *distance* matrices M_{nm}, and the differential algebraic structure is almost equivalent.

2.1. *Differential calculus*

On the n-sheeted Minkowski space $M^4 \times Z_n$, the coordinates are represented by $(x^\mu, y = 1 - N)$, and a function on the nth sheet is expressed as $f(x,n) \equiv f_n$. In this space, the exterior derivative is generalized to $\boldsymbol{d} = d + d_\chi$ introducing extra exterior derivatives d_χ and the differential forms χ_m. They are defined as follows;

$$\begin{aligned} \boldsymbol{d} f_n &= (d + d_\chi) f_n, \quad d f_n = \partial_\mu f_n dx^\mu, \\ d_\chi f &= \sum_{m \neq n} d_{\chi_m} f = \sum_{m \neq n} [M_{nm} f_m - f_n M_{nm}] \chi_m, \end{aligned}$$

where the matrices $M^\dagger_{nm} = M_{mn} (n \neq m)$ represent the *distances* between two sheets. In order to keep the nilpotency of the generalized exterior derivative $\boldsymbol{d}^2 = 0$, the differential forms dx^μ and χ_m should be anti-commutative $dx^\mu \wedge \chi_m = -\chi_m \wedge dx^\mu$, and the d_χ should also be nilpotent $d_\chi^2 = 0$. To prove $d_\chi^2 = 0$, and to keep the differential algebra consistent, there are several "index shifting rules" between f_n, M_{nm}, and χ_m. These shifting rules are source of the noncommutativity that corresponds to the relation $ydy = -dyy$ in other formulations [5–8]. The precise form of these rules and a proof of $d_\chi^2 = 0$ is presented in Ref. [12].

2.2. *Generalized connection and the Lagrangian*

The generalized connection one-form $\boldsymbol{A}$, which contains usual gauge bosons and Higgs bosons is defined as

$$\boldsymbol{A}(x,n) \equiv \sum_i a_i^\dagger(x,n) \boldsymbol{d} a_i(x,n) \equiv A(x,n) + \sum_{m \neq n} \Phi_{nm}(x) \chi_m. \tag{1}$$

Here, the "constituent field" $a_i(x,n)$ is square-matrix-valued continuous function defined on the nth sheet, and the summation over i is assumed to be a finite sum.

The field strength two-form is defined as usual $\boldsymbol{F}_n = \boldsymbol{dA}_n + \boldsymbol{A}_n \wedge \boldsymbol{A}_n$, where $\boldsymbol{dA}_n = \sum_i \boldsymbol{d}a_n^{i\dagger} \wedge \boldsymbol{d}a_n^i$, because of $\boldsymbol{d}^2 = 0$. After some calculation, the explicit form of the $\boldsymbol{F}_n$ contains the following three pieces:

$$\boldsymbol{F}_n = F_{n\,\mu\nu} dx^\mu \wedge dx^\nu + \sum_{m\neq n} D_\mu H_{nm}\, dx^\mu \wedge \chi_m + \sum_{\substack{m\neq n,\\ l\neq m}} (H_{nm}H_{ml} - X'_{nml})\chi_m \wedge \chi_l,$$

where

$$\begin{aligned} F_{\mu\nu}(x,n) \equiv F_{n\,\mu\nu} &= \frac{1}{2}\left(\partial_\mu A_\nu(x,n) - \partial_\nu A_\mu(x,n) + [A_\mu(x,n), A_\mu(x,n)]\right) \\ D_\mu H_{nm}(x) &= \partial_\mu H_{nm}(x) + A_\mu(x,n)H_{nm}(x) - H_{nm}(x)A_\mu(x,m), \\ X'_{nml}(x) &= \sum_i a_i^\dagger(x,n) M_{nm} M_{ml} a_i(x,l). \end{aligned}$$

It contains the gauge boson, the *back-shifted* Higgs field $H_{nm} \equiv \Phi_{nm} + M_{nm} = \sum_i a_n^{i\dagger} M_{nm} a_m^i$ and the new field $X'_{nml}(x)$. The treatment of X'_{nml} is decided by whether X'_{nml} is a dependent function of the Higgs fields H_{nm} or not. If X'_{nml} can be written as a some function of the Higgs fields $X'_{nml} = f(H_{nm})$, X'_{nml} is treated as a Higgs interaction terms. If not, X'_{nml} is regarded as an *auxiliary field* that does not have a kinetic term, and then it will be eliminated from Lagrangian by the equation of motion $\partial\mathcal{L}/\partial X'_{nml} = 0$. In consequence, the gauge-invariant Lagrangian is formulated by

$$\mathcal{L}_{\text{YMH}} = -\frac{1}{4}\sum_n \frac{1}{g_n^2}\text{tr}\langle \boldsymbol{F}_n, \boldsymbol{F}_n\rangle, \tag{2}$$

where independent coupling constants g_n are introduced for gauge fields on each nth sheet. In order to calculate Eq. (2), the metric of the space $M^4 \times Z_n$ is specified as

$$\langle dx^\mu, dx^\nu\rangle = g^{\mu\nu}, \quad \langle \chi_n, dx^\mu\rangle = 0, \quad \langle \chi_n, \chi_m\rangle = -\delta_{nm}\alpha_n^2,$$

where $g^{\mu\nu} = \text{diag}(+,-,-,-)$. and similar relation holds for the inner products of two-forms.

3. *SU*(5) GUT breaking by parity assignment

In this section, we review a $SU(5)$ GUT in the NHM shortly, and implement the orbifold GUT mechanism in this toy model. In the formalism presented the previous section, $SU(5)$ GUT model requires $N \geqq 3$. This is because that $N \geqq 3$ realizes more than two independent M_{nm}'s, that corresponds to two energy scales of $SU(5)$ symmetry breaking, $SU(5) \to SU(3)_c \times SU(2)_L \times U(1)_Y$ and $SU(3)_c \times SU(2)_L \times U(1)_Y \to SU(3)_c \times U(1)_{\text{em}}$. Then we choose $N = 3$ to describe $SU(5)$ GUT[10,12].

At first, "constituent fields" a_1^i, a_2^i are assumed to be complex 5×5 matrices and a_3^i has a real-valued continuous function, that satisfies the Hermitian condition

$\sum_i a_n^{i\dagger} a_n^i = 1$. Additionally, we consider a parity symmetry between $n=1$ and $n=2$ sheets of spacetime, and impose the following parity condition for the "constituent fields";

$$a_2^i = P a_1^i P, \tag{3}$$

where $P = \mathrm{diag}\,(-1,-1,-1,+1,+1)$.

The $SU(5)$ gauge fields at each three copies of M^4 are calculated as follows

$$A(x,1) = \sum_i a_1^{i\dagger} d a_1^i = iT^a A_1^a(x) \equiv A, \tag{4}$$

$$A(x,2) = \sum_i a_2^{i\dagger} d a_2^i = iT^a A_2^a(x) \equiv PAP, \tag{5}$$

$$A(x,3) = \sum_i a_3^{i\dagger} d a_3^i = 0, \tag{6}$$

where $T^a(a=1,\cdots,24)$ are the generators of $SU(5)$, provided that the traceless condition $\mathrm{Tr}A(x,1) = \mathrm{Tr}A(x,2) = 0$ is imposed.

In order to determine the Higgs fields, we set the matrix M_{nm} as

$$M_{12} = M_{21} = M\mathrm{diag}(1,1,1,1,1) \equiv \Sigma_0, \tag{7}$$

$$M_{13} = M_{23} = M_{31}^{\dagger} = M_{32}^{\dagger} = \mu \left(0\;0\;0\;0\;1\right)^T \equiv H_0, \tag{8}$$

where $M(\mu)$ is the energy scale at the stage of GUT (SM) symmetry breaking $SU(5) \to$ SM (SM $\to SU(3)_c \times U(1)_{\mathrm{em}}$). Thus the back-shifted Higgs fields are found to be

$$\Sigma(x) + \Sigma_0 = H_{12}(x) = \sum_i a_1^{i\dagger} M P a_1^i P = P H_{21}(x) P, \tag{9}$$

$$H(x) + H_0 = H_{13}(x) = \sum_i a_1^{i\dagger} M_{13} a_3^i = P H_{23}(x). \tag{10}$$

From Eqs. (9) and (10), it is found that H_{13} is a 5×1 matrix-valued field transforming like 5 representation under $SU(5)$, and H_{12} is a 5×5 matrix-valued field that is a linear combination of the field like 1 and 24 representation. The discussion on the field $\Sigma(x)$ is presented later. Note that both the parity assignments of the gauge boson (5) and the Higgs boson (10) are determined by the condition of the "constituent fields" (3). They are independent conditions in the original orbifold GUT model.

In order to calculate the Lagrangian (2), it is necessary to consider which $X'_{nml}(x)$ are independent of Higgs field and which $X'_{nml}(x)$ are not. We found that only $X'_{121}, X'_{313}, X'_{123}$, and X'_{312} are dependent function of Higgs fields. Other $X'_{nml}(x)$ are independent that are entirely eliminated in Eq. (2) by $\partial\mathcal{L}/\partial X'_{nml} = 0$.

Substituting these results into Eq. (2), we found the final form of the Lagrangian [16]

$$\begin{aligned}\mathcal{L}_{SU(5)} =& -\frac{1}{4g^2}\mathrm{tr}F^\dagger_{\mu\nu}F^{\mu\nu} + \frac{\alpha^2}{2g^2}\mathrm{tr}|D_\mu(\Sigma+\Sigma_0)|^2 + \left(\frac{\alpha^2}{2g'^2}+\frac{\beta^2}{2g^2}\right)|D_\mu(H+H_0)|^2 \\ & -\frac{\alpha^4}{2g^2}\mathrm{tr}||\Sigma+\Sigma_0|^2 - M^2|^2 - \frac{\alpha^2\beta^2}{2g'^2}||H+H_0|^2-\mu^2|^2 \\ & -\left(\frac{\alpha^2\beta^2}{2g^2}+\frac{\alpha^4}{2g'^2}\right)|(\Sigma P + MP - M)(H+H_0)|^2. \end{aligned} \tag{11}$$

Here, $F_{\mu\nu} = \partial_\mu A_\nu - \partial_\nu A_\mu + [A_\mu, A_\nu]$, and we have assumed the $1 \leftrightarrow 2$ symmetry $g_1^2 = g_2^2 = g^2$, $\alpha_1^2 = \alpha_2^2 = \alpha^2$, and set $g_3^2 = g'^2, \alpha_3^2 = \beta^2$.

In particular, the mass term of the 5 representation Higgs is computed as

$$|(MP - M)H|^2 = M^2\mathrm{diag}(4,4,4,0,0)\, H^\dagger H. \tag{12}$$

Similarly the gauge boson masses are

$$|D_\mu H_{12}|^2 \ni (A_\mu M - MPA_\mu P)^2 = (MA_\mu^{\hat{a}}T^{\hat{a}})^2, \tag{13}$$

where $\hat{a}$ runs the broken generators except those of $SU(3)_c \times SU(2)_L \times U(1)_Y$. Eqs. (12) and (13) shows that the parity assignment condition (3) $a_2^i = Pa_1^iP$ invokes $SU(5)$ symmetry breaking, and provides the colored triplet Higgs and broken gauge bosons with heavy mass of order M. Therefore, it is reasonable to consider that the symmetry breaking by the condition (3) corresponds to the orbifold GUT mechanism [17] in the NHM. This is our main result. Of course, this result holds only at tree level. However, if the parity symmetry imposed between the first and second sheet is not broken by the quantum correction, a mixing between the Higgs doublet and triplet is prohibited at the quantum level.

4. Conclusions

In this study, we have implemented the orbifold GUT mechanism in the noncommutative Higgs model. An assignment of Z_2 parity to the "constituent fields" induces the parity assignments of both the gauge and Higgs bosons, because these bosons are treated as some kind of composite fields in this formalism. This is the main difference from the original orbifold GUT model that treats the assignments of bosons as independent conditions. As a result, some of the gauge bosons and the colored triplet Higgs boson receive heavy mass comparable to GUT scale, and the $SU(5)$ gauge symmetry is broken. No particles appear other than the SM ones in the massless states. In fact, the nucleon decay is a problem in this model. However, there is a possibility that couplings between fermions and X_μ, Y_μ, H^c bosons could be forbidden by the proper parity assignment of fermions. We leave the construction of the fermionic sector for future work.

Acknowledgements

The author would like to appreciate R. Kitano for useful discussions and valuable comments. This study is supported by JSPS Research Fellowships for Young Scientists, No 24 · 8357.

References

1. ATLAS Collaboration, G. Aad *et al.*, Observation of a new particle in the search for the Standard Model Higgs boson with the ATLAS detector at the LHC, Phys. Lett. B **716**, 1 (2012), arXiv:1207.7214.
2. CMS Collaboration, S. Chatrchyan *et al.*, Observation of a new boson at a mass of 125 GeV with the CMS experiment at the LHC, Phys. Lett. B **716**, 30 (2012), arXiv:1207.7235.
3. A. Connes and J. Lott, Particle Models and Noncommutative Geometry (Expanded Version), Nucl. Phys. Proc. Suppl. **18B**, 29 (1991).
4. A. Connes, Noncommutative geometry, Academic Press, London (1994).
5. M. Dubois-Violette, R. Kerner, and J. Madore, Noncommutative Differential Geometry of Matrix Algebras, J. Math. Phys. **31**, 316 (1990).
6. R. Coquereaux, G. Esposito-Farese, and G. Vaillant, Higgs fields as Yang-Mills fields and discrete symmetries, Nucl. Phys. B **353**, 689 (1991).
7. A. Sitarz, Higgs mass and noncommutative geometry, Phys. Lett. B **308**, 311 (1993), arXiv:hep-th/9304005.
8. K. Morita and Y. Okumura, Weinberg-Salam theory in noncommutative geometry, Prog. Theor. Phys. **91**, 959 (1994).
9. A. H. Chamseddine, Connection between space-time supersymmetry and noncommutative geometry, Phys. Lett. B **332** (1994) 349 arXiv:hep-th/9404138.
10. A. H. Chamseddine, G. Felder, and J. Frohlich, Unified gauge theories in noncommutative geometry, Phys. Lett. B**296**, 109 (1992).
11. A. H. Chamseddine, G. Felder, and J. Frohlich, Grand unification in noncommutative geometry, Nucl. Phys. B **395**, 672 (1993), arXiv:hep-ph/9209224.
12. K. Morita and Y. Okumura, Gauge theory on discrete space M(4) x Z(N) and SU(5) GUT in noncommutative geometry, Prog. Theor. Phys. **91**, 975 (1994).
13. I. Sogami, Minimal SU(5) grand unified theory based on generalized covariant derivatives with gauge and Higgs fields, Prog. Theor. Phys. **95**, 637 (1996).
14. A. H. Chamseddine and J. Frohlich, SO(10) unification in noncommutative geometry, Phys. Rev. D **50**, 2893 (1994), arXiv:hep-th/9304023.
15. Y. Okumura, SO(10) grand unified theory in noncommutative differential geometry on the discrete space M(4) X Z(N), Prog. Theor. Phys. **94**, 607 (1995).
16. M. J. S. Yang, A symmetry breaking mechanism by parity assignment in the noncommutative Higgs model, PTEP **2015**, no. 4, 043B10 (2015) [arXiv:1501.03888 [hep-ph]].
17. Y. Kawamura, Triplet doublet splitting, proton stability and extra dimension, Prog. Theor. Phys. **105**, 999 (2001), arXiv:hep-ph/0012125.

Topology, the Meson Spectrum and the Scalar Glueball: Three Probes of Conformality and the Way It Is Lost

E. Pallante
Van Swinderen Institute, University of Groningen,
Groningen, 9747 AG, The Netherlands
E-mail: e.pallante@rug.nl

We discuss properties of non-Abelian gauge theories that change significantly across the lower edge of the conformal window. Their probes are the topological observables, the meson spectrum and the scalar glueball operator. The way these quantities change tells about the way conformal symmetry is lost.

Keywords: Conformal symmetry, QCD.

1. Introduction

It is a compelling task to understand the role played by conformal symmetry in the yet to be discovered unified theory for gravity, electroweak and strong interactions. Instrumental to this aim and relevant for phenomenological constructions beyond the Standard Model is the study of the breaking patterns of this symmetry in each component of the complete theory taken in isolation as well as when coupled to each other.

In this work we mainly focus on the conformal window of non-Abelian gauge theories in isolation, and in order to simplify the discussion we consider the case of $SU(N)$ Yang-Mills theory with massless Dirac fermions in the fundamental representation, while explicitly commenting on other realizations whenever useful. Towards the end, in Sec. 4, we come back to some considerations about the embedding in the complete theory.

The phase diagram of massless QCD in the plane temperature versus the number of flavors is shown in Fig. 1. We are interested in the change of properties of the system across the lower edge of the conformal window; the latter defines a family of theories that has a stable infrared fixed point (IRFP) where the theory is conformal with nonzero anomalous dimensions. Theories inside the conformal window are deconfined and chiral symmetry is exact. In Sec. 2 we discuss properties of the scalar glueball operator and how its anomalous dimension probes different mechanisms for the closing of the conformal window. Sec. 3 discusses the topology, the eigenvalue distribution of the Dirac operator and the role of the $U(1)$ axial anomaly. In Sec. 4 we entertain the possibility of a spontaneous breaking of conformal symmetry in the complete theory and discuss under which conditions a dilaton could manifest in the QCD sector.

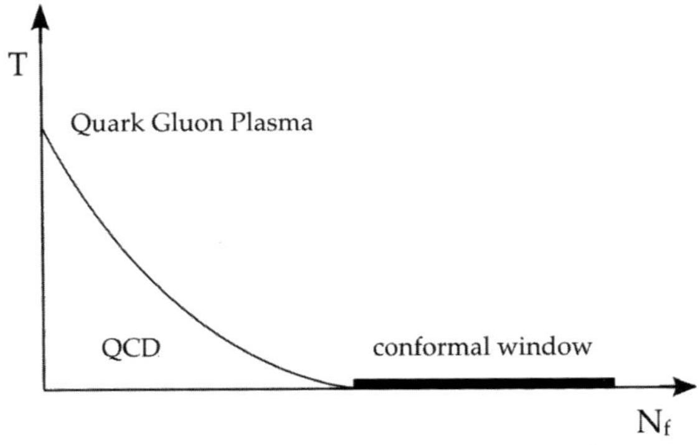

Fig. 1. Temperature (T) – flavor (N_f) phase diagram of $SU(N)$ Yang-Mills theory with N_f massless Dirac fermions in the fundamental representation. A line of thermal chiral phase transitions, the chiral phase boundary, separates QCD from the high-temperature quark-gluon plasma phase. The endpoint of the chiral phase boundary is also where the theory deconfines and it coincides with the lower edge N_f^c of the conformal window; the addition of supersymmetry and/or different fermion representations can change the latter property and make the phase diagram richer.

2. The scalar glueball and its anomalous dimension

The lower edge of the conformal window of QCD as depicted in Fig. 1 can be seen as the point in the space of relevant couplings where the system at the same time deconfines and realizes exact chiral symmetry at zero temperature. There is no physical reason, nor evidence for a separation of where the system deconfines and where chiral symmetry is restored; if it were not so, a new phase would appear below the conformal window, separated from QCD by an additional change of symmetry patterns[a]. In fact, the recent study in Ref. 2 provides close agreement between the lattice determination of the lower edge of the conformal window using chiral observables $6 < N_f^c < 8$ and the prediction in Ref. 3, where large-N QCD in the Veneziano limit ($N \to \infty$, $N_f/N = \text{const}$) is solved and the lower edge is identified as the threshold for the quantum instability of the glueball kinetic term $N_f/N = 5/2$.

The properties of gluodynamics must change across the lower edge of the conformal window, where conformal symmetry is restored and the system realizes an exact IRFP. In particular, the gluon condensate, i.e., the vacuum expectation value of the scalar glueball operator $G^a_{\mu\nu}G^a_{\mu\nu}$ must identically vanish at the fixed point and the two-point correlator of the scalar glueball acquires conformal scaling with nonzero anomalous dimension γ_G. In Ref. 2 we realized that γ_G, for varying N_f along the IRFP line, carries information on the nature of the endpoint and the zero-temperature phase that precedes the conformal window. At a fixed point the

[a]Differently from QCD, supersymmetric QCD (SQCD) develops a free non-Abelian magnetic phase in terms of the dual magnetic variables for $N_c + 2 \leq N_f \leq 3N_c/2$, just below the conformal window where the fundamental electric theory is infinitely strongly coupled. This can be seen as a consequence of exact electric-magnetic duality[1].

beta-function vanishes $\beta(\alpha_*) = 0$ and one finds[2]

$$\gamma_G = -\beta'(\alpha_*)\,. \tag{1}$$

The latter is a physical property of the system, renormalization scheme independent. In Ref. 2 we have determined γ_G in perturbation theory to a given loop order, $n = 2, 3, 4$ loops, inside the conformal window and compared this result with the Veneziano limit of large-N QCD[3]. One salient aspect is the fact that perturbation theory as well as the large-N solution predict an *increasing* anomalous dimension $|\gamma_G|$ ($|\beta'(\alpha_*|$) for *decreasing* $N_f \searrow N_f^c$. This behavior turns out to be opposite[2] to the one implied by a UV-IR fixed-point merging phenomenon at N_f^c. In this case we can assume without loss of generality

$$\beta(\alpha, N_f) = (N_f - N_f^c) - (\alpha - \alpha^c)^2 \tag{2}$$

close to N_f^c where the merging would occur, see also Ref. 4. This beta-function has pairs of zeros $\alpha_\pm = \alpha^c \pm \sqrt{N_f - N_f^c}$ and it develops a local maximum $\beta'(\alpha^c) = 0$ for $N_f = N_f^c$, with $|\beta'|$ *decreasing* for *decreasing* $N_f \searrow N_f^c$. Both zeros $\alpha_\pm$ acquire an imaginary component for $N_f < N_f^c$.

A UV-IR merging mechanism is the simplest, maybe not unique, framework where the so called walking phenomenon is realized just below the conformal window. Instead of using the running coupling — a renormalisation scheme dependent quantity — this phenomenon is best identified by the realization of preconformal Miransky/Berezinsky-Kosterlitz-Thouless (BKT) scaling[5–7] of observables in the vicinity of the conformal window[b], for $N_f \nearrow N_f^c$. From a statistical mechanics point of view, the realization of this scaling is a direct consequence of a so-called conformal phase transition that would occur at N_f^c, treating N_f as a continuous parameter. The inverse of the correlation length ξ of the system

$$\frac{1}{\xi} \sim \Lambda_{UV}\, \theta(N_f^c - N_f)\, e^{-c/\sqrt{N_f^c - N_f}}\,, \qquad c > 0 \tag{3}$$

decreases exponentially in the broken phase $N_f < N_f^c$ and it vanishes identically in the symmetric phase $N_f > N_f^c$, leaving an imprint in the spectrum on both sides. We come back to this point in Sec. 4 when studying the spectrum across the endpoint and the role of conformal symmetry.

A scenario where no UV-IR fixed point merging occurs remains a plausible one. It is actually favored by perturbation theory and it more closely resembles what happens in SQCD, where no physical UVFP emerges — differences can still reside in the detailed structure of the beta-function, as noticed in Ref. 3 and further discussed in Ref. 2, and in the occurrence of the free magnetic phase in SQCD.

[b]The comparison of theories with different N_f requires the setting of a common reference scale. A common UV scale is to be chosen in order to trace a preconformal Miransky/BKT scaling in the IR-dominated observables.

A nonperturbative, e.g., lattice determination of the anomalous dimension γ_G is therefore desirable, in order to discriminate between realizations of loss of conformality with and without UV-IR merging. One obvious way to extract γ_G is by a nonperturbative measurement of the two-point correlator of the scalar glueball operator; on the lattice one should separate its universal scaling law from nonuniversal violations of scaling. Another intriguing possibility may be suggested by the study of the Wilson flow of the vacuum expectation value of the scalar glueball operator (the gauge energy density) in Ref. 8. There, the evolution of this expectation value away from the ultraviolet has been derived in perturbative QCD as the solution of a Wilson flow equation, an approximately trivializing map realized by a first order ordinary differential equation in field space with the Wilson gauge action as the evolution kernel. Trivializing maps have been discussed in the context of field theories in the continuum and on the lattice[9], while Ref. 8 discusses explicit examples and proposes their use for high precision scale setting in lattice QCD studies, see also Refs. 10,11. By defining the quantity $t^2\langle E\rangle$, with t the flow "time" with dimensions $[\text{energy}^{-2}]$, in the $\overline{MS}$ scheme and at a renormalization scale $\mu = 1/\sqrt{8t}$ one obtains for $SU(3)$[8]

$$\begin{aligned} t^2\langle E\rangle &= \frac{3}{4\pi}\alpha(\mu)\left\{1 + k_1\alpha(\mu) + O(\alpha^2)\right\} \\ k_1 &= \frac{1}{4\pi}\left\{11\gamma_E + \frac{52}{3} - 9\ln 3 - N_f\left(\frac{2}{3}\gamma_E + \frac{4}{9} - \frac{4}{3}\ln 2\right)\right\} \\ &= 1.0978 + 0.0075\,N_f \end{aligned} \tag{4}$$

to next-to-leading order in the running coupling $\alpha(\mu) = g^2(\mu)/(4\pi)$. The quantity in Eq. (4) exhibits the important property of being finite once expressed in terms of the renormalized gauge coupling, at least at next-to-leading order in perturbation theory. In other words, the flow of $t^2\langle E\rangle \to 0$ on a Euclidean spacetime lattice as a function of t is dictated by dimensional arguments, while its proportionality to the running coupling in perturbation theory encodes asymptotic freedom. In particular, the residual t-scale dependence in the running coupling encodes the specific and highly nontrivial way in which asymptotic freedom realizes the breaking of conformal symmetry in the quantized theory; asymptotic freedom also guarantees that $t^2\langle E\rangle \to 0$ in the ultraviolet limit $t \to 0$. Beyond perturbation theory, we could think to infer the evolution towards the infrared of a properly renormalized $t^2\langle E\rangle$ based, again, on dimensional reasoning and underlying symmetries. However, constraints on one-point functions are usually not as powerful as those on other $n-$point functions. To illustrate this we use an argument that we argue should be valid for both $\bar{\psi}\psi$ and GG at an IRFP. Both VEVs $\langle\bar{\psi}\psi\rangle$ and $\langle GG\rangle$ must vanish at the IRFP. One would like to know how they approach zero when a scale perturbation — it being the fermion mass or t, respectively — is applied. Chiral symmetry is exact in the conformal window, and only at the lower endpoint of the IRFP line a chiral restoration phase transition occurs. In other words, the chiral order parameter $\langle\bar{\psi}\psi\rangle$ along this line, except its endpoint, does not undergo a phase transition and it must obey all constraints of exact chiral symmetry. One constraint that will also be later useful is the one on the chiral cumulant $R_\pi = \chi_\sigma/\chi_\pi = (\partial\langle\bar{\psi}\psi\rangle/\partial m)/(\langle\bar{\psi}\psi\rangle/m)$, or equivalently the ratio of the renormalized pseudoscalar and scalar meson masses. At the chiral restoration transition the relevant magnetic EoS[12]

$$m = A\langle\bar{\psi}\psi\rangle^\delta + B\langle\bar{\psi}\psi\rangle + \langle\bar{\psi}\psi\rangle_0 \tag{5}$$

has $B = 0$, $\langle\bar{\psi}\psi\rangle_0 = 0$ and it implies $R_\pi = 1/\delta$, with critical exponent δ. Chiral symmetry implies that $\lim_{m\to 0} R_\pi = 0$ in the broken phase, while $\lim_{m\to 0} R_\pi = 1$ in the symmetric phase; thus $0 \leq R_\pi \leq 1$. Away from the endpoint along the IRFP line no chiral phase transition occurs, and exact chiral symmetry suggests $\lim_{m\to 0} R_\pi = 1$; thus $B \neq 0$. Differently from the endpoint, the only constraint imposed by conformal symmetry on the scaling of this and other one-point functions with a scale perturbation is that their VEVs vanish at the IRFP; this is guaranteed for $\langle\bar{\psi}\psi\rangle$ also with $B \neq 0$. On the other hand, given that $0 \leq 1/\delta \leq 1$, the contribution of the A term is dominant w.r.t. the linear term close to the chiral limit. An analogous reasoning can be applied to $\langle GG\rangle$ ($\langle E\rangle$), where now chiral symmetry is no longer imposing constraints. The t-evolution can contain a term carrying its anomalous dimension $t^{-2-\gamma_G/2}$, but it is not the only one. Also note that this term goes to zero for $t \to \infty$ only if $\gamma_G > -4$.

Obviously, the infrared behavior realized on the lattice by the quantity $t^2\langle E\rangle$ will depend on the chosen kernel of the Wilson-flow differential equation. It may still be possible to device a useful evolution equation able to distinguish between the confining and the conformal behavior of the theory in the infrared.

3. Topology and the $U(1)$ axial anomaly

In order to identify all the relevant properties of the lower edge of the conformal window we need to determine the role that each symmetry plays in this context. In particular, three symmetries or lack thereof and their order parameters should be considered: conformal symmetry, chiral (flavored) symmetry $SU(N_f)_L \times SU(N_f)_R$ and the $U(1)$ axial anomaly. Despite the fact that a true order parameter for confinement is absent in QCD, we can still identify confinement with the presence of a mass-gap, i.e, a glueball mass in quenched QCD. The lower edge of the conformal window is then also the point where the properties of glueball operators change and the mass-gap disappears, as in Ref. 3.

In this section we discuss the topology, the way the $U(1)$ axial anomaly manifests and its interplay with chiral and conformal symmetry, in the conformal window and across its lower edge. Two quantities that are relevant to understand the fate of the $U(1)$ axial anomaly are the topological charge Q

$$Q = \frac{1}{32\pi^2}\int_V d^4x\, G^a_{\mu\nu}\tilde{G}^a_{\mu\nu}\,, \tag{6}$$

with $\tilde{G}$ the dual of the gauge-field strength tensor $G_{\mu\nu}$, and the topological susceptibility χ_t, i.e., the second moment of the topological charge distribution

$$\chi_t = \frac{\langle Q^2 \rangle}{V} = -\frac{1}{V}\left(\frac{1}{Z}\frac{\partial^2 Z}{\partial \theta^2}\right)\bigg|_{\theta=0} \tag{7}$$

with the QCD partition function Z dependent on the QCD θ-angle. The $U(1)$ axial anomaly can be defined as the nonzero contribution to the divergence of the flavor singlet axial current

$$\partial_\mu J_\mu^5 = \frac{g^2 N_f}{16\pi^2} G^a_{\mu\nu}\tilde{G}^a_{\mu\nu} + \text{explicit fermion mass contribution.} \tag{8}$$

This is an operator identity and it implies that the axial current is conserved, i.e., no anomaly in two limits: $N_f = 0$ and $N = \infty$. The latter limit becomes evident in the Witten-Veneziano formula, where the anomaly contribution to the mass of the η' meson only appears at $O(1/N)$. We add that results in the large-N limit are directly relevant in the formulation of AdS/CFT arguments.

What is the fate of the $U(1)$ axial anomaly, what is the distribution of the topological charge Q, and what is the value of χ_t in the conformal window? To give what is maybe a straightforward answer, we first clarify how many relevant, in the renormalisation group sense, order parameters are at the lower edge of the conformal window. By dimensional arguments and the effective lagrangian description[13] the $U(1)$ axial order parameter

$$\langle \det \bar{\psi}_L^f \psi_R^f \rangle \sim \Lambda^{3N_f} \tag{9}$$

becomes irrelevant for infrared physics for $N_f > 2$, while the chiral symmetry order parameter $\langle \bar{\psi}_L \psi_R \rangle$ is always relevant. This already suggests that one single phase transition should be associated with the restoration of chiral symmetry and the effective restoration of the $U(1)$ axial symmetry[c] for $N_f > 2$. This holds true for the chiral phase transition of finite temperature QCD with $N_f > 2$, as well as the conformal window with $N_f^c > 2$. We can even add that the approximate statements that apply to the near-conformal system at $T > T_c$ become exact for the massless theory at the IRFP inside the conformal window.

We now recall the relation of the topological charge Q to the eigenmodes and eigenvalues of the Dirac operator, the relation of the latter to the chiral condensate and the decomposition of the meson two-point functions in terms of the Dirac eigenvalues.

The Atiyah-Singer index theorem[14] expresses the index of an elliptic operator of any compact oriented differentiable manifold in topological terms. While Eq. (6) is in terms of the gauge fields, the index theorem provides the topological charge in terms of the index of the Dirac operator, part of the QCD fermion sector: Q can

[c]While the $U(1)$ axial anomaly in Eq. (8) is an operator identity and does not vanish in QCD for any finite N_f and N, its effect may be absent in the n-point functions of the theory, at least for $n = 2$, and one talks about the effective restoration of the symmetry.

be written as the sum of the chiralities of the eigenmodes of the Dirac operator

$$Q = \sum_s \chi_s = n_+ - n_- \,, \tag{10}$$

where $n_\pm$ indicates the number of zero modes with chirality $+1$ and -1, respectively. In fact, since the zero modes of the (massless) theory have chirality $\chi_0 = \pm 1$, while the nonzero modes have chirality $\chi_{(s\neq 0)} = 0$, one should conclude that the topological charge is nonzero (and integer) only in the presence of zero modes. To summarize, the index theorem tells that any field configuration with $Q \neq 0$ has at least one eigenfunction of vanishing eigenvalue, and thus vanishing fermion determinant.

The Banks-Casher relation[15]

$$\langle 0|\bar{\psi}\psi|0\rangle \;=\; -\lim_{m\to 0} m \int_{-\infty}^{\infty} d\lambda \, \frac{\rho(\lambda)}{\lambda^2 + m^2} \begin{cases} = -\pi\,\rho(0), & \text{if } \rho(0)\neq 0 \\ \propto m^\alpha, & \text{if } \rho(\lambda)\propto \lambda^\alpha \end{cases} \tag{11}$$

expresses the order parameter of chiral symmetry $\langle 0|\bar{\psi}\psi|0\rangle$ in terms of the Dirac eigenvalue distribution $\rho(\lambda)$. The limit $V \to \infty$ taken before the limit $m \to 0$ in the r.h.s. is mandatory in the presence of spontaneously broken chiral symmetry; it is not so if chiral symmetry is exact. Eq. (11) tells that the vanishing/appearance of the chiral condensate corresponds to the vanishing/appearance of $\rho(0)$, i.e., the Dirac eigenvalue level density at $\lambda = 0$. Note that the Banks-Casher relation carries no explicit information on the N_f dependence.

The third useful ingredient is the decomposition of the meson two-point functions in terms of Dirac eigenvalues; the analysis is based on Ref. 16, see also Ref. 17. The degeneracy pattern of the pseudoscalar isovector (π), scalar isosinglet (σ), scalar isovector (δ) and pseudoscalar isosinglet (η') two-point functions when $SU(N_f)_L \times SU(N_f)_R$ or $U_A(1)$ are restored

$$\begin{array}{ccc} C_{\vec{\pi}} & \xleftrightarrow{\;SU(N_f)\;} & C_\sigma \\ U_A(1)\updownarrow & & \updownarrow \\ C_{\vec{\delta}} & \longleftrightarrow & C_{\eta'} \end{array}$$

is a useful vademecum. When chiral symmetry is exact, for $T > T_c$ or in the conformal window, the ordering of the infinite volume and zero mass limits is irrelevant and the spectral decomposition of the meson correlators reads[16]

$$\begin{aligned} C_{\vec{\pi}} &= -(Q=0) + (Q=\pm 1) \\ C_{\vec{\delta}} &= -(Q=0) - (Q=\pm 1) \\ C_\sigma &= -(Q=0) - (Q=\pm 1) + N_f(Q=\pm 1) \\ C_{\eta'} &= -(Q=0) + (Q=\pm 1) - N_f(Q=\pm 1)\,, \end{aligned} \tag{12}$$

where the $Q = 0, \pm 1$ terms are a shorthand notation for the zero and ± 1 topological charge sector contributions, and only the terms proportional to N_f in C_σ and $C_{\eta'}$ come from disconnected contributions. The case $N_f = 2$ is special: exact chiral symmetry, i.e. the degeneracy of the chiral partners $C_{\vec{\pi}} = C_\sigma$ and $C_{\vec{\delta}} = C_{\eta'}$ is automatically verified, while the degeneracy of the $U(1)$ axial partners, i.e., $C_{\vec{\pi}} = C_{\vec{\delta}}$ and $C_\sigma = C_{\eta'}$ requires the vanishing of the contributions from the nonzero topological charge sector. For $N_f > 2$, restored chiral symmetry itself requires the absence of $(Q = \pm 1)$ contributions and it implies the degeneracy of the $U(1)$ axial partners. This is consistent with the fact that the chiral and $U(1)$ axial order parameters are both relevant for $N_f = 2$, and their restoration can thus occur separately, while only the chiral order parameter is relevant for $N_f > 2$. One crucial difference between theories below the conformal window with $T > T_c$, and theories in the conformal window is that all contributions of irrelevant operators[d] to n-point functions vanish *exactly* at the IRFP in the conformal window. In other words, the $U(1)$ axial anomaly is not manifest at the IRFP.

To understand the fate of Q and χ_t we observe that the formulae valid for the symmetry-restoration region $V\Sigma m \ll 1$ of QCD with $N_f \geq 3$ degenerate flavors[e] can to a certain extent be applied to the IRFP. In practice, we suggest that one can take the results in Ref. 18 in the limit $V\Sigma m \to 0$[f]. The chiral condensate of QCD for $V\Sigma m \ll 1$ reads[18]

$$\langle \bar{\psi}_L \psi_R \rangle = -\frac{V\Sigma^2}{4N_f} m + \ldots \quad (N_f \geq 3)\,, \tag{13}$$

in contrast with $\langle \bar{\psi}_L \psi_R \rangle = -(\Sigma/2)e^{-i\theta}$ in the opposite limit $V\Sigma m \gg 1$. The θ-vacuum angle dependence has disappeared in Eq. (13), the condensate is proportional to the fermion mass[g] and it decreases as N_f increases, a screening effect. Also, the chiral condensate is exclusively due to nonzero modes in the trivial topology sector $Q = 0$.

From Ref. 18 we also learn that the partition function at fixed winding number is $\propto (V\Sigma m)^{|Q|N_f}$ for $V\Sigma m \ll 1$, while Gaussian for $V\Sigma m \gg 1$. In both cases the topological susceptibility $\chi_t = \Sigma m/N_f$ to leading order in m, is proportional to the fermion mass and inversely proportional to N_f, consequence of the fact that fermions drastically modify the gluodynamics by introducing a leading order fermion mass dependence in the topological quantities. Also, the suppression of large Q sectors increases with N_f, i.e., the shift towards larger eigenvalues increases with N_f.

The interplay of gluodynamics and fermions is even more evident when taking the $N \to \infty$ limit with finite N_f in mass-degenerate QCD[18]. In this limit,

---

[d]It does not apply to dangerously irrelevant operators.
[e]The low-energy constant Σ is related to the infinite volume fermion condensate in the chiral limit $\langle 0|\bar{\psi}\psi|0\rangle = -\Sigma$ at $\theta = 0$.
[f]Given that the ordering of the infinite volume and zero mass limits does not matter in the restored phase, we can take $m \to 0$ first.
[g]Note that we cannot obtain the power-law dependence with exponent δ of Eq. (5) valid at a critical point.

fermion degrees of freedom strongly affect the properties of the partition function in the symmetry-restoration region $V\Sigma m \ll 1$. On the contrary, outside this region the probability for finding field configurations of large winding number and the topological susceptibility are the same as in gluodynamics. Gluodynamics itself changes drastically for $T > T_c$. Lattice results[19,20] are consistent with an exponential suppression at finite N of the Yang-Mills topological susceptibility $\chi_t \sim \exp(-\gamma(T)N)\cos\theta$; it vanishes in the large-N limit[h]. This suppression is also consistent with the exponential suppression of instantons, which carry nonzero topological charge. Refs. 21,22 have suggested that an analogous suppression occurs at the lower edge of the conformal window.

Our discussion uses instead the fact that topology is drastically affected by fermions, that chiral symmetry is exact in the conformal window, that the $U(1)$ axial order parameter is infrared irrelevant, the absence of zero modes of the Dirac operator, and that all VEVs of dimensionful quantities should vanish at the IRFP[i]. All these properties suggest that the topological charge is zero at the IRFP, as well as the topological susceptibility. This means that there are no instantons (in contrast to instantons of all sizes in QCD). Cluster decomposition is satisfied, differently from what happens when taking a fixed topological charge sector in QCD[j]. In some sense, the same result is obtained in the infinite temperature limit $T \to \infty$ of QCD, where instantons are squeezed until none contributes and $U(1)$ axial symmetry becomes exact.

A lattice formulation of the theory inside the conformal window will only approximately reproduce the properties so far discussed, and a numerical study will need to consider corrections to the physics of the IRFP induced by finite lattice spacing, finite volume and a non vanishing fermion mass.

4. The spectrum and conformal symmetry

Understanding the breaking pattern of conformal symmetry close to the Planck scale, i.e., at an energy scale where gravity is relevant and it is to be included in the complete lagrangian of the universe, is likely to shed light on the way short distances relate to large distances within a unified description of physical forces. It is also an appealing possibility that answers are in large part to be found in the quantum field theoretical framework of which many features are familiar to us.

In which manner is conformal symmetry broken? The spontaneous breaking of

---

[h] For $T < T_c$ the topological susceptibility of pure Yang-Mills remains nonzero at large-N and it guarantees the presence of the anomaly contribution to the η' mass, $m^2_{\eta'} = O(1/N)$ via the Witten-Veneziano formula $f^2_\pi m^2_{\eta'} = 2N_f\chi_t$.

[i] This is true also for $\langle G\tilde{G}\rangle$. In QCD, instantons contribute to the θ-vacuum expectation value as $\langle\theta|G\tilde{G}|\theta\rangle \propto K \sin\theta e^{-S_0}$, $S_0 = 8\pi^2/g^2$ and K contains the fermion determinant. This VEV vanishes in QCD because $K = 0$ due to the presence of vanishing eigenvalues.

[j] Zero topological charge in QCD can be made by a widely separated instanton with $+n$ and anti-instanton with $-n$ which would spoil cluster decomposition.

conformal symmetry is appealing in many respects, though we do not easily find examples of it in known physical systems. QCD in its present formulation is never conformal, it deviates from the free (conformal) theory in a highly nontrivial way, encoded in the logarithms of asymptotic freedom; in this example, as many others, conformal symmetry is explicitly broken due to the presence of conformally non invariant terms in the renormalized lagrangian. The situation can be different once we couple matter fields to gravity; in the end conformal symmetry is a symmetry of spacetime itself. It is also plausible to think that we need to first formulate the complete theory in curved spacetime and eventually recover known physics in flat spacetime, with possible corrections. To what extent these corrections modify known low-energy theories is unknown, while they should not spoil the agreement of the Standard Model as presently formulated with established experimental results.

In this context, one interesting question is: What are the signatures that allow to determine which way conformal symmetry is broken? We can choose among explicit breaking, the spontaneous breaking of a global symmetry and the spontaneous breaking of a local, i.e., gauge symmetry; we have examples of all three in nature. The spontaneous breaking of conformal symmetry can be implemented analogously to known examples, and it carries the imprint of dynamical scalar degree(s) of freedom[k]. If conformal symmetry is treated as a global symmetry, then a massless Goldstone boson, the dilaton would arise analogously to pions in QCD, the Goldstone bosons of global chiral symmetry. We could also promote the conformal symmetry group, or a subgroup of it, to a local gauge symmetry. This is a highly non trivial change, whose imprint would be a Brout-Englert-Higgs mechanism and the associated scalar boson. The idea of local gauge symmetry (local scale invariance) has been pursued recently.[23,24]

We can conceive that the scalar field of the spontaneously broken local or global conformal symmetry will couple to matter fields of the complete theory for gravity, electroweak and strong interactions as dictated by the underlying symmetries, which now include conformal symmetry. Such scalar is therefore a dynamical degree of freedom which may leave its imprint at all scales. The breaking happens only once, in the complete theory, and one needs to uncover the renormalisation-group (RG) flow of all couplings from the Planck scale down to the electroweak symmetry breaking scale, and further down to low-energy QCD. It is not yet clear to which extent one should allow for additional explicit breakings at multiple energy scales below the Planck energy. Certainly, all physical quantum anomalies responsible for explicit breakings should be preserved; an exception can be the conformal anomaly.

This work focusses on the conformal window of QCD, and non-Abelian gauge theories in general, and it is legitimate to ask under which conditions is conformal symmetry lost at the lower edge of the conformal window and if a scalar degree of freedom related to its breaking may arise. Massless QCD in isolation, i.e. not

---

[k]This scalar may carry some different properties as compared to ordinary matter scalar fields.

embedded in the complete theory, seems to suggest a straightforward solution: the breaking of conformal symmetry is in this case a consequence of the spontaneous breaking of chiral symmetry and the appearance of a mass-gap, i.e., confinement. The induced conformal symmetry breaking is thus explicit, no dilaton is needed; in other words, the vacuum expectation value of the trace of the energy-momentum tensor T^{μ}_{μ} can become nonzero due to the contributions from the condensates $\langle GG \rangle$ and $\langle \bar{\psi}\psi \rangle$. This is not different in nature from the conformal symmetry breaking realized by the quantum corrections to classical chromodynamics.

Perhaps physically more interesting is the fate of the conformal window, once QCD is embedded in the complete theory. We have said that the scalar field(s) of the spontaneously broken global or local conformal symmetry at the Planck scale leaves its imprint in the dynamics, hence it affects the beta-functions of the complete theory. The RG flow and fixed-point structure of the complete theory must be reconsidered in the presence of the new scalar degree(s) of freedom. A priori, the beta-function of the QCD gauge coupling itself $\beta(N, N_f)$ becomes a function of the scalar contributions, i.e., matter scalar fields and the dilaton. From the general construction in Ref. 24 and the well known one-loop results[25] we deduce some relevant features in flat spacetime:

- The dilaton does not couple directly to $SU(N)$ Yang-Mills fields. It couples to matter scalar fields (the Higgs field and others) through the scalar potential and to fermions through a purely imaginary Yukawa term $(im_f/\Lambda_{Pl})\bar{\psi}\eta(x)\psi$. This term disappears if fermions are massless, $m_f = 0$. If matter scalar fields are also absent, low-energy strong interactions do not talk to the dilaton.
- In the presence of scalar fields the gauge coupling beta-function $\beta(g)$ is modified; in particular, to find the fixed points one needs to consider the system of differential equations $\beta(g, \lambda_i, y_i) = 0$ that includes all scalar fields self-couplings λ_i and the Yukawa couplings y_i.
- Matter scalar fields usually screen gauge forces as fermions do. Thus, the coefficient of the one-loop beta-function $\beta(g) = -b_0 g^3$ becomes $b_0 = 11/3N - 2/3N_f - 1/6N_s$[25–27], where N_s does *not* include the dilaton which does not couple directly to YM fields. One concludes that the loss of asymptotic freedom, $b_0 = 0$, is anticipated by matter scalars to a lower N_f and the upper edge of the conformal window is shifted, unless asymptotic freedom has been fully destroyed by the scalar self-couplings λ_i.
- At two-loops in QCD $\beta(g) = -b_0 g^3 - b_1 g^5$ and an IRFP arises at $g_*^2 = -b_0/b_1$, as long as $b_0 > 0$ and $b_1 < 0$. A sign change of b_1 implies the disappearance of the IRFP and would signal the lower edge of the conformal window N_f^c. Corrections to b_1 in the presence of matter scalar fields and the dilaton are more involved and depend on the details of the theory, its group and representation structure. We can still say that matter scalar fields will act analogously to fermions in screening gauge forces; all in all

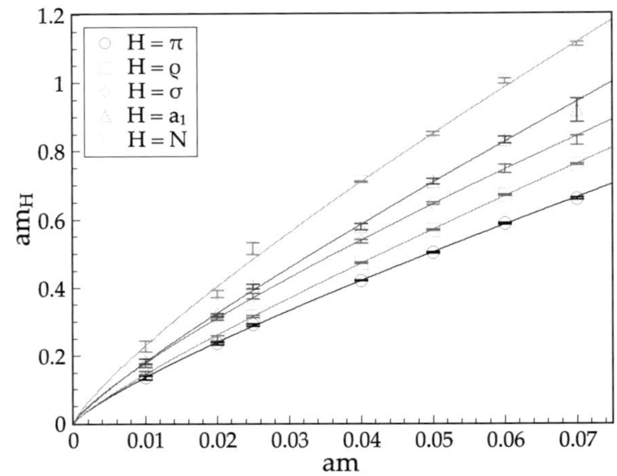

Fig. 2. Spectrum of the $N_f = 12$ theory from Ref. 28.

the conformal window would be shifted to lower N_f and its width changed, or disappear due to scalar self-couplings.

- The dilaton is special. Interestingly, its Yukawa coupling y to fermions can contribute to $\beta(g)$ at $O(y^2g)$[24] and it being imaginary, it carries a sign opposite to the other Yukawa contributions. On the other hand, if y is proportional to the fermion mass its contribution vanishes if fermions are massless.

To summarize, if fermion mass terms are turned on via, e.g., a Brout-Englert-Higgs mechanism, then the Yukawa coupling of the dilaton to fermions is turned on. It becomes then phenomenologically interesting to study small mass perturbations to a fixed point of the massless theory.

Turning back to massless QCD in isolation, we comment on its spectrum when adding a small fermion mass perturbation and varying N_f across N_f^c. QCD with $N_f = 12$ massless flavors can be taken as a prototype of a theory inside the conformal window. Its particle spectrum in the presence of a fermion mass perturbation has been theoretically discussed in Ref. 28 with analogies to quantum critical phenomena. A perturbative mass deformation at the IRFP allows to parametrize violations of universal scaling and to determine on the lattice the anomalous dimension of the fermion mass operator at the IRFP. We find $\gamma_m \sim 0.25$[28]; note that a value $\gamma_m \ll 1$ already suggests that $N_f = 12$ is not close to the lower edge of the conformal window, in agreement with our recent results[2] that place the endpoint between $N_f = 8$ and $N_f = 6$. The ordering of states, pseudoscalar (PS), vector (V), scalar (S), axial (PV) mesons and the nucleon (N) for twelve flavors does not lead to surprises, see Fig. 2. All masses approach zero in the chiral limit with a universal power-law, corrected by mass-induced small violations of universal scaling[28]. This behavior

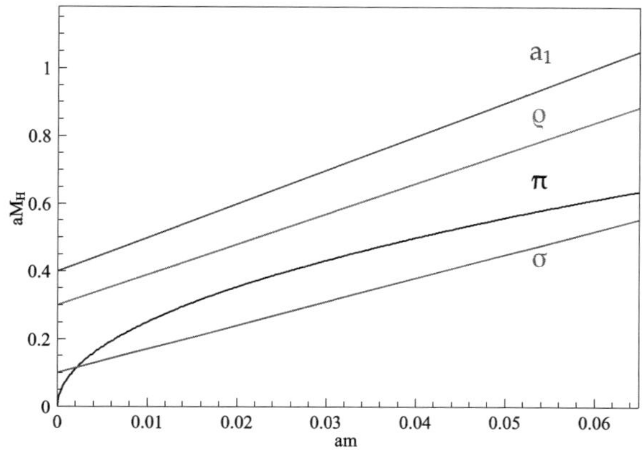

Fig. 3. Illustration of $\pi - \sigma$ inversion near the conformal window, away from the chiral limit.

signals exact chiral symmetry and the presence of a fixed point where conformal scaling holds. One obvious constraint below the conformal window, $N_f < N_f^c$, is that the spectrum of the zero-temperature theory at a given N_f cannot qualitatively differ from that of zero-temperature QCD, because they share the same underlying symmetries. Massless pseudoscalar mesons, the pions, are the Goldstone bosons of the spontaneously broken chiral symmetry, while the vector, axial and scalar mesons have a nonzero mass in the chiral limit. Away from the chiral limit, an ordering of states different from real world QCD can occur, provided all constraints implied by the spontaneously broken chiral symmetry and the $U(1)$ axial anomaly are satisfied. Some of these constraints, of immediate use in lattice studies, are i) the non degeneracy of chiral partners, which translates in the (pseudo)scalar sector into a constraint on the chiral cumulant $R_\pi \underset{m \to 0}{\to} 0$, ii) the Gell-Mann-Oakes-Renner (GMOR) relation[29], iii) the non degeneracy of $U(1)$ axial partners. Generally, even an inversion of the pseudoscalar (π) and scalar (σ) states should be possible away from the chiral limit as shown in Fig. 3, without violating any of the constraints. No dilaton will be present for QCD in isolation, while a scalar state numerically, but not parametrically, lighter can occur just below the conformal window. We add that recent AdS/CFT constructions[30] also fail to produce a parametrically lighter scalar. If a conformal phase transition occurs at the lower edge of the conformal window, the corresponding preconformal scaling would leave its imprint also in the spectrum, still ensuring that its properties smoothly merge with those of the QCD spectrum — because no change of underlying symmetries separates the preconformal region and QCD.

5. Conclusions

Some observables change drastically across the lower edge of the conformal window of non-Abelian gauge theories. We have identified them and discussed their properties inside and below the conformal window; nonperturbative (lattice) studies of many of these properties would be desirable. Throughout, we have used the idea that the drastic change from QCD to theories inside the conformal window results from the interplay of chiral symmetry and confinement, and we discussed the somewhat different role of the $U(1)$ axial anomaly and conformal symmetry itself.

Within the gauge sector, the scalar glueball operator exhibits the nice feature that its anomalous dimension at the IRFP is constrained by the nonrenormalization of the energy-momentum tensor. The way this anomalous dimension varies along the IRFP line is a direct probe of the mechanism for the loss of conformal symmetry. Interestingly, perturbation theory and large-N arguments predict a variation with N_f opposite to the one associated with a UV-IR merging mechanism. In addition to the study of the two-point function of the scalar glueball operator in order to determine its anomalous dimension at the IRFP and its variation with N_f, we commented on the possibility that a Wilson-flow equation with an appropriately devised evolution kernel could discriminate between a conformal and a confining behavior in the infrared.

Topology is a powerful probe of the conformal window. We have suggested a close analogy between high-temperature massless QCD with $N_f > 2$, $T > T_c$ and a theory at the IRFP in the conformal window. One crucial difference between the two is the actual existence in the latter of an IRFP, which makes approximate statements exact. It is also observed that the commutativity of the infinite volume and zero mass limits in the chirally symmetric phase allows to a certain extent to establish an analogy with the symmetry-restoration region of QCD, where one works in the limit $V\Sigma m \ll 1$. The restoration of chiral symmetry, the infrared irrelevance of the $U(1)$ axial order parameter for $N_f > 2$, the exact disappearance of irrelevant operators at the IRFP, the absence of zero eigenmodes of the Dirac operator and the vanishing of all vacuum expectation values suggest consistently that the topological charge and the topological susceptibility vanish at the IRFP; in other words, the effect of the anomaly is irrelevant for infrared physics which is all contained in the trivial topology sector. Fermions provide the dominant effect inside the conformal window, canceling all effects from gluodynamics. On a lattice, mass corrections jointly to lattice spacing effects and finite spacetime volume force the system to be never exactly at the IRFP, so that topological quantities will not be exactly zero, though drastically suppressed as compared to QCD and theories just below the conformal window.

If conformal symmetry is spontaneously broken at the energy scale where gravity matters, then its associated scalar degree of freedom plays a role in the complete theory at all scales. For QCD in isolation it is plausible that conformal symmetry is explicitly broken as a consequence of the spontaneous breaking of chiral symmetry and confinement, and no dilaton arises. The spectrum just below the conformal window allows for inversion in the ordering of states away from the chiral limit, but it does not allow for a parametrically lighter scalar meson due to the constraints imposed by the spontaneously broken chiral symmetry, i.e., the vanishing of the chiral cumulant in the massless limit due to the pseudoscalar Goldstone bosons of chiral symmetry. It remains interesting and challenging to consider QCD as well as the complete Standard Model for strong and electroweak interactions embedded in a hypothesized complete theory that couples them to gravity, and study which way a spontaneously broken conformal symmetry modifies low-energy interactions.

Acknowledgments

I thank the organizers of SCGT15 for the interesting workshop and warm hospitality and my collaborators A. Deuzeman, M.P. Lombardo, K. Miura, T. Nunes da Silva and L. Robroek for all the inspiring discussions and work together.

References

1. N. Seiberg, Electric - magnetic duality in supersymmetric nonAbelian gauge theories, *Nucl. Phys.* **B435**, 129 (1995).
2. T. N. da Silva, E. Pallante and L. Robroek, Conformal or Confining, arXiv:1506.06396 [hep-th].
3. M. Bochicchio, Yang-Mills mass gap, Floer homology, glueball spectrum, and conformal window in large-N QCD, arXiv:1312.1350 [hep-th].
4. D. B. Kaplan, J.-W. Lee, D. T. Son and M. A. Stephanov, Conformality Lost, *Phys. Rev.* **D80**, 125005 (2009).
5. V. A. Miransky, Dynamics of Spontaneous Chiral Symmetry Breaking and Continuum Limit in Quantum Electrodynamics, *Nuovo Cim.* **A90**, 149 (1985).
6. V. A. Miransky and K. Yamawaki, Conformal phase transition in gauge theories, *Phys. Rev.* **D55**, 5051 (1997).
7. J. M. Kosterlitz, The Critical properties of the two-dimensional x y model, *J. Phys.* **C7**, 1046 (1974).
8. M. Luescher, Properties and uses of the Wilson flow in lattice QCD, *JHEP* **1008**, 071 (2010).
9. R. Narayanan and H. Neuberger, Infinite N phase transitions in continuum Wilson loop operators, *JHEP* **0603**, 064 (2006).
10. M. Luescher and S. Schaefer, Lattice QCD without topology barriers, *JHEP* **1107**, 036 (2011).
11. S. Borsanyi, S. Durr, Z. Fodor, C. Hoelbling, S. D. Katz *et al.*, High-precision scale setting in lattice QCD, *JHEP* **1209**, 010 (2012).
12. A. Deuzeman, M. Lombardo and E. Pallante, Evidence for a conformal phase in SU(N) gauge theories, *Phys. Rev. D* **82**, 074503 (2010).
13. R. D. Pisarski and F. Wilczek, Remarks on the Chiral Phase Transition in Chromodynamics, *Phys. Rev.* **D29**, 338 (1984).
14. M. F. Atiyah and I. M. Singer, The index of elliptic operators on compact manifolds, *Bull. Am. Math. Soc.* **69**, 422 (1969).

15. T. Banks and A. Casher, Chiral Symmetry Breaking in Confining Theories, *Nucl. Phys.* **B169**, 103 (1980).
16. J. B. Kogut, J. F. Lagae and D. K. Sinclair, Topology, fermionic zero modes and flavor singlet correlators in finite temperature QCD, *Phys. Rev.* **D58**, 054504 (1998).
17. A. Deuzeman, M. P. Lombardo, T. Nunes Da Silva and E. Pallante, The bulk transition of QCD with twelve flavors and the role of improvement, *Phys. Lett.* **B720**, 358 (2013).
18. H. Leutwyler and A. V. Smilga, Spectrum of Dirac operator and role of winding number in QCD, *Phys. Rev.* **D46**, 5607 (1992).
19. L. Del Debbio, H. Panagopoulos and E. Vicari, Topological susceptibility of SU(N) gauge theories at finite temperature, *JHEP* **09**, 028 (2004).
20. C. Bonati, M. DElia, H. Panagopoulos and E. Vicari, Change of theta Dependence in 4D SU(N) Gauge Theories Across the Deconfinement Transition, *Phys. Rev. Lett.* **110**, 252003 (2013).
21. A. Parnachev and A. R. Zhitnitsky, Phase Transitions, theta Behavior and Instantons in QCD and its Holographic Model, *Phys. Rev.* **D78**, 125002 (2008).
22. A. R. Zhitnitsky, Conformal window in QCD for large numbers of colors and flavors, *Nucl. Phys.* **A921**, 1 (2014).
23. G. T. Hooft, Local Conformal Symmetry: the Missing Symmetry Component for Space and Time, arXiv:1410.6675 [gr-qc].
24. G. 't Hooft, A class of elementary particle models without any adjustable real parameters, *Found. Phys.* **41**, 1829 (2011).
25. D. J. Gross and F. Wilczek, Asymptotically Free Gauge Theories. 1, *Phys. Rev.* **D8**, 3633 (1973).
26. D. J. Gross and F. Wilczek, Ultraviolet Behavior of Nonabelian Gauge Theories, *Phys. Rev. Lett.* **30**, 1343 (1973).
27. H. D. Politzer, Reliable perturbative results for strong interactions?, *Phys. Rev. Lett.* **30**, 1346 (1973).
28. M. Lombardo, K. Miura, T. J. N. da Silva and E. Pallante, On the particle spectrum and the conformal window, *JHEP* **1412**, 183 (2014).
29. M. Gell-Mann, R. J. Oakes and B. Renner, Behavior of current divergences under SU(3) x SU(3), *Phys. Rev.* **175**, 2195 (1968).
30. D. Kutasov, J. Lin and A. Parnachev, Conformal Phase Transitions at Weak and Strong Coupling, *Nucl. Phys.* **B858**, 155 (2012).

Strong Coupling Limit of Lattice QCD with Many Staggered Quarks

Seyong Kim

Department of Physics, Sejong University, Seoul 05000, Republic of Korea

Using Monte Carlo simulation, we study the strong coupling limit of lattice QCD with many staggered quark flavors. After identifying an existence of chiral symmetric and color deconfined phase, we investigate the properties of this phase via meson spectrum, gluonic observable and Dirac eigenvalue spectrum for various lattice volume sizes and then speculate the phase diagram of QCD with many quark flavors.

Old Wine in a New Bottle: Technidilaton as the 125 GeV Higgs

Koichi Yamawaki

Kobayashi-Maskawa Institute for the Origin of Particles and the Universe (KMI), Nagoya University, Nagoya, 464-8602, Japan

Dedicated to the Late Professor Yoichiro Nambu*

– One theme that emerged from these observations is that some concepts have a long life. They may not always be right, but they can undergo reincarnations and become relevant later ("old wine in a new bottle", a remark we have heard).—Y. Nambu, Concluding Remarks at SCGT 88.—

The first Nagoya SCGT workshop back in 1988 (SCGT 88) was motivated by the walking technicolor and technidilaton. Now at SCGT15 I returned to the "old wine" in "a new bottle", the recently discovered 125 Higgs boson as the technidilaton. We show that the Standard Model (SM) Higgs Lagrangian is identical to the nonlinear realization of both the scale and chiral symmetries ("scale-invariant nonlinear sigma model"), and is further gauge equivalent to the "scale-invariant Hidden Local Symmetry (HLS) model" having possible new vector bosons as the HLS gauge bosons with scale-invariant mass: SM Higgs is nothing but a (pseudo) dilaton. The effective theory of the walking technicolor has precisely the same type of the scale-invariant nonlinear sigma model, thus further having the scale-invariant HLS gauge bosons (technirho's, etc.). The technidilaton mass M_ϕ comes from the trace anomaly, which yields $M_\phi^2 F_\phi^2 \simeq [\frac{8}{N_F}\frac{4}{N_C}] \cdot (2.5)^2 \cdot v^4$ via PCDC, in the underlying walking $SU(N_C)$ gauge theory with N_F massless flavors, where F_ϕ is the the decay constant and $v = 246$ GeV. This implies $F_\phi \simeq 5\, v$ for $M_\phi \simeq 125\,\text{GeV} \simeq \frac{1}{2} v$ in the one-family walking technicolor model ($N_C = 4, N_F = 8$), in good agreement with the current LHC Higgs data. In the anti-Veneziano limit, $N_C \to \infty$, with $N_C\alpha =$ fixed and $N_F/N_C =$ fixed ($\gg 1$), we have a result: $M_\phi^2/v^2 \sim M_\phi^2/F_\phi^2 \sim 1/(N_F N_C) \to 0$. Then the technidilaton is a naturally light composite Higgs out of the strongly coupled conformal dynamics, with its couplings even weaker than the SM Higgs. Related holographic and lattice results are also discussed. In particular, such a light flavor-singlet scalar does exists in the lattice simulations in the walking regime.

1. Introduction

This talk is an impromptu substitute filling the hole of the cancellation of the talk to be given by Volodya Miransky, my old friend, so this is a kind of personal sentiments style rather than a scientific presentation. Sorry for that in advance.

History repeats itself: At the first SCGT workshop in 1988 motivated by our work[1,2] proposing the walking technicolor based on his paper[3], the Volodya's talk was also cancelled, although he was a frequent repeater to many (out of ten) SCGT workshops during the long period more than a quarter century 1988 - 2015.

*Deceased on July 5, 2015

Volodya discovered[3] so-called Miransky scaling:

$$m_F \simeq 4\Lambda \cdot \exp\left(-\frac{\pi}{\sqrt{\frac{\alpha}{\alpha_{\rm cr}}-1}}\right) \ll \Lambda \quad \left(\frac{\alpha}{\alpha_{\rm cr}}-1 \ll 1\right) \tag{1}$$

in the scale-invariant gauge model (ladder model), an essential singularity scaling analogous to the Berezinsky-Kosterlitz-Thouless phase transition (what we called "conformal phase transition"[4]), where $m_F (\neq 0)$ is the spontaneous chiral symmetry breaking (SχSB) solution of the ladder Schwinger-Dyson (SD) gap equation for the fermion and Λ is the ultraviolet cutoff to regularize the theory to act as an intrinsic scale $\Lambda_{\rm TC}$ of the walking technicolor, similarly to the $\Lambda_{\rm QCD}$ in QCD, responsible for the trace anomaly $\theta^\mu_\mu \propto \Lambda^4$ as the explicit breaking of the scale symmetry manifesting itself as the running of the coupling in the ultraviolet region far bigger than the SχSB scale m_F.

The SχSB solution exists only in the *strong coupling* phase $\alpha > \alpha_{\rm cr} \neq 0$, where the *non-zero critical coupling* $\alpha_{\rm cr} \neq 0$ discovered by Maskawa-Nakajima[5] is the characterization of the *strong coupling gauge theories (SCGT)*. It is a gauge analogue of the *non-zero critical coupling,* $G^{\rm (NJL)} > G_{\rm cr} \neq 0$, *of the Nambu-Jona-Lasinio (NJL) model*[6], which, despite the resemblance, should actually be *distinguished from the BCS dynamics having zero critical coupling* $G^{\rm (BCS)}_{\rm cr} = 0$ (weak coupling theory in broken phase even for infinitesimal coupling).[a]

Volodya further invented nonperturbative renormalization[3] taking $\Lambda \to \infty$ as $m_F =$ fixed for the Miransky scaling Eq.(1), which yields a nonperturbative beta function:[8]

$$\beta^{(NP)}(\alpha) = \Lambda \frac{\partial \alpha(\Lambda)}{\partial \Lambda}\bigg|_{m_F} = -\frac{2\pi^2 \alpha_{\rm cr}}{\ln^3(\frac{4\Lambda}{m_F})} = -\frac{2\alpha_{\rm cr}}{\pi}\left(\frac{\alpha}{\alpha_{\rm cr}}-1\right)^{\frac{3}{2}}, \tag{2}$$

$$\alpha(\mu) = \alpha_{\rm cr}\left[1 + \frac{\pi^2}{\ln^2(\frac{\mu}{m_F})}\right], \quad (\alpha > \alpha_{\rm cr} \neq 0) \tag{3}$$

Thus the coupling actually starts running nonperturbatively once the $m_F \neq 0$ is generated, even if the initial ladder coupling is scale-invariant (nonrunning). Besides the explicit breaking by the intrinsic scale Λ, the scale symmetry is now *explicitly broken* also by m_F which was generated by the *spontaneous breaking* of the scale symmetry, with the *new trace anomaly (nonperturbative trace anomaly)* $|\theta^{\mu\,(NP)}_\mu| = \mathcal{O}(m_F^4) (\ll \Lambda^4)$. From Eq.(2) we can see that the $\alpha_{\rm cr} \neq 0$ is regarded as a *nontrivial ultraviolet fixed point,* although the input ladder coupling $\alpha(\mu^2) = \alpha$ in the infrared region $\mu^2 < \Lambda^2 = \Lambda^2_{\rm TC}$ is regarded as the *nontrivial infrared fixed point.* This opened

[a]The characteristic weak coupling of the BCS theory is due to the existence of the Fermi surface where the fermions are effectively in the 2-dimensional brane instead of 4 dimensional free space bulk, whereas the Cooper pairs and Nambu-Goldstone (NG) modes are bosons and hence live in the bulk to trigger the Higgs mechanism (Meissner effect), escaping the Mermin-Wagner-Coleman theorem in genuine 2 dimensional theories.

a new phase of the SCGT for the composite model and turned out to be the basics of the whole walking dynamics.

We applied this Volodya's result to the technicolor theory, and proposed what we called "Scale-invariant Technicolor" (now called Walking Technicolor)[1,2], where we found a large anomalous dimension

$$\gamma_m = 1 \qquad (\alpha > \alpha_{\rm cr}) \tag{4}$$

in the broken phase[b], as a solution of the Flavor-Changing-Neutral Currents (FCNC) problem of the original Technicolor[7],[c] and predicted a technidilaton, a pseudo Nambu-Goldstone boson of the approximate scale symmetry, which is spontaneously broken at the same time as the SχSB due to the fermionic condensate responsible for the electroweak symmetry breaking. *The technidilaton was predicted as a flavor-singlet technifermion bound state* (not a glueball-type bound state), behaving similarly to the standard model (SM) Higgs itself.

Below I will explain[11] how the technidilaton is a *naturally light and weakly coupled composite Higgs* out of *strongly coupled* underlying conformal gauge theory, the walking technicolor, in the light of the anti-Veneziano limit $N_C, N_F \to \infty$ with N_F/N_C =fixed $\gg 1$ for $SU(N_C)$ gauge theory with N_F massless flavors. The technidilaton particularly for the one-family walking technicolor with $N_F = 8$ and $N_C = 4$ is nicely fit to the current 125 GeV Higgs data at LHC.[11–13]

The present SCGT15 workshop is entitled "Sakata Memorial $\cdots$". Composite model is indeed Nagoya University tradition trace back to the late Professor Shoichi Sakata, who invented the Sakata model[14], a composite hadron model, in 1955 (published in 1956), which turned out to be the forerunner of the quark model in 1964[15]. In the context of the extended Sakata model with the four constituents corresponding to the four lepton flavors, the famous Maki-Nakagawa-Sakata (MNS) neutrino mixing was proposed in 1962[16]. This motivated M. Kobayashi and T. Maskawa, both disciples of Sakata, to believe in four quarks, with quark flavor mixing counter to the MNS lepton flavor mixing, which was the initial motivation of the paper of Kobayashi-Maskawa[17], stepping eventually further to the six quarks in 1972 (published in 1973), well before the J/ψ discovery in 1974.

The present SCGT15 workshop is also entitled "Origin of Mass $\cdots$". The origin of mass of all the SM particles is the Higgs VEV $v = \sqrt{\frac{-\mu_0^2}{\lambda}} = 246\,\text{GeV}$ or the Higgs mass $M_\phi^2 = 2\lambda v^2 = -2\mu_0^2$ read from the Lagrangian:

$$\mathcal{L}_{\rm Higgs} = |\partial_\mu h|^2 - \mu_0^2 |h|^2 - \lambda |h|^4 \,. \tag{5}$$

[b]The anomalous dimension in the unbroken phase ($\alpha < \alpha_{\rm cr}$) was known[8] to be $\gamma_m = 1 - \sqrt{1 - \alpha/\alpha_{\rm cr}}$, which is irrelevant to the dynamical mass m_F ($\neq 0$ only for $\alpha > \alpha_{\rm cr}$) and hence to the nonperturbative running of the coupling in Eq.(2).

[c]The FCNC solution by the large anomalous dimension was proposed without concrete dynamical model nor concrete value of the anomalous dimension at a hypothetical nontrivial fixed point in the asymptotically non-free theory.[9] Subsequently to our paper, a similar FCNC solution based on the ladder-type SD equation was discussed[10], without notion of the anomalous dimension nor scale symmetry, and hence without technidilaton.

Then the origin of mass is attributed to the mysterious input mass parameter of the *tachyon* with the mass μ_0 such that $\mu_0^2 < 0$ *as a free parameter.* But why tachyon? How is the tachyon mass determined? SM cannot answer to these questions.

Here we should recall that *the spontaneous symmetry breaking was born as a dynamical symmetry breaking, thanks to Professor Y. Nambu*[6], where the tachyon is in fact generated as a dynamical consequence of the strong dynamics, but not an ad hoc input. Before we discuss the dynamical origin of mass a la Nambu, we first show[18] that *the SM Higgs Lagrangian itself possesses "hidden" symmetries (scale symmetry and gauge symmetry, both spontaneously broken, i.e., nonlinearly realized)*, in addition to the well-known symmetry (chiral symmetry, also spontaneously broken) to be gauged by the electroweak symmetry.

2. SM Higgs as a Dilaton: Hidden Scale Symmetry and Hidden Local Symmetry in the SM Higgs Lagrangian[18]

Here we show[18] that the SM Higgs Lagrangian Eq.(5) in the form of the *linear sigma model* is rewritten into precisely the form equivalent to the *scale-invariant* version of the chiral $SU(2)_L \times SU(2)_R$ *nonlinear sigma model*, as far as it is in the broken phase, with *both the chiral and scale symmetries spontaneously broken* due to the same Higgs VEV $v \neq 0$, and thus are *both nonlinearly realized.* The SM Higgs Lagrangian is further shown to be *gauge equivalent* to the *scale-invariant* version[19] of the Hidden Local Symmetry (HLS) Lagrangian[20–22], which contains *possible new vector bosons*, analogue of the ρ mesons, as the gauge bosons of the (spontaneously broken) HLS *hidden behind the SM Higgs Lagrangian.*

Let us rewrite Eq.(5) as

$$\begin{aligned}\mathcal{L}_{\text{Higgs}} &= \frac{1}{2}\left[(\partial_\mu \hat{\sigma})^2 + (\partial_\mu \hat{\pi}_a)^2\right] - \frac{1}{2}\mu_0^2\left[\hat{\sigma}^2 + \hat{\pi}_a^2\right] - \frac{\lambda}{4}\left[\hat{\sigma}^2 + \hat{\pi}_a^2\right]^2 \\ &= \frac{1}{2}\text{tr}\left(\partial_\mu M \partial^\mu M^\dagger\right) - \left[\frac{\mu_0^2}{2}\text{tr}\left(MM^\dagger\right) + \frac{\lambda}{4}\left(\text{tr}\left(MM^\dagger\right)\right)^2\right] . \end{aligned} \tag{6}$$

where we have noted

$$h = \begin{pmatrix} \phi^+ \\ \phi^0 \end{pmatrix} = \frac{1}{\sqrt{2}}\begin{pmatrix} i\hat{\pi}_1 + \hat{\pi}_2 \\ \hat{\sigma} - i\hat{\pi}_3 \end{pmatrix} , \tag{7}$$

and 2×2 matrix M reads

$$M = (i\tau_2 h^*, h) = \frac{1}{\sqrt{2}}\left(\hat{\sigma} \cdot 1_{2\times 2} + 2i\hat{\pi}\right) \quad \left(\hat{\pi} \equiv \hat{\pi}_a \frac{\tau_a}{2}\right) , \tag{8}$$

which transforms under $G = SU(2)_L \times SU(2)_R$ as:

$$M \to g_L\, M\, g_R^\dagger , \quad (g_{R,L} \in SU(2)_{R,L}) . \tag{9}$$

Note first that any complex matrix M can be decomposed into the Hermitian (always diagnonalizable) matrix H and unitary matrix U as $M = HU$ ("polar decomposition"):

$$M = H \cdot U\,, \quad H = \frac{1}{\sqrt{2}} \begin{pmatrix} \sigma & 0 \\ 0 & \sigma \end{pmatrix}\,, \quad U = \exp\left(\frac{2i\pi}{F_\pi}\right) \quad (F_\pi = v = \langle\sigma\rangle)\,. \tag{10}$$

The chiral transformation of M is inherited by U, while H is a chiral singlet such that:

$$U \to g_L\, U\, g_R^\dagger\,, \quad H \to H\,, \tag{11}$$

and $U\,U^\dagger = 1$ implies $\langle U\rangle = \langle \exp\left(\frac{2i\pi}{F_\pi}\right)\rangle = 1 \neq 0$, namely the spontaneous breaking of the chiral symmetry is taken granted in the polar decomposition. Note that the radial mode σ is a chiral-singlet in contrast to $\hat{\sigma}$ which is chiral non-singlet.

We further parametrize σ as

$$\sigma = v \cdot \chi\,, \quad \chi = \exp\left(\frac{\phi}{F_\phi}\right)\,, \tag{12}$$

where $F_\phi = v$ is the decay constant of the dilaton ϕ as the Higgs. The scale (dilatation) transformations for these fields are

$$\delta_D \sigma = (1 + x^\mu \partial_\mu)\sigma\,, \qquad \delta_D \chi = (1 + x^\mu \partial_\mu)\chi\,, \qquad \delta_D \phi = F_\phi + x^\mu \partial_\mu \phi\,. \tag{13}$$

Note that $\langle\sigma\rangle = v\langle\chi\rangle = v \neq 0$ breaks spontaneously the scale symmetry, but not the chiral symmetry, since σ (χ as well) is a chiral singlet. This is a nonlinear realization of the scale symmetry: the ϕ is a dilaton, NG boson of the spontaneously broken scale symmetry. Although χ is a dimensionless field, it transforms as that of dimension 1, while ϕ having dimension 1 transforms as the dimension 0, instead.

Then the SM Higgs Lagrangian reads:[18]

$$\begin{aligned} \mathcal{L}_{\text{Higgs}} &= \left[\frac{F_\phi^2}{2}\left(\partial_\mu \chi\right)^2 + \frac{F_\pi^2}{4}\chi^2 \cdot \mathrm{tr}\left(\partial_\mu U \partial^\mu U^\dagger\right)\right] - V(\phi) \\ &= \chi^2 \cdot \left[\frac{1}{2}\left(\partial_\mu \phi\right)^2 + \frac{F_\pi^2}{4}\mathrm{tr}\left(\partial_\mu U \partial^\mu U^\dagger\right)\right] - V(\phi)\,, \\ V(\phi) &= \frac{\lambda}{4}v^4\left[\left(\chi^2 - 1\right)^2 - 1\right] = \frac{M_\phi^2 F_\phi^2}{8}\left[\left(\chi^2 - 1\right)^2 - 1\right]\,, \end{aligned} \tag{14}$$

which is nothing but *the scale-invariant nonlinear sigma model*[11,12], with $F_\phi = F_\pi = v$, plus the explicit scale-symmetry breaking potential $V(\phi)$ such that $\delta_D V(\phi) = \lambda v^4 \chi^2 = -\theta^\mu_\mu$ whose scale dimension $d_\theta = 2$ (originally the tachyon mass term) instead of 4: namely, the scale symmetry is broken only by the dimension 2 operator.[d] This yields the mass of the (pseudo-)dilaton as the Higgs

[d]Note that *mass of all the SM particles except the Higgs is scale-invariant.* By the electro-weak gauging as usual; $\partial_\mu U \Rightarrow \mathcal{D}_\mu U = \partial_\mu U - ig_2 W_\mu U + ig_Y U B_\mu$ in Eq.(14), we see that the mass of W/Z is scale-invariant thanks to the dilaton factor χ, and so is the mass of the SM fermions f: $g_Y \bar{f} h f = (g_Y v/\sqrt{2})(\chi \bar{f} f)$, all with the scale dimension 4.

$M_\phi^2 = 2\lambda v^2$, which is in accord with the Partially Conserved Dilatation Current (PCDC):

$$M_\phi^2 F_\phi^2 = -\langle 0|\partial^\mu D_\mu|\phi\rangle F_\phi = -d_\theta\langle\theta_\mu^\mu\rangle = 2\lambda v^4\langle\chi^2\rangle = 2\lambda v^4\,, \tag{15}$$

with $F_\phi = v$, where D_μ is the dilatation current: $\langle 0|D_\mu(x)|\phi\rangle = -iq_\mu F_\phi e^{-iqx}$.

Hence the SM Higgs as it stands is a (pseudo) dilaton, with the *mass arising from the dimension 2 operator* in the potential,

$$M_\phi^2 = 2\lambda v^2 \to 0 \qquad \left(\lambda \to 0\,,\ v = \sqrt{\frac{-\mu_0^2}{\lambda}} = \text{fixed} \neq 0\right) \tag{16}$$

("conformal limit"[18]).[e] In fact the Higgs mass 125 GeV would imply $\lambda = M_\phi^2/(2v^2) \simeq 1/8 \ll 1$. It should be noted that $\lambda \ll 1$ *(with v = fixed $\neq 0$) can be realized even when the underlying theory is strong coupling*, particularly when the *scale symmetry is operative*, as we discuss below.

On the other hand, if we take the limit $\lambda \to \infty$, then the SM Higgs Lagrangian goes over to the usual nonlinear sigma model *without scale symmetry*:

$$\mathcal{L}_{\text{NL}\sigma} = \frac{F_\pi^2}{4}\,\text{tr}\left(\partial^\mu U \partial_\mu U^\dagger\right)\,, \tag{17}$$

where the potential is decoupled and $\chi(x) \equiv 1$ is frozen, so that the scale symmetry breaking is transferred from the potential to the kinetic term, which is no longer transform as the dimension 4 operator. This is known to be a good effective theory (chiral perturbation theory) of the ordinary QCD which in fact lacks the scale symmetry at all, perfectly consistent with the nonlinear sigma model, Eq.(17). However, it cannot be true for the walking technicolor which does have the scale symmetry, and the effective theory must respect the symmetry of the underlying theory, in a form of the scale-invariant nonlinear sigma model Eq.(14) in the conformal limit $\lambda \to 0$.

Once rewritten in the form of Eq.(14), it is easy to see[18] that the *SM Higgs Lagrangian is gauge equivalent to the "scale-invariant HLS model" (s-HLS)*[19], a scale-invariant version of the HLS model[20–22] [f], which *contains massive spin-1 states*, spontaneously broken HLS gauge bosons, as possible yet other composite states in some underlying theory hidden behind the SM Higgs:

$$\begin{aligned} \mathcal{L}_{\text{s-HLS}} &= \mathcal{L}_{\text{Higgs}} + \mathcal{L}_{\text{Kinetic}}(V_\mu)\,, \\ \mathcal{L}_{\text{Higgs}} &= \chi^2\cdot\left(\frac{1}{2}(\partial_\mu\phi)^2 + \mathcal{L}_A + a\mathcal{L}_V\right) - V(\phi)\,, \end{aligned} \tag{18}$$

with a being an arbitrary parameter, and $\mathcal{L}_{\text{Kinetic}}(V_\mu)$ is the kinetic term of the HLS gauge boson V_μ which is obviously scale-invariant. The mass term of the HLS

[e] This limit should be distinguished from the popular limit $\mu_0^2 \to 0$ with λ =fixed $\neq 0$, where the Coleman-Weinberg potential as the explicit scale symmetry breaking is generated by the trace anomaly (dimension 4 operator) due to the quantum loop.

[f] The s-HLS model was also discussed in a different context, ordinary QCD in medium.[23]

gauge bosons is given by $\chi^2 \cdot a\mathcal{L}_V = \chi^2 \cdot a(F_\pi^2/4)\mathrm{tr}(g_{_{HLS}} V_\mu + \cdots)^2$, which yields the mass $M_V^2 = a g_{_{HLS}}^2 F_\pi^2$, $(F_\pi = v)$, where $g_{_{HLS}}$ is the gauge coupling of the HLS. For the low energy $p^2 < M_V^2$ where the kinetic term can be ignored, the HLS gauge boson V_μ becomes just an auxiliary field to be solved away to yield $\mathcal{L}_V = 0$, and we are left with $\chi^2\mathcal{L}_A = \chi^2 \cdot (F_\pi^2/4)\mathrm{tr}(\partial_\mu U \partial^\mu U^\dagger)$. Hence $\mathcal{L}_{\mathrm{s-HLS}}$ is reduced back to the original SM Higgs Lagrangian $\mathcal{L}_{\mathrm{Higgs}}$ in nonlinear realization, Eq.(14). Note that *the HLS gauge boson acquires the scale-invariant mass thanks to the dilaton factor* χ^2, the nonlinear realization of the scale symmetry, in sharp contrast to the *Higgs (dilaton) which acquires mass only from the explicit breaking of the scale symmetry.*

This form of the Lagrangian Eq.(18) is the same as that of the effective theory of the walking technicolor, except for the shape of the scale-violating potential $V(\phi)$ which has a scale dimension 4 (trace anomaly) in the case of the walking technicolor instead of 2 of the SM Higgs case (Lagrangian mass term). We shall come back to this later.

3. Composite Higgs from the NJL Model

Let us now elaborate the composite Higgs model based on the strong coupling theory $G > G_{\mathrm{cr}} \neq 0$ pioneered by Nambu. In the Nambu-Jona-Lasinio (NJL) model[6] for the N_C−component 2-flavored fermion ψ the Lagrangian takes the form:

$$\begin{aligned}\mathcal{L}_{\mathrm{NJL}} &= \bar\psi i\gamma^\mu\partial_\mu\psi + \frac{G}{2}\left[(\bar\psi\psi)^2 + (\bar\psi i\gamma_5\tau^a\psi)^2\right] \\ &= \bar\psi\left(i\gamma^\mu\partial_\mu + \hat\sigma + i\gamma_5\tau^a\hat\pi_a\right)\psi - \frac{1}{2G}\left(\hat\sigma^2 + \hat\pi_a^2\right), \end{aligned} \tag{19}$$

where equation of motion of the auxiliary fields $\hat\sigma \sim G\bar\psi\psi$ and $\hat\pi^a \sim G\bar\psi i\gamma_5\tau^a\psi$ are plugged back in the Lagrangian to get the original Lagrangian. In the large N_C limit ($N_C \to \infty$ with $N_C G \neq 0$ fixed), after rescaling the induced kinetic term to the canonical one, $Z_\phi^{1/2}\hat\sigma \to \hat\sigma$, the quantum theory for $\hat\sigma$ and $\hat\pi$ sector yields precisely the same form as the SM Higgs Eq.(6), with

$$\begin{aligned}\mu_0^2 &= \left(\frac{1}{G} - \frac{1}{G_{\mathrm{cr}}}\right) Z_\phi^{-1} = -2m_F^2 = -v^2 Z_\phi^{-1} = -\lambda v^2 < 0 \quad (G > G_{\mathrm{cr}} = \frac{4\pi^2}{N_C\Lambda^2}) \\ \lambda &= Z_\phi Z_\phi^{-2} = Z_\phi^{-1} = \left[\frac{N_C}{8\pi^2}\ln\frac{\Lambda^2}{m_F^2}\right]^{-1} \sim \left[\frac{N_C}{8\pi^2}\ln\frac{\Lambda^2}{v^2}\right]^{-1}, \end{aligned} \tag{20}$$

where the gap equation has been used:

$$\frac{1}{G} - \frac{1}{G_{\mathrm{cr}}} = -\frac{N_C}{4\pi^2} m_F^2 \ln\frac{\Lambda^2}{m_F^2} = -2m_F^2 Z_\phi = -F_\pi^2 = -v^2 . \tag{21}$$

Eq.(20) shows that the *tachyon with* $\mu^2 < 0$ *is in fact generated* by the dynamical effects for the *strong coupling* $G > G_{\mathrm{cr}} \neq 0$, corresponding to the generation of mass $m_F \neq 0$ in the gap equation. Or, we can explicitly see it by computing the $\bar\psi\psi$ bound state using the $m_F = 0$ solution (wrong solution) of the gap equation at $G > G_{\mathrm{cr}}$. The correct spectrum $M_\pi^2 = 0, M_\sigma^2 = 2\lambda v^2 = -2\mu_0^2 = 4m_F^2$ can be obtained when we use the correct solution $m_F \neq 0$ in the gap equation. The last equality $M_\sigma^2 = 4m_F^2$ is specific to the $N_C \to \infty$ (with $N_C G \neq 0$ fixed) limit of the NJL model (*"weak coupling" limit* $G > G_{\rm cr} \sim 1/N_C \to 0$ in the *strong coupling phase*), but not the general outcome of the NJL model nor the generic linear sigma model.

There are two extreme limits for λ in Eq.(20) : $\lambda \to 0$ ($N_C \gg 1$ and/or $\Lambda/v^2 \gg 1$) reproduces precisely the conformal limit, or scale-invariant nonlinear sigma model limit, Eq.(14), of the SM Higgs Lagrangian, while $\lambda \to \infty$ ($N_C, \Lambda^2/v^2 = \mathcal{O}(1)$) does the nonlinear sigma model limit without scale symmetry, Eq.(17).

We are interested in the limit $\lambda = [N_C \ln(\Lambda^2/v^2)/(8\pi^2)]^{-1} \to 0$ (conformal limit in Eq.(16)) realized for $\Lambda/v \to \infty$ and/or $N_C \to \infty$, with $v = F_\pi = F_\phi \neq 0$ fixed.[g] Then the effective Lagrangian in the large N_C limit takes precisely the same as the SM Higgs Lagrangian, which is further equivalent to the scale-invariant nonlinear sigma model, Eq.(14), as mentioned before. Now the SM Higgs is identified with the composite (pseudo-)dilaton with mass vanishing $M_\phi^2 = 2\lambda v^2 \to 0$.

The limit theory gives an *interacting low energy effective theory even in the* $\Lambda/v \to \infty$ *limit*: a scale-invariant nonlinear sigma model [11,12] where massless π and ϕ are *interacting each other* with the (derivative) couplings $\sim (1/F_\pi, 1/F_\phi) \neq 0$. It is actually the basis for the *scale-invariant chiral perturbation theory* (sChPT) with the derivative expansion as a loop expansion[24], although the Yukawa couplings of π, ϕ to the fermions are vanishing $g_Y \sim m_F/F_\pi, m_F/F_\phi \to 0$ (The composite particles are still interacting due to the loop divergence compensation of the vanishing Yukawa coupling). This limit should be sharply distinguished from a similar limit $\Lambda/m_F \to \infty$, m_F =fixed (not $\Lambda/v \to \infty$, v =fixed), which is the famous triviality limit (Gaussian fixed point) where the theory becomes a free theory: free massive scalar for $G < G_{\rm cr}$ and free tachyon for $G > G_{\rm cr}$, with not just the Yukawa couplings but all the couplings vanishing.

One might wonder why dilaton in NJL model? Obviously the NJL model has the explicit scale-breaking coupling G having dimension $[M]^{-2}$. But this scale is an ultraviolet scale to which the low energy effective theory is insensitive. This is in exactly the same sense as in the walking technicolor where the intrinsic scale $\Lambda_{\rm TC}$ generated by the trace anomaly can be far bigger than the infrared scale of spontaneous symmetry breaking $F_\pi, F_\phi = \mathcal{O}(v) \ll \Lambda_{\rm TC}$ thanks to the approximate scale symmetry due to the almost nonnruning coupling.

Actually we can formulate the nonperturbative running of the (dimensionless) four-fermion coupling $g = \frac{N_C \Lambda^2}{4\pi^2} G$ in the same way as the Miransky nonperturbative

[g]If Λ is regarded as a physical cutoff in contrast to the nonperturbative renormalization arguments below, this argument would not be realistic for the 125 GeV Higgs with $\lambda \simeq 1/8$, corresponding to $\Lambda \simeq v \cdot e^{32\pi^2/N_C} \gg 10^{19}$ GeV. For the NJL model with N_D doublets, however, we would have $\Lambda \simeq v \cdot e^{32\pi^2/(N_D N_C)} \sim 10^{11}$ GeV for $N_D = N_C = 4$.

renormalization:[h] The gap equation Eq. (21) reads

$$\left(\frac{1}{g_{\rm cr}}-\frac{1}{g}\right)\Lambda^2 = m_F^2 \ln\frac{\Lambda^2}{m_F^2} \simeq \frac{4\pi^2}{N_C}v^2\,, \quad g_{\rm cr}=1\,, \tag{22}$$

which leads to a nonperturbative beta function

$$\beta(g) = \Lambda\frac{\partial g(\Lambda)}{\partial\Lambda}\bigg|_{v={\rm fixed}} = -\frac{2}{g_{\rm cr}}g\cdot(g-g_{\rm cr})\,, \quad g(\mu) = g_{\rm cr}\frac{1}{1-\frac{4\pi^2}{N_C g_{\rm cr}}\frac{v^2}{\mu^2}} \tag{23}$$

by *fixing* v = *constant* and taking $\Lambda \to \infty$. Thus $g = g_{\rm cr} = 1$ is the ultraviolet fixed point that the running coupling $g(\mu)$ reaches even much faster than the walking coupling in Eq.(3). Then the scale symmetry is operative $g(\mu) \approx g_{\rm cr}$ for the wide region $\frac{4\pi^2}{N_C g_{\rm cr}}v^2 < \mu^2 < \Lambda^2$, with the *explicit scale symmetry breaking in the dimension 2 operator*: $\theta^\mu_\mu = \frac{\beta(g)}{g}\frac{G}{2}\left[\left(\bar\psi\psi\right)^2 + \left(\bar\psi i\gamma_5\tau^a\psi\right)^2\right] = -\lambda v^4\chi^2$, [i] where $\lambda = 8\pi^2/[N_C\ln(\Lambda^2/v^2)] \to 0$ as in Eq.(20). The PCDC follows precisely the same way as in the SM Higgs as $M_\phi^2 F_\phi^2 = -d_{(\bar\psi\psi)^2}\langle\theta^\mu_\mu\rangle = 2\lambda v^4$ (See below Eq.(14)). In any case the trace of energy-momentum tensor vanishes in the limit $\lambda \sim 1/[N_C\ln(\Lambda^2/v^2)] \to 0$, and the dilaton mass should come from the trace anomaly in the $1/N_C$ sub-leading loop effects, or the chiral loops of the effective theory Eq.(14).

Again the spin 1 composites can also be introduced via HLS, precisely in the same way as Eq.(18) for the SM Higgs Lagrangian. This time it can be done more explicitly by introducing the vector/axialvector type four-fermion coupling which in fact become the "explicit" composite HLS gauge bosons.(See section 5.3 of Ref.[21]).

One of the concrete composite Higgs models as the straightforward application of the NJL type theory is the top quark condensate model (Top-Mode Standard Model)[25–27]. The crucial ingredient of the model is again the *non-zero critical coupling in sharp contrast to the weakly-coupled BCS theory which has* $g_{\rm cr} = 0$ as already mentioned.: *only the top quark coupling is strong coupling larger than the critical coupling* $G_t > G_{\rm cr}$ while others are less, $G_{b,c,s,d,u} < G_{\rm cr}$, so that only the top acquires the dynamical mass of order of weak scale $\mathcal{O}(v)$ to produce only three NG bosons to be absorbed into the W/Z bosons[25,27]. The large N_C limit relation $M_\phi = 2m_t$ is modified by the effects of the SM gauge interactions, $M_\phi \simeq \sqrt{2}m_t$[27,28], and further reduced by the non-leading order in $1/N_C$ expansion[27]. Different reductions

---

[h]The argument here is somewhat similar to the renormalizability arguments of the D-dimensional NJL model ($2 < D < 4$)[32] and the gauged NJL model[33], although the explicit scale-breaking from the Lagrangian parameters, i.e., the four-fermion interaction and fermion mass term (if present), depend on the renormalization point (vanish at the UV limit).

[i]Note that $-\langle\bar\psi_i\psi_j\rangle = \delta_{i,j}\Lambda^2 m_F N_C/(4\pi^2) = Z_m^{-1}\,\delta_{i,j}v^3 N_C/(4\pi^2)$, where $Z_m^{-1} = Z_m^{-1}(\Lambda/v) = (\Lambda/v)^2[N_C\ln(\Lambda^2/v^2)/(4\pi^2)]^{-1/2}$ is the mass renormalization constant. This implies $\gamma_m = Z_m\Lambda\frac{\partial Z_m^{-1}}{\partial\Lambda} = 2 - 1/\ln(\Lambda^2/v^2) \to 2$, and hence the operators have scale dimension $d_{\bar\psi\psi} = 1$ and $d_{(\bar\psi\psi)^2} = 2$. Thus we may write $\bar\psi_i\psi_j = -Z_m^{-1}\delta_{i,j}v^3 N_C/(4\pi^2)\cdot\chi$, or $(G/2)(\bar\psi\psi)^2 = 2g\Lambda^2 v^2\cdot\chi^2/[\ln(\Lambda^2/v^2)]$. The gap equation implies $\beta(g)/g = -g(4\pi^2/N_C)(v^2/\Lambda^2)$. Putting all together, we have $\beta(g)/g\cdot(G/2)\cdot(\bar\psi\psi)^2|_{g\to g_{\rm cr}=1} = -\lambda v^4\chi^2$.

have been considered, top seesaw[29], and its NG boson Higgs version[30]. Updated detailed discussions are given in the talk by H. Fukano in this meeting.[31]

4. Composite Higgs in Walking Technicolor: Technidilaton as the 125 GeV Higgs

We have already seen that the SM Higgs is a dilaton in the conformal limit. Here I come to the main subject, the technidilaton in the Walking Technicolor, the SCGT. Details are given in the talk by S. Matsuzaki in this meeting.[36]

Let us discuss a typical walking technicolor, the QCD-like vector-like $SU(N_C)$ gauge theory with N_F massless technifermions, particularly near the "anti-Veneziano limit" (in distinction to the original Veneziano limit with $N_F/N_C \ll 1$):[11]

$$N_C \to \infty \quad \text{and} \quad \lambda \equiv N_C \cdot \alpha = \text{fixed}, \quad \text{with} \quad r \equiv N_F/N_C = \text{fixed} \gg 1\,. \tag{24}$$

Such a limit is ideal for the ladder approximation with nonrunning coupling, since the coupling growth in the infrared region by the (anti-screening) gluon loops is cancelled by the opposite (screening) effects of the increasing fermion loops as N_F increases, and eventually levels off by balance at certain large number $N_F > N_F^*(\gg 2)$ $(N_F < 11N_C/2)$, as realized by the infrared (IR) fixed point α_* such that $\beta^{(\text{perturbative})}(\alpha = \alpha_*) = 0$ as demonstrated in the two-loop beta function (Caswell-Banks-Zaks (CBZ) IR fixed point)[37], with $N_F^* \simeq 8(N_C = 3)$. Thus the input perturbative coupling in the SD equation becomes almost nonrunning, $\alpha(\mu^2) \simeq \alpha_*$, in the infrared region $\mu^2 < \Lambda_{\text{TC}}^2$ (infrared conformality), where Λ_{TC} is the intrinsic scale, analogous to the Λ_{QCD}, generated by the perturbative trace anomaly responsible for the perturbative (asymptotically-free) running of the coupling in the UV region $\mu^2 > \Lambda_{\text{TC}}^2$. It plays the role of the cutoff Λ in the ladder approximation.

Then the anti-Veneziano limit Eq.(24) corresponds to the original walking technicolor[1,2] based on the ladder SD equation in Landau gauge for the fermion propagator $iS_F^{-1}(p) = \gamma^\mu p_\mu - \Sigma(-p^2)$, with the gauge coupling $\alpha(p^2) \equiv g(p^2)^2/(4\pi^2) \equiv \alpha$ =constant for $|p^2| < \Lambda^2 = \Lambda_{\text{TC}}^2$, which has the broken solution, $\Sigma(p^2 = -m_F^2) = m_F \neq 0$, in the strong coupling phase $C_2\alpha_* > C_2\alpha_{\text{cr}} = \pi/3 \neq 0$, where $C_2 = (N_C^2 - 1)/(2N_C)$ is the quadratic Casimir. All the ladder results are intact in the limit: Eqs.(1-4) and the technidilaton as a composite Higgs. While in the weak coupling phase ("conformal window") $C_2\alpha_* < C_2\alpha_{\text{cr}}$, there remains the unbroken approximate scale symmetry, and no bound states exist ("unparticle").

When $\alpha \simeq \alpha_* > \alpha_{\text{cr}}$, such that $N_F > N_F^{\text{cr}}(\simeq 4N_C > N_F^*)$[38], the SD equation has spontaneous breaking solution $m_F \neq 0$, arising from the technifermion condensate $\langle \bar{F}F \rangle \neq 0$, which obviously *breaks both chiral symmetry and the scale symmetry spontaneously.* The scale symmetry is also *broken explicitly by the same origin* $m_F \neq 0$: the would-be IR fixed point actually is washed out, with the coupling starting running (walking) as in Eq.(3), according to the nonpertubative beta function Eq.(2) responsible for a tiny *nonperturbative trace anomaly of dimension 4 operator*: $\theta_\mu^\mu = \beta^{(\text{NP})}(\alpha)/(4\alpha) \cdot G_{\mu\nu}^2 = \mathcal{O}(m_F^4)(\ll \Lambda_{\text{TC}}^4)$, where $G_{\mu\nu}$ is the technigluon field strength with $G^2_{\mu\nu}$ also induced by the m_F. (Usual (perturbative) trace anomaly corresponding to the asymptotically free running of order $\mathcal{O}(\Lambda^4_{\rm TC})$, has been subtracted out.) In accord with the smallness of the trace anomaly, the coupling is still almost nonrunning as in Eq.(3) for a wide infrared region $m_F < \mu < \Lambda_{\rm TC}$ ($m_F \ll \Lambda_{\rm TC}$).

Note[11] that Eq.(2) has a multiple zero at $\alpha = \alpha_{\rm cr} \simeq \alpha_*$ (not linear zero) and is completely different from the two-loop beta function having the CBZ IR fixed point (having linear zero at α_*) which is no longer valid in the broken phase $\alpha > \alpha_* \simeq \alpha_{\rm cr}$, where $\alpha(\mu) \searrow \alpha_{\rm cr}$ ($\mu \nearrow$). Then the would-be *IR fixed point* $\alpha_* \simeq \alpha_{\rm cr}$ *is also regarded as the UV fixed point* of the nonperturbative running (walking) coupling $\alpha(\mu) \approx \alpha_{\rm cr}$[1] in the wide IR region $m_F < \mu < \Lambda = \Lambda_{\rm TC}$ for the characteristic large hierarchy $m_F \ll \Lambda_{\rm TC}$[39,40]. (See also Ref.[41] for a similar observation.) See Fig. 1.[j]

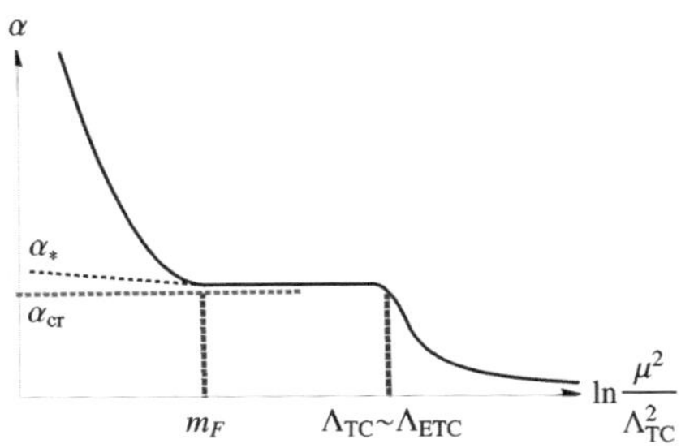

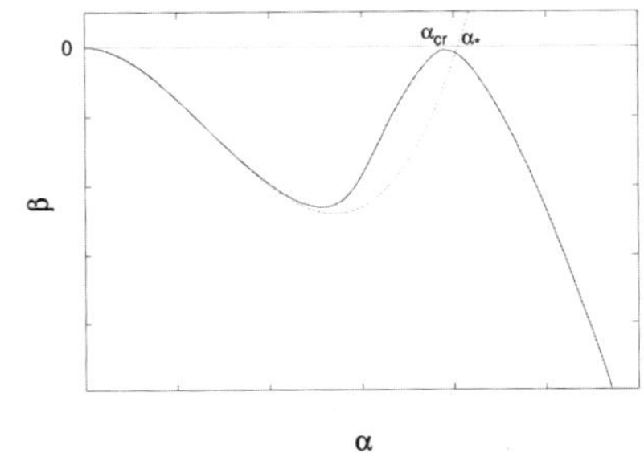

Fig. 1. Possible perturbative running coupling (left) and the beta function (right) in the region $\alpha < \alpha_{\rm cr}$, in comparison with the nonperturbative region $\alpha > \alpha_{\rm cr}$.

Now we come to our core result, ladder evaluation of the nonperturbative trace anomaly in the anti-Veneziano limit, which then yields the mass M_ϕ and decay constant F_ϕ of the technidilaton ϕ through PCDC[2] as in Eq.(15):[11,40]

$$M_\phi^2 F_\phi^2 = -d_\theta \langle \theta^\mu_\mu \rangle = -\frac{\beta^{(\rm NP)}(\alpha(\mu^2))}{\alpha(\mu^2)} \langle G^2_{\mu\nu}(\mu^2)\rangle \simeq N_C N_F \frac{16}{\pi^4} m_F^4 \quad (d_\theta = 4) \tag{25}$$

$$\simeq 2.5 \left[\frac{8}{N_F}\frac{4}{N_C}\right] v^4 . \quad (v = 246\,{\rm GeV}) \tag{26}$$

Firstly, the rightmost value in Eq.(25) can be obtained by two different ladder calculations: one through direct evaluation of the vacuum energy by the effective potential at the stationary point (Solution of the SD equation, $\Sigma = \Sigma_{\rm sol}$)[44], $E = V_{\rm eff}(\Sigma = \Sigma_{\rm sol}) = \langle \theta^0_0 \rangle = (1/4)\langle \theta^\mu_\mu \rangle$, the other through the ladder evaluation of the

[j] Note also that the walking technicolor in the UV region $\mu^2 > \Lambda^2_{\rm TC}$ must be changed into only a part of some larger picture, such as the Extended Technicolor (ETC)[42] or models having both technifermions and SM fermions as composites on the equal footing[43], to provide the mass to the SM fermions through communicating technifermion condensate to the SM fermions, so that discussing the walking technicolor in isolation does not make sense for $\mu^2 > \Lambda^2_{\rm TC}$.

trace anomaly[11,40], i.e., the technigluon condensate $\langle G_{\mu\nu}^2 \rangle$ times the nonperturbative beta function Eq.(2), both in precise agreement with each other. The agreement is in highly nontrivial manner, being *independent of the renormalization point* μ as it should be: $\langle G_{\mu\nu}^2(\mu^2)\rangle \sim \ln^3(\mu^2/m_F^2)$, while $\beta^{(\rm NP)}(\alpha(\mu^2))/\alpha(\mu^2) \sim 1/\ln^3(\mu^2/m_F^2)$, precisely cancelled by each other.[11]

Secondly, Eq.(26) is obtained by use of the Pagels-Stokar formula:

$$v^2 = (246\,{\rm GeV})^2 = N_D F_\pi^2 \simeq N_F N_C \frac{1}{4\pi^2}\, m_F^2 \simeq m_F^2 \left[\frac{N_F}{8}\frac{N_C}{4}\right], \tag{27}$$

and the result indicates *important* N_F, N_C- *dependence of* $M_\phi^2 F_\phi^2$ *in the anti-Veneziano limit when* $v =$ *fixed*[11]. Since the technidilaton is a flavor-singlet bound state, its decay constant by definition scales like $F_\phi^2 \propto N_F N_C m_F^2 (\propto v^2)$ (Actually $F_\phi^2 \simeq N_F N_C m_F^2$). Then $M_\phi^2/F_\phi^2, M_\phi^2/v^2 \sim 1/(N_F N_C) \to 0$ in the anti-Veneziano limit, where the technidilaton becomes NG boson although no exact massless limit exists: the situation is in the same sense as the η' meson in the original Veneziano limit $N_C \to \infty$ with $N_C\alpha$ =fixed, and $N_F/N_C \ll 1$.[k]

Numerically, Eq.(26) implies that

$$F_\phi \simeq 5\,v \quad \text{for } M_\phi \simeq \frac{v}{2} \simeq 125\,{\rm GeV} \quad (N_F = 8, N_C = 4)\ , \tag{28}$$

in the one-family model, which is best fit to the current LHC data of the 125 GeV Higgs.[11,12] Similar results are also obtained in the holographic model for the walking technicolor.[13]

5. Discovering Walking Technicolor at LHC

Now to the LHC phenomenology of the walking technicolor. Details are given in the talk by S. Matsuzaki in this meeting.[36] The model has in general a larger chiral symmetry $SU(N_F)_L \times SU(N_F)_R$ $(N_F > 2)$ spontaneously broken by the technifermion condensate $\langle \bar{F}F\rangle \neq 0$, or the technifermion dynamical mass $m_F \neq 0$, down to the diagonal $SU(N_F)_V$. Also *spontaneously broken by the same technifermion condensate* is the *approximate scale symmetry* due to the almost nonruning (walking) coupling $\alpha(\mu^2) \approx \alpha_{\rm cr}$ in the wide infrared region $m_F^2 < \mu^2 < \Lambda_{\rm TC}^2$ $(m_F \ll \Lambda_{\rm TC}^2)$. Moreover the scale symmetry is *simultaneously broken explicitly by the same origin* $m_F \neq 0$, an emergent infrared mass scale, resulting in the *nonperturbative trace anomaly of dimension 4 operator* as was given in Eq.(25).

The effective theory should have the same symmetry structure as the underlying theory. Namely, the *nonlinear realization of both the chiral symmetry and the scale symmetry*, which implies the *scale-invariant nonlinear sigma*

[k] There exists no exact massless limit in the conformal phase transition at $\alpha = \alpha_{\rm cr}$, with $m_F = 0$, where no massless spectrum exists (conformal phase), in sharp contrast to the Ginzburg-Landau phase transition where the spectrum continuously passes through the phase transition point with massless particles.[4]

model[11,12]. Then the effective theory of the walking technicolor with N_F massless flavors takes *precisely the same scale-invariant form as the nonlinearly realized SM Higgs Lagrangian* in Eq.(14), with $U = e^{i\pi^a T^a}$ being $N_F \times N_F$ unitary matrix ($\mathrm{tr}\, T^a = 0, \mathrm{tr}(T^a T^b) = \delta^{ab}/2$, $a = 1, \cdots, N_F^2 - 1$), except that the explicit scale breaking comes from the different potential $V^{(4)}(\phi)$ [1] responsible for the *trace anomaly of dimension 4 operator* this time:[11,12]

$$
\begin{aligned}
\mathcal{L}_{\mathrm{WTC}} &= \chi^2 \cdot \left[\frac{1}{2}\left(\partial_\mu \phi\right)^2 + \frac{F_\pi^2}{4}\mathrm{tr}\left(\mathcal{D}_\mu U \mathcal{D}^\mu U^\dagger\right)\right] - V^{(4)}(\phi) - V^{(\mathrm{SM})}(\phi)\,, \\
V^{(4)}(\phi) &= -\ln\chi \cdot \frac{\beta^{(\mathrm{NP})}(\alpha)}{4\alpha} G_{\mu\nu}^2 = \frac{M_\phi^2 F_\phi^2}{4}\chi^4\left(\ln\chi - \frac{1}{4}\right) \\
V^{(\mathrm{SM})}(\phi) &= -\chi^{2-\gamma_m}\left(m_f \chi \bar{f} f\right) - \ln\chi\left[\frac{\beta_F(\alpha_s)}{4\alpha_s} G_{\mu\nu}^{(\mathrm{gluon})\,2} + \frac{\beta_F(\alpha_e)}{4\alpha_e} F_{\mu\nu}^{(\gamma)\,2}\right], \\
\chi &= \exp\left(\frac{\phi}{F_\phi}\right), \qquad (29)
\end{aligned}
$$

where $F_\phi \neq F_\pi = v/\sqrt{N_D} = v/\sqrt{N_F/2}$ *in general* in contrast to the SM Higgs case $F_\phi = F_\pi = v$, the electroweak gauging was done as usual $\partial_\mu U \Rightarrow \mathcal{D}_\mu U = \partial_\mu U - i g_2 W_\mu U + i g_Y U B_\mu$, and we have added $V^{(\mathrm{SM})}(\phi)$, the scale symmetry breaking terms related to the SM particles arising from the technifermion contributions: mass term of the SM fermion f, (one loop) technifermion contributions to the trace anomaly for the gluon and photon operators, with $\beta_F(g_s) = \frac{g_s^3}{(4\pi)^2}\frac{4}{3}N_C$ and $\beta_F(e) = \frac{e^3}{(4\pi)^2}\frac{16}{9}N_C$. It is obvious that $\theta_\mu^{\mu\,(\mathrm{TC})} = -\delta_D V^{(4)}(\phi) = \beta^{(\mathrm{NP})}(\alpha)/(4\alpha)\cdot G_{\mu\nu}^2 = -(M_\phi^2 F_\phi^2/4)\chi^4$ up to total derivative, corresponding to the PCDC with $d_\theta = 4$ ($\langle\chi\rangle = 1$), Eq.(25).

The technidilaton potential $V^{(4)}(\phi)$ is expanded in ϕ/F_ϕ:

$$
V^{(4)}(\phi) = -\frac{M_\phi^2 F_\phi^2}{16} + \frac{1}{2}M_\phi^2 \phi^2 + \frac{4}{3}\frac{M_\phi^2}{F_\phi}\phi^3 + 2\frac{M_\phi^2}{F_\phi^2}\phi^4 + \cdots\,, \qquad (30)
$$

which shows a remarkable fact that in the anti-Veneziano limit the technidilaton self couplings (trilinear and quartic couplings) are highly suppressed: $\lambda_{\phi^3} = 4M_\phi^2/(3F_\phi) \sim 1/\sqrt{N_F N_C}$, $\lambda_{\phi^4} = 2M_\phi^2/F_\phi^2 \sim 1/(N_F N_C)$, by $M_\phi/F_\phi \sim 1/\sqrt{N_F N_C}$ and $M_\phi \sim N_F^0 N_C^0$. Numerically, we may compare the technidilaton self couplings with those of the SM Higgs with $m_h = M_\phi = 125$ GeV for $v/F_\phi \simeq 1/5$ in the

[1]This potential is indeed obtained by the explicit ladder computation of the effective potential at the conformal phase transition point: $V^{(4)}(\phi) = -(4N_F N_C m_F^4/\pi^4)\chi^4(\ln\chi - 1/4)$, in precise agreement with Eq.(29) through Eq.(25). See Eq.(65) in Ref.[4]

one-family model ($N_F = 8, N_C = 4$):[11]

$$\left.\frac{\lambda_{\phi^3}}{\lambda_{h^3_{\rm SM}}}\right|_{M_\phi=m_h} = \left.\frac{\frac{4M_\phi^2}{3F_\phi}}{\frac{m_h^2}{2v}}\right|_{M_\phi=m_h} \simeq \frac{8}{3}\left(\frac{v}{F_\phi}\right) \simeq 0.5\,,$$

$$\left.\frac{\lambda_{\phi^4}}{\lambda_{h^4_{\rm SM}}}\right|_{M_\phi=m_h} = \left.\frac{\frac{2M_\phi^2}{F_\phi^2}}{\frac{m_h^2}{8v^2}}\right|_{M_\phi=m_h} = 16\left(\frac{v}{F_\phi}\right)^2 \simeq 0.6\,. \tag{31}$$

This shows that the *technidilaton self couplings, although generated by the strongly coupled interactions, are even smaller than those of the SM Higgs*, a salient feature of the approximate scale symmetry in the ant-Veneziano limit !!

The coupling of the technidilaton ($M_\phi = 125$ GeV) to the SM particles can be seen by expanding $\chi = 1 + \phi/F_\phi + (1/2!)(\phi/F_\phi)^2 + \cdots$ in Eq.(29):

$$\frac{g_{\phi WW/ZZ}}{g_{h_{\rm SM}WW/ZZ}} = \frac{g_{\phi ff}}{g_{h_{\rm SM}ff}} = \frac{v}{F_\phi}\,. \tag{32}$$

$$\frac{g_{\phi gg}}{g_{h_{\rm SM}gg}} \simeq \frac{v}{F_\phi}\cdot(1+2N_C)\,,\quad \frac{g_{\phi\gamma\gamma}}{g_{h_{\rm SM}\gamma\gamma}} \simeq \frac{v}{F_\phi}\cdot\left(\frac{63-16}{47} - \frac{32}{47}N_C\right)\,, \tag{33}$$

where besides the technifermion, only the top and W of the SM contributions were included at one-loop. Note the couplings in Eq.(32) with $v/F_\phi \sim 1/5$ are even weaker than the SM Higgs, which are however compensated by those in Eq.(33) for gg and $\gamma\gamma$ rather enhanced by the extra loop contributions of the technifermions other than the SM particles, particularly for large N_C, resulting in signal strength similar to the SM Higgs within the current experimental accuracy.

In fact the current LHC data for 125 GeV Higgs are fit by the technidilaton as good as by the SM Higgs, particularly for $N_F = 8, N_C = 4$, i.e., near the anti-Veneziano limit.[12] Most recent detailed analyses are given in Ref.[11]. It should be mentioned here that the one-family model will be most naturally imbedded into the ETC in the case for $N_C = 4$[47]. More precise data at LHC Run II will discriminate among them, SM Higgs or technidilaton. We will see.

Next to the technipions: in the walking technicolor with $N_D = N_F/2 > 1$, the spontaneous breaking of the chiral symmetry larger than $SU(2)_L \times SU(2)_R$ produces NG bosons (technipions) more than 3 to be absorbed into W/Z. Let us take the one-family model with $N_F = 8$, which has colored techniquarks (3 weak doublets) Q_i^a and non-colored technileptons (one weak-doublet) L_i ($a = 1, 2, 3; i = 1, 2$), the resultant chiral symmetry being $SU(8)_L \times SU(8)_R$[45]. There are 63 technipions, 60 of which are unabsorbed technipions. All of them acquire the mass from the explicit chiral symmetry breaking due to the SM gauge and ETC gauge interactions. Due to the large anomalous dimension $\gamma_m \simeq 1$, the mass of them are all enhanced to TeV region[46], which will be discovered at LHC Run II.

Another signatures of the walking technicolor are higher resonances such as the spin 1 boson, the walking techni-ρ, walking techni-a_1, etc.. The straightforward

N_F extension of Eq.(18) is also obvious: Eq.(29) is gauge equivalent to the *scale-invariant HLS Lagrangian* explicitly constructed for one-family walking technicolor with $N_F = 8$[19]:

$$\mathcal{L}_{\text{s-HLS}} = \mathcal{L}_{\text{WTC}} + \mathcal{L}_{\text{Kinetic}}(V_\mu) ,$$
$$\mathcal{L}_{\text{WTC}} = \chi^2 \cdot \left(\frac{1}{2} (\partial_\mu \phi)^2 + \mathcal{L}_A + a\mathcal{L}_V \right) - V^{(4)}(\phi) , \tag{34}$$

where the HLS gauge bosons V_μ in the mass term $a\chi^2 \mathcal{L}_V = a e^{2\phi/F_\phi} \text{tr}(g_{_{\text{HLS}}} V_\mu + \cdots)^2$ are the bound states of the walking technicolor, the walking techni-ρ, with *mass* $M_V^2 = a g_{_{\text{HLS}}}^2 F_\pi^2$ *being scale-invariant* thanks to the overall technidilaton factor χ^2, as mentioned before. The loop expansion is formulated as the scale-invariant HLS perturbation theory[18] in the same way as the scale-invariant chiral perturbation theory[24], a straightforward extension of the (non-scale-invariant) HLS perturbation theory[22]. The HLS is readily extendable to include techni-a_1, etc.[20–22], with a infinite set of the HLS tower being equivalent to the deconstructed extra dimension[34] and/or the holographic models[35], and the scale-invariant version of them are also straightforward and *mass of all the higher HLS vector bosons are scale-invariant*, an outstanding characterization in sharp contrast to other formulations for the spin 1 bosons. We will see at LHC Run II.

6. Walking on the Lattice

Finally, I briefly mention the walking technicolor on the lattice focussing on our own results by LatKMI. Updated details are given in the talks by Y. Aoki, K-i. Nagai and H. Ohki in this meeting[48]. The LatKMI Collaboration started in 2010 for the lattice simulations on the possible candidate for the walking technicolor by systematic studies of the $N_F = 16, 12, 8, 4$ degenerate fermions in $SU(3)$ gauge theories (dubbed "Large N_F QCD"), using the HISQ (Highly Improved Staggered Quarks) action with tree-level Symanzik gauge action. We have mainly focused on the low-lying fermionic bound states (plus some gluonic ones), i,e., pseudoscalar (denoted as π), scalar (σ, a_0), vector (ρ), axialvector mesons (a_1), and nucleon-like states (N, N^*), particularly the flavor-singlet scalar σ as a candidate for the technidilaton.

We found[49] that $N_F = 12$ is consistent with the conformal window indicating no spontaneous chiral symmetry breaking, which is in agreement with the results of many other groups. We further found[50] that $N_F = 8$ is consistent with the spontaneously broken phase with remnants of the conformality with large anomalous dimension $\gamma_m \simeq 1$, namely the walking theory, in accord with other groups[51].

The most remarkable result of the LatKMI Collaboration is the *discovery of a light flavor-singlet scalar on the lattice* in both $N_F = 12$[52] and $N_F = 8$[53]. The $N_F = 12$ results are also consistent with other groups[54]. Since $N_F = 8$ seems to be a walking theory in the broken phase, the light flavor-singlet scalar is particularly attractive as a candidate for the technidilaton. Also $N_F = 8$ is of phenomenological relevance to the LHC data as well as direct relevance to the one-family model as the most natural model building. Future confirmation of our results is highly desired. Also $N_C = 4$ simulations should be studied for various reasons as mentioned before.

I am really proud of the LatKMI Collaboration, although it will loose funding soon.

7. Summary

I have argued that the 125 GeV Higgs is a (pseudo-)dilaton even if it is described by the Standard Model Higgs Lagrangian (!!) which is actually shown to be equivalent to the scale-invariant nonlinear sigma model with both chiral and scale symmetries are nonlinearly realized. The SM Higgs Lagrangian is further gauge equivalent to the scale-invariant Hidden Local Symmetry (HLS) Lagrangian which include new massive vector bosons as the gauge bosons of the (spontaneously broken) HLS, with the mass being scale-invariant.

All these features of the SM Higgs Lagrangian are reminiscence of the conformal UV completion behind the Higgs, the existence of the underlying theory with (approximate) scale symmetry, so strong coupling as to produce composite states such as the Higgs (dilaton), new vector bosons (HLS gauge bosons), etc.. We have seen that even the NJL model, though not gauge theory, can be regarded as such a conformal UV completion. The walking technicolor, conformal SCGT, is such a typical underlying theory, where the 125 GeV Higgs is a composite dilaton, the technidilaton.

The walking technicolor in the anti-Veneziano limit $N_C \to \infty$ with $N_C\alpha$ =fixed $= \mathcal{O}(1)$ and N_F/N_C = fixed ($\gg 1$) makes the ladder approximation reasonable, which yields a naturally light and weakly coupled technidilaton through the PCDC:

$$M_\phi^2 F_\phi^2 = -4\langle\theta_\mu^\mu\rangle = -\frac{\beta(\alpha(\mu^2))}{\alpha(\mu^2)}\langle G_{\mu\nu}^2(\mu^2)\rangle \simeq N_C N_F \frac{16}{\pi^4} m_F^4 \,, \tag{35}$$

independently of the renormalization point μ, where the scale symmetry is explicitly broken by the trace anomaly of the dimension 4 operator $G_{\mu\nu}^2$, which is induced by m_F the dynamical mass of the technifermion arising from the simultaneous spontaneous breaking of the scale symmetry and the chiral symmetry.

I have defined "strong coupling theories" as "those having non-zero critical coupling" $N_C g_{\rm cr} = \mathcal{O}(1)$, even though its value could be small $g \sim 1/N_C \ll 1$ in the typical large N_C limit. The NJL model pioneered by Professor Nambu is the first and a typical example of such, to be distinguished from its preceding, the BCS theory, which has a zero critical coupling $g_{\rm cr} = 0$. Existence of such a non-zero critical coupling in gauge theory was discovered by Maskawa and Nakajima in the scale-invariant dynamics, ladder approximation, and became crucial for the walking technicolor with the coupling $N_C\alpha > N_C\alpha_{\rm cr} = \mathcal{O}(1)$ in the spontaneous broken phase of the scale symmetry as well as the chiral symmetry.

The existence of the non-zero critical coupling is actually "hidden" even in the QCD which is regarded to have only one phase in the ordinary situation without signal of the no-zero critical coupling: it manifests itself in the extreme condition, such as the large number of fermions $N_F \gg N_C$ (so as to keep the asymptotic freedom), high temperature, high density, etc..

Indeed, it is the large N_F QCD that models the walking technicolor where the large number of fermions give the screening effects and level off of the infrared coupling which otherwise brows up due to the gluon anti-screening effects (Caswell-Banks-Zaks infrared fixed point). For large N_F with the fixed point value smaller than the critical coupling, the spontaneously broken phase is gone (what we called conformal phase transition). Then the infrared scale invariance becomes manifest.

I would infer existence of the similar "hidden" nonzero critical coupling even for ordinary QCD at high temperature and/or high density, where the infrared coupling cannot blow up in the region below the temperature/density scale, effectively frozen in the infrared region (effective scale symmetry similar to the walking technicolor). If the frozen coupling is smaller than the hidden "critical coupling", then the spontaneously broken phase would be gone, resulting in the quark-gluon plasma and/or the alternative color superconductor (genuine BCS weak coupling for the Fermi surface (only for quark, not anti-quark)). Then the effective scale invariance would manifest itself.

I am retiring from KMI, Nagoya University as of March 2015. This meeting will probably be the last one of the series of the Nagoya SCGT workshops. I hope that future of the Strong Coupling Gauge Theories will be fruitful ever, no matter how the SCGT workshop might be over. LHC and lattice activity will tell us something. Thank you everybody, with my never-ending dream to toast to the old wine in a new bottle.

Acknowledgements

I would like to express my hearty thanks to all the participants in SCGT15 as well as those in the previous SCGT workshops. This work was supported by the JSPS Grant-in-Aid for Scientific Research (S) # 22224003 and (C) #23540300.

[Note added] After the SCGT15 workshop, there appeared an interesting report on the 2 TeV excess in the diboson channels [55]. We have argued [56] that this would be a most natural candidate for the walking techni-ρ as a gauge boson of the Hidden Local Symmetry described by the scale-invariant HLS model in Eq.(34) [19]. We further found [18,57] that a salient feature of this possibility is the scale symmetry which forbids the decay of the walking techni-ρ to the 125 GeV Higgs (technidilaton) plus W/Z (what we called "conformal barrier"), in sharp contrast to the popular "equivalence theorem". This applies not only to the techni-ρ but also to all the higher vector/axialvector resonances as the HLS gauge bosons, having scale-invariant mass. The LHC Run II will tell us whether or not it is the case.

References

1. K. Yamawaki, M. Bando and K. Matumoto, Phys. Rev. Lett. **56**, 1335 (1986); M. Bando, T. Morozumi, H. So and K. Yamawaki, Phys. Rev. Lett. **59**, 389 (1987).
2. M. Bando, K. Matumoto and K. Yamawaki, Phys. Lett. B **178**, 308 (1986).
3. V. A. Miransky, Nuovo Cim. A **90** (1985), 149.
4. V. A. Miransky and K. Yamawaki, Phys. Rev. D **55**, 5051 (1997); Errata, **56**, 3768 (1997).
5. T. Maskawa and H. Nakajima, Prog. Theor. Phys. **52** (1974), 1326; **54** (1975), 860.
6. Y. Nambu, Superconductor Model of Elementary Particles and its Consequencies - Talk given at a conference at Purdue (1960); Y. Nambu and G. Jona-Lasinio, Phys. Rev. **122**, 345 (1961).
7. S. Weinberg, Phys. Rev. D **13** (1976), 974; D **19** (1979), 1277; L. Susskind, Phys. Rev. D **20** (1979), 2619: See for a review of earlier literature, E. Farhi and L. Susskind, Phys. Rep. **74** (1981), 277.
8. C. N. Leung, S. T. Love and W. A. Bardeen, Nucl. Phys. B **273**, 649 (1986).
9. B. Holdom, Phys. Rev. D **24**, 1441 (1981).
10. T. Akiba and T. Yanagida, Phys. Lett. B **169** (1986), 432; T. W. Appelquist, D. Karabali and L. C. R. Wijewardhana, Phys. Rev. Lett. **57** (1986), 957; For an earlier work on this line see B. Holdom, Phys. Lett. B **150** (1985), 301.
11. S. Matsuzaki and K. Yamawaki, arXiv:1508.07688 [hep-ph], to be published in JHEP.
12. S. Matsuzaki and K. Yamawaki, Phys. Lett. B **719**, 378 (2013); Phys. Rev. D **86**, 035025 (2012); Phys. Rev. D **85**, 095020 (2012).
13. S. Matsuzaki and K. Yamawaki, Phys. Rev. D **86**, 115004 (2012).
14. S. Sakata, Prog. Theor. Phys. **16**, 686 (1956).
15. M. Gell-Mann, Phys. Lett. **8**, 214 (1964); G. Zweig, CERN-TH-401 (1964);PRINT-64-170 (1964).
16. Z. Maki, M. Nakagawa and S. Sakata, Prog. Theor. Phys. **28**, 870 (1962).
17. M. Kobayashi and T. Maskawa, Prog. Theor. Phys. **49**, 652 (1973).
18. H. S. Fukano, S. Matsuzaki, K. Terashi and K. Yamawaki, arXiv:1510.08184 [hep-ph].
19. M. Kurachi, S. Matsuzaki and K. Yamawaki, Phys. Rev. D **90**, no. 5, 055028 (2014).
20. M. Bando, T. Kugo, S. Uehara, K. Yamawaki and T. Yanagida, Phys. Rev. Lett. **54** (1985), 1215; M. Bando, T. Kugo and K. Yamawaki, Nucl. Phys. B **259** (1985), 493.
21. M. Bando, T. Kugo and K. Yamawaki, Phys. Rept. **164** (1988), 217.
22. M. Harada and K. Yamawaki, Phys. Rept. **381**, 1 (2003).
23. H. K. Lee, W. G. Paeng and M. Rho, arXiv:1504.00908 [nucl-th]; M. Rho, arXiv:1507.05173 [hep-ph], in this Proceedings; W. G. Paeng, T. T. S. Kuo,

H. K. Lee and M. Rho, arXiv:1508.05210 [hep-ph].
24. S. Matsuzaki and K. Yamawaki, Phys. Rev. Lett. **113**, no. 8, 082002 (2014)
25. V. A. Miransky, M. Tanabashi and K. Yamawaki, Phys. Lett. B **221**, 177 (1989); Mod. Phys. Lett. A **4**, 1043 (1989);
26. Y. Nambu, Chicago preprint EFI 89-08 (Feb., 1989). Earlier related discussions are given in H. Terazawa, K. Akama and Y. Chikashige, Phys. Rev. D **15**, 480 (1977).
27. W. A. Bardeen, C. T. Hill and M. Lindner, Phys. Rev. D **41**, 1647 (1990).
28. S. Shuto, M. Tanabashi and K. Yamawaki, in *Proc. 1989 Workshop on Dynamical Symmetry Breaking*, Dec. 21-23, 1989, Nagoya, eds. T. Muta and K. Yamawaki (Nagoya Univ., Nagoya, 1990) 115-123; M. S. Carena and C. E. M. Wagner, Phys. Lett. B **285**, 277 (1992); M. Hashimoto, Phys. Lett. B **441**, 389 (1998).
29. B. A. Dobrescu and C. T. Hill, Phys. Rev. Lett. **81**, 2634 (1998); R. S. Chivukula, B. A. Dobrescu, H. Georgi and C. T. Hill, Phys. Rev. D **59**, 075003 (1999)
30. H. S. Fukano, M. Kurachi, S. Matsuzaki and K. Yamawaki, Phys. Rev. D **90**, no. 5, 055009 (2014); H. C. Cheng, B. A. Dobrescu and J. Gu, JHEP **1408**, 095 (2014).
31. H. S. Fukano, arXiv:1507.08003 [hep-ph]. to appear in the SCGT15 Proceedings.
32. Y. Kikukawa and K. Yamawaki, Phys. Lett. B **234**, 497 (1990).
33. K. i. Kondo, S. Shuto and K. Yamawaki, Mod. Phys. Lett. A **6**, 3385 (1991); N. V. Krasnikov, Mod. Phys. Lett. A **8**, 797 (1993); K. i. Kondo, M. Tanabashi and K. Yamawaki, Prog. Theor. Phys. **89**, 1249 (1993); M. Harada, Y. Kikukawa, T. Kugo and H. Nakano, Prog. Theor. Phys. **92**, 1161 (1994).
34. N. Arkani-Hamed, A. G. Cohen and H. Georgi, Phys. Rev. Lett. **86**, 4757 (2001); C. T. Hill, S. Pokorski and J. Wang, Phys. Rev. D **64**, 105005 (2001).
35. D. T. Son and M. A. Stephanov, Phys. Rev. D **69**, 065020 (2004); T. Sakai and S. Sugimoto, Prog. Theor. Phys. **113**, 843 (2005).
36. S. Matsuzaki, arXiv:1510.04575 [hep-ph], to be published in the SCGT15 Proceedings.
37. W. E. Caswell, Phys. Rev. Lett. **33**, 244 (1974); T. Banks and A. Zaks, Nucl. Phys. B **196**, 189 (1982).
38. T. Appelquist, J. Terning and L. C. Wijewardhana, Phys. Rev. Lett. **77**, 1214 (1996) T. Appelquist, A. Ratnaweera, J. Terning and L. C. Wijewardhana, Phys. Rev. D **58**, 105017 (1998).
39. K. Yamawaki, Prog. Theor. Phys. Suppl. **167**, 127 (2007).
40. M. Hashimoto and K. Yamawaki, Phys. Rev. D **83**, 015008 (2011).
41. D. B. Kaplan, J. W. Lee, D. T. Son and M. A. Stephanov, Phys. Rev. D **80**, 125005 (2009).
42. S. Dimopoulos and L. Susskind, Nucl. Phys. B **155**, 237 (1979); E. Eichten and K. D. Lane, Phys. Lett. B **90**, 125 (1980).

43. K. Yamawaki and T. Yokota, Phys. Lett. B **113** (1982), 293; Nucl. Phys. B **223** (1983), 144.
44. V. P. Gusynin and V. A. Miransky, Phys. Lett. B **198**, 79 (1987) [Ukr. Fiz. Zh. (Russ. Ed.) **33**, 485 (1988)].
45. S. Dimopoulos, Nucl. Phys. B **168**, 69 (1980); E. Farhi and L. Susskind, Phys. Rept. **74**, 277 (1981).
46. M. Kurachi, S. Matsuzaki and K. Yamawaki, Phys. Rev. D **90**, no. 9, 095013 (2014), and references cited therein.
47. M. Kurachi, R. Shrock and K. Yamawaki, Phys. Rev. D **91**, no. 5, 055032 (2015).
48. Y. Aoki, T. Aoyama, E. Bennett, M. Kurachi, T. Maskawa, K. Miura, K. -i. Nagai, H. Ohki and E. Rinaldi, A. Shibata, K. Yamawaki and T. Yamazaki (the LatKMI Collaboration), arXiv:1510.05863 [hep-lat]; arXiv:1510.07373 [hep-lat]. To be published in SCGT15 Proceedings.
49. Y. Aoki, T. Aoyama, M. Kurachi, T. Maskawa, K. -i. Nagai, H. Ohki, A. Shibata, K. Yamawaki and T. Yamazaki (the LatKMI Collaboration), Phys. Rev. D **86**, 054506 (2012).
50. Y. Aoki, T. Aoyama, M. Kurachi, T. Maskawa, K. -i. Nagai, H. Ohki, A. Shibata, K. Yamawaki and T. Yamazaki (the LatKMI Collaboration), Phys. Rev. D **87**, no. 9, 094511 (2013).
51. T. Appelquist *et al.* [LSD Collaboration], Phys. Rev. D **90**, 114502 (2014); A. Hasenfratz, D. Schaich and A. Veernala, arXiv:1410.5886 [hep-lat].
52. Y. Aoki, T. Aoyama, M. Kurachi, T. Maskawa, K. -i. Nagai, H. Ohki and E. Rinaldi, A. Shibata, K. Yamawaki and T. Yamazaki (the LatKMI Collaboration), Phys. Rev. Lett. **111**, no. 16, 162001 (2013).
53. Y. Aoki, T. Aoyama, M. Kurachi, T. Maskawa, K. Miura, K. -i. Nagai, H. Ohki and E. Rinaldi, A. Shibata, K. Yamawaki and T. Yamazaki (the LatKMI Collaboration), Phys. Rev. D **89**, 111502(R) (2014); PoS LATTICE 2013 (2013) 070.
54. Z. Fodor, K. Holland, J. Kuti, D. Nogradi and C. H. Wong, PoS LATTICE **2013**, 062 (2014); R. Brower, A. Hasenfratz, C. Rebbi, E. Weinberg and O. Witzel, arXiv:1411.3243 [hep-lat].
55. G. Aad *et al.* [ATLAS Collaboration], arXiv:1506.00962 [hep-ex]. See also V. Khachatryan *et al.* [CMS Collaboration], JHEP **1408**, 173 (2014).
56. H. S. Fukano, M. Kurachi, S. Matsuzaki, K. Terashi and K. Yamawaki, Phys. Lett. B **750**, 259 (2015).
57. H. S. Fukano, S. Matsuzaki and K. Yamawaki, arXiv:1507.03428 [hep-ph].

Summary Talk — SCGT15

Howard Georgi

Center for the Fundamental Laws of Nature,
The Physics Laboratories,
Harvard University, USA
E-mail: hgeorgi@fas.harvard.edu

In this brief talk, I will try to focus on the things that happened at the conference that seemed very important, but that I didn't understand.

Keywords: Conformal, Strong-couple gauge theories, Higgs, Dilaton, Pseudo-Goldstone.

1. Introduction

I am honored to give the summary talk at this historic conference.[1,2] Of course, it is impossible to summarize everything in all of the interesting talks that we have heard in the last four days. But I will try to highlight a couple of major themes. I should say from the beginning that there was a lot of beautiful work presented at the conference that I could not summarize even if I had lots of time, because I don't understand it. I should say from the beginning that there was a lot of beautiful work presented at the conference that I could not summarize even if I had lots of time, because I don't understand it. My favorite comment related to my ignorance

$$V = \frac{e^{-2(g+2(p+x))}}{128}\left[16\left(a^4 + 2\left(\left(e^g - e^{6p+2x}\right)^2 - 1\right)a^2 + e^{4g} - 4e^{g+6p+2x}\left(1+e^{2g}\right) + 1\right) + e^{12p+2x+\phi}\left(2e^{2g}(a-b)^2 + e^{4g} + \left(a^2 - 2ba + 1\right)^2\right)N_c^2\right],$$

$$G_{ab} = \mathrm{diag}\left(\frac{1}{2}, 6, 1, \frac{1}{4}, \frac{e^{-2g}}{2}, \frac{e^{-2x+\phi}N_c^2}{32}\right)$$

Fig. 1. Daniel Elander's complicated slide.

came from Daniel Elander (see figure 1) "If I had had a choice, I would not have made it this complicated." I have felt that way frequently in the last four days.

There are also some things that I won't discuss because I don't understand why I should care. For example, I won't mention quadratic divergences (as in figure 2). I don't care about them because continuum quantum field theory doesn't care about them either. I will come back to this later when I talk about Francesco's fascinating contribution.

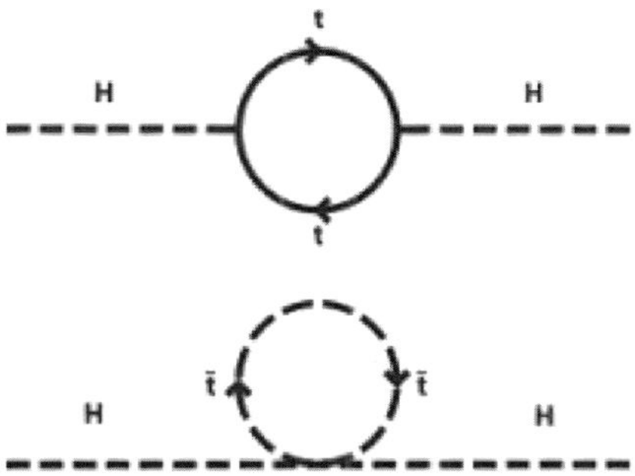

Fig. 2. The classic SUSY quadratic divergence diagram which I don't understand at all.

What I will do is to try to fit what I have heard into my own simple-minded language of perturbative effective field theory (PEFT), and to suggest when this is not possible, and why.[3–7]

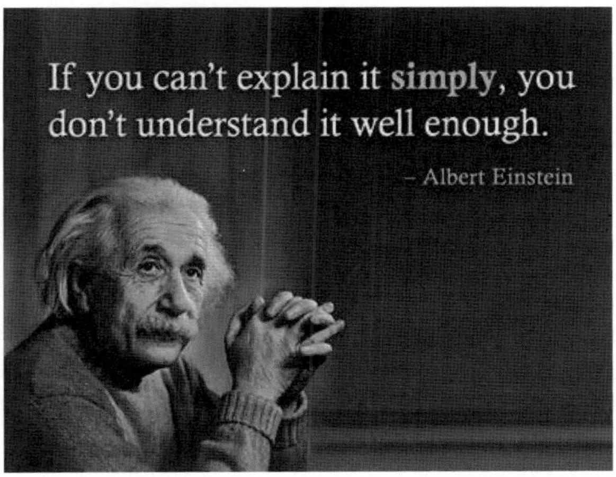

Fig. 3. Words to live by!

In PEFT one tries to hide from the hard problems of strong dynamics by choosing different degrees of freedom that are appropriate at different scales.

The classic example is chiral perturbation theory, cartooned in figure 4, which

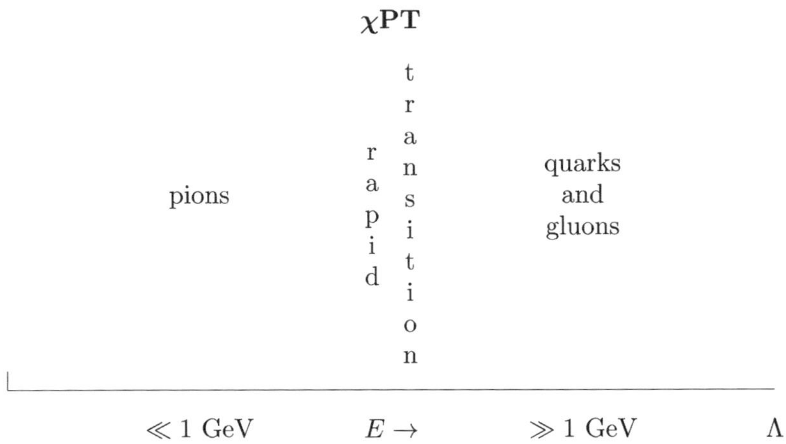

Fig. 4. Chiral perturbation theory — the classic PETF

allows us to parameterize the physics of pions at low energies in terms of a manageable number of parameters.[8] The QCD gauge theory of quarks and gluons is, we think, an accurate description of their interactions up to some scale very large compared to 1 GeV (Liz & Sekhar's colorons?). And below 1 GeV it is not wrong — **but it is useless.** Fortunately, we can make progress by focusing on the right degrees of freedom.

The pseudo-Goldstone Bosons are parametrically light, and we can describe their low energy interactions accurately. χPT works because the transition between the high-energy and low-energy descriptions of QCD is rather abrupt. This is another way of saying that the number of parameters in the χPT is manageable. The effective scale of the higher dimension operators in χPT is relatively large. χPT is my model for understanding complicated physics. When I am trying to understand something, I try to fit it into this paradigm, and if I can't I get confused, and try to understand why I can't. With that introduction, let me review the complex of ideas that is the major subject of the conference. I hope to give a thought-provoking but also a short talk so that we have time to discuss!

In the effective theory that we call the standard model, we assume that we know the physics up to some very high scale, Λ. Everything is perturbative at low energies. We know the source of EWSB so we know exactly what the Goldstone bosons and the Higgs are. Of course we know that this works beautifully. Furthermore, the only sources of scale violation in this theory are the scale anomaly from quantum renormalization effects (small at the 100 GeV scale) and the doublet mass parameter that is very small compared to whatever the high scale is. So there is a fairly precise sense in which the violation of scale invariance is small and the Plain Old Higgs is also a pseudo-Dilaton.[9]

In a classic technicolor theory, figure 6, there is a new QCD-like strong interac-

Plain Old Higgs

standard
model
+Higgs

100 GeV $E \to$ Λ

Fig. 5. The standard model PEFT

Plain Old Technicolor

standard model w/o Higgs	new strong QCD-like TC	perturbative ETC from unknown physics	
100 GeV	Λ_{TC}	Λ_{ETC}	Λ

Fig. 6. The classic technicolor theory

tion with U and D techniquarks analgous to the u and d quarks of QCD, but with the chiral symmetry breaking scale scaled up to about 1 TeV. This is a beautiful dynamical way of giving mass to the W and Z. In Plain Old Technicolor,[10–12] there is no Higgs and there is no pseudo-Dilaton. Let me be more precise about what I mean. What I mean by a Higgs (and I will defend this definition to my last breath) is an object that is part of an electroweak $SU(2) \times U(1)$ doublet with the longitudinal components of the W and Z, and therefore plays the dominant role of canceling the bad high-energy behavior of WW scattering. In Plain Old Technicolor, the longitudinal components of the W and Z are the technipions. They are well-described at the W, Z scale by an effective chiral theory, and, analogous to pions in χPT, the bad high-energy behavior in their scattering is controlled by the contribution of techni-resonances like the techni-rho. Plain Old Technicolor also does not have a pseudo-Dilaton. Except for the Goldstone bosons, the resonances have masses (and/or widths) of the order of the technicolor scale. At this this scale, where the technicolor interaction is strong and scale invariance is very badly broken. There is no reason for a techni-Dilaton to be lighter than anthing else, and no sense in talking about such an object. Plain Old Technicolor also doesn't give the right structure of the W, Z masses at the level of the radiative corrections. We know this because it is QCD like and we can just scale up parameters from χPT. We also know a lot about these things from detailed studies on the lattice. But nobody should be too upset about that, because the ETC sector is the really ugly part of Plain Old Technicolor. Robert Shrock[13] reminded us of a very old but very beautiful complex of ideas — Extended technicolor[14,15] - self-breaking of chiral gauge theories — but it is never enough. My personal belief is that nobody has ever constructed a completely realistic model without explicit scalars.

Composite Higgs

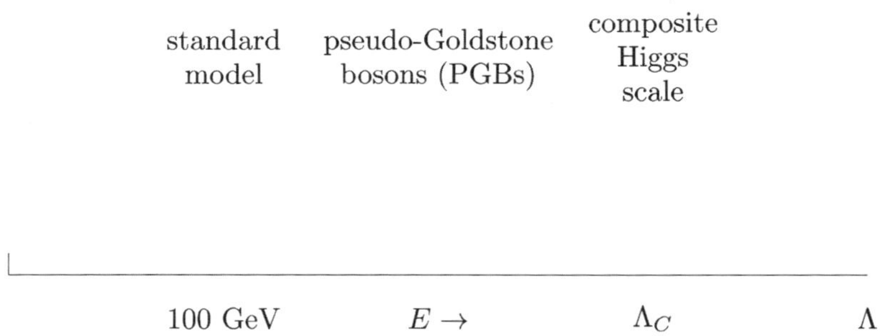

Fig. 7. The classic composite Higgs theory

The composite Higgs idea, figure 7, originally developed out of χPT. You can build a Higgs doublet out of PGBs produced by some strong interaction that gets strong at Λ_C above a TeV.[16] To generate a Higgs potential you have to break the global symmetries and finely tune the parameters to keep the mass of the Higgs doublet well below Λ_C (which you typically have to do to satisfy electroweak constraints). There are clever ideas like "little higgs" to make the required tuning less severe.[17]

In principle, a composite Higgs need not be a PGB. But there are two good reasons why most composite Higgs model building has gone the PGB route. The first reason is simply that we don't know enough about strong interactions. If we can argue that a symmetry is spontaneously broken, we know that spinless PGBs are present, independent of the details of the strong dynamics that we typically don't understand. Once you have spinless particles, you can start model building and trying to make one look like a Higgs. But if you don't even know that your strong dynamics produces scalar bound states at all, it is hard to get started and you may just be whistling in the dark. And even if somehow you know that have composite scalars, either you or your theory at Λ_C still has to somehow make a scalar doublet that is much lighter than Λ_C. Otherwise, there is no reason for such a model to give anything like the standard model. And if you do succeed in making your composite Higgs light, then it is just a Higgs! So like the Plain Old Higgs, a composite Higgs is both a Higgs and a techni-Dilaton.

You have all been very patient with my summary of the last 40 years of particle theory, and finally, I am getting to the main subject of many (if not most) of the talks at this conference. What happens as our technicolor theory approaches walking/conformality?[18] The naive picture that most of us naturally take as our starting point is illustrated in figure 8. Now I want to claim that in this case, the PEFT picture is very misleading.

Walking/nearly conformal/... Technicolor

standard model + scalar?	new nearly conformal TC	perturbative ETC from unknown physics	
100 GeV	Λ_{TC}	Λ_{ETC}	Λ

Fig. 8. Naive walking/nearly conformal Technicolor

From here on, I will be speculating. But I hope that these speculations will be thought provoking, and will help us make sense of some of the amazing amount of work that has gone into studying theories of this kind, as exemplified by many of the beautiful talks we have heard at this conference.[19,20] It may also be that these speculations are so obvious to everyone else, that they are not worth discussing. But I found them useful for my own thinking, so I am going to subject all of you to them. Walking/Conformal TC is in many ways just the opposite of Plain Old Technicolor and QCD. Rather than having a sharp transition from a high-energy theory to a low-energy theory, WCTC spreads the transition out over the entire region between the ETC scale and the electroweak scale, as suggested in figure 9. Given the importance of a sharp transition from the high scale to the low scale for the success of χPT, this should make you worry a lot about the applicability of a conventional low energy EFT at the weak scale in the WCTC scenario. There is another way of arriving at the same conclusion. We have heard a lot about light dilatons in WCTC theories. It is not clear to me whether a light dilaton is an inescapable consequence of this picture. Scale invariance is tricky and this might be model dependent. But what is clear is that there must be some small dynamical scales in WCTC. Otherwise, the dynamics could not be spread even appoximately uniformly between Λ_{TC} and Λ_{ETC}. For this you need small energy splittings.

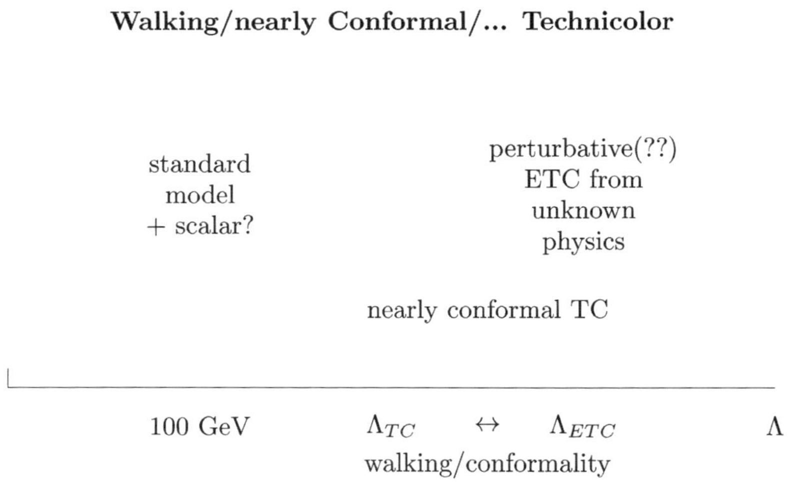

Fig. 9. A more suggesting picture of walking/nearly conformal Technicolor

Approximate walking/conformal invariance is a very strong constraint. There may not be small masses, **but there must be small mass scales.** The dilaton mass, or whatever the small dynamical scale is in your favorite WCTC theory is not at all like the Higgs mass. The Higgs self-coupling can give us a light Higgs independent of the scale of symmetry breaking. The small scale of WCTC is built into the theory in a much more dynamical way and I believe that it will surely appear in the denominator of any effective field theory description of the physics of WCTC. Thus my tentative conclusion is that one reason I have found all this so confusing is that it is impossible (or at least not useful) to fit WCTC into a perturbative EFT at low energies in the trivial way that usually pops into my head. A different description may be required.

The same, I think will be true of the interesting theories that Francesco described for us (see figure 10).[21] The big question here is what to say about scalar masses.

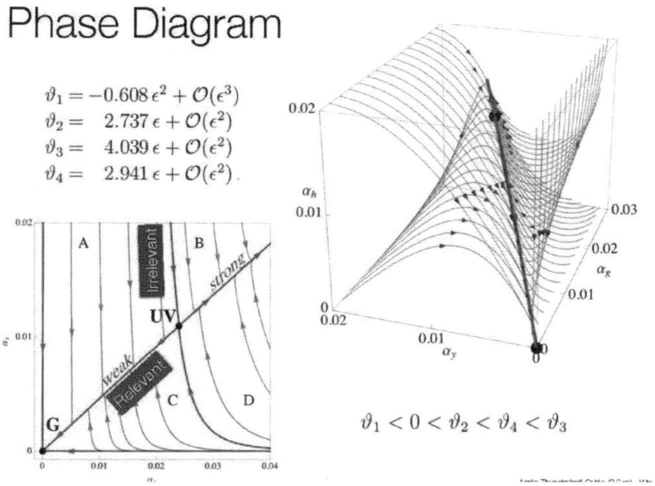

Fig. 10.

As I mentioned above, I agree with Francesco that the naive treatment of quadratic divergences in the RG is not appropriate. But the issue was never about quadratic divergences. The question is whether these new theories help us with the issues of fine tuning at the large scale. This would be great, but it is not obvious.

Returning to WCTC, I don't really have a conclusion but I have a question and a challenge. I don't think know what the appropriate description of the low energy physics in such theories is. What is it? How do the small mass scales that must exist in the WCTC appear in the low energy physics? Maybe this is what holography is trying to get for us, if I could only understand and believe it. But I hope that knowing the nature of the problem will help, even if I don't know the solution.

Let me close by thanking the organizers, particularly Koichi Yamawaki, for this amazing conference. I say thanks in spite of the fact that I have never been so tired at the end of a conference before. Koichi has worked us very hard. But one thing you know about Koichi is that no matter how hard he works you, he is working even harder himself. I am sure that this will continue even after as he moves on to a new phase of his career. A couple of hundred years ago, the English poet and print maker William Blake wrote a wonderful poem called "The Marriage of Heaven and Hell" and there is a line in that poem that I would like to recall as we thank Koichi not just for this conference, but for many many years of physics.

Energy is Eternal Delight!

References

1. V. I. Zakharov, Some nondynamical relations in SCGT, in *In *Nagoya 1990, Proceedings, Strong coupling gauge theories and beyond* 407-414. (see HIGH ENERGY PHYSICS INDEX 29 (1991) No. 12260)*, 1990.
2. J. Nishimura and K. Yamawaki (eds.), *Perspectives of strong coupling gauge theories. Proceedings, International Workshop, SCGT'96, Nagoya, Japan, November 13-16, 1996* 1997.
3. A. G. Cohen, Selected topics in effective field theories for particle physics, in *Theoretical Advanced Study Institute (TASI 93) in Elementary Particle Physics: The Building Blocks of Creation - From Microfermius to Megaparsecs Boulder, Colorado, June 6-July 2, 1993*, 1993.
4. H. Georgi, Effective field theory, *Ann. Rev. Nucl. Part. Sci.* **43**, 209 (1993).
5. M. E. Luke and A. V. Manohar, Bound states and power counting in effective field theories, *Phys. Rev.* **D55**, 4129 (1997).
6. D. B. Kaplan, Five lectures on effective field theory2005.
7. I. Z. Rothstein, TASI lectures on effective field theories2003.
8. H. Leutwyler, Chiral dynamics (2000).
9. W. D. Goldberger, B. Grinstein and W. Skiba, Distinguishing the Higgs boson from the dilaton at the Large Hadron Collider, *Phys. Rev. Lett.* **100**, p. 111802 (2008).
10. L. Susskind, Dynamics of Spontaneous Symmetry Breaking in the Weinberg-Salam Theory, *Phys. Rev.* **D20**, 2619 (1979).
11. S. Weinberg, Implications of Dynamical Symmetry Breaking, *Phys. Rev.* **D13**, 974 (1976).
12. S. Weinberg, Implications of Dynamical Symmetry Breaking: An Addendum, *Phys. Rev.* **D19**, 1277 (1979).
13. M. Kurachi, R. Shrock and K. Yamawaki, Determination of $SU(4)_{TC}$ Technicolor Gauge Group from Embedding in Extended Technicolor, *Phys. Rev.* **D91**, p. 055032 (2015).
14. S. Dimopoulos and L. Susskind, Mass Without Scalars, *Nucl. Phys.* **B155**, 237 (1979).
15. E. Eichten and K. D. Lane, Dynamical Breaking of Weak Interaction Symmetries, *Phys. Lett.* **B90**, 125 (1980).
16. R. Contino, The Higgs as a Composite Nambu-Goldstone Boson, in *Physics of the large and the small, TASI 09, proceedings of the Theoretical Advanced Study Institute in Elementary Particle Physics, Boulder, Colorado, USA, 1-26 June 2009*, 2011.
17. M. Schmaltz, Little Higgs goes to TASI, in *Physics in D ¿= 4. Proceedings, Theoretical Advanced Study Institute in elementary particle physics, TASI 2004, Boulder, USA, June 6-July 2, 2004*, 2004.
18. B. Holdom, Raising the Sideways Scale, *Phys. Rev.* **D24**, p. 1441 (1981).
19. Z. Fodor, K. Holland, J. Kuti, S. Mondal, D. Nogradi and C. H. Wong, The

running coupling of the minimal sextet composite Higgs model, *JHEP* **09**, p. 039 (2015).
20. V. Drach, M. Hansen, A. Hietanen, C. Pica and F. Sannino, Conformal symmetry vs. chiral symmetry breaking in the SU(3) sextet model2015.
21. D. F. Litim and F. Sannino, Asymptotic safety guaranteed, *JHEP* **12**, p. 178 (2014).

Deviation of Yukawa Coupling and Higgs Decay in Gauge-Higgs Unification

Y. Adachi

Department of Sciences, Matsue College of Technology,
Matsue 690-8518, Japan
E-mail: y-adachi@matuse-ct.ac.jp

N. Maru

Department of Mathematics and Physics, Osaka City University,
Osaka 558-8585, Japan
E-mail: nmaru@sci.osaka-cu.ac.jp

We study the deviation of yukawa coupling in the gauge-Higgs unification scenario from the Standard Model one. Applying the obtained results to the tau and bottom yukawa couplings, we numerically calculate the signal strength of $gg \to H \to b\bar{b}, \tau\bar{\tau}$ in the gauge-Higgs unification.

Keywords: Gauge-Higgs unification.

1. Introduction

Gauge-Higgs unification (GHU)[1] is one of the attractive scenarios beyond the Standard Model(SM), which provides a possible solution to the hierarchy problem without supersymmetry[2]. In this scenario, the SM Higgs boson and the gauge fields are unified into the higher dimensional gauge fields, *i.e.* Higgs boson is identified with extra spatial components of higher dimensional gauge fields. Due to this, the quantum correction to Higgs mass is UV-finite and calculable due to the higher dimensional gauge symmetry though the theory is the non-renormalizable[3,4]. The finiteness of other physical observables have been investigated[5–7].

The fact that the yukawa couplings are governed by the gauge principle may deviate from the SM one[8,9]. Actually, in the flat extra dimensional case, the ratio of yukawa coupling of GHU and the SM one is derived as

$$\frac{f_{\rm GHU}}{f_{\rm SM}} \simeq \frac{g_4}{2} v\pi R \cot\left(\frac{g_4}{2} v\pi R\right). \tag{1}$$

It is because the Higgs fields $A_y^{(0)}$ appears the Wilson line phase which is defined by

$$W = P\exp\left[i\frac{g}{2}\oint_{S^1} A_y {\rm d}y\right] = \exp\left[ig_4\pi R A_y^{(0)}\right], \tag{2}$$

where g, g_4 are 5D and 4D gauge couplings, respectively. Namely, W is periodic with respect to A_y under $A_y \to A_y + 2/(gR)$ so that the yukawa couplings $f_{\rm GHU}$ in this scenario becomes periodic as shown in eq. (1).

Although the above result is general feature in such a scenario, but their model[8] is not realistic. A crucial point is that the brane mass terms of the fermions are not included. They are necessary for generating the flavor mixing[10] and removing the exotic massless fermions absent in the SM. Therefore, it is important to study the deviation of Yukawa coupling in a realistic model to incorporate the appropriate brane mass terms.

2. Deviation of yukawa couplings in gauge-Higgs unification

We consider an $SU(3) \times U(1)'$ GHU model in a five-dimensional flat space-time compactified on S^1/Z_2 with the radius R of S^1. The up-type quarks except for the top quark, the down-type quarks and the charged leptons are embedded into **3** and $\overline{\mathbf{6}}$ representations of $SU(3)$, respectively[11]. In order to realize the large top Yukawa coupling, the top quark is embedded into $\overline{\mathbf{15}}$ representation of $SU(3)$[12].

The boundary conditions are assigned to reproduce the SM fields as the zero modes. A periodic boundary condition with respect to S^1 is taken for all of the bulk SM fields, and the Z_2 parity is assigned for the gauge fields and fermions in the representation $\mathcal{R}$ by using the parity matrix $P = \mathrm{diag}(-,-,+)$ in the following.

$$A_\mu(-y) = P^\dagger A_\mu(y) P, \quad A_y(-y) = -P^\dagger A_y(y) P, \quad \psi(-y) = \mathcal{R}(P)\gamma^5\psi(y) \tag{3}$$

where the subscripts μ (y) denotes the four (the fifth) dimensional component. With this choice of parities, zero-mode vector bosons in the model are only the SM gauge fields.

This parity assignment also leaves exotic massless fermions which is not included in the SM. Such exotic fermions are made massive by introducing brane localized fermions with conjugate $SU(2) \times U(1)$ charges and an opposite chirality to the exotic fermions, allowing us to write brane-localized Dirac mass terms. These brane localized mass terms are also very important to generate the flavor mixing in the context of GHU[10].

In this context, the deviation of the tau and the bottom yukawa couplings in GHU from the SM one has been obtained as[13]

$$\begin{aligned}\frac{f_{\rm GHU}}{f_{\rm SM}} =& \frac{M^2 - m^2_{\tau(b)}}{M^2 - \pi R m^2_{\tau(b)}\sqrt{M^2 - m^2_{\tau(b)}}\coth(\pi R\sqrt{M^2 - m^2_{\tau(b)}})} \\ &\times \pi R M_W \frac{\sin(2\pi R M_W) - \left[\sin(2\pi R M_W) - \sqrt{2}\sin\left(2\sqrt{2}\pi R M_W\right)\right]\sin^2\theta}{1 - \cos(2\pi R M_W) - \left[\cos(2\sqrt{2}\pi R M_W) - \cos(2\pi R M_W)\right]\sin^2\theta}.\end{aligned} \tag{4}$$

The parameter θ is a mixing angle between two $SU(2)$ doublet zero modes existing per generation[10]. We can easily show that they reduce to the eq. (1) or the result in the previous work[8] for the case $\theta = 0$.

3. Calculation of the signal strength of $gg \to H \to b\bar{b}, \tau\bar{\tau}$

In this section, we calculate the signal strength of $gg \to H \to b\bar{b}, \tau\bar{\tau}$. The Higgs production is dominated by the gluon fusion process at the LHC and calculated from the coefficient of the following dimension five operator between the Higgs and the digluon,

$$\mathcal{L}_{\text{eff}} = C_g H G^a_{\mu\nu} G^{a\mu\nu} \tag{5}$$

where $G^a_{\mu\nu}$ $(a = 1 - 8)$ is the gluon field strength.

In GHU,we have to take into account the KK top loop contributions, which is found to be

$$C_g^{\text{KK}} = F(m_1) \times \frac{1}{2} \times 2 + F(m_2) \times 1 \times 2 + F(m_3) \times \frac{3}{2} \times 1 + F(m_4) \times \frac{4}{2} \times 1 \tag{6}$$

where the first factor behind $F(m_a)$ denotes the ratio for the top yukawa coupling and the second factor is a multiplicity of the same KK mass spectrum.

$$F(m_a) \equiv -\frac{1}{16\pi}\frac{m_t}{v}\alpha_s \sum_{n=1}^{\infty}\left[\frac{1}{m_{a+}^{(n)}} F_{1/2}\left(\left(\frac{2m_{a+}^{(n)}}{m_h}\right)^2\right) - (+ \to -)\right], \tag{7}$$

$m_t = 2M_W$ and $(m_{a\pm}^{(n)})^2 = (n/R \pm aM_W)^2$.

We perform a numerical calculation of the deviation of the yukawa couplings and Higgs production $C_g^{\text{KK}}/C_g^{\text{SM}}$. Combining them, the signal strength μ of the process $gg \to H \to b\bar{b}, \tau\bar{\tau}$

$$\mu = \left|\frac{C_g^{\text{KK}}}{C_g^{\text{SM}}}\right|^2 \times \left|\frac{f_{\text{GHU}}}{f_{\text{SM}}}\right|^2 \tag{8}$$

are shown in Figure 1. Our prediction is that the signal strength is always smaller

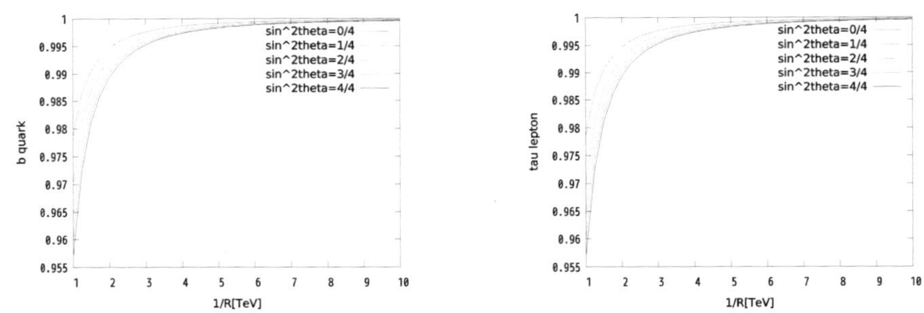

Fig. 1. The left (right) plot is the signal strength of $gg \to H \to b\bar{b}(\tau\bar{\tau})$. The horizontal line denotes the compactification scale. The results does not almost depend on the parameter θ.

than the unity, namely the process $gg \to H \to b\bar{b}, \tau\bar{\tau}$ in GHU is always suppressed comparing to the SM prediction. This is because the suppression is dominantly due to the suppression of the Higgs production via the gluon fusion[14], while the deviation of yukawa coupling is known to be very small[13].

4. Summary

In this presentation, we study the deviation of the yukawa coupling in GHU scenario from SM one and calculate the signal strength of the bottom and tau decays of the Higgs boson produced via the gluon fusion at the LHC $gg \to H \to b\bar{b}, \tau\bar{\tau}$ is suppressed. Our generic prediction is that the deviation of yukawa coupling is quite small in the realistic parameter space and the signal strength is always smaller than the unity, namely the process $gg \to H \to b\bar{b}, \tau\bar{\tau}$ in GHU is always suppressed comparing to the SM prediction. This is because the suppression is dominantly due to the suppression of the Higgs production via the gluon fusion, while the deviation of yukawa coupling is known to be very small[13].

References

1. N. S. Manton, Nucl. Phys. B **158**, 141 (1979); D. B. Fairlie, Phys. Lett. B **82**, 97 (1979), J. Phys. G **5**, L55 (1979); Y. Hosotani, Phys. Lett. B **126**, 309 (1983), Phys. Lett. B **129**, 193 (1983), Annals Phys. **190**, 233 (1989).
2. H. Hatanaka, T. Inami and C. S. Lim, Mod. Phys. Lett. A **13**, 2601 (1998).
3. I. Antoniadis, K. Benakli and M. Quiros, New J. Phys. **3**, 20 (2001); G. von Gersdorff, N. Irges and M. Quiros, Nucl. Phys. B **635**, 127 (2002); R. Contino, Y. Nomura and A. Pomarol, Nucl. Phys. B **671**, 148 (2003); C. S. Lim, N. Maru and K. Hasegawa, J. Phys. Soc. Jap. **77**, 074101 (2008); C. S. Lim, N. Maru and T. Miura, PTEP 2015 (2015) 4, 043B02.
4. N. Maru and T. Yamashita, Nucl. Phys. B **754**, 127 (2006); Y. Hosotani, N. Maru, K. Takenaga and T. Yamashita, Prog. Theor. Phys. **118**, 1053 (2007).
5. C. S. Lim and N. Maru, Phys. Rev. D **75**, 115011 (2007).
6. N. Maru, Mod. Phys. Lett. A **23**, 2737 (2008).
7. Y. Adachi, C. S. Lim and N. Maru, Phys. Rev. D **76**, 075009 (2007); Phys. Rev. D **79**, 075018 (2009); Phys. Rev. D **80**, 055025 (2009).
8. K. Hasegawa, N. Kurahashi, C. S. Lim and K. Tanabe, Phys. Rev. D **87**, 016011 (2013).
9. Y. Hosotani and Y. Kobayashi, Phys. Lett. B **674**, 192 (2009); Y. Hosotani, P. Ko and M. Tanaka, Phys. Lett. B **680**, 179 (2009).
10. Y. Adachi, N. Kurahashi, C. S. Lim and N. Maru, JHEP **1011**, 150 (2010); JHEP **1201**, 047 (2012); Y. Adachi, N. Kurahashi, N. Maru and K. Tanabe, Phys. Rev. D **85**, 096001 (2012); Y. Adachi, N. Kurahashi and N. Maru, arXiv:1404.4281 [hep-ph].
11. C. A. Scrucca, M. Serone and L. Silvestrini, Nucl. Phys. B **669**, 128 (2003).
12. G. Cacciapaglia, C. Csaki and S. C. Park, JHEP **0603**, 099 (2006).
13. Y. Adachi and N. Maru, arXiv:1501.4281 [hep-ph].
14. N. Maru and N. Okada, Phys. Rev. D **77**, 055010 (2008); Phys. Rev. D **87**, no. 9, 095019 (2013); Phys. Rev. D **88**, no. 3, 037701 (2013); arXiv:1310.3348 [hep-ph].

Emergent Geometries from Strong Coupling Gauge Theories

Yuhma Asano[1,2)*], Goro Ishiki[3)] and Shinji Shimasaki[4)]

[1)] *Department of Physics, Kyoto University, Kyoto, 606-8502, Japan*
[2)] *School of Theoretical Physics, Dublin Institute for Advanced Studies, 10 Burlington Road, Dublin 4, Ireland*
[3)] *Graduate School of Pure and Applied Sciences, University of Tsukuba, Tsukuba, Ibaraki 305-8571, Japan*
[4)] *Theory Center, High Energy Accelerator Research Organization (KEK), Ibaraki 305-0801, Japan*
* *E-mail: yuhma@stp.dias.ie*

We find a new evidence of a gauge/gravity duality: an explicit correspondence between eigenvalue densities and geometries. We examine the duality between IIA supergravity solutions and $SU(2|4)$ symmetric gauge theories, which are BMN matrix model, $\mathcal{N}=8$ super Yang-Mills theory on $R\times S^2$ and $\mathcal{N}=4$ super Yang-Mills theory on $R\times S^3/Z_k$. We show that the corresponding geometries are realized as the solution to the saddle point equations in the gauge theories, and that typical scales of the geometries are characterized by the range of the eigenvalue distribution.

Keywords: Emergent geometry; Matrix theory; Gauge/gravity; Localization.

1. Introduction and Summary

For many years, string theories have been considered to be somehow described by certain large-N gauge theories[1]. In this duality the strong coupling regimes in the gauge theories are mapped to the regimes where the semi-classical approximation is valid. Degrees of freedom of strings are considered to be described by field configurations in the corresponding gauge theory. The background geometries in string theories are also expected to emerge as typical configurations in the strong coupling limit of the gauge theories (see for example Ref. 2).

As an example of gauge/gravity dualities, we analyze the duality for $SU(2|4)$ symmetric gauge theories[4], which consists of BMN matrix model[5], $\mathcal{N}=8$ super Yang-Mills theory (SYM) on $R\times S^2$ and $\mathcal{N}=4$ SYM on $R\times S^3/Z_k$. These theories have a lot of discrete vacua which preserve sixteen supersymmetries. For each vacuum, the corresponding background geometry on the string theory side was constructed in Refs. 3 and 4. The geometries are written purely in terms of solutions to the Laplace equation in the three-dimensional axially symmetric system with some conducting disks and are labeled by the disk configurations as well as the background potential. The disk configuration specifies the corresponding vacuum on the gauge theory side and the identification of the D-brane charges and the electrostatic quantity such as the total charges on the disks have been obtained.

Recently, we found[6,7] that these background geometries emerge as eigenvalue distributions of adjoint scaler fields in strong coupling limits on the gauge theory side. In obtaining the eigenvalue distributions on the gauge theory side, we applied the localization technique to the gauge theories[8], which makes exact calculations possible for a BPS sector even in the strong coupling limit. The procedure of our computation consists of the following steps: (I) Identify the appropriate BPS sector. (II) Perform the localization computation. (III) Derive the saddle point equation of the partition function. (IV) Compare the saddle point equation with the equation of motion satisfied by the charge densities on the gravity side. Through this procedure, we verified that the eigenvalue distributions agree with the charge densities on the disk on the gravity side, which completely determines the background geometry. We also found that typical scales of the geometry are exactly reproduced as ranges of the eigenvalues. This result gives not only a new evidence of the gauge/gravity duality but also a concrete example of emergent geometry.

2. Gravity Side

The corresponding geometry also has $SU(2|4)$ symmetry, whose bosonic subgroup is $R\times SO(3)\times SO(6)$, so that the geometry locally contains $R\times S^2\times S^5$. By solving the Killing spinor equation, one finds that the metric and fluxes are written by a single function[4] $V(r,z)$, which satisfies the Laplace equation: $r^{-1}\partial_r(r\partial_r V)+\partial_z^2 V=0$. Here, r and z are the coordinates for directions perpendicular to $R\times S^2\times S^5$.

The positivity and regularity of the metric impose some conditions on $V(r,z)$. The positivity restricts the asymptotic behavior of $V(r,z)$. For the dual geometry of BMN matrix model, the space is restricted to $z>0$, and from the positivity of the metric, the background part of $V(r,z)$ is determined as $r^2z-\frac{2}{3}z^3$ multiplied by a positive constant V_0. The regularity at the boundary where the radius of S^2 or S^5 shrinks requires either boundary condition, (i) $\partial_r V\sim r$ at $r=0$ or (ii) $z=$ constant if $\partial_r V=0$. Hence, the problem of finding the form of $V(r,z)$ is reduced to solving the Laplace equation in the axisymmetric electrostatic system with conducting disks in the background potential $V_{\text{b.g.}}(r,z)=V_0\left(r^2z-\frac{2}{3}z^3\right)$. The geometries for the other gauge theories are described in a similar way.

The z-coordinate and charge of the s-th disk denoted by d_s and Q_s, respectively, are quantized since they are related with the brane charges. The explicit relations are $d_s-d_{s-1}=\frac{\pi}{2}N_5^{(s)}$ and $Q_s=\frac{\pi^2}{8}N_2^{(s)}$, where $N_5^{(s)}$ and $N_2^{(s)}$ are NS5-brane and D2-brane charges, respectively. Therefore, the geometries are labeled by those integers as well as the parameter V_0.

The potential $V(r,z)$ generally takes the form, $V(r,z)=V_{\text{b.g.}}+\sum_{s=1}^{\Lambda}\phi_s$, where Λ is the number of disks and ϕ_s is the potential produced by the s-th disk. ϕ_s is determined by a charge density of the s-th disk $f_s(u)$ satisfying $Q_s=\frac{1}{\pi}\int_{-R_s}^{R_s}du\,f_s(u)$, where R_s is the radius of the s-th disk. The radius of a disk is related to the S^5 radius at the tip of the disk as $R_s=R_{S^5}^{(s)2}/4$. In order to make the geometry well-defined, one should impose[9] $f_s(\pm R_s)=0$. The Laplace equation leads to the following integral equations:

$$\frac{1}{\pi}\sum_{t=1}^{\Lambda}\int_{-R_t}^{R_t} du'\left[-\frac{d_s+d_t}{(d_s+d_t)^2+(u-u')^2}+\frac{|d_s-d_t|}{(d_s-d_t)^2+(u-u')^2}\right]f_t(u')$$
$$+f_s(u) = \frac{1}{\pi}\left(\Delta_s+\frac{2}{3}V_0d_s^3-2V_0d_su^2\right), \qquad (1)$$

where Δ_s is the constant value of $V(r,z)$ on the s-th disk.

3. Gauge Theory Side

Let us consider BMN matrix model[5] which is a mass-deformation of BFSS model[11]. The deformation breaks the original symmetry $R\times SO(9)$ to $R\times SO(3)\times SO(6)$. The action is written as

$$S=\frac{1}{g^2}\int d\tau\,\mathrm{Tr}\Big(\frac{1}{2}(D_\tau X_a)^2+\frac{1}{2}(D_\tau X_m)^2+\frac{1}{4}(2\epsilon_{abc}X_c-i[X_a,X_b])^2$$
$$-\frac{1}{2}[X_a,X_m]^2-\frac{1}{4}[X_m,X_n]^2+\frac{1}{2}X_mX^m+\text{fermions}\Big), \qquad (2)$$

where X_a $(a=1,2,3)$ are $SO(3)$ scalar fields and X_m $(m=4,\cdots,9)$ are $SO(6)$ scalar fields. All the fields are $SU(N)$ adjoint. D_τ is the gauge covariant derivative.

The vacua are characterized by general $SU(2)$ representations. General vacuum configuration is given by $\hat{X}_m=0$ and

$$\hat{X}_a=-2\bigoplus_{s=1}^{\Lambda}\mathbf{1}_{N_2^{(s)}}\otimes L_a^{[D_s]}, \qquad (3)$$

where $L_a^{[D_s]}$ is an $SU(2)$ generator in the spin $(D_s-1)/2$ representation, and $N_2^{(s)}$ is a multiplicity of the representation. The sum $\sum_{s=1}^{\Lambda}N_2^{(s)}D_s$ should be equal to N. According to the gauge/gravity dictionary of $SU(2|4)$ symmetric theories[4,12], the integers $N_5^{(s)}=D_s-D_{s-1}$ and $N_2^{(s)}$ are identified with the NS5-brane and D2-brane charges on the gravity side. Also, the gauge coupling is identified[6,7] with V_0 by $V_0=2/(\pi^2g^2)$.

Now, let us consider how the corresponding geometries are realized. Emergence of S^2 and S^5 is rather trivial, since the field theories have $SO(3)\times SO(6)$ symmetry. A scalar field corresponding to the non-trivial directions, r and z, perpendicular to S^2 and S^5, is found[6] as $\phi(\tau)=2(-X_3(\tau)+i(-i\sinh(\tau)X_8(\tau)+\cosh(\tau)X_9(\tau)))$. This field preserves a quarter of the supersymmetry. Insertions of operators $\mathrm{Tr}\,\phi^j$ into the path integral, where j is small enough, make small perturbations around a vacuum one considers. In contrast, without any insertion, the field $\phi(\tau)$ takes the vacuum configuration. We compute this configuration by using the localization method. The localizing locus is expressed as the same equations as (3) for $SO(3)$ scalars and $\hat{X}_9(\tau)=\bigoplus_{s=1}^{\Lambda}M_s\otimes\mathbf{1}_{D_s}/\cosh\tau$ with the other $SO(6)$ fields zero, where M_s is a constant $N_2^{(s)}\times N_2^{(s)}$ matrix. We found that the saddle point equations is given by[6,7]

$$\frac{1}{\pi}\sum_{t=1}^{\Lambda}\int_{-q_m^{(t)}}^{q_m^{(t)}} dq' \left[-\frac{D_s+D_t}{(D_s+D_t)^2+(q-q')^2}+\frac{|D_s-D_t|}{(D_s-D_t)^2+(q-q')^2}\right]\rho^{(t)}(q')$$
$$+\rho^{(s)}(q) = \frac{\mu_s}{\pi}-\frac{2D_s}{\pi g^2}q^2, \tag{4}$$

in the limit in which the classical gravity approximation is valid: large-$N_2^{(s)}$, large-$N_5^{(s)}$, and $g^2N_2^{(s)} \gg D_s$. Here, $\rho^{(s)}(q)$ is the eigenvalue density of M_s, $q_m^{(s)}$ is an upper limit of the support of $\rho^{(s)}(q)$, and μ_s is a Lagrange multiplier for $\int dq\, \rho^{(s)}(q) = N_2^{(s)}$.

The saddle point equations (4) is equivalent to (1) under the identification: $f_s(x) = \frac{\pi}{4}\rho^{(s)}(\frac{2}{\pi}x)$ and $R_s = \frac{\pi}{2}q_m^{(s)}$. Furthermore, it gives the S^5 radii as $R_{S^5}^{(s)2} = 2\pi q_m^{(s)}$. For $\Lambda = 1$, one can consider two scaling limits where the D2- or NS5-brane solution is realized[9]. In these limits, one can solve the saddle point equation and find that the relation $R_{S^5}^2 = 2\pi q_m$ gives correct values for the S^5 radius[6]. Hence, the agreement between (1) and (4) is not accidental, but the identification correctly reproduces supergravity quantities.

It is known that the other $SU(2|4)$ symmetric theories are realized as the theories around certain vacua of BMN matrix model[10]. By using this relationship, we also found the similar equivalences between eigenvalue densities and the disk charge densities for those gauge theories.

References

1. J. M. Maldacena, Adv. Theor. Math. Phys. **2** (1998) 231.
2. D. Berenstein, JHEP **0407**, 018 (2004) [hep-th/0403110].
3. H. Lin, O. Lunin and J. M. Maldacena, JHEP **0410**, 025 (2004)
4. H. Lin and J. M. Maldacena, Phys. Rev. D **74**, 084014 (2006)
5. D. Berenstein, J. M. Maldacena and H. Nastase, JHEP **0204** (2002) 013
6. Y. Asano, G. Ishiki, T. Okada and S. Shimasaki, JHEP **1405**, 075 (2014)
7. Y. Asano, G. Ishiki and S. Shimasaki, JHEP **1409**, 137 (2014)
8. Y. Asano, G. Ishiki, T. Okada and S. Shimasaki, JHEP **1302**, 148 (2013)
9. H. Ling, A. R. Mohazab, H. -H. Shieh, G. van Anders and M. Van Raamsdonk, JHEP **0610**, 018 (2006)
10. G. Ishiki, S. Shimasaki, Y. Takayama and A. Tsuchiya, JHEP **0611** (2006) 089
11. T. Banks, W. Fischler, S. H. Shenker and L. Susskind, Phys. Rev. D **55** (1997) 5112.
12. J. Maldacena, M. M. Sheikh-Jabbari and M. Van Raamsdonk, JHEP **0301** (2003) 038.

Descriptions of Inflation in Scalar Field Theories, Perfect Fluid, and $F(R)$ Gravity

Kazuharu Bamba[1,2,3]

[1] *Leading Graduate School Promotion Center, Ochanomizu University, 2-1-1 Ohtsuka, Bunkyo-ku, Tokyo 112-8610, Japan*

[2] *Department of Physics, Graduate School of Humanities and Sciences, Ochanomizu University, Tokyo 112-8610, Japan*

[3*] *Division of Human Support System, Faculty of Symbiotic Systems Science, Fukushima University, Fukushima 960-1296, Japan*

E-mail: bamba@sss.fukushima-u.ac.jp

We explore the descriptions for the three observables of the inflationary universe: the spectral index of curvature perturbations, the tensor-to-scalar ratio, and the running of the spectral index, in the context of scalar field theories, perfect fluid models, and $F(R)$ gravity theories with the methods of reconstruction. By comparing the descriptions of these three observables with the recent cosmological observations, we explicitly show that the inflationary universe represented in scalar field theories, perfect fluid models, and $F(R)$ gravity theories can explain the Planck results.

Keywords: Modified theories of gravity, Cosmology, Particle-theory and field-theory models of the early Universe, Models beyond the standard model.

1. Introduction

Recently, the inflationary cosmology have been investigated much more extensively, thanks to the observational data acquired from the Planck satellite[1,2] on the anisotropy of the cosmic microwave background (CMB) radiation. The spectrum of the primordial density perturbations lead to the information of the inflaton potential[3].

In this paper, we review our main results in Refs. 4 and 5. We propose the descriptions of inflation in scalar field theories, perfect fluid models, and $F(R)$ gravity theories. In particular, with the quantities in scalar field theories, perfect fluid models[6], and $F(R)$ gravity theories[7,8], we represent the three observables of the inflationary universe, namely, the spectral index of curvature perturbations, the tensor-to-scalar ratio, and the running of the spectral index. Furthermore, we compare these descriptions in scalar field theories, perfect fluid models, and $F(R)$ gravity theories with the recent the Planck results. We use the units $k_{\mathrm{B}} = c = \hbar = 1$ and describe the Newton's constant by $G = 1/M_{\mathrm{P}}^2$, where $M_{\mathrm{P}} = 2.43 \times 10^{18}$ GeV is the reduced Planck mass.

---

*The author's current affiliation

2. Slow-roll parameters

The action for a scalar field ϕ is given by $S = \int d^4x\sqrt{-g}\left[R/\left(2\kappa^2\right) - (1/2)\,\partial_\mu\phi\partial^\mu\phi - V(\phi)\right]$, where R is the scalar curvature. The so-called slow-roll parameters ϵ, η and ξ are defined as

$$\epsilon \equiv \frac{1}{2\kappa^2}\left(\frac{V'(\phi)}{V(\phi)}\right)^2\,, \quad \eta \equiv \frac{1}{\kappa^2}\frac{V''(\phi)}{V(\phi)}\,, \quad \xi^2 \equiv \frac{1}{\kappa^4}\frac{V'(\phi)V'''(\phi)}{\left(V(\phi)\right)^2}\,, \tag{1}$$

where the prime shows the derivative with respect to the argument of $V'(\phi) \equiv dV(\phi)/d\phi$. The spectral index n_{s} of the curvature perturbations, the tensor-to-scalar ratio r of the density perturbations, and the running of the spectral index α_{s} are represented as

$$n_{\mathrm{s}} \sim 1 - 6\epsilon + 2\eta\,, \quad r = 16\epsilon\,, \quad \alpha_{\mathrm{s}} \equiv \frac{dn_{\mathrm{s}}}{d\ln k} \sim 16\epsilon\eta - 24\epsilon^2 - 2\xi^2\,, \tag{2}$$

with k the wave number. We take the flat Friedmann-Lemaître-Robertson-Walker (FLRW) metric $ds^2 = -dt^2 + a^2(t)\sum_{i=1,2,3}\left(dx^i\right)^2$, where $a(t)$ is the scale factor, $H \equiv \dot{a}/a$ is the Hubble parameter, and the dot means the time derivative.

2.1. *Descriptions in scalar field theories*

From the action of ϕ shown above, we can obtain the gravitational equations in the FLRW space-time. By introducing the other scalar field φ, we redefine the scalar field ϕ as $\phi = \phi(\varphi)$. We also identify φ with the number of e-folds N and express H as a function of N. Eventually, the gravitational equations read

$$\frac{3}{\kappa^2}\left(H(N)\right)^2 = \frac{1}{2}\omega(\varphi)\left(H(N)\right)^2 + V\left(\phi\left(\varphi\right)\right)\,, \tag{3}$$

$$-\frac{1}{\kappa^2}\left(3\left(H(N)\right)^2 + 2H'(N)H(N)\right) = \frac{1}{2}\omega(\varphi)\left(H(N)\right)^2 - V\left(\phi\left(\varphi\right)\right)\,, \tag{4}$$

with $\omega(\varphi) \equiv \left(d\phi/d\varphi\right)^2$. As a result, we obtain

$$\omega(\varphi) = -\left.\frac{2H'(N)}{\kappa^2 H(N)}\right|_{N=\varphi}\,, \quad V(\varphi) = \left.\frac{1}{\kappa^2}\left[3\left(H(N)\right)^2 + H(N)H'(N)\right]\right|_{N=\varphi}\,. \tag{5}$$

We find $\varphi = N$ as a solution for the field equation ϕ or φ and the Einstein equation. Thus, we can express the slow-roll parameters ϵ, η and ξ in terms of H (for the concrete expressions, see (2.8) in Ref. 4).

2.2. *Perfect fluid descriptions*

Next, we represent the slow-roll parameters in the perfect fluid descriptions. In the FLRW background, the gravitational field equations are expressed as $\left(3/\kappa^2\right)\left(H(N)\right)^2 = \rho$ and $-\left(2/\kappa^2\right)H(N)H'(N) = \rho + P$ with where ρ and P the energy density and pressure of the perfect fluid, respectively. With an arbitrary function $f(\rho)$ of ρ, the equation of state (EoS) can be represented as $P(N) = -\rho(N) + f(\rho)$. In this case, the conservation law becomes $0 = \rho'(N) + 3f(\rho)$.

Consequently, we find $(2/\kappa^2)\left[(H'(N))^2 + H(N)H''(N)\right] = 3f'(\rho)f(\rho)$, where $f'(\rho) \equiv df(\rho)/d\rho$, $H'(N) \equiv dH(N)/dN$, and $\rho'(N) \equiv d\rho(N)/dN$. It is seen that all of the slow-roll parameters can be described only with $\rho(N)$ and $f(\rho(N))$. Accordingly, we have the representations of observables for the inflationary universe of n_{s}, r, and α_{s} (the explicit forms of which are given in Appendix A of Ref. 4).

2.3. *F(R) gravity descriptions*

The action of $F(R)$ gravity is $S_{F(R)} = \int d^4x\sqrt{-g}\left[F(R)/\left(2\kappa^2\right) + \mathcal{L}_{\mathrm{matter}}\right]$, where g the determinant of the metric tensor $g_{\mu\nu}$, and $\mathcal{L}_{\mathrm{matter}}$ is the matter Lagrangian. The gravitational field equation is derived as $(1/2)\, g_{\mu\nu}F(R) - R_{\mu\nu}F'(R) - g_{\mu\nu}\Box F'(R) + \nabla_\mu\nabla_\nu F'(R) = -\left(\kappa^2/2\right) T_{\mathrm{matter}\,\mu\nu}$, where ∇_μ is the covariant derivative operator, $\Box \equiv g^{\mu\nu}\nabla_\mu\nabla_\nu$ is the covariant d'Alembertian, and $T_{\mathrm{matter}\,\mu\nu}$ is the energy-momentum tensor of matter. In the flat FLRW universe, when the matter component is negligible, the Friedmann equation leads to $H^2 = \left(-F(R) + RF'(R)\right) / \left[6\left(F'(R) + R'(N)F''(R)\right)\right]$ with $R = 12H^2 + 6\dot{H}$, $F'(R) \equiv dF(R)/dR$, and $R'(N) \equiv dR/dN$. In what follows, simplicity, the arguments of $H(N)$, $R(N)$, and $F(R)$ are omitted. The prime operating H and R denotes the derivative with respect to N, whereas the one operating F show that with respect to R. Thus, all of the slow-roll parameters are represented in terms of R, R', R'', R''', and $F(R)$, so that the expressions of n_{s}, r, and α_{s} can be obtained (for their explicit forms, see Appendix B in Ref. 4).

3. Comparison of each description with the observations

First, as an inflation model, we study the action for the scalar field ϕ with the following inflaton potential proposed in Ref. 9, given by $V(\phi) = \bar{V}_{\mathrm{n}} \cosh\left(\gamma_{\mathrm{n}}\kappa\phi\right)$ with $\bar{V}_{\mathrm{n}}$ and γ_{n} constants. If $(N, \gamma_{\mathrm{n}}, \kappa\phi_{\mathrm{f}}) = (50, 3.00 \times 10^{-2}, 1.04 \times 10^{-20})$ and $(60, 4.00 \times 10^{-2}, 3.01 \times 10^{-25})$, we have $(n_{\mathrm{s}}, r, \alpha_{\mathrm{s}}) = (0.999, 6.17 \times 10^{-3}, 5.98 \times 10^{-7})$ and $(0.999, 9.90 \times 10^{-3}, 2.69 \times 10^{-6})$, respectively. Here, ϕ_{f} is the amplitude of ϕ at the end of inflation.

Next, as a model of the perfect fluid, we consider the case that the time variation of $f(\rho)$ and ρ during inflation is sufficiently small and therefore exponential inflation can be realized as $w(N) \equiv P(N)/\rho(N) = -1 + f(\rho)/\rho(N) \approx -1$, namely, $|f(\rho)/\rho(N)| \ll 1$. In this case, we get $n_{\mathrm{s}} \sim 1 - 6f(\rho)/\rho(N)$, $r \approx 24f(\rho)/\rho(N)$, and $\alpha_{\mathrm{s}} \approx -9\left(f(\rho)/\rho(N)\right)^2$. If the condition $f(\rho)/\rho(N) = 6.65 \times 10^{-3}$ is satisfied during inflation, we acquire $(n_{\mathrm{s}}, r, \alpha_{\mathrm{s}}) = (0.960, 0.160, -3.98 \times 10^{-4})$.

Finally, we explore $F(R)$ gravity descriptions. We investigate the exponential form of H^2 as $(H(N))^2 = G_2 \exp\left(\beta N\right) + G_3$, where $G_2(< 0)$, $G_3(> 0)$, and $\beta(> 0)$ are constants. In this case, when $(N, G_2, G_3) = (50.0, -1.10, 10.0)$ and $(60.0, -1.20, 15.0)$, we acquire $(n_{\mathrm{s}}, r, \alpha_{\mathrm{s}}) = (0.963, 6.89 \times 10^{-2}, -5.06 \times 10^{-5})$ and $(0.965, 5.84 \times 10^{-2}, -4.51 \times 10^{-5})$, respectively.

As a consequence, in these descriptions, there could be the cases consistent with the Planck results[1,2] of $n_s = 0.968 \pm 0.006$ (68% CL), $r < 0.11$ (95% CL), and $\alpha_s = -0.003 \pm 0.007$ (68% CL).

4. Summary

In the present paper, we have reconstructed the descriptions for the three observables of inflation, i.e., the spectral index n_s of curvature perturbations, the tensor-to-scalar ratio r, and the running of the spectral index α_s, in scalar field theories, perfect fluid models, and $F(R)$ gravity theories. We have demonstrated that the values of (n_s, r, α_s) in these descriptions can be consistent with the Planck analysis.

Acknowledgments

The author would like to sincerely appreciate the collaborations with Professor Shin'ichi Nojiri, Professor Sergei D. Odintsov and Dr. Diego Sáez-Gómez in Refs. 4, 5. This work was partially supported by the JSPS Grant-in-Aid for Young Scientists (B) # 25800136 and the research-funds supplied from Fukushima University.

References

1. P. A. R. Ade *et al.* [Planck Collaboration], Planck 2015 results. XIII. Cosmological parameters, arXiv:1502.01589 [astro-ph.CO].
2. P. A. R. Ade *et al.* [Planck Collaboration], Planck 2015 results. XX. Constraints on inflation, arXiv:1502.02114 [astro-ph.CO].
3. J. E. Lidsey, A. R. Liddle, E. W. Kolb, E. J. Copeland, T. Barreiro and M. Abney, Reconstructing the inflation potential : An overview, *Rev. Mod. Phys.* **69**, 373 (1997).
4. K. Bamba, S. Nojiri and S. D. Odintsov, Reconstruction of scalar field theories realizing inflation consistent with the Planck and BICEP2 results, *Phys. Lett. B* **737**, 374 (2014).
5. K. Bamba, S. Nojiri, S. D. Odintsov and D. Sáez-Gómez, Inflationary universe from perfect fluid and $F(R)$ gravity and its comparison with observational data, *Phys. Rev. D* **90**, 124061 (2014).
6. K. Bamba, S. Capozziello, S. Nojiri and S. D. Odintsov, Dark energy cosmology: the equivalent description via different theoretical models and cosmography tests, *Astrophys. Space Sci.* **342**, 155 (2012).
7. S. Nojiri and S. D. Odintsov, Unified cosmic history in modified gravity: from F(R) theory to Lorentz non-invariant models, *Phys. Rept.* **505**, 59 (2011).
8. S. Nojiri and S. D. Odintsov, Introduction to modified gravity and gravitational alternative for dark energy, *eConf C* **0602061**, 06 (2006) [*Int. J. Geom. Meth. Mod. Phys.* **4**, 115 (2007).].
9. J. Yokoyama and K. I. Maeda, On the Dynamics of the Power Law Inflation Due to an Exponential Potential, *Phys. Lett. B* **207** (1988) 31.

Hydrodynamics in $1+1$ Dimensions from Maxwell-Chern-Simons Theory in AdS_3

Han-Chih Chang[a], Mitsutoshi Fujita[b*†] and Matthias Kaminski[c]

[a]*Department of Physics University of Virginia 382 McCormick Rd. PO Box 400714, Charlottesville, VA, USA*

[b]*Yukawa Institute for Theoretical Physics, Kyoto University, Kyoto, Japan*

[c]*Department of Physics and Astronomy, University of Alabama, Tuscaloosa, AL 35487, USA*

[†]*E-mail: mitsutoshi.fujita@yukawa.kyoto-u.ac.jp*

In this presentation we review our work on Abelian Maxwell-Chern-Simons theory in three-dimensional AdS black brane backgrounds, with both integer and non-integer Chern-Simons coupling. Such theories can be derived from several string theory constructions, and we found exact solutions in the low frequency, low momentum limit ($\omega, k \ll T$, the hydrodynamic limit). Our results are translated into correlation functions of vector operators in the dual strongly coupled $1+1$-dimensional quantum field theory with a chiral anomaly at non-zero temperature T, via the holographic correspondence. The applicability of the hydrodynamic limit is discussed, together with the comparison between an exact field theoretic computation and the found holographic correlation functions in the conformal case.

Keywords: Hydrodynamics, 1+1 Dimensions, Anomalies, Holography.

1. Introduction

The hydrodynamic behavior of the $1+1$ dimensional theory dual to the $1+2$-dimensional Maxwell-Chern-Simons (MCS) theory on the AdS_3 black brane has not been well-understood with the presence of the massive sector. In our work[9], we specifically focus on this hydrodynamic limit, that is when temperature is much larger than frequency and momentum. Related work on the Maxwell-Chern-Simons theory includes analysis in the zero temperature limit[1,2], and in the hydrodynamic limit with the flat sector[3–5]. Pure Yang-Mills theory (without Chern-Simons term) in the AdS_3 black brane background is also interesting, as it provides an example for a $1+1$ dimensional holographic superconductor at large charge density by virtue of the large N limit[6]. It is known that only the special (Neumann) boundary condition[7] is allowed for the gauge field in such a theory.

It was suggested that the Maxwell-Chern-Simons theory was dual to the chiral Luttinger liquid, namely, a theory of electrons in the gapless edge state in the context of the fractional quantum Hall effect[8]. In the chiral Luttinger theory, excitations travel into one particular direction, which is the drift motion in the presence of a strong magnetic field. In this presentation, our holographic model of the probe

*Speaker.

Maxwell-Chern-Simons action is conjectured to be dual to a chiral Luttinger liquid coupled to $1+1$-dimensional thermal CFT (AdS_3 black brane background)[9].

2. 1+1 dimensional anomalous hydrodynamics

In order to compare with the gravity dual, on the field theory side we consider the leading terms of the hydrodynamic expansion in the presence of a anomaly

$$T^{\mu\nu} = \epsilon_I u^\mu u^\nu + P_I \Delta^{\mu\nu} + \ldots, \qquad J^\mu = \rho_I u^\mu + \chi \mu \epsilon^{\mu\nu} u_\nu + \ldots, \tag{1}$$

$$D_\mu J^\mu = -\frac{\chi}{2}\epsilon^{\mu\nu} F_{\mu\nu}, \quad D_\mu T^{\mu\nu} = J_\mu F^{\nu\mu}, \tag{2}$$

where u^μ denotes the fluid velocity ($u_\mu u^\mu = -1$), and $\Delta^{\mu\nu} = g^{\mu\nu} + u^\mu u^\nu$. Here, ϵ_I, P_I, ρ_I are the energy density, the pressure, and the charge density in equilibrium, respectively. In the second line, D_μ is the covariant derivative and $-\chi\epsilon^{\mu\nu}F_{\mu\nu}/2$ is the anomaly term associated with the anomaly in the global axial $U(1)$ symmetry. The terms in Eq. (1) are the leading contributions to the conserved currents $T^{\mu\nu}$ and J^μ in the hydrodynamic limit[3,4]. Note that it has been shown that the anomalous transport terms do not lead to any entropy production[4], and all the terms in Eq. (1) are non-dissipative. Transport peaks (associated with dissipation) are not observed in the holonomy current (dual to a pure Chern-Simons theory), because the holomorphic decomposition does not mix the two sectors of different chirality. We are going to argue that the Maxwell-Chern-Simons theory is holographically dual to a field theory whose low-energy limit is described by anomalous hydrodynamics including terms with the parity breaking epsilon symbol in Eq. (1) and (2). Note also that temperature does not break conformal invariance in $1+1$ dimensions.

3. Analysis of the gravity dual: Maxwell-Chern-Simons action

On the gravity side, the Maxwell-Chern-Simons theory is analyzed on the AdS_3 black brane background. The θ Maxwell-Chern-Simons action ($\theta-$MCS) is given by

$$S = T\int d^3x(\sqrt{-g} f_{\mu\nu} f^{\mu\nu} + \theta\epsilon^{\mu\nu\rho} a_\mu f_{\nu\rho}), \tag{3}$$

where $T\theta$ is the Chern-Simons level and T is a tension. There are many string theory embeddings of the Maxwell-Chern-Simons theory, yielding both integer and non-integer θ[11]. A rather famous one is the dimensional reduction of a type IIB string theory embedding on $AdS_3 \times S^3 \times M_4$ ($M_4 = S^3 \times S^1$ or T^4)[10]. The resulting Maxwell-Chern-Simons theory has two Chern-Simons terms with levels of opposite sign. Moreover, a string theory reduction on $AdS_3 \times S^3 \times T^4$ yields $\theta = 2-$MCS, and leads to the presence of a scalar potential in addition to the cosmological constant[12]. Maxwell-Chern-Simons theories can also be obtained from the probe brane actions on the supergravity backgrounds. More specifically, a Chern-Simons term arises from the Wess-Zumino term of the probe brane actions. The D3/D7 system with ♯ND= 8, which has chiral fermions as zero modes from $3-7$ strings, leads to $\theta = -4$–MCS[13].

To derive the renormalized retarded Green's function, counter-terms and boundary terms need to be added to the action Eq. (3) as follows[14,15]:

$$\begin{aligned}S_{reg} = S + T\int d^2x\sqrt{-\gamma}\Big(&2K_0 a_i f^i + K_1 f_{ij}f^{ij} + K_3 f_i f^i + 2K_0' f^i\Delta a_i \\ &+K_3' f_i\Delta f^i + K_p a_i a^i + R_1(n^\mu\partial_\mu f_{ij})^2 + R_2(n^\mu\partial_\mu f_i)^2 + R_3(n^\mu\partial_\mu a_i)^2 \\ &+Q_1\gamma^{ij}a_i\Delta a_j + Q_2\gamma^{ii'}\gamma^{jj'}f_{ij}\Delta f_{i'j'} + Q_3\gamma^{ij}a_i\Delta f_j\Big) \\ &+T\int[K_2\epsilon^{ij}f_i a_j + K_2'\epsilon^{ij}a_j\Delta f_i + Q_4\epsilon^{ij}a_i\Delta a_j + Q_5\epsilon^{ij}f_i\Delta f_j \\ &+Q_6\epsilon^{ij}f_i\Delta a_j] + \mathcal{O}(\log u) + \ldots,\end{aligned} \tag{4}$$

where $i,j = 1,2$, $f^i = f^{\mu i}n_\mu$, n_μ is the vector normal to the boundary satisfying $n_\mu n^\mu = 1$, ϵ^{ij} is the Levi-Civita symbol, and $\Delta = \gamma^{ij}\partial_i\partial_j$ is the boundary Laplacian in terms of the induced boundary metric γ. Note that counter-terms multiplied by $\mathcal{O}(\log)$ terms are to be included, although not explicitly displayed here. Assuming $\theta > 0$, solutions of the gauge field are expanded near the AdS boundary ($u \sim 0$)

$$\begin{aligned}a_t &= B_t^{(-\theta)}u^{-\theta} + \cdots + a_t^{(0)} + \cdots + B_t^{(\theta)}u^\theta \\ &\quad + \log(u)\left(B_t^{(0),\log} + B_t^{(1),\log}u + B_t^{(2),\log}u^2 + \ldots\right), \\ a_x &= B_x^{(-\theta)}u^{-\theta} + \cdots + a_x^{(0)} + \cdots + B_x^{(\theta)}u^\theta \\ &\quad + \log(u)\left(B_x^{(0),\log} + B_x^{(1),\log}u + B_x^{(2),\log}u^2 + \ldots\right).\end{aligned} \tag{5}$$

Imposing the EOM, one constraint implies $B_t^{(-\theta)} = B_x^{(-\theta)}$, where $B_t^{(-\theta)}$ is the source term and $B_t^{(\theta)}$, $B_x^{(\theta)}$ are VEV terms of the vector operator $\langle\mathcal{O}_1\rangle$. The VEV terms are related with the source terms via the IR boundary conditions. $a_i^{(0)}$ is dual to the chiral current $\langle\mathcal{O}_\pm\rangle$, and it is a flat connection satisfying $da^{(0)} = 0$.

Let's focus on the case of θ = 2-MCS which corresponds to the string theory embedding on $AdS_3\times S^3\times M_4$. The on-shell action is holographically renormalized and becomes finite, with the following choice of counter-term coefficients:

$$\begin{aligned}&K_0 = 1, \quad K_1 = -\frac{1}{4}, \quad K_2 = 0, \quad K_2' = 2K_0' + \frac{1}{2} - Q_1, \quad K_3 = \frac{1}{4}, \\ &K_{3l}' = \frac{1}{2}, \quad K_p = 0, \quad R_1 = \frac{1}{32}, \quad R_2 = -\frac{1}{16}.\end{aligned} \tag{6}$$

Then the renormalized Green's functions are derived in the hydrodynamic limit by performing the variation of the finite on-shell action, yielding the vector operator correlators and chiral current correlator

$$\langle\mathcal{O}_1\mathcal{O}_1\rangle = \frac{\delta^2 S_{\text{reg}}}{\delta B_t^{(-2)}\delta B_t^{(-2)}} = Ti4\omega + \mathcal{O}(2), \tag{7}$$

$$\langle\mathcal{O}_\pm\mathcal{O}_\pm\rangle = \frac{\delta^2 S_{\text{reg}}}{\delta a_\mp^{(0)}\delta a_\mp^{(0)}} = \pm 8T\frac{q_\pm}{q_\mp}, \tag{8}$$

with $a_\pm^{(0)} = (a_x^{(0)} \pm a_t^{(0)})/\sqrt{2}$, where $q_\pm \propto (k \pm \omega)$. It is found that the hydrodynamic two-point functions exactly agree with the appropriate CFT correlation function at the leading order. Note that the scaling behavior of temperature is omitted and can be recovered by dimensional analysis.

4. Summary

We have outlined our obtained results[9]: First, hydrodynamic solutions of MCS theory were obtained on AdS_3 black branes for $\theta = -2, -4, \ldots$, implying that analytic solutions exist for all even Chern-Simons levels. Second, the on-shell action of the MCS theory was holographically renormalized for $\theta = 2$ (and non-integer θ), satisfying a well-defined variational principle. Details in our field theory calculation of the current correlators can be found in our original publication[9].

References

1. P. Minces and V. O. Rivelles, *Phys.Lett.* **B455** (1999) 147–154.
2. T. Andrade, J. I. Jottar, and R. G. Leigh, *JHEP***1205** (2012) 071.
3. S. Jain and T. Sharma, `arXiv:1203.5308`.
4. R. Loganayagam, `arXiv:1106.0277`.
5. K. Jensen, *JHEP* **1101** (2011) 109, `arXiv:1012.4831`.
6. X. Gao, M. Kaminski, H.-B. Zeng, and H.-Q. Zhang, `arXiv:1204.3103`.
7. D. Marolf and S. F. Ross, *JHEP* **11** (2006) 085, `hep-th/0606113`.
8. M. Fujita, W. Li, S. Ryu, and T. Takayanagi, *JHEP* **0906** (2009) 066, `arXiv:0901.0924`.
9. H. C. Chang, M. Fujita and M. Kaminski, *JHEP* **1410**, 118 (2014) `arXiv:1403.5263`.
10. S. Gukov, E. Martinec, G. W. Moore, and A. Strominger, *Adv.Theor.Math.Phys.* **9** (2005) 435–525, `hep-th/0403090`.
11. P. Karndumri and E. O. Colgáin, *JHEP* **1310** (2013) 094, `arXiv:1307.2086`.
12. S. Detournay and M. Guica, *JHEP* **1308** (2013) 121, `arXiv:1212.6792`.
13. J. A. Harvey and A. B. Royston, *JHEP* **0808** (2008) 006, `arXiv:0804.2854`.
14. S. de Haro, S. N. Solodukhin, and K. Skenderis, *Commun.Math.Phys.* **217** (2001) 595–622, `hep-th/0002230`.
15. H.-U. Yee and I. Zahed, *JHEP* **1107** (2011) 033, `arXiv:1103.6286`.

The Born-Infeld Gravity in the Palatini Formalism and the Condition of the Black Hole Formation

Meguru Komada* and Shin'ichi Nojiri†

Department of Physics, Nagoya University,
Nagoya, 464-8602, Japan
* *E-mail: komada@gravity.phys.nagoya-u.ac.jp*
† *E-mail: nojiri@gravity.phys.nagoya-u.ac.jp*

In recent years, many kinds of modified gravity theories are suggested to get over problems in the general relativity. We consider a kind of modified gravity theories "Born-Infeld gravity in the Palatini formalism" and apply this theory to the black hole formation by considering the spherically symmetric collapse of the dust ball. We find the condition for the mass that the black hole horizon is formed.

Keywords: Born-Infeld gravity, Black hole formation.

1. Introduction

In recent years, many kinds of modified gravity theories have been suggested and investigated in several motivations. First motivation is the quantum gravity theory. Although Einstein's general relativity is a successful theory of the classical theory of gravity, we cannot obtain the quantum theory based on the general relativity due to the non-renormalizability. On the other hand, there are problems such as the accelerating expansion of the present universe and the presence of the dark matter in cosmology. Many kinds of gravity theories beyond the Einstein gravity are considered to solve above problems.

As a model of such modified gravities, in this paper, we consider about the Born-Infeld gravity[1] in the Palatini formalism[2 3](also called Eddington inspired Born-Infeld gravity).

$$S = \frac{1}{\kappa^2 b}\int d^4x \left\{\sqrt{|\det\left(g_{\mu\nu} + bR_{\mu\nu}\right)|} - \sqrt{|\det\left(g_{\mu\nu}\right)|}\right\} + S_{\text{matter}}\,. \tag{1}$$

Where, κ is the gravitational constant corresponding to the Einstein gravity and we introduce new parameter b, which has the dimension of the square of the length.

We add an explanation about the Palatini formalism[4]. In usual gravitational theory, only metric is the degree of freedom. Palatini formalism adopts the metric and the connection as the degree of freedom. In the metric formalism of the Born-Infeld gravity, theory includes a ghost in general and we need to tune the action by adding the higher derivative terms so that the ghost does not appear[1]. In the Palatini formalism, however, there does not appear ghost[2 3] and the only propagating mode is massless graviton.

The Born-Infeld type theory has a characteristic. This theory has a different behavior from general relativity in the strong gravitational field region. Because the action includes the square root, there appears the upper limit in the strength given by the scale, which may have suggested that there might not appear the divergence. Therefore, it is expected that the divergence in general relativity like the singularity is removed in this theory.

We apply this theory to FLRW universe and gravitational collapse in next section to confirm the effect of the avoiding the singularity.

2. FLRW cosmology and black hole formation

2.1. *FLRW cosmology*

We consider the FRW space-time with flat spacial part:

$$ds^2 = -dt^2 + a(t)^2 \sum_{i=1,2,3} \left(dx^i\right)^2 , \tag{2}$$

and assume that the non-vanishing components of the connection is given by

$$\Gamma^t_{tt} = A(t) , \quad \Gamma^t_{ij} = a(t)^2 B(t) \delta ij , \quad \Gamma^i_{jt} = \Gamma^i_{tj} = C(t) {\delta^i}_j . \tag{3}$$

In the Einstein gravity, we have $A = 0$, $B = C = H \equiv \dot{a}/a$. We now assume that the matter is given by the dust whose pressure p vanishes and the energy density is denoted by ρ. Substituting this assumption to the equation derived from action (1), we obtain following equation.

$$A = C - H = -\frac{3}{4} b\kappa^2 H\rho \left(1 + b\kappa^2\rho\right)^{-1} , \tag{4}$$

$$B = H + \frac{b\kappa^2}{4} H\rho , \tag{5}$$

$$C = H - \frac{3}{4} b\kappa^2 H\rho \left(1 + b\kappa^2\rho\right)^{-1} , \tag{6}$$

$$b\kappa^2\rho = \left\{1 + b\left(\dot{H} + 3H^2 + \frac{b\kappa^2}{2}\dot{H}\rho\right)\right\}^2 - 1 . \tag{7}$$

With solving this equation, we revealed the behavior of the scale factor and discovered that the shrinking universe doesn't collapse to the one point. that stops and bounces unlike the general relativity. Therefore, the initial singularity is removed in this theory. Furthermore we derived the value of the scale factor when the universe bounces.

$$e^{3N_{\min}} \sim \frac{3}{8} b\kappa^2\rho_0 \quad \text{for } \mathrm{b}\kappa^2\rho_0 \gg 1. \tag{8}$$

$$e^{3N_{\min}} \sim \frac{9}{64} b\kappa^2\rho_0 \quad \text{for } \mathrm{b}\kappa^2\rho_0 \ll 1. \tag{9}$$

Where, N is thr e-folding $a = e^N$ and ρ_0 is the initial dust density.

2.2. *Black hole formation*

With using above results, we consider the black hole formation. We now assume there is a spherically symmetric and uniform ball made of dust and consider the collapse of ball in the Schwarzschild spacetime. Inside the ball, the space-time can be regarded with the shrinking FRW universe as in the last section. The results in the previous section tell that there could be a bouncing. If the radius of the ball at the bouncing is larger than the Schwarzschild radius, the black hole cannot be formed. With using the junction condition of spacetime, we can estimate the condition that black hole can be formed from (8) (9).

$$M > \frac{18\pi b}{\kappa^4} \quad \text{for } \mathrm{b}\kappa^2\rho_0 \gg 1. \tag{10}$$

$$M > \frac{27\pi b}{\kappa^4} \quad \text{for } \mathrm{b}\kappa^2\rho_0 \ll 1. \tag{11}$$

3. Summary

In the Born-Infeld gravity by using the Palatini formalism, we investigated the FRW cosmology where the matter is dust and we have shown that when $b > 0$, there occurs the bouncing. And by applying the results in the FRW universe, we also investigated the collapse of the sphere of dust and the black hole formation. Then we have shown that although the large black hole might be formed but the small black holes are prohibited to be formed.

Acknowledgements

The work by S.N. is supported by the JSPS Grant-in-Aid for Scientific Research (S) # 22224003 and (C) # 23540296.

References

1. S. Deser and G. W. Gibbons, Class. Quant. Grav. **15** (1998) L35 [hep-th/9803049].
2. D. N. Vollick, Phys. Rev. D **69** (2004) 064030 [gr-qc/0309101].
3. M. Banados, Phys. Rev. D **77** (2008) 123534 [arXiv:0801.4103 [hep-th]];
   M. Banados and P. G. Ferreira, Phys. Rev. Lett. **105** (2010) 011101 [arXiv:1006.1769 [astro-ph.CO]];
   J. H. C. Scargill, M. Banados and P. G. Ferreira, Phys. Rev. D **86** (2012) 103533 [arXiv:1210.1521 [astro-ph.CO]].
4. A. Palatini, Rend Circ. Mat, Palermo 43 (1919) 203

Lattice Study of SU(3) Gauge Theory with Four Fundamental Fermions

Yasumichi Aoki[a], Tatsumi Aoyama[a], Ed Bennett[b], Masafumi Kurachi[c], Toshihide Maskawa[a], Kohtaroh Miura[d], Kei-ichi Nagai[a], Hiroshi Ohki[e], Enrico Rinaldi[f], Akihiro Shibata[g], Koichi Yamawaki[a] and Takeshi Yamazaki[h]

(LatKMI Collaboration)

[a] *Kobayashi-Maskawa Institute for the Origin of Particles and the Universe (KMI), Nagoya University, Nagoya 464-8602, Japan*

[b] *Department of Physics, Swansea University, Singleton Park, Swansea SA2 8PP, UK*

[c] *Institute of Particle and Nuclear Studies, High Energy Accelerator Research Organization (KEK), Tsukuba 305-0801, Japan*

[c] *Nuclear and Chemical Sciences Division, Lawrence Livermore National Laboratory, Livermore, CA 94550, USA*

[d] *Centre de Physique Theorique(CPT), Aix-Marseille University, Campus de Luminy, Case 907, 163 Avenue de Luminy, 13288 Marseille cedex 9, France*

[e] *RIKEN/BNL Research Center, Brookhaven National Laboratory, Upton, NY, 11973, USA*

[f] *Lawrence Livermore National Laboratory, Livermore, California, 94550, USA*

[g] *Computing Research Center, High Energy Accelerator Research Organization (KEK), Tsukuba 305-0801, Japan*

[h] *Graduate School of Pure and Applied Sciences, University of Tsukuba, Tsukuba, Ibaraki 305-8571, Japan*

LatKMI Collaboration has been studying SU(3) gauge theories with large numbers of fermion flavors, N_f. Here, we report results of lattice simulations for SU(3) gauge theory with four fundamental fermions. We first show fermion mass dependence of F_π, $\langle\bar{q}q\rangle$, M_π, M_ρ and chiral extrapolations of them, showing the evidence of chiral symmetry breaking. Then, we show preliminary result of the measurement of the mass of the flavor-singlet scalar bound state.

Keywords: Lattice gauge theory, Large N_f QCD, Walking technicolor.

1. Introduction

As a part of the project of studying large N_f QCD, the LatKMI Collaboration has been studying the SU(3) gauge theory with four fundamental fermions (four-flavor QCD). The main purpose of studying four-flavor QCD is to provide a qualitative comparison to $N_f = 8$, 12, 16 QCD; however, a quantitative comparison to real-world QCD is also interesting. To make such comparisons more meaningful, it is desirable to use the same kind of lattice action consistently, so that qualitative difference of different theories are less affected by artifacts of lattice discretization. Here, we adopt the highly-improved staggered quark action with the tree-level Symanzik gauge action (HISQ/tree), which is exactly the same as the setup for our simulations for $SU(3)$ gauge theories with $N_f = 8$, 12 and 16 fundamental fermions[1–3]. In the next section, we show the fermion mass dependence of F_π, $\langle\bar{q}q\rangle$, M_π, M_ρ and chiral extrapolations of them. In section 3, preliminary result of the measurement

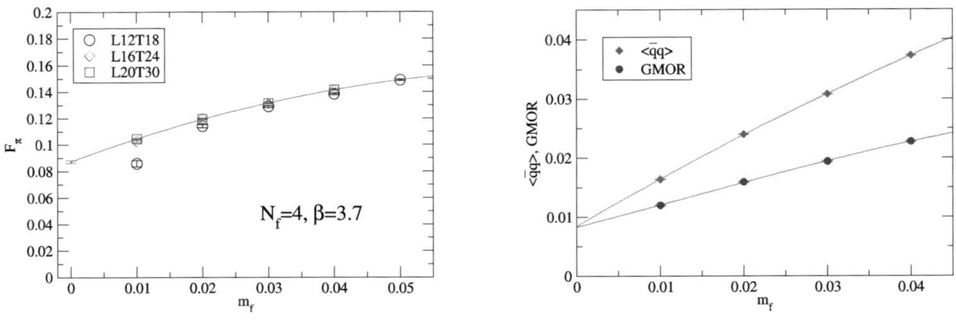

Fig. 1. The decay constant of the NG boson, F_π, (left panel) and the chiral condensation, $\langle \bar{q}q \rangle$, (right panel) as functions of m_f for $N_f = 4$, $\beta = 3.7$. On the right panel, only the largest volume data is plotted. The chiral condensate estimated by using the GMOR relation is also plotted in the right panel. All the curves appearing in figures are obtained by fitting the quadratic function to the largest volume data.

of the mass of the flavor-singlet scalar bound state will be reported.

2. Chiral property and the vector meson mass

For our study of four-flavor QCD, we generated gauge configurations for $\beta = 3.7$ with three volumes, $(L, T) = (12, 18)$, $(16, 24)$, $(20, 30)$, and four or five bare mass parameter, m_f, in the range of $0.01 \leq m_f \leq 0.05$ depending on the volume.

In Fig. 1, we show the decay constant of the Nambu–Goldstone (NG) boson, F_π, and the chiral condensation, $\langle \bar{q}q \rangle$, as functions of m_f. Quadratic fits of F_π and $\langle \bar{q}q \rangle$ show a non-zero value in the chiral limit, which can be considered as evidence for four-flavor QCD being in the chirally broken phase. As a consistency check, we also calculated the chiral condensate by using the GMOR relation, which is shown in the plot of $\langle \bar{q}q \rangle$. From the figure, we can see that the value of the chiral condensate obtained from the GMOR relation agrees with that of $\langle \bar{q}q \rangle$ in the chiral limit.

In the left panel of Fig. 2, we show the plot of the mass-squared (m_π^2) of the NG boson obtained from our simulations as a function of m_f. m_π^2 can be fitted well by a liner ansatz; if the quadratic term is allowed, its coefficient is small. This is consistent with predictions from chiral perturbation theory.

The right panel of Fig. 2 shows the ratio of the mass of the vector meson, M_ρ, to F_π for various values of m_f. The curve in the figure is the quadratic fit to the data of the largest volume, from which we can see the value of the ratio in the chiral limit being $M_\rho / F_\pi \simeq 6$. This value is very close to the ratio in the real-world QCD, $775\,\text{MeV}/130\,\text{MeV} \simeq 6$. In this sense, though the number of light fermion flavors is slightly larger than that in the real-world QCD, the two theories are very similar as far as the mass of the vector meson is concerned.

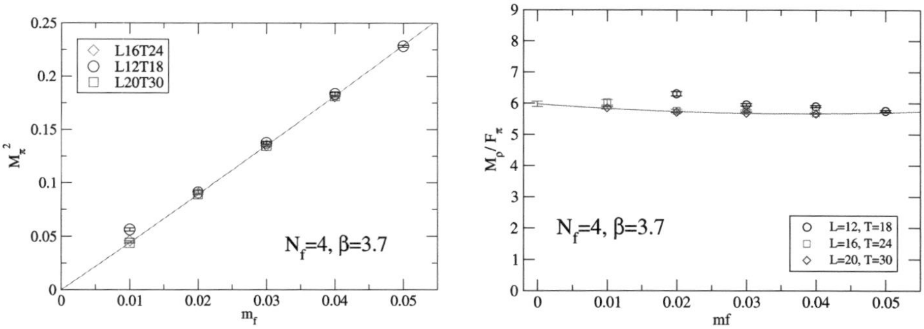

Fig. 2. **Left:** Mass-squared of the NG boson as a function of m_f. Quadratic fit to the largest volume data is also shown. **Right:** The ratio of the mass of the vector meson, M_ρ, to F_π for various values of m_f. Quadratic fit to the largest volume data is also shown.

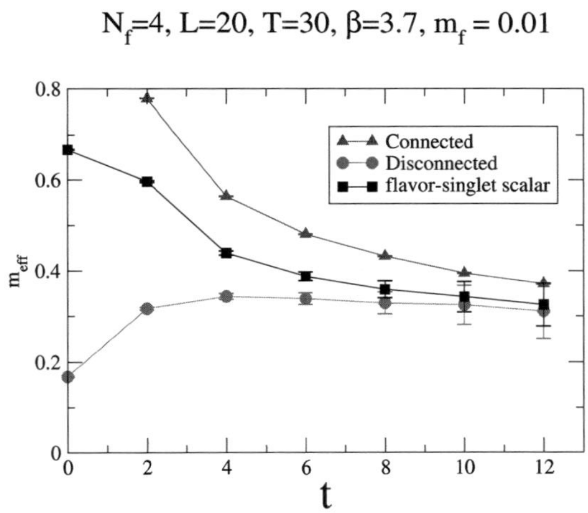

Fig. 3. The effective mass plot of the flavor-singlet scalar for $L = 20$, $T = 30$, $m_f = 0.01$. The (black) square symbols represent effective mass calculated from the full scalar correlator, while (blue) triangle and (red) circle symbols represent those obtained from the connected and disconnected correlators, respectively.

3. Flavor-singlet scalar bound state

The flavor-singlet scalar meson is very interesting object both in large N_f QCD and in real-world QCD. In the former case, it could be related to the 126 GeV Higgs boson discovered at the LHC, while for the latter case, it is related to the σ meson. We calculate the flavor-singlet scalar correlator to extract the bound state mass by using the same technique we adopted for $N_f = 8$ and 12 QCD. (See Refs.[4,5] for details.) The preliminary effective mass plot $L = 20$, $T = 30$, $m_f = 0.01$ is shown in Fig. 3. The plateau is observed in the disconnected channel, and the mass extracted from it is about 0.3, which is heavier than the pseudo NG boson ($M_\pi \sim 0.2$) in the same parameter. This is quite different from $N_f = 8$ and 12 QCD, in which it was shown that the mass of the flavor-singlet scalar bound state is comparable to or smaller than the pseudo NG boson[4,5].

4. Summary

In this report, we showed our lattice simulation results for four-flavor QCD, which was performed by using the HISQ action with the tree-level Symanzik gauge action. It was shown that the theory is in the chiral symmetry breaking phase, and the ratio of M_ρ to F_π in the chiral limit is quite close to that of real-world QCD. Preliminary results for flavor-singlet bound state mass were also shown, and the qualitative difference between four-flavor QCD and (near-)conformal theories such as $N_f = 8$ and 12 QCD was pointed out. More detailed analysis with higher statistics and similar investigation in multiple mass parameter would help to understand the physical interpretation of the result, which will be pursued in the future.

Acknowledgments

Numerical computations have been carried out on the high-performance computing systems at KMI (φ) and at the Research Institute for Information Technology in Kyushu University (CX400 and HA8000). This work is supported by the JSPS Grant-in-Aid for Scientific Research (S) No.22224003, (C) No.23540300 (K.Y.), for Young Scientists (B) No.25800139 (H.O.) and No.25800138 (T.Y.), and also by the MEXT Grants-in-Aid for Scientific Research on Innovative Areas No.23105708 (T.Y.), No. 25105011 (M.K.). E.R. acknowledges the support of the U.S. Department of Energy under Contract DE-AC52- 07NA27344 (LLNL).

References

1. Y. Aoki *et al.* [LatKMI Collaboration], "Walking signals in $N_f = 8$ QCD on the lattice," Phys. Rev. D **87**, no. 9, 094511 (2013) [arXiv:1302.6859 [hep-lat]].
2. Y. Aoki *et al.*, "Lattice study of conformality in twelve-flavor QCD," Phys. Rev. D **86**, 054506 (2012) [arXiv:1207.3060 [hep-lat]].
3. Y. Aoki *et al.* [LatKMI Collaboration], "Non-Perturbative Study of 16-Flavor QCD," Proceedings of "KMI-GCOE Workshop on Strong Coupling Gauge Theories in the LHC Perspective (SCGT 12)" : Nagoya, Japan, December 4-7, 2012
4. Y. Aoki *et al.* [LatKMI Collaboration], "Light composite scalar in twelve-flavor QCD on the lattice," Phys. Rev. Lett. **111**, no. 16, 162001 (2013) [arXiv:1305.6006 [hep-lat]].
5. Y. Aoki *et al.* [LatKMI Collaboration], "Light composite scalar in eight-flavor QCD on the lattice," Phys. Rev. D **89**, 111502 (2014) [arXiv:1403.5000 [hep-lat]].

GUT Scale Threshold Effect on Proton Decay

Takumi Kuwahara

Department of Physics, Nagoya University, Nagoya 464-8602, Japan
E-mail: kuwahara@th.phys.nagoya-u.ac.jp

The supersymmetric grand unified theories (SUSY GUTs) are the promising models beyond the standard model for particle physics. The standard SUSY GUTs predict that there exist the baryon-number violating processes, such as proton decay. It is important to estimate RGE effects on these processes in order to predict precise nucleon lifetime since there exists a large scale running between the GUT scale and 1GeV where the nucleon matrix elements are calculated. In this work, we have evaluated the GUT scale threshold corrections to the Wilson coefficients of the dimension-six baryon-number violating operators, and then we have estimated the effect of the threshold corrections on the decay rate in the minimal SUSY SU(5) GUT setup and its vector-like extensions.

Keywords: Supersymmetric grand unified theories, Proton decay.

1. Introduction

The supersymmetric grand unified theories (SUSY GUTs) are the promising extensions of the Standard Model (SM). The gauge groups of the SM are unified into the larger group, such as $SU(5)$, and the SM fields are embedded into the GUT multiplets. SUSY also plays a crucial role in the gauge coupling unification as well as the natural explanation of the gauge hierarchy problem, and we are looking forward to the discovery of the SUSY particles at the LHC experiment. In 2012, a scalar particle, which may be consistent with the SM Higgs, was discovered around 125 GeV[1]. The SM is established and we expect the discovery of new physics predicted by the SUSY GUT in near future, although no signal for new physics has been found yet.

It is true that there are several issues which should be carefully studied in the SUSY GUTs. One of the issues is how to achieve the 125 GeV Higgs boson. The minimal supersymmetric standard model (MSSM) is considered as the low-energy effective theory of the SUSY GUTs. It is known, however, that the MSSM predicts the upper bound on the Higgs mass at tree-level, and the observed Higgs mass may require the extension of MSSM or the specific mass spectrum (*e.g.*, the quantum correction to the Higgs mass is enhanced by adding the additional matters[2]).

Another is from the experimental constraints on the baryon-number violating ($\not{B}$) processes, such as nucleon decay, which are induced through the interaction between the GUT and MSSM fields. The current status of the nucleon decay experiments is as follows: $\tau(p \to \pi^0 e^+) > 1.4 \times 10^{34}$ years[3] for the dominant decay mode via a massive gauge boson (X boson), and $\tau(p \to K^+\overline{\nu}) > 5.9 \times 10^{33}$ years for the dominant mode by the additional color-triplet Higgs exchange[4]. The minimal

SUSY $SU(5)$ GUT is excluded by the short partial lifetime of $p \to K^+\bar{\nu}$ if the SUSY scale is close to the electroweak scale[5]. If the SUSY scale is much higher, the constraint from the color-triplet Higgs becomes mild and the dominant decay mode $p \to K^+\bar{\nu}$ may be detected at the future detectors[6]. This dangerous decay mode is also suppressed by extending the models, such as the missing-partner model[7]. Thus, we do not consider the terrible decay mode in this work.

In this talk, we focus on the extra vector-like matter extended models[2]. Since, in these models, the additional fields couple to the SM Higgs field, there is additional contribution to the quantum correction to the Higgs mass. Furthermore, if we introduce the additional fields as complete multiplets of $SU(5)$, the unification of gauge couplings is manifest and these couplings become larger at the GUT scale. Then the nucleon decay through the X-boson exchange is enhanced[8].

This talk is based on our work[9]. The analytic expression for the threshold corrections to the Wilson coefficients of the $\not{B}$ operators is given in the original paper[9]. The GUT mass spectrum dependence of these correction is also shown in the original paper, which is not shown in this talk.

2. SUSY $SU(5)$ GUTs

In the SUSY extensions of the SM, matter fields, Higgs fields, and their superpartners are embedded in chiral superfields and their conjugation. Gauge bosons and gauginos are described by vector superfields. In the minimal SUSY $SU(5)$ GUT, the matter fields are given by the $\bar{\mathbf{5}}$ and $\mathbf{10}$ representational superfields which are denoted by Φ and Ψ as follows:

$$\Phi_A(\bar{\mathbf{5}}) = \begin{pmatrix} D^C_\alpha \\ \epsilon_{rs} L^s \end{pmatrix}, \qquad \Psi^{AB}(\mathbf{10}) = \frac{1}{\sqrt{2}} \begin{pmatrix} \epsilon^{\alpha\beta\gamma} U^C_\gamma & Q^{r\alpha} \\ -Q^{s\beta} & \epsilon^{sr} E^C \end{pmatrix}, \tag{1}$$

where $A, B, \cdots = 1, 2, \cdots, 5$ are the indices of the $SU(5)$, $\alpha, \beta, \cdots = 1, 2, 3$ and $r, s, \cdots = 1, 2$ are the indices of the $SU(3)_C$ and $SU(2)_L$, respectively. All the chiral superfields include the left-handed fermions in the flavor basis. Q and L denote the weak-doublet chiral superfields for left-handed quarks and left-handed leptons, respectively. U^C, D^C, and E^C denote the chiral superfields for the charge-conjugation of right-handed up-type and down-type quarks, and right-handed charged lepton, respectively. In the Higgs sector, there are $\mathbf{5}$, $\bar{\mathbf{5}}$, and $\mathbf{24}$ representational superfields, especially the $\mathbf{5}$ and $\bar{\mathbf{5}}$ Higgs multiplets are decomposed as

$$H^A_{\mathbf{5}}(\mathbf{5}) = \begin{pmatrix} H^\alpha_C \\ H^r_u \end{pmatrix}, \quad H_{\bar{\mathbf{5}}A}(\bar{\mathbf{5}}) = \begin{pmatrix} H_{\overline{C}\alpha} \\ \epsilon_{rs} H^s_d \end{pmatrix}. \tag{2}$$

$H_{\mathbf{5}}(\mathbf{5})$ and $H_{\bar{\mathbf{5}}}(\bar{\mathbf{5}})$ include the MSSM Higgs doublets, H_u and H_d. In order to embed the MSSM Higgs multiplets in the $SU(5)$ multiplets, we have to introduce the color-triplet Higgs multiplets H_C and $H_{\overline{C}}$. The adjoint Higgs multiplet ($\mathbf{24}$) is introduced to cause the spontaneous symmetry breaking of the $SU(5)$ gauge symmetry according to the non-zero vacuum expectation value (VEV).

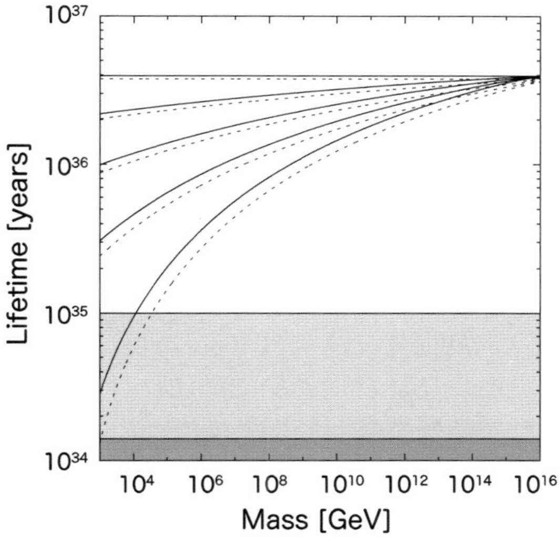

Fig. 1. Partial proton lifetime $\tau(p \to \pi^0 + e^+)$ in the vector-like extended SUSY SU(5) models. The solid (dotted) lines describe the lifetime with (without) threshold corrections at GUT scale. The number of vector-like multiplets is set to be $N_{\mathbf{5}} = 0, 1, \cdots, 4$ from top to bottom. The dark gray (gray) region corresponds to the region excluded by Super-Kamiokande (expected the future sensitivity by the Hyper-Kamiokande).

3. Results

Let us consider the finite corrections to the $\not{B}$ operators via gauge interactions since the gauge couplings become strong if there exist additional vector-like multiplets. Above the GUT scale, there are three-type finite one-loop corrections to the $\not{B}$ operators; the vacuum polarization of the X boson, the vertex correction to the gauge interactions between the X boson and matter multiplets, and the box corrections. On the other hand, below the GUT scale, the finite corrections arise from the vertex corrections of the dimension-six operators. We calculate the matrix elements for the $\not{B}$ processes with the one-loop finite corrections above and below the GUT scale, and then we determine the threshold corrections to the Wilson coefficients by matching these matrix elements.

In Fig. 1, we show the partial proton lifetime $\tau(p \to \pi^0 + e^+)$ in the vector-like extended SUSY SU(5) models as a function of the mass of vector-like multiplets. For simplicity, all vector-like multiplets are degenerate in mass and the SUSY breaking scale is set to be 1 TeV. The solid and dotted lines correspond in this figure to the case with and without the threshold corrections at GUT scale, respectively.

This figure shows that the threshold corrections at GUT scale correct the matrix elements for nucleon decay maximally 50%, which is correspond to the case that there are four pairs of vector-like multiplets and these masses lie around 1 TeV. On the other hand, in the minimal SUSY $SU(5)$ GUT, the effect of these corrections make the decay rate suppressed about 5%.

4. Conclusion and Discussion

In this talk, we showed the effect of threshold corrections to the Wilson coefficients for the dimension-six baryon-number violating operators in the minimal SUSY $SU(5)$ GUT and its vector-like matter extensions. Since the gauge couplings at the GUT scale become strong in the vector-like extensions, the higher-order corrections may be significant in order to precisely estimate the lifetime of nucleon. In the minimal SUSY $SU(5)$ GUT, the proton lifetime becomes longer about 5%. On the other hand, in the vector-like extended models, we find the proton decay rate receives $\mathcal{O}(10)\%$ suppression due to the large unified coupling at the GUT scale, maximally about 50% suppression. We note that the precise proton lifetime depends on the nucleon matrix elements. For the calculation of nucleon matrix elements by lattice simulation, there remains about 30% uncertainty[10].

We also note that we neglected the threshold corrections at the SUSY scale and the effect of the additional multiplets at the GUT scale in this work. In order to complete the two-loop level estimation, we need to evaluate the threshold corrections at the SUSY scale. Furthermore, for instance, the additional multiplets are required to solve the doublet-triplet splitting problem in the SUSY $SU(5)$ GUTs. In this work, we estimated all vertex corrections via gauge interactions, so that we need only to calculate the vacuum polarization of X boson in order to apply our work in other GUT models. These effects will be shown in the future work.

References

1. G. Aad *et al.* [ATLAS and CMS Collaborations], Phys. Rev. Lett. **114** (2015) 191803.
2. S. P. Martin, Phys. Rev. D **81** (2010) 035004.
3. K. S. Babu *et al.*, arXiv:1311.5285 [hep-ph].
4. K. Abe *et al.* [Super-Kamiokande Collaboration], Phys. Rev. D **90** (2014) 7, 072005.
5. N. Sakai and T. Yanagida, Nucl. Phys. B **197** (1982) 533; S. Weinberg, Phys. Rev. D **26** (1982) 287.
6. J. Hisano, D. Kobayashi, T. Kuwahara and N. Nagata, JHEP **1307** (2013) 038; N. Nagata and S. Shirai, JHEP **1403** (2014) 049.
7. J. Hisano, T. Moroi, K. Tobe and T. Yanagida, Phys. Lett. B **342** (1995) 138.
8. J. Hisano, D. Kobayashi and N. Nagata, Phys. Lett. B **716** (2012) 406.
9. J. Hisano, T. Kuwahara and Y. Omura, Nucl. Phys. B **898** (2015) 1.
10. Y. Aoki, E. Shintani and A. Soni, Phys. Rev. D **89** (2014) 1, 014505.

Thermodynamics in 8-Flavor QCD

Yasumichi Aoki[a], Tatsumi Aoyama[a], Ed Bennett[b], Masafumi Kurachi[c], Toshihide Maskawa[a], Kohtaroh Miura*[,a,d], Kei-ichi Nagai[a], Hiroshi Ohki[e], Enrico Rinaldi[f], Akihiro Shibata[g], Koichi Yamawaki[a] and Takeshi Yamazaki[h]

(LatKMI Collaboration)

[a] *Kobayashi-Maskawa Institute for the Origin of Particles and the Universe (KMI), Nagoya University, Nagoya, 464-8602, Japan*
[b] *Department of Physics, Swansea University, Singleton Park, Swansea SA2 8PP, UK*
[c] *Institute of Particle and Nuclear Studies, High Energy Accelerator Research Organization (KEK), Tsukuba 305-0801, Japan*
[d] *Centre de Physique Theorique(CPT), Aix-Marseille University, Campus de Luminy, Case 907, 163 Avenue de Luminy, 13288 Marseille cedex 9, France*
* *E-mail: Kohtaroh.Miura@cpt.univ-mrs.fr*
[e] *RIKEN/BNL Research Center, Brookhaven National Laboratory, Upton, NY, 11973, USA*
[f] *Lawrence Livermore National Laboratory, Livermore, California 94550, USA*
[g] *Computing Research Center, High Energy Accelerator Research Organization (KEK), Tsukuba 305-0801, Japan*
[h] *Graduate School of Pure and Applied Sciences, University of Tsukuba, Tsukuba, Ibaraki 305-8571, Japan*

Motivated by the scenario of electroweak baryogenesis in the walking technicolor (WTC) model, we investigate thermodynamic properties of 8-flavor QCD by using Monte Carlo simulations in the lattice gauge theory framework. We adopt HISQ and tree-level Symanzik improved actions. We show that the chiral susceptibility has a peak structure, and that the ratio of chiral to pion susceptibilities starts showing an inflection point with decreasing fermion mass. These results suggest the existence of a chiral phase transition in 8-flavor QCD.

Keywords: Lattice gauge theory, Conformal/Walking dynamics, Chiral and electroweak phase transitions.

1. Introduction

The baryon number asymmetry in the Universe (BAU) indicates that baryogenesis took place in the early stage of the Universe. The baryogenesis requires processes satisfying Sakharov's three conditions, 1) baryon number violation, 2) C and CP violation, 3) out-of-equilibrium. In the standard model of particle physics, the only electroweak phase transition has a possibility to satisfy 1) and 2). However, lattice studies of the standard model at finite temperature (T)[1] have shown that a rapid crossover instead of a first-order phase transition, which is necessary for 3), occurs in the electroweak gauge-Higgs sector. Thus, the standard model fails to explain BAU. If a walking technicolor (WTC) model[2] replaces the Higgs sector of the standard model, the finite-T chiral phase transition of that model is of great interest for electroweak baryogenesis. Recently, the LatKMI Collaboration has shown that 8-flavor ($N_f = 8$) QCD is a candidate for WTC[3]. That theory is expected to have a first-order phase transition, *a la* Pisarski and Wilczek[4].

First-principle studies of 8-flavor QCD thermodynamics using lattice gauge theory simulations are demanded and the subject of this proceedings. The peak of ($\chi_{\rm chiral}$) gives the most important signal of the chiral phase transition, while it has not been shown yet in the previous works[5,6]. Therefore, we particularly focus on the ($\chi_{\rm chiral}$) and related quantities.

2. Lattice Setup

The Monte Carlo simulations are organized to be consistent with zero temperature simulations[3]: we have utilized the MILC-code[7] with some modifications, and adopted tree-level Symanzik gauge action and HISQ fermion action.

We used lattice box sizes of $(L_s, L_t) = (24, 12)$ and $(32, 16)$ with fixed aspect ratio $L_s/L_t = 2$, and the temperature $T = 1/(L_t a(\beta_{\rm L}))$ ($a(\beta_{\rm L})$ = lattice spacing at lattice coupling $\beta_{\rm L}$) is controlled by varying $\beta_{\rm L} \in [2.8, 4.0]$. We will find the critical coupling $\beta_{\rm L}{}^{\rm c}$ associated with a chiral phase transition temperature $T_c = 1/(L_t a(\beta_{\rm L}{}^{\rm c}))$. In the strong coupling region $\beta_{\rm L} \lesssim 3.5$, the HISQ fermion cannot achieve a good acceptance ratio in Hybrid Monte Carlo (HMC) algorithm with Metropolis tests, and we alternatively used Hybrid Moleculer Dynamics (HMD). Systematic errors due to the lack of Metropolis tests have not been investigated yet in this proceedings.

The lattice fermion mass is set to be $m_f \times a(\beta_{\rm L}) = 0.02$ or 0.04. We shall consider the fermion mass normalized by temperature as follows: $\tilde{m}_f(\beta_{\rm L}) \equiv [m_f \cdot a(\beta_{\rm L})] \times L_t = m_f \times [a(\beta_{\rm L}) \cdot L_t] = m_f/T$. At the critical coupling, we have $\tilde{m}_f(\beta_{\rm L}{}^{\rm c}) = m_f/T_c$ which measures the strength of fermion mass effects. In the present studies, we have three values on $\tilde{m}_f(\beta_{\rm L}{}^{\rm c}) = \{0.02 \cdot 12,\ 0.02 \cdot 16,\ 0.04 \cdot 12\}$ at the transition. We have generated $\mathcal{O}(1000)$ gauge configurations with $\mathcal{O}(3000 - 10000)$ trajectories at each point of the parameter space explained above. Using those configurations, we have evaluated the chiral condensate, Polyakov loops, and their susceptibilities.

3. Results

We investigate the ratio of chiral to pion susceptibility $R_\pi = \chi_{\rm chiral}/\chi_\pi$. In the chiral limit, the R_π vanishes (becomes 1) in the chiral broken (restored) phase and thus is very sensitive to the chiral phase transition[8]. Even for the case with a finite fermion mass $m_f > 0$, R_π shows a rapid increase near the chiral transition/crossover[5]. Since the R_π is dimensionless, it is renormalization invariant under the $\beta_{\rm L}$ variation.

In order to compute the R_π, we evaluate the pion susceptibility from the chiral condensate, $\chi_\pi a^2(\beta_{\rm L}) = \langle\bar{\psi}\psi\rangle a^3(\beta_{\rm L})/(m_f a(\beta_{\rm L}))$. This relation becomes relevant in the vicinity of the second-order chiral phase transition[8]. We will discuss the validity of the assumption just later.

In the left panel of Fig. 1, we show R_π as a function of $\beta_{\rm L}$. A larger $\beta_{\rm L}$ corresponds to a higher temperature, $T = 1/(L_t a(\beta_{\rm L}))$. As the fermion mass $\tilde{m}_f(\beta)$ gets smaller, R_π becomes step-function-like. As a result, an inflection point appears for $\tilde{m}_f = 0.02 \cdot 12$ (blue circles/squares) as indicated by the peak of the R_π-increasing rate (middle panel). Such a mass dependence of R_π would result from the chiral dynamics but not from the conformal one. In the right panel of Fig. 1, we show the chiral susceptibility $\chi_{\rm chiral}$ as a function of $\beta_{\rm L}$ for $\tilde{m}_f = 0.02 \cdot 12$. The decreasing trend is due to the scale variation with $\beta_{\rm L}$. We find the peak at $\beta_{\rm L} = 3.08$, which suggests the chiral phase transition $\beta_{\rm L}{}^{\rm c} \simeq 3.08$. We note that the real parts of Polyakov loops start deviating from zero at $\beta_{\rm L} = 3.16$, which is slightly larger than $\beta_{\rm L}{}^{\rm c}$.

Above observations support that the 8-flavor QCD at zero temperature is in the chiral broken phase rather than the conformal phase. In the vicinity of the inflection point, the chiral condensates are found to change smoothly, which suggests a crossover or second-order transition. Even in the crossover case, the peak observed in $\chi_{\rm chiral}$ suggests that the present system is feeling the critical phenomena near the second-order transtion, and our assumption for the pion susceptibility $\chi_\pi a^2(\beta_{\rm L}) = \langle\bar{\psi}\psi\rangle a^3(\beta_{\rm L})/(m_f a(\beta_{\rm L}))$ should be reasonable.

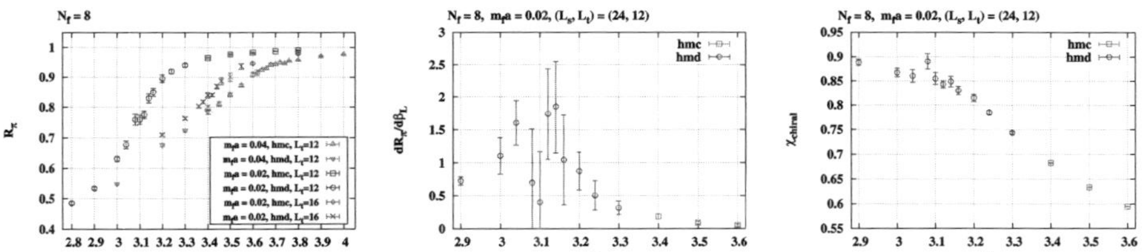

Fig. 1. The susceptibility ratio R_π (left), its increasing rate $dR_\pi/d\beta_{\rm L}$ (middle), and the chiral susceptibility $\chi_{\rm chiral}$ (right) as a function of lattice bare couplings $\beta_{\rm L}$ associated with a temperature via $T = 1/(L_t a(\beta_{\rm L}))$.

4. Summary and Outlook

Motivated by the electroweak baryogenesis in WTC scenario, we have investigated thermodynamic properties of 8-flavor QCD using Monte Carlo simulations in the lattice gauge theory. In contrast to the other group activities[5,6], we have shown a *Peak Structure in the Chiral Susceptibility*, which suggests the physical chiral transition in 8-flavor QCD. The susceptibility ratio $R_\pi = \chi_{\rm chiral}/\chi_\pi$ becomes more step-function-like with decreasing fermion mass m_f, implying the growing signal of the chiral phase transition/crossover as m_f becomes smaller.

We should note that the aforementioned results are still preliminary and the existence of the chiral phase transition in 8-flavor QCD would not be conclusive. It is mandatory to estimate the systematic errors coming from the use of HMD in the strong coupling region as a next step.

Moreover, there are several subjects to be investigated in future. First, it is desirable to investigate a phase transition of 8-flavor QCD for sufficiently small fermion masses without a contamination of un-physical bulk transitions. Second, it is interesting to study Dirac eigen values of 8-flavor QCD at finite T by utilizing Wilson flow smearing technique as performed in[6]. Third, it is necessary to set a scale by using the Wilson flow at zero temperature.

Acknowledgments

Numerical calculations have been carried out on the high-performance computing systems at KMI (φ), at the Information Technology Center in Nagoya University (CX400), and at the Research Institute for Information Technology in Kyushu University (CX400 and HA8000). This work is supported by the JSPS Grant-in-Aid for Scientific Research (S) No.22224003, (C) No.23540300 (K.Y.), for Young Scientists (B) No.25800139 (H.O.) and No.25800138 (T.Y.), and also by the MEXT Grants-in-Aid for Scientific Research on Innovative Areas No.23105708 (T.Y.) and No.25105011 (M.K.). E.R. acknowledges the support of the U.S. Department of Energy under Contract DE-AC52-07NA27344 (LLNL).

References

1. For a review including lattice results and references, see V. A. Rubakov and M. E. Shaposhnikov, Usp. Fiz. Nauk **166** (1996) 493 [Phys. Usp. **39** (1996) 461].
2. K. Yamawaki, M. Bando, and K.-i. Matumoto, Phys. Rev. Lett. **56**, 1335 (1986); B. Holdom, Phys. Lett. **B150**, 301 (1985); T. Akiba and T. Yanagida, Phys. Lett. **B169**, 432 (1986); T. W. Appelquist, D. Karabali, and L. Wijewardhana, Phys. Rev. Lett. **57**, 957 (1986).
3. Y. Aoki *et al.* [LatKMI Collaboration], Phys. Rev. D **87**, no. 9, 094511 (2013); Phys. Rev. D **89**, no. 11, 111502 (2014).
4. R. D. Pisarski and F. Wilczek, Phys. Rev. D **29** (1984) 338. Phys. Rev. D **77** (2008) 015014.
5. A. Deuzeman, M. P. Lombardo and E. Pallante, Phys. Lett. B **670**, 41 (2008); K. Miura and M. P. Lombardo, Nucl. Phys. B **871**, 52 (2013).
6. D. Schaich, A. Cheng, A. Hasenfratz and G. Petropoulos, PoS LATTICE **2012**, 028 (2012); D. Schaich *et al.* [LSD Collaboration], arXiv:1506.08791 [hep-lat].
7. MILC Collaboration, http://www.physics.indiana.edu/~sg/milc.html.
8. A. Kocic, J. B. Kogut and M. P. Lombardo, Nucl. Phys. B **398**, 376 (1993).

Tree Level Unitarity and Finiteness of Electroweak Oblique Corrections*

Ryo Nagai

Department of Physics, Nagoya University,
Nagoya 464-8602, Japan
E-mail: nagai@th.phys.nagoya-u.ac.jp

We study perturbative unitarity and electroweak oblique corrections in the electroweak symmetry breaking models including an arbitrary number of neutral Higgs bosons. Requiring the perturbative unitarity of the high energy scattering amplitudes of weak gauge bosons and the neutral Higgs bosons at tree level, we obtain a set of conditions among the Higgs coupling strengths (unitarity sum rules). It is shown that the one-loop finiteness of the electroweak oblique corrections is automatically ensured once the unitarity sum rules are imposed among the Higgs coupling strengths. Applying the unitarity sum rules, we obtain severe constraints on the mass of the second lightest neutral Higgs boson and the lightest neutral Higgs (a 125GeV Higgs) coupling strength from the results of the electroweak precision tests.

Keywords: Perturbative unitarity, Electroweak precision tests.

1. Introduction

In 2012, the Large Hadron Collider (LHC) experiments confirmed the existence of a 125 GeV Higgs boson which interacts with the Standard Model (SM) particles. The current experimental results of the measured couplings of the Higgs boson are consistent with the SM predictions. However, the experimental uncertainties are still large. (The 68% confidence level (CL) uncertainty of the coupling scale factor for weak gauge bosons is approximately 10% [2].) The deviation of the Higgs coupling is therefore allowed at the present stage. From the theoretical point of view, the deviation of the Higgs coupling strength causes two problems if there is no new physics beyond the SM up to the Plank scale. The first one is the violation of the perturbative unitarity of the high energy weak gauge bosons' scattering amplitudes. The other one is the appearance of non-renormalizable ultraviolet (UV) divergences in radiative corrections. The deviation of the Higgs coupling strength is therefore evidence for new physics which makes the theory unitary and renormalizable. Can we predict the property of the required new physics if the measured value of the Higgs coupling strength turns out to deviate from the SM prediction? In order to answer this question, we study perturbative unitarity and electroweak oblique corrections in the electroweak symmetry breaking (EWSB) scenarios only with neutral Higgs bosons in a model independent manner. We find the precision measurements

*This report is based on Ref. 1.

of the discovered Higgs boson are a great probe for investigating the property of the extra neutral Higgs bosons.

2. Unitarity sum rules and Finiteness of electroweak oblique corrections

In this section, considering the EWSB scenarios only with neutral Higgs particles, we (re)derive two sets of conditions among the Higgs coupling strengths. The first set is known as "unitarity sum rules", which are required for the cancelation of the perturbatively unitary violating high energy scattering amplitudes of weak gauge bosons and the neutral Higgs bosons[3]. The other set is needed to guarantee the finiteness of electroweak oblique corrections.

We apply the electroweak chiral Lagrangian technique to describe interactions between the weak gauge bosons and the neutral Higgs bosons, ϕ_n^0, in an $SU(2)\times U(1)$ gauge invariant manner. The Lagrangian $\mathcal{L}$ is explicitly given by Ref. 1.

Let us first investigate perturbative unitarity in the present framework. Evaluating the high energy scattering amplitudes of the longitudinally polarized weak gauge bosons and the neutral Higgs bosons, we obtain a set of conditions among the Higgs coupling strengths which are needed to cancel the unitarity violating high energy scattering amplitudes. The unitarity sum rules at tree level are given by

$$WW \to WW \text{ scattering}: \quad -4 + \frac{3}{\rho_0} + \sum_{n=1}^{N_0} \kappa_{WW}^{\phi_n^0} \kappa_{WW}^{\phi_n^0} = 0, \tag{1}$$

$$WW \to ZZ \text{ scattering}: \quad \frac{1}{\rho_0} - \rho_0 \sum_{n=1}^{N_0} \kappa_{ZZ}^{\phi_n^0} \kappa_{WW}^{\phi_n^0} = 0, \tag{2}$$

$$WW \to \phi_n^0 Z \text{ scattering}: \quad \kappa_{WW}^{\phi_n^0} - \rho_0 \kappa_{ZZ}^{\phi_n^0} = 0, \quad \text{and} \quad \sum_{m=1}^{N_0} \kappa_Z^{\phi_n^0 \phi_m^0} \kappa_{WW}^{\phi_m^0} = 0, \tag{3}$$

$$WW \to \phi_n^0 \phi_m^0 \text{ scattering}: \quad \kappa_{WW}^{\phi_n^0 \phi_m^0} - \kappa_{WW}^{\phi_n^0} \kappa_{WW}^{\phi_m^0} = 0, \text{ and } \kappa_Z^{\phi_n^0 \phi_m^0} = 0, \tag{4}$$

$$ZZ \to \phi_n^0 \phi_m^0 \text{ scattering}: \quad \kappa_{ZZ}^{\phi_n^0 \phi_m^0} - \rho_0 \kappa_{ZZ}^{\phi_n^0} \kappa_{ZZ}^{\phi_m^0} - \sum_{l=1}^{N_0} \kappa_Z^{\phi_n^0 \phi_l^0} \kappa_Z^{\phi_m^0 \phi_l^0} = 0. \tag{5}$$

Here κs and ρ_0 denote the Higgs coupling strengths and tree level ρ parameter, respectively. Notice that if we combine Eq.(1), Eq.(2), and Eq.(3), we find a condition, $\rho_0^{-1}(\rho_0 - 1) = 0$. Thus, it is found that the unitarity sum rules require ρ_0 to be unity or infinity in any EWSB model if it only possesses neutral Higgs bosons.

Thanks to the gauge invariance of the non-linear sigma model Lagrangian we use, we can also study radiative corrections in the present framework. Let us next focus on electroweak oblique corrections, which can be expressed by S, T and U parameters for the model satisfying $\rho_0 = 1$[4]. Since the non-linear sigma model Lagrangian is not renormalizable, non-renormalizable UV divergences appear in the electroweak oblique corrections. The finiteness of the S, T and U parameters at one-loop level requires the following conditions among the Higgs coupling strengths,

Finiteness condition of the S parameter :

$$1-\sum_{n=1}^{N_0}\kappa_{ZZ}^{\phi_n^0}\kappa_{ZZ}^{\phi_n^0}-\frac{1}{2}\sum_{n=1}^{N_0}\sum_{m=1}^{N_0}\kappa_Z^{\phi_n^0\phi_m^0}\kappa_Z^{\phi_n^0\phi_m^0}=0, \tag{6}$$

Finiteness condition of the T parameter :

$$\sum_{n=1}^{N_0}\left[\kappa_{WW}^{\phi_n^0\phi_n^0}-2\kappa_{WW}^{\phi_n^0}\kappa_{WW}^{\phi_n^0}-\kappa_{ZZ}^{\phi_n^0\phi_n^0}+2\kappa_{ZZ}^{\phi_n^0}\kappa_{ZZ}^{\phi_n^0}+\sum_{m=1}^{N_0}\kappa_Z^{\phi_n^0\phi_m^0}\kappa_Z^{\phi_n^0\phi_m^0}\right]=0,$$

$$\sum_{n=1}^{N_0}\left(-\kappa_{WW}^{\phi_n^0\phi_n^0}+\kappa_{WW}^{\phi_n^0}\kappa_{WW}^{\phi_n^0}+\kappa_{ZZ}^{\phi_n^0\phi_n^0}-\kappa_{ZZ}^{\phi_n^0}\kappa_{ZZ}^{\phi_n^0}-\sum_{m=1}^{N_0}\kappa_Z^{\phi_n^0\phi_m^0}\kappa_Z^{\phi_n^0\phi_m^0}\right)\frac{M_{\phi_n^0}^2}{v^2}$$

$$-\frac{3}{4}g^2\sum_{n=1}^{N_0}(\kappa_{WW}^{\phi_n^0}\kappa_{WW}^{\phi_n^0}-\kappa_{ZZ}^{\phi_n^0}\kappa_{ZZ}^{\phi_n^0})-\frac{3}{4}g_Y^2(1-\sum_{n=1}^{N_0}\kappa_{ZZ}^{\phi_n^0}\kappa_{ZZ}^{\phi_n^0})=0, \tag{7}$$

Finiteness condition of the U parameter :

$$\sum_{n=1}^{N_0}(-\kappa_{WW}^{\phi_n^0}\kappa_{WW}^{\phi_n^0}+\kappa_{ZZ}^{\phi_n^0}\kappa_{ZZ}^{\phi_n^0})+\frac{1}{2}\sum_{n=1}^{N_0}\sum_{m=1}^{N_0}\kappa_Z^{\phi_n^0\phi_m^0}\kappa_Z^{\phi_n^0\phi_m^0}=0, \tag{8}$$

with $M_{\phi_n^0}$ being the mass of the neutral Higgs boson ϕ_n^0. It should be noted that the above conditions Eq.(6)-Eq.(8) are automatically satisfied if the Higgs coupling strengths satisfy the unitarity sum rules Eq.(1)-Eq.(5) in the present framework.

3. Constraints on heavy Higgs bosons

As discussed in the previous section, the one-loop finiteness of electroweak oblique corrections is ensured by the tree level unitarity sum rules. This allows us to evaluate finite contributions to the electroweak oblique corrections at one-loop level in any perturbatively unitary model only with neutral Higgs bosons. The finite corrections to the S and T parameters are given by

$$S \gtrsim \frac{1-\kappa_V^2}{12\pi}\left[\ln\frac{M_H^2}{M_h^2}+0.86\right]>0, \tag{9}$$

$$T \lesssim -\frac{3(1-\kappa_V^2)}{4v^2}\frac{(g^2+g_Y^2)}{g^2g_Y^2}(M_Z^2-M_W^2)\left[\ln\frac{M_H^2}{M_h^2}-1.05\right]<0, \tag{10}$$

for sufficiently the heavy extra Higgs bosons. Here κ_V denotes the coupling strength for the lightest Higgs boson h (a 125GeV Higgs boson), $\kappa_V \equiv \kappa_{WW}^h = \kappa_{ZZ}^h$. M_H denotes the mass of the second lightest Higgs boson. On the other hand, the EWPTs constraints the value of these parameters[6]. These constraints are summarized in Figure. 1. The 95% (99%) CL excluded areas from the EWPTs are shown by gray (dark-gray) area. The hatched area is disfavored from the perturbative unitarity[5]. We then find, for $\Delta\kappa_V \leq -0.008$, the constraints on the S and T parameters at 95% CL give more stringent bound on M_H than the unitarity bound.

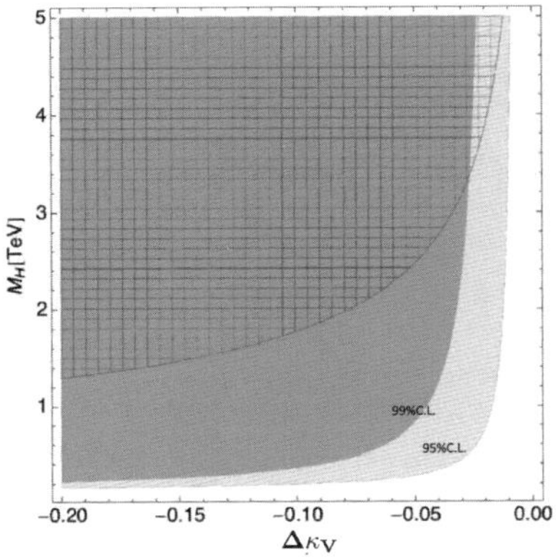

Fig. 1. Constraints on the second lightest Higgs boson mass as function of the deviation of the discovered Higgs coupling, $\Delta\kappa_V \equiv \kappa_V - 1$.

4. Summary

We study perturbative unitarity and electroweak oblique corrections in the EWSB model only with neutral Higgs bosons in a model independent manner. It is found that the one-loop finiteness of the electroweak oblique corrections is automatically guaranteed by the tree level perturbative unitarity sum rules in our framework. Applying the unitarity sum rules, we can test any perturbatively unitary model with neutral Higgs bosons extension by the results of the electroweak precision tests. The electroweak precision tests give us a severe bound on the second lightest neutral Higgs boson's mass as a function of the deviation of the discovered Higgs coupling strength. Once the deviation of the discovered Higgs coupling strengths with the weak gauge bosons is experimentally confirmed in future experiments, our results can be used to predict properties of the extra neutral Higgs boson.

Acknowledgments

Author is grateful to M. Tanabashi and K. Tsumura for fruitful collaborations. This work is supported by Research Fellowships of the Japan Society for the Promotion of Science (JSPS) for Young Scientists No.26·3947.

References

1. R. Nagai, M. Tanabashi and K. Tsumura, Phys. Rev. D **91**, no.3, 034030 (2015)
2. The ATLAS collaboration, ATLAS-CONF-2015-007, ATLAS-COM-CONF-2015-011. V. Khachatryan *et al.* [CMS Collaboration],
3. J. F. Gunion, H. E. Haber and J. Wudka, Phys. Rev. D **43** (1991) 904. B. Grinstein and M. Trott, Phys. Rev. D **76** (2007) 073002
4. M. E. Peskin and T. Takeuchi, Phys. Rev. Lett. **65** (1990) 964.
5. B. W. Lee, C. Quigg and H. B. Thacker, Phys. Rev. D **16** (1977) 1519.
6. M. Baak *et al.* [Gfitter Group Collaboration], Eur. Phys. J. C **74** (2014) 9, 3046

3d Seiberg Duality with an Adjoint Matter from 4d Duality

K. Nii

Department of Physics,
Nagoya University, Nagoya 464-8602, Japan

KEK Theory Center,
1-1 Oho, Tsukuba, Ibaraki 305-0801 Japan
E-mail: nii@post.kek.jp

We study the Seiberg dualities with an adjoint matter for the $SU(N)$ gauge groups in three- and four-spacetime dimensions with four supercharges. We derive the three-dimensional duality from four-dimensional one by the dimensional reduction including the non-perturbative effect of the $\mathbb{S}^1$-compactification, where the twisted instantons play a crucial role.

Keywords: Seiberg duality, Supersymmetry.

1. Introduction

The gauge theory is a fundamental framework to describe the particle physics. Up to the present date, the standard model is compatible with the experimental result. However the non-perturbative or strongly coupled regime of the gauge theory is difficult to calculate and to predict analytically about the physics. The understanding of the strongly coupled dynamics of the gauge theories is very important because it explains the confinement of the quarks, the chiral symmetry breaking and the mysterious duality between the gauge theory and the string theory.

The supersymmetry helps us overcome these difficulties. The holomorphy [1] of the supersymmetric gauge theories constrains the dynamics and we can exactly analyze the low-energy dynamics. Especially, for the $\mathcal{N} = 1$ supersymmetric QCD which is a supersymmetric version of the ordinary QCD, the most fascinating property, so-called Seiberg duality [2], is believed to hold. The Seiberg duality is a conjecture that the two seemingly different superymmetric gauge theories are identical in the infrared region, where these two theories have the different ranks of the gauge groups.

The Seiberg duality was applied to the more complicated gauge thories and found in three-dimensional gauge theories. However the proof of the Seiberg duality is now not accomplished and the relation between the 3d and 4d Seiberg dualities was not well understood.

Recently, Aharony et al. [3,4] have given the procedure to derive the 3d duality from the 4d duality. They consider the 4d theory on $\mathbb{S}^1 \times \mathbb{R}^3$. The important point is that when we put the theory on $\mathbb{S}^1 \times \mathbb{R}^3$, we need to care about the nonpertutbative superpotential from the twisted-instantons. With the presence of this potential they correctly derived the 3d duality. Although their technique in principle can be applied to the generic 4d dualities, the procedure becomes more difficult and unknown. In this paper we investigate the 4d Seiberg duality with fundamental matters and an adjoint matter and derive the 3d Seiberg duality form the 4d duality by using the dimensional reduction technique[5].

The organization of this paper is as follows. In section 2, we will briefly introduce the 4d duality which we would like to reduce to the 3d. In section 3, we consider the 4d theory on $\mathbb{S}^1 \times \mathbb{R}^3$ and show the effects of the compactification. In section 4 we will reduce the 4d duality to the 3d one. Finally in section 5 we summarize.

2. 4d Kutasov-Schwimmer duality

Here we briefly show the 4d $\mathcal{N} = 1$ SUSY gauge theory with an adjoint matter and its Seiberg duality which is known as the Kutasov-Schwimmer duality[6,7].

The "electric side" is a $SU(N_c)$ supersymmetric gauge theory with N_f fundamental matters $Q, \tilde{Q}$ and with an adjoint matter X, containing the superpotential $W = \operatorname{tr} X^{k+1}$. The "magnetic side" is a $SU(kN_f - N_c)$ supersymmetric gauge theory with N_f dual fundamental matters $q, \tilde{q}$, an adjoint matter Y and the gauge singlets M_j $(j = 0, \cdots, k-1)$. The superpotential of the magnetic theory is as follows.

$$W = \operatorname{tr} Y^{k+1} + \sum_{j=0}^{k-1} M_j \tilde{q} Y^{k-1-j} q. \tag{1}$$

The Seiberg duality conjectures that these two theories are equivalent in the far infrared region.

3. 4d theory on $\mathbb{S}^1 \times \mathbb{R}^3$

In order to derive the 3d duality from the 4d duality, we need to put the theory on a circle and take the low-energy limit $E \ll 1/r$, where r is a compactification radius[3,4]. The theory on $\mathbb{S}^1 \times \mathbb{R}^3$ is almost the same as the 4d one. However the non-perturbative effects play a important role. The twisted-instanton[8] generates the dynamical superpotential which drastically modifies the low-energy structure of the moduli space.

In this theory containing the adjoint matter, the twisted-instantons produces the following non-perturbative superpotential[5].

$$W_{\mathsf{electric}} = \sum_{i+j=k-1} \eta\, v_{i,+} v_{j,-} \tag{2}$$

$$W_{\mathsf{magnetic}} = \sum_{i+j=k-1} \tilde{\eta}\, \tilde{v}_{i,+} \tilde{v}_{j,-} \tag{3}$$

where η, $\tilde{\eta}$ are the some constants and v, $\tilde{v}$ are the Coulomb moduli which is the flat direction of the massless scalar fields. This superpotential changes the structure of the flat directions.

4. Dimensional Reduction of the 4d Seiberg duality with adjoint matter

Including the non-perturbative superpotential discussed above, we can correctly derive the 3d duality. The 3d duality reduced from the 4d Kutasov-Schwimmer duality is as follows (This duality was first conjectured by Park and Park[9]):

The electric side is a three-dimensional $\mathcal{N}=2$ supersymmetric $SU(N_c)$ gauge theory with N_f fundamental matters $Q, \widetilde{Q}$s and an adjoint matter X with a superpotential $W = \mathrm{tr}\, X^{k+1}$ which is completely the same as the 4d theory except for the dimensionality. This is a simple dimensional reduction of the 4d electric theory.

The "magnetic" theory is a $\mathcal{N}=2$ supersymmetric $U(kN_f - N_c) \times U(1)$ gauge theory with N_f fundamental dual matters $q, \tilde{q}$, one adjoint field Y, k $U(1)$-charged fields b_i, $\tilde{b}_i$ and the gauge singlets M_j. The superpotential is as follows.

$$W = \mathrm{Tr} Y^{k+1} + \sum_{j=0}^{k-1} M_j \tilde{q} Y^{k-1-j} q + \sum_{j=0}^{k-1} \left(b_{k-1-j} V_{j,+}^{U(kN_f - N_c)} + \tilde{b}_{k-1-j} V_{j,-}^{U(kN_f - N_c)} \right) \tag{4}$$

where $V_{j,\pm}^{U(kN_f - N_c)}$ $(j = 0, \cdots, k-1)$ are the Coulomb branch coordinates of the magnetic side and b_j, $\tilde{b}_j$ are identified with the Coulomb branch coordinates of the electric side. In the magnetic side, b_j, $\tilde{b}_j$ are the elementary fields while in the electric side these operators are the non-local and composite ones.

5. Summary

In this paper we considered the supersymmetric gauge theory with the fundamental matters and one adjoint matter and derived the 4d Seiberg duality with the adjoint matter to the 3d Seiberg duality. It is important to take into account the non-perturbative superpotential from the twisted instantons.

It will be worth extending this analysis to the more complicated 4d Seiberg duality. For example, we can do the same analysis to the $O(N)$ or $Sp(N)$ gauge theories. We can also consider the Seiberg duality with the tensorial matter fields.

References

1. N. Seiberg, hep-th/9408013.
2. N. Seiberg, Nucl. Phys. B **435**, 129 (1995) [hep-th/9411149].
3. O. Aharony, S. S. Razamat, N. Seiberg and B. Willett, JHEP **1307**, 149 (2013) [arXiv:1305.3924 [hep-th]].

4. O. Aharony, S. S. Razamat, N. Seiberg and B. Willett, JHEP **1308**, 099 (2013) [arXiv:1307.0511 [hep-th]].
5. K. Nii, JHEP **1502**, 024 (2015) [arXiv:1409.3230 [hep-th]].
6. D. Kutasov and A. Schwimmer, Phys. Lett. B **354**, 315 (1995) [hep-th/9505004].
7. D. Kutasov, Phys. Lett. B **351**, 230 (1995) [hep-th/9503086].
8. K. M. Lee and P. Yi, Phys. Rev. D **56**, 3711 (1997) [hep-th/9702107].
9. J. Park and K. J. Park, JHEP **1310**, 198 (2013) [arXiv:1305.6280 [hep-th]].

Pion Condensation in Asymmetric Nuclear Matter Using a Holographic QCD Model

Hiroki Nishihara* and Masayasu Harada
Department of Physics, Nagoya University,
Nagoya 464-8602, Japan
**E-mail: h248ra@hken.phys.nagoya-u.ac.jp*

We summarize the main points of our work on the study of the pion condensation in asymmetric nuclear matter by using a holographic QCD model. We show that the phase transition from the normal hadron phase to the pion condensation phase is delayed in asymmetric matter.

Keywords: Pion condensation, asymmetric nuclear matter, holographic QCD model.

1. Introduction

It is expected that studying hadron physics in extreme conditions provides some clues to understand Quantum ChromeDynamics (QCD). Related to the recent discovery of very heavy neutron star[2], it is very interesting to investigate the asymmetric nuclear matter.

The existence of various phases are expected at finite isospin chemical potential μ_I as well as at finite baryon chemical potential. When the isospin chemical potential is increased at zero temperature, the pion condensation phase appears for $\mu_I > m_\pi$ as shown in Ref. 4 by the $\mathcal{O}(p^2)$ chiral Lagrangian. In Ref. 3, it was shown that, using a holographic QCD model including mesons only, the chiral symmetry breaking is enhanced in the pion condensation phase for zero baryon number chemical potential. Recently in Ref. 1, this enhancement also occurs even when the effects of baryons are included. There it was shown that the critical isospin chemical potential is larger than the pion mass for non-zero baryon number chemical potential. In this write-up, we summarize main points of the results shown in Ref. 1.

2. Model and Analysis

The model which we use is a hard-wall type holographic QCD model[5–7] which describes dynamics of mesons. The model is constructed by requiring the $U(2)_L \times U(2)_R$ chiral symmetry[a] in five-dimensional space-time. Fifth direction denoted by z is bounded as $\epsilon \le z \le z_m$, where ϵ and z_m are UV- and IR-boundary,

[a]Our analysis is done in $L_5 = R_5 = 0$ gauge. Here L_5 and R_5 are fifth components of gauge fields corresponding to $U(2)_L \times U(2)_R$.

respectively. The explicit chiral symmetry braking due to the current quark mass and the spontaneous symmetry braking are introduced in the model.

To take account of the effect from a nuclear matter, we add the following term to the action:

$$S_{\text{int}} = \int d^4x \int_{\epsilon}^{z_m} dz \left[V_0^0 n_q + V_0^3 n_I^{\text{Baryon}} \right] \delta\left(z - z_m + \delta z\right) \tag{1}$$

where n_q and n_I^{Baryon} are the quark number density and the baryonic contribution to the isospin density of the nuclear matter, respectively. V_0^0 is the time component of the $\mathrm{U}(1)_V$ gauge field, and V_0^3 is the time component of the gauge field corresponding to the generator T^3 of $\mathrm{SU}(2)_{L+R}$. The baryon number density n_B is defined as $n_B = n_q/N_c$. The profile with respect to z direction of the introduced source is the δ-function which has a peak at IR-boundary[8].

Here we assume that proton (neutron) does not exist when the proton (neutron) chemical potential μ_p (μ_n) is smaller than its mass. Then, it is convenient to perform the analysis by introducing three cases as in Table 1. Here μ_p and μ_n are written

Table 1. Three cases

Case-(i)	$-m_N \leq \mu_p < m_N$,	$-m_N \leq \mu_n < m_N$
Case-(ii)	$m_N \leq \mu_p$,	$-m_N \leq \mu_n < m_N$
Case-(iii)	$m_N \leq \mu_p$,	$m_N \leq \mu_n$

in terms of the baryon number chemical potential μ_B and the isospin chemical potential μ_I as $\mu_p = \mu_B + \mu_I/2$ and $\mu_n = \mu_B - \mu_I/2$, respectively. The present assumption means that $2n_I^{\text{Baryon}} = n_B = 0$ in Case-(i) and $2n_I^{\text{Baryon}} = n_B = n_p$ in Case-(ii)[b].

Let us introduce the quark number chemical potential μ_q and the isospin chemical potential μ_I. When we add S_{int} to the Lagrangian, we have to fix the gauge of four-dimensional part. Therefore we introduce μ_q and μ_I into the time component of the gauge field at UV-boundary as $V_0^0\big|_\epsilon = \mu_q - c_{(0)}$, $V_0^3\big|_\epsilon = \mu_I - c_{(3)}$,[c] where $c_{(0)}$ and $c_{(3)}$ correspond to the degrees of freedom of the geuge transformation. We take $c_{(3)} = 0$ in order that two charged pions are degenerate at $\mu_I = 0$. The value of $c_{(0)}$ for Case-(ii) and Case-(iii) in Table 1 is determined as $c_{(0)} = m_N/N_c$, while we do not need to determine the value for Case-(i).

There exist five parameters in this model. One of them is determined by comparing to the vector current correlation function in QCD. The values of the other parameters are obtained by using the pion mass $m_\pi = 139.6\,\text{MeV}$, the pion decay constant $f_\pi = 92.4\,\text{MeV}$, the ρ meson mass $m_\rho = 775.8\,\text{MeV}$, and the a_0 meson

[b] n_I^{Baryon} and n_B are given by sum and difference between the proton number density n_p and the neutron number density n_n, respectively: $n_I^{\text{Baryon}} = \frac{n_p - n_n}{2}$, $n_B = n_p + n_n$.
[c] The baryon number chemical potential μ_B relates to μ_q through $\mu_B \equiv N_c \mu_q$.

mass. Although we use $m_{a_0} = 980\,\mathrm{MeV}$ as the a_0 meson mass, our results are almost independent of its value.

The grand potential Ω is obtained as $\Omega = -\int_{\epsilon}^{z_m} dz\mathcal{L}$ from the five-dimensional Lagrangian. By minimizing the Ω, we study the phase transition. We note that the isospin density is given as $n_I = -\frac{\partial\Omega}{\partial\mu_I}$.

3. Result

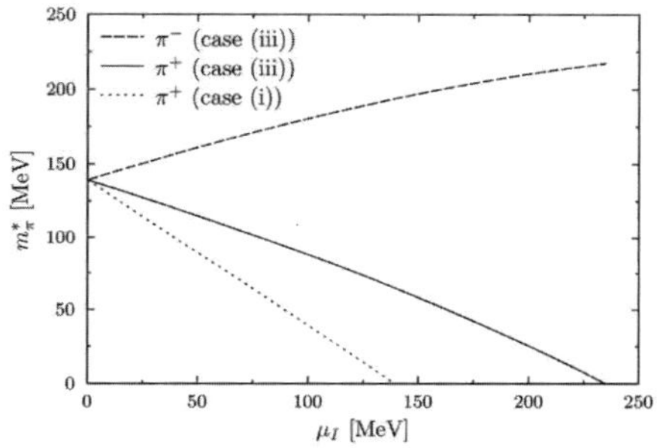

Fig. 1. μ_I dependence of the charged pion mass. The solid line and the dashed line express the π^+ and π^- mass in Case-(iii), respectively. The dotted line is the π^+ mass in Case-(i).

Figure 1 shows the μ_I dependence of the pion mass m_π^* in the normal hadron phase. The π^- mass increases with the increasing μ_I. On the other hand, the π^+ mass decreases, and its mass vanishes at $\mu_I = 235\,\mathrm{MeV}$, which implies that the π^+ condensation occurs. We find that the decrease of the π^+ mass in the asymmetric nuclear matter (Case-(iii)) is slower than that in the pure mesonic case (Case-(i)). Furthermore, the value of the critical chemical potential for the asymmetric nuclear matter is larger than that for the pure mesonic case.

We can get the symmetry energy $E_{\mathrm{sym}}(n_0)$ at the saturation density $n_0 = 0.16\mathrm{fm}^{-3}$ and its slope parameter L. Our results are $E_{\mathrm{sym}}(n_0) = 29\,\mathrm{MeV}$ and $L = 87\,\mathrm{MeV}$, which are comparable with experimental data[9], $E_{\mathrm{sym}}(n_0) = 32.3 \pm 1.0\,\mathrm{MeV}$ and $L = 45.2 \pm 10.0\,\mathrm{MeV}$.

4. Summary

We showed our results for studying the pion condensation in asymmetric nuclear matter by introducing a baryonic source to a holographic QCD model. Figure 2 is the QCD phase diagram obtained by our analysis. In this figure, the pion condensation phase and the normal hadron phase are shown by the dark area and the white area, respectively. Although, the phase transition in the pure mesonic case (Case-(i)) occurs at $\mu_I = m_\pi$, the existence of asymmetric nuclear matter (Case-(iii))

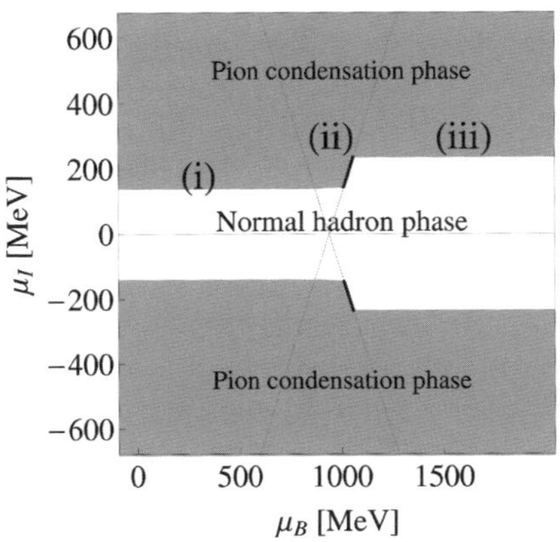

Fig. 2. Phase diagram: μ_B vs. μ_I.

delays the transition.

One can reproduce delay of the phase transition by using a model with Hidden Local Symmetry which includes the ρ meson. This implies that existence of the ρ meson causes delay of the phase transition.

In addition, by solving coupled equations of motion, we can discuss the equation of state and the value of the chiral condensate in the pion condensation phase. Its detail is shown in Ref. 1.

Acknowledgements

This work was supported in part by Grant-in-Aid for Scientific Research on Innovative Areas (No. 2104) "Quest on New Hadrons with Variety of Flavors" from MEXT, and the JSPS Grant-in-Aid for Scientific Research (S) No. 22224003, (c) No. 24540266.

References

1. H. Nishihara and M. Harada, Phys. Rev. D **90**, no. 11, 115027 (2014)
2. P. Demorest, T. Pennucci, S. Ransom, M. Roberts and J. Hessels, Nature **467**, 1081 (2010).
3. H. Nishihara and M. Harada, Phys. Rev. D **89**, 076001 (2014).
4. D. T. Son and M. A. Stephanov, Phys. Rev. Lett. **86**, 592 (2001).
5. J. Erlich, E. Katz, D. T. Son and M. A. Stephanov, Phys. Rev. Lett. **95**, 261602 (2005).
6. L. Da Rold and A. Pomarol, Nucl. Phys. B **721**, 79 (2005).
7. L. Da Rold and A. Pomarol, JHEP **0601**, 157 (2006).
8. K. -Y. Kim, S. -J. Sin and I. Zahed, JHEP **0801**, 002 (2008).
9. Z. Zhang and L. -W. Chen, Phys. Lett. B **726**, 234 (2013).

Unification of SUSY Breaking and GUT Breaking

Yuji Omura
Kobayashi-Maskawa Institute for the Origin of Particles and the Universe (KMI), Nagoya University, Nagoya 464-8602, Japan
E-mail: yujiomur@eken.phys.nagoya-u.ac.jp

We introduce an explicit supersymmetric unification model where grand unified gauge symmetry breaking and supersymmetry (SUSY) breaking are caused by the same field. Besides, the SM-charged particles are also predicted by the symmetry breaking sector, and their loop corrections induce the soft SUSY breaking terms. Especially, nonzero A-term and B-term are generated at one-loop level according to the mediation via the vector superfields, so that the electro-weak symmetry breaking and 125 GeV Higgs mass could be achieved even if the stop mass is around 1 TeV.

Keywords: SUSY GUT, SUSY breaking, gauge mediation.

1. Introduction

We are sure that there are several mysteries in our nature, although the Standard Model (SM) successfully describes the experimental results. For instance, the structure of the fermion generations, the hyper-charge assignment and the Higgs mass in the SM seem to suggest unknown physics behind the SM. One of the candidates for the Beyond Standard Models (BSMs) is the supersymmetric grand unified theory (GUT), which reveals the origin of the Higgs mass and the fermion charges. The SM gauge group, $SU(3)_c \times SU(2)_L \times U(1)_Y$, is naturally unified into a $SU(5)$ gauge group, and supersymmetry (SUSY) can explain why the Higgs mass is tachyonic and around the electro-weak (EW) scale. The SUSY $SU(5)$-GUT has been studied for a long time, and several issues, as well as interesting features, are pointed out so far: for instance, the too short life time of proton, how to realize the Yukawa coupling, and how to achieve 125-GeV Higgs mass as a recent topics. Futhermore, we should explicitly mention how to break SUSY and induce the soft SUSY breaking terms in the Minimal Supersymmetric Standard Model (MSSM).

In this paper, we propose an explicit supersymmetric GUT with $SU(5)_F \times SU(2) \times U(1)_\phi$ gauge symmetry. There are two sectors in this model: the MSSM sector charged under only $SU(5)_F$, and the breaking sector which causes SUSY breaking and GUT symmetry breaking spontaneously. The SM fields belong to the MSSM sector, so that the charge quantization can be realized. The breaking sector consists of one $SU(5)_F$ adjoint plus singlet filed (Φ) and $SU(5)_F$ fundamental and anti-fundamental fields $(\phi,\ \widetilde{\phi})$. The vector-like pairs $(\phi,\ \widetilde{\phi})$ are also charged under $SU(2) \times U(1)_\phi$. In this setup, the nonzero vacuum expectation values (VEVs) of (Φ) and $(\phi,\ \widetilde{\phi})$ break $SU(5)_F \times SU(2) \times U(1)_\phi$ symmetry to the SM gauge groups, $SU(3)_c \times SU(2)_L \times U(1)_Y$, where $SU(3)_c$ is from the subgroup of $SU(5)_F$, and $SU(2)_L \times U(1)_Y$ are the linear combinations of the subgroup of $SU(5)_F$ and $SU(2) \times U(1)_\phi$. Besides, SUSY is broken by the F-component of the part of Φ. After the symmetry breaking, additional SM-charged particles are generated by the fluctuation of Φ and $(\phi, \widetilde{\phi})$ around the VEVs. One interesting point is that the fermionic partner of the massive gauge boson of $SU(5)_F$ could mediate the SUSY breaking effect through the gauge coupling with Φ, and play a crucial role in generating the non-zero A-term and B-term as discussed in Refs.[1,2]. It is well-known that large A-term could shift the upper bound on the lightest Higgs mass in the MSSM, even if squark is light, and the sizable B-term is required to realize the EW symmetry breaking. Our A-term and B-term are given at one-loop level, so that they are the same order as the squark masses and gaugino masses. In fact, we will see that Higgs mass could be around 125 GeV, even if the SUSY scale (Λ_{SUSY}) is less than $O(1)$ TeV, and the B-term could be consistent with the EW symmetry breaking in our model.

2. Setup

We briefly explain our setup of this model which causes SUSY breaking together with GUT gauge symmetry breaking, based on Ref.[3].

The MSSM chiral superfields are only charged under $SU(5)_F$ gauge symmetry, and they belong to the **10** and $\overline{\mathbf{5}}$ representations of $SU(5)_F$, based on the Georgi-Glashow $SU(5)$ GUT[4]. The charge assignment and matter contents are summarized below. **5**-representation Higgs, $(H, \overline{H})$, are also introduced to write the Yukawa couplings in the visible sector.

	$\overline{\mathbf{5}}_i$	$\mathbf{10}_i$	H	$\overline{H}$	ϕ	$\widetilde{\phi}$	Φ
$SU(5)_F$	$\overline{\mathbf{5}}$	$\mathbf{10}$	$\mathbf{5}$	$\overline{\mathbf{5}}$	$\mathbf{5}$	$\overline{\mathbf{5}}$	$\mathbf{adj}_5\mathbf{+1}$
$SU(2)$	$\mathbf{1}$	$\mathbf{1}$	$\mathbf{1}$	$\mathbf{1}$	$\mathbf{2}$	$\mathbf{2}$	$\mathbf{1}$
$U(1)_\phi$	0	0	0	0	Q_ϕ	$-Q_\phi$	0

Φ is the $SU(5)_F$ adjoint plus singlet field and $(\phi, \widetilde{\phi})$ pair is the vector-like under $SU(5)_F \times SU(2) \times U(1)_\phi$ gauge group. The superpotential for the breaking sector is given by

$$W_{SB} = W_R + W_{RB} \tag{1}$$

$$W_R = -hTr_2(\widetilde{\phi}\Phi\phi) + h\Lambda_G \mathrm{Tr}_5(\Phi), \tag{2}$$

$$W_{RB} = m_\phi Tr_2(\widetilde{\phi}\phi) + c. \tag{3}$$

W_R respects the R-symmetry and breaks the SUSY spontaneously[5]. W_{RB} breaks the R-symmetry explicitly, and gives mass of R-axion. Some ideas to induce W_{RB} have been proposed in Ref.[6], where the small wave-function factor of Φ suppresses

Φ^2 and Φ^3 terms according to the strong dynamics or the profile in the extra dimension. Here, we simply start the discussion from the superpotential W_{SB} assuming that such a mechanism, as discussed in Ref.[6], works in underlying theories above the GUT scale, and study the symmetry breaking.

In the global SUSY with canonical Kähler potential, the scalar potential is given by $V = |\partial_\Phi W_{SB}|^2 + |\partial_\phi W_{SB}|^2 + |\partial_{\tilde{\phi}} W_{SB}|^2$, and SUSY vacua satisfy $\partial_\Phi W_{SB} = \partial_\phi W_{SB} = \partial_{\tilde{\phi}} W_{SB} = 0$. In this model, $\partial_\Phi W_{SB}$ is given by $\partial_{\Phi_{ji}} W_{SB} = -h(\phi\tilde{\phi})_{ij} + h\Lambda_G \delta_{ij}$ $(i, j = 1, \ldots, 5)$, and all elements cannot be vanishing, because 5×5 matrix $(\phi\tilde{\phi})$ has the rank 2^5. This means that SUSY is broken by the F-components of the remnant 3 elements in Φ and $SU(5)_F$ is also broken at the same time. There are two scales in W_{SB}: m_ϕ and Λ_G. The former relates to the $SU(5)_F$ breaking scale and the letter is concerned with the SUSY breaking scale[3].

As well studied in Ref.[3], the fluctuations around the SUSY and $SU(5)_F$ breaking vacuum correspond to the fields charged under the SM gauge symmetry, whose masses are below/at the GUT scale. Then, the extra chiral superfields works as the mediator to induce the soft SUSY breaking terms in the MSSM. Futhermore, we could find that the massive vector superfields, which are generated by the $SU(5)_F$ symmetry breaking, could also mediate the SUSY breaking effect, because the $SU(5)_F$-adjoint field, Φ, breaks both of SUSY and GUT gauge symmetry. Then we will have some additional contributions to the soft SUSY breaking terms compared to the ordinary gauge-mediation scenario[1–3]. The soft SUSY breaking terms, including A-term and B-term, could be estimated as

$$m_{\rm soft} \approx \frac{\alpha_{SM}}{4\pi} \frac{M_p}{\Lambda_{GUT}} m_{3/2}, \tag{4}$$

where M_p, Λ_{GUT}, and $m_{3/2}$ are the Planck scale, the GUT scale and the gravitino mass respectively. α_{SM} denotes the SM gauge coupling, and controls the soft SUSY breaking parameters, so that our model has very explicit predictions for the mass spectrums of the SUSY particles.

One issue in supersymmetric models is how to realize the μ and B terms which are consistent with the EW scale. Especially, the fine-tuning of μ might be required by the recent Higgs discovery around the 125 GeV mass region. In fact, 125 GeV Higgs mass seems to suggest $\Lambda_{SUSY} \gtrsim O(10)$ TeV in the simple scenarios as discussed in Ref.[7,8]. $O(10)$-TeV SUSY scale corresponds to at least 0.01% fine-tuning against μ without any cancellation in $m_{H_u}^2$. As pointed out in Refs.[9,10], it is known that a special relation between A_t and squark mass relaxes the fine-tuning, maximizing the loop corrections in the Higgs mass in the MSSM. This relation is so-called "maximal mixing" and described as $X_t/m_{stop} = \sqrt{6}$, where $X_t = A_t - \mu/\tan\beta$ and $m_{stop}^2 = \sqrt{m_Q^2 m_U^2}$ are defined. If this relation is satisfied, the 125 GeV Higgs mass could be achieved even if the stop is less than 1 TeV. In our model, the predicted A-term is quite large, so that the maximal mixing might be possible.

3. Summary

We proposed an explicit and simple supersymmetric model, where the spontaneous SUSY breaking and GUT breaking are achieved by the same sector. The origin of the hyper-charge assignment in the MSSM is also explained by the analogy with the Georgi-Glashow $SU(5)$ GUT[4]. The SM-charged particles are also introduced by the breaking sector, so that we could also predict the soft SUSY breaking terms via the gauge mediation with the gauge and chiral messenger superfields. The crucial role of the gauge-messenger mediation is to induce large A-terms and B-terms at the one-loop level. We investigated the scenario with light superpartners that such a large A-term realizes the maximal mixing and shifts the lightest Higgs mass.

Our light SUSY particles are wino, bino, and gravitino, and the mass difference is not so big. The lightest particle is bino, and wino is heavier than bino. The mass difference is $O(0.1) \times m_{3/2}$ GeV.

The author would like to thank T. Kobayashi for fruitful collaboration. This work is supported by Grant-in-Aid for Scientific research from the Ministry of Education, Science, Sports, and Culture (MEXT), Japan, No. 23104011.

References

1. R. Dermisek, H. D. Kim and I. W. Kim, JHEP **0610**, 001 (2006) [hep-ph/0607169].
2. L. Matos, JHEP **1012**, 042 (2010) [arXiv:1007.3616 [hep-ph]].
3. T. Kobayashi and Y. Omura, JHEP **1502**, 114 (2015) [arXiv:1410.5173 [hep-ph]].
4. H. Georgi and S. L. Glashow, Phys. Rev. Lett. **32**, 438 (1974).
5. K. Intriligator, N. Seiberg and D. Shih, JHEP **0604**, 021 (2006) [arXiv:hep-th/0602239].
6. H. Abe, T. Kobayashi and Y. Omura, Phys. Rev. D **77**, 065001 (2008) [arXiv:0712.2519 [hep-ph]].
7. G. F. Giudice and A. Strumia, Nucl. Phys. B **858**, 63 (2012) [arXiv:1108.6077 [hep-ph]].
8. G. Degrassi, S. Di Vita, J. Elias-Miro, J. R. Espinosa, G. F. Giudice, G. Isidori and A. Strumia, JHEP **1208**, 098 (2012) [arXiv:1205.6497 [hep-ph]].
9. M. S. Carena, M. Quiros and C. E. M. Wagner, Nucl. Phys. B **461**, 407 (1996) [hep-ph/9508343].
10. T. Hahn, S. Heinemeyer, W. Hollik, H. Rzehak and G. Weiglein, Phys. Rev. Lett. **112**, 141801 (2014) [arXiv:1312.4937 [hep-ph]].

Finite-Temperature Study of Eight-Flavor SU(3) Gauge Theory

David Schaich,[1*] Anna Hasenfratz[2] and Enrico Rinaldi[3]
for the Lattice Strong Dynamics (LSD) Collaboration
[1]*Department of Physics, Syracuse University, Syracuse, New York 13244, USA*
[2]*Department of Physics, University of Colorado, Boulder, Colorado 80309, USA*
[3]*Lawrence Livermore National Laboratory, Livermore, California 94550, USA*
**E-mail: dschaich@syr.edu*

We present new lattice investigations of finite-temperature transitions for SU(3) gauge theory with $N_f = 8$ light flavors. Using nHYP-smeared staggered fermions we are able to explore renormalized couplings $g^2 \lesssim 20$ on lattice volumes as large as $48^3 \times 24$. Finite-temperature transitions at non-zero fermion mass do not persist in the chiral limit, instead running into a strongly coupled lattice phase as the mass decreases. That is, finite-temperature studies with this lattice action require even larger $N_T > 24$ to directly confirm spontaneous chiral symmetry breaking.

SU(3) gauge theory with $N_f = 8$ massless fundamental flavors is currently the subject of considerable interest, both as a quantum field theory exhibiting strongly coupled dynamics significantly different from QCD, and also as the basis for models of new strong dynamics producing a standard-model-like 125-GeV Higgs particle. We can only include here an incomplete collection of references to recent investigations employing both continuum and lattice methods.[1–11] In this work we attempt to confirm the conventional wisdom that chiral symmetry breaks spontaneously for $N_f = 8$, which would rule out the existence of an 8-flavor conformal IR fixed point (IRFP). However, this requires extrapolating to the massless chiral limit, and we are unable to establish that chiral symmetry breaking persists in that limit.

Previous lattice studies have explored the discrete β function of the 8-flavor system, denoted $\beta_s(g^2)$ for scale change s.[1,10,11] An IRFP would correspond to $\beta_s(g_\star^2) = 0$, and no such zero has been observed. In fact, the β function was found to be monotonic throughout the ranges of couplings explored: $g_{SF}^2 \lesssim 6.6$ in the Schrödinger functional scheme,[1] $g_c^2 \lesssim 6.3$ in a gradient flow scheme with $c = 0.3$,[11,12] and $g_c^2 \lesssim 14$ in gradient flow schemes with $c = 0.25$ and $c = 0.3$.[10] The last result is the most relevant to this work, since we use the same lattice action with nHYP-smeared staggered fermions and both fundamental and adjoint plaquette terms with couplings related by $\beta_A/\beta_F = -0.25$.[13] This enables us to relate our bare lattice couplings β_F to renormalized g_c^2. This work is part of ongoing investigations by the Lattice Strong Dynamics (LSD) Collaboration, using this action to extend the USBSM project.[5]

The non-observation of an IRFP by lattice studies of the discrete β function does not guarantee that the 8-flavor theory exhibits spontaneous chiral symmetry breaking. On their own such calculations cannot rule out the possibility that the system flows to an IRFP at a stronger coupling. In fact, the non-perturbative $\beta_s(g_c^2 \approx 14)$ is comparable to the four-loop $\overline{\text{MS}}$ prediction,[10] which does possess an IRFP at $g^2_{\overline{\text{MS}}} \approx 19.5$. To exclude such behavior one must demonstrate that the massless system spontaneously breaks chiral symmetry at some g^2 for which the β function is still non-zero. Here we attempt to do this by studying chiral symmetry breaking at finite temperature $T = 1/(aN_T)$ and non-zero fermion mass am, where "a" is the lattice spacing and the lattice volume is $L^3 \times N_T$ with $L/N_T = 2$.[2,4] We work with fixed $N_T = 20$ and 24 for small $0.0025 \leq am \leq 0.01$, and extrapolate $am \to 0$ to investigate the massless chiral limit.

The chiral extrapolation is crucial since $am > 0$ explicitly breaks chiral symmetry and can even produce QCD-like scaling that disappears as am decreases. This behavior was observed in previous studies (using the same lattice action) that considered $am \geq 0.005$ and $N_T \leq 20$.[2,4] As shown in the left plot of Fig. 1, QCD-like scaling between $12 \leq N_T \leq 16$ for $am \geq 0.01$ is lost at $am = 0.005$ where the finite-temperature transitions merge with bulk (zero-temperature) transitions into the "$\not{S}^4$" lattice phase in which the single-site shift symmetry (S^4) of the staggered action is spontaneously broken.[13] In IR-conformal systems such behavior persists in the chiral limit as $N_T \to \infty$, whereas for chirally broken systems the massless transitions must move to $\beta_F^{(c)} \to \infty$ as $N_T \to \infty$.

The right plot of Fig. 1 shows our new results, which include smaller $am = 0.0025$ on $40^3 \times 20$ and $48^3 \times 24$ lattices. These $N_T = 20$ and 24 are extraordinarily large compared to typical lattice QCD calculations. Unfortunately they do not suffice to establish spontaneous chiral symmetry breaking. At $am = 0.0025$ the $N_T = 20$ finite-temperature transition also merges with the bulk transition into the $\not{S}^4$ lattice phase. Even the $N_T = 24$ finite-temperature transitions will clearly run into the $\not{S}^4$ phase at non-zero mass, rather than reaching the chiral limit.

In Fig. 2 we present some results for observables that have proven useful to

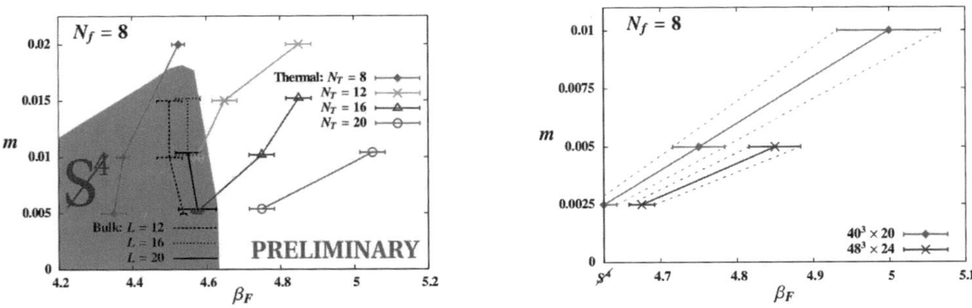

Fig. 1. **Left:** Previous $N_f = 8$ studies found that the finite-temperature transitions merge with zero-temperature bulk transitions into the $\not{S}^4$ lattice phase as the fermion mass am decreases.[2,4] **Right:** Our new results produce the same behavior at smaller masses on larger lattice volumes, implying that $N_T > 24$ is required to establish spontaneous chiral symmetry breaking for $N_f = 8$. This plot zooms in on the weak-coupling regime to the right of the $\not{S}^4$ phase.

identify both bulk and finite-temperature transitions: the Wilson-flowed Polyakov loop PL_W and the massless Dirac eigenvalue spectrum $\rho(\lambda)$. PL_W is a modern adaptation of the RG-blocked Polyakov loop investigated in previous studies, which significantly improves signals of the finite-temperature transition without altering its location.[2,4] It is trivial to measure the Polyakov loop as a function of Wilson flow time t, and sufficiently large t produces a clear contrast between confined systems with small PL_W and deconfined systems with large PL_W. This is shown in the left plot of Fig. 2 for $N_T = 20$ and $t = (0.3N_T)^2/8$ corresponding to $c = \sqrt{8t}/N_T = 0.3$. As the fermion mass decreases from $am = 0.01$ the finite-temperature transition in PL_W steadily sharpens and moves to stronger coupling, merging with the bulk transition into the $\not{S}^4$ phase at $am = 0.0025$ as in Fig. 1.

The right plot of Fig. 2 shows the eigenvalue spectrum $\rho(\lambda)$ for a subset of the $N_T = 20$ ensembles with $am = 0.005$, clearly contrasting the three different phases we can observe for this mass. At the weakest coupling shown, $\beta_F = 4.8$, the system is deconfined and chirally symmetric, with $\rho(0) = 0$ and a gap below the smallest eigenvalue $\lambda_0 > 0$. The gap grows at even weaker couplings that are not included in this plot. Moving to stronger couplings, at $\beta_F = 4.7$ we observe the expected chiral symmetry breaking, with $\rho(0) \neq 0$ and a small slope $\frac{d\rho}{d\lambda}$. That is, we find $4.7 < \beta_F^{(c)} < 4.8$ for the $N_T = 20$ transition with $am = 0.005$, slightly sharper than the signal in PL_W. However, at the strongest coupling shown, $\beta_F = 4.5$, chiral symmetry breaking is lost ($\lambda_0 > 0$) and the system exhibits the "soft edge" $\rho(\lambda) \propto \sqrt{\lambda - \lambda_0}$ characteristic of the $\not{S}^4$ phase.[13] Finally, $\beta_F = 4.6$ appears to exhibit partial features of both the chirally broken and $\not{S}^4$ phases, with $\rho(0) \neq 0$ but a much larger slope approaching the square-root behavior of the soft edge.

While this finite-temperature study with $N_T = 20$ and 24 is not able to establish spontaneous chiral symmetry breaking for $N_f = 8$, it is only part of ongoing inves-

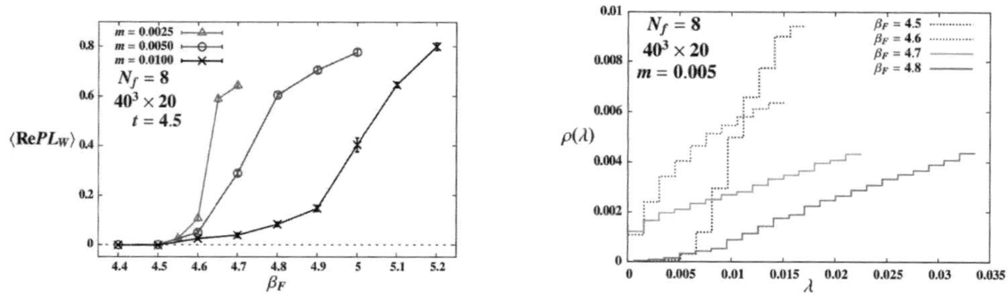

Fig. 2. **Left:** The $c = 0.3$ Wilson-flowed Polyakov loop PL_W for $40^3 \times 20$ lattices with $am = 0.01$, 0.005 and 0.0025 vs. the bare lattice coupling β_F. As the fermion mass decreases the transitions sharpen and move to stronger coupling, eventually merging with the zero-temperature bulk transition into the $\not{S}^4$ phase at $\beta_F \approx 4.625$. **Right:** The massless Dirac eigenvalue spectrum $\rho(\lambda)$ contrasts the three phases encountered for $40^3 \times 20$ lattices with $am = 0.005$: the chirally symmetric phase at weak coupling ($\beta_F = 4.8$), the chirally broken phase at intermediate coupling ($\beta_F = 4.7$) and the $\not{S}^4$ lattice phase at strong coupling ($\beta_F = 4.5$). We directly measure 200 eigenmodes to produce each histogram.

tigations by the LSD Collaboration that primarily focus on the zero-temperature hadron spectrum, and in particular the scalar Higgs particle. When that work is finalized, combining these complementary studies of the discrete β function, finite-temperature transitions, and hadron spectrum, all using the same lattice action, will shed further light on $N_f = 8$ and its phenomenological viability as the basis for new strong dynamics beyond the standard model.

Acknowledgments: We thank the other members of LSD for continuing collaboration on $N_f = 8$. AH and DS are grateful for the hospitality of the Aspen Center for Physics, supported by the U.S. National Science Foundation under Grant No. PHYS-1066293. This work was supported by Lawrence Livermore National Laboratory (LLNL) LDRD 13-ERD-023 (ER) and by the U.S. Department of Energy (DOE), Office of Science, Office of High Energy Physics, under Award Numbers DE-SC0010005 (AH), DE-SC0008669 (DS) and DE-SC0009998 (DS). Numerical calculations were carried out through LLNL Institutional Computing Grand Challenge program allocations on the LLNL BlueGene/Q (rzuseq and vulcan) supercomputer and on the DOE-funded USQCD facilities at Fermilab.

References

1. T. Appelquist, G. T. Fleming and E. T. Neil, *Phys. Rev.* **D79**, 076010 (2009).
2. D. Schaich, A. Cheng, A. Hasenfratz and G. Petropoulos, *Proc. Sci.* **Lattice 2012**, 028 (2012).
3. A. Bashir, A. Raya and J. Rodriguez-Quintero, *Phys. Rev.* **D88**, 054003 (2013).
4. A. Hasenfratz, A. Cheng, G. Petropoulos and D. Schaich, *Strong Coupling Gauge Theories in the LHC Perspective (SCGT12)*, 44–50 (2014).
5. D. Schaich, *Proc. Sci.* **LATTICE 2013**, 072 (2013).
6. Y. Aoki, T. Aoyama, M. Kurachi, T. Maskawa, K. Miura, K.-i. Nagai, H. Ohki, E. Rinaldi, A. Shibata, K. Yamawaki and T. Yamazaki, *Phys. Rev.* **D89**, 111502 (2014).
7. M. Kurachi, S. Matsuzaki and K. Yamawaki, *Phys. Rev.* **D90**, 055028 (2014).
8. T. Appelquist, R. C. Brower, G. T. Fleming, J. Kiskis, M. F. Lin, E. T. Neil, J. C. Osborn, C. Rebbi, E. Rinaldi, D. Schaich, C. Schroeder, S. Syritsyn, G. Voronov, P. Vranas, E. Weinberg and O. Witzel, *Phys. Rev.* **D90**, 114502 (2014).
9. M. P. Lombardo, K. Miura, T. J. Nunes da Silva and E. Pallante, *Int. J. Mod. Phys.* **A29**, 1445007 (2014).
10. A. Hasenfratz, D. Schaich and A. Veernala, *JHEP* **1506**, 143 (2015).
11. Z. Fodor, K. Holland, J. Kuti, S. Mondal, D. Nogradi and C. H. Wong, *JHEP* **1506**, 019 (2015).
12. Z. Fodor, K. Holland, J. Kuti, D. Nogradi and C. H. Wong, *JHEP* **1211**, 007 (2012).
13. A. Cheng, A. Hasenfratz and D. Schaich, *Phys. Rev.* **D85**, 094509 (2012).

Schwinger-Dyson Study for Walking/Conformal Dynamics with IR Cutoffs

Kohtaro Miura[a,b], Kei-ichi Nagai[a] and Akihiro Shibata[c]

[a] *Kobayashi-Maskawa Institute for the Origin of Particles and the Universe (KMI), Nagoya University, Nagoya, 464-8602, Japan*

[b] *Centre de Physique Theorique(CPT), Aix-Marseille University, Campus de Luminy, Case 907, 163 Avenue de Luminy, 13288 Marseille cedex 9, France*

[c] *Computing Research Center, High Energy Accelerator Research Organization (KEK), Tsukuba 305-0801, Japan*

Motivated by the recent progress on many flavor QCD on a lattice, we investigate conformal/walking dynamics by using Schwinger-Dyson (SD) equation within an improved ladder approximation for two-loop running coupling. By numerically solving the SD equation, we obtain a pole mass (m_p), pion decay constant (f_π), and investigate the chiral symmetry breaking and mass anomalous dimension (γ_m) in the presence of IR cutoffs ($\Lambda_{\rm IR}$). We find that the chiral symmetry breaking is suppressed if IR cutoff ($\Lambda_{\rm IR}$) becomes larger than the critical value near the dynamical mass ($\Lambda^*_{\rm IR} \sim m_D$). In the conformal phase the γ_m is strongly suppressed by IR cutoffs for $\Lambda_{\rm IR} \gtrsim m_p$. We, then, obtain finite size hyperscaling (FSS) relation by adapting a linearized approximation for the SD equation, and examine the γ_m. The results offer valuable insight and suggestion for analyses in lattice gauge theories.

Keywords: Many-flavor QCD, Conformal/walking theory, Schwinger-Dyson equation.

1. Introduction

When a fermion multiplicity (flavor) exceeds a critical value $N_f = N_f^*$ in non-Abelian gauge theories, a quantum phase transition from a chiral broken phase to the conformal window is anticipated to take place. In the conformal window, strong coupling gauge theories are asymptotically-free in ultra violet (UV) region, while dominated by a non-trivial infrared fixed point (IRFP) in infrared (IR) region. The IRFP-conformal dynamics can be probed via a response to a small fermion mass (m_0) perturbation, which gives rise to a bound state with a mass gap ($M_{\rm H}$) satisfying the conformal hyperscaling relation with a mass anomalous dimension γ_m. In the chiral broken phase near to the conformal window $N_f \lesssim N_f^*$, it is advocated that the system shows an approximate conformality (walking) with a "would-be" γ_m, which is of great interest in physics beyond the Standard Model[1]. In recent lattice studies, the precise determination of γ_m in many flavor QCD is reported, while there exist tensions among the values[2]. The discrepancy presumably originates to scale violations by UV/IR cutoffs (lattice spacing and finite box size) as well as a probe fermion mass. In this proceedings, we focus on the IR cutoff effects by using Schwinger-Dyson (SD) equation.

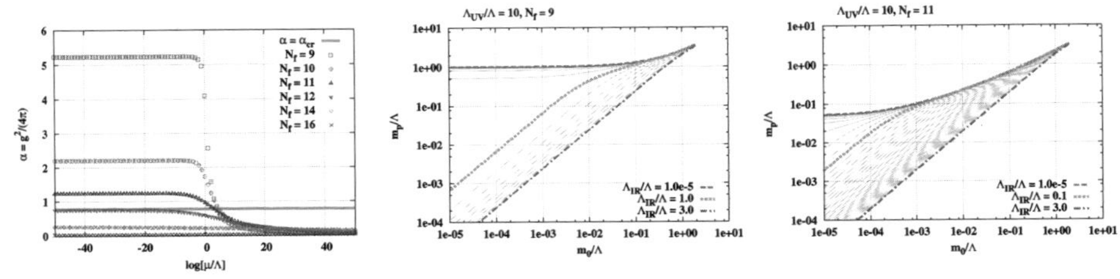

Fig. 1. Left: The running coupling for the various fermion multiplicity $9 \leq N_f \leq 16$ in the color $SU(N_c = 3)$ gauge theory. Middle and Right: Pole masses m_p as a function of a bare fermion mass m_f for various IR cutoffs ($\Lambda_{IR}/\Lambda \in [10^{-5}, 3.0]$, blue-thin-solid lines) with the UV cutoff $\Lambda_{UV}/\Lambda = 10.$ in the chiral broken phase: (Middle panel) $N_f = 9$, (Right) $N_f = 11$.

2. Schwinger-Dyson equation analysis

We consider the SD equation for the fermion propagator $iS_F(q) \equiv \left[\not{p} - \Sigma(p^2)\right]^{-1}$ within improved ladder approximation in the Landau gauge. After integrating over angular degrees of freedom in a momentum space, the SD equation is expressed as

$$\Sigma(p^2) = m_0 + \int_{\Lambda_{IR}}^{\Lambda_{UV}} dq^2 \frac{3C_2[F]\alpha(p^2+q^2)}{4\pi} \left[\frac{q^2}{p^2}\theta_{p^2-q^2} + \theta_{q^2-p^2}\right] \frac{\Sigma(q^2)}{q^2 + \Sigma^2(q^2)}, \quad (1)$$

where m_0, θ_{p^2}, $\alpha(\mu) = g^2(\mu)/(4\pi^2)$, and $C_2[F]$ represent a bare fermion mass, step function, running coupling constant in the two-loop perturbation theory, and quadratic Casimir operator, respectively. The UV/IR cutoffs $\Lambda_{UV/IR}$ are introduced in the momentum space integral. For a given fermion mass m_0 and in the presence of $\Lambda_{UV/IR}$, we numerically solve the SD equation (1) and evaluate the physical pole mass m_p and pion decay constant f_π as

$$m_p \equiv \Sigma(p^2 = m_p)\ , \quad f_\pi^2 = \frac{N_c}{4\pi^2} \int_0^{\Lambda_{UV}^2} dz\ z\ \frac{\left(1 - \frac{1}{4} z \frac{d}{dz}\right)\Sigma^2(z)}{\left(z + \Sigma^2(z)\right)^2}\ . \quad (2)$$

We take account of full momentum dependences of the two-loop running coupling without recourse to the usual step-function type approximation for the coupling. We then perform the hyperscaling fit analyses for the obtained numerical data similarly to analyses in lattice gauge theories. Moreover, we derive the SD-based finite-size hyperscaling formula (FSS), which allows us to handle the IR cutoff artifacts, even in the case of $\Lambda_{IR} \gtrsim m_p, f_\pi$. These are the advantages to the previous work[3].

3. Chiral broken and conformal phases with IR cutoff

First, we investigate the chiral broken phase in the presence of IR cutoffs. The middle and right panels of Fig. 1 show the pole mass m_p against m_0 for $N_f = 9$ and 11, respectively. Each blue-thin-solid line represents a result with a different IR cutoff Λ_{IR}/Λ [a]. For a small Λ_{IR}/Λ, the pole masses remain finite at small m_0,

[a] The scale Λ denotes the intrinsic scale which is renormalization invariant and associated with the onset of non-perturbative dynamics[4].

indicating the spontaneous chiral symmetry breaking. We find a critical value of the IR cutoff above which the dynamical fermion mass m_D is strongly suppressed as shown in the figures:

$$\Lambda_{\rm IR}^{*} = m_p(\Lambda_{\rm IR} = 0, m_0 = 0) =: m_D. \tag{3}$$

Thus, a *Fake Conformality* results from a large $\Lambda_{\rm IR}$. As a system becomes closer to the conformal window ($N_f = 9 \to 11$), the fake conformality appears for a smaller $\Lambda_{\rm IR}$. This gives a caveat for lattice works studying the border region of chiral broken and conformal phases.

Next, we investigate a conformal phase with IR cutoffs by using the SD results for $N_f = 12$. The mass anomalous dimension γ_m is obtained by the scaling property of m_p (f_π) against m_0. The left panel of Fig. 2 shows a logarithm plot of the pole mass m_p as a function of fermion bare masses m_0 for various IR cutoffs $\Lambda_{\rm IR}$. For the smallest IR cutoff $\Lambda_{\rm IR} = 10^{-8}$ (red-squares), the data points align on a straight line, therefore, the hyperscaling relation $m_p = C m_0^{1/(1+\gamma_m)}$ is satisfied. For the fit range of $m_0 \in [0.001, 0.1]$ (non-shaded region in the figure), we obtain $\gamma_m = 0.56$. While, for larger IR cutoffs, the data points are on a polygonal line with a bend around $\Lambda_{\rm IR} \simeq m_p$. As seen in the figure, the fit range $m_0 \in [0.001, 0.1]$ starts affected by the bend down with increasing $\Lambda_{\rm IR}$. As a result, the fit with the ansatz $m_p = C m_0^{1/(1+\gamma_m)}$ results in a γ_m strongly suppressed, as shown in the middle panel of Fig. 2. When the suppression sets in, the γ_m evaluated by m_p (red-squares) and f_π (blue-circles) get impaired, and thus the universality of γ_m does not hold any more.

4. Finite size hyperscaling (FSS) and mass anomalous dimension

We shall now derive a SD-based FSS which allows us to handle the IR cutoff artifacts explained above. To this end, we further approximate the SD equation (1) with a linearized fermion propagator and $\alpha(\mu) \to \alpha_* \theta_{\Lambda-\mu}$ where α_* denotes the running coupling at IRFP, and obtain the analytic expression of the renormalization factor $Z_M^{-1} := m_0/m_p$. The Z_M is a function of dimensionless variables, $\hat{m}_p := m_p/\Lambda$, $\hat{m}_0 := m_0/\Lambda$, $\hat{\Lambda} := \Lambda_{\rm IR}/\Lambda$, and $\gamma = 1 - \sqrt{1 - \alpha_*/\alpha_{\rm cr}}$, and the analytic expression allows us to derive the SD-based FSS formula[7]:

$$\hat{m}_p/\hat{\Lambda}_{\rm IR} = C_{\rm X} \cdot X \ , \quad X \equiv \hat{m}_0^{1/(1+\gamma_m)}/\hat{\Lambda}_{\rm IR} \ , \quad \Lambda_{\rm IR} \ll m_p \ll \Lambda \ , \tag{4}$$

$$\hat{m}_p/\hat{\Lambda}_{\rm IR} = C_{\rm Y} \cdot Y \ , \quad Y \equiv \hat{m}_0/\hat{\Lambda}_{\rm IR}^{1+\gamma_m} \ , \quad m_p \ll \Lambda_{\rm IR} \ll \Lambda \ . \tag{5}$$

The equation (4) is same as what dictated by the renormalization group equation[5]. Remarkably, the SD predicts another FSS formula (5) which is characterized by the scaling variable Y rather than X for larger $\Lambda_{\rm IR}$.

We apply the SD-based FSS (4) and (5) for the data obtained by solving Eq. (1) numerically. Right panel of Fig. 2 shows the results. The various symbols (colors) show data with different IR cutoffs. For the case of $\Lambda_{\rm IR} \ll m_p \ll \Lambda$, we adopt the ansatz (4). The fit works well (blue-solid line) and we obtain $\gamma_m \simeq 0.60 =: \gamma_1$ which is fairly consistent to $\gamma_m = 0.56$ obtained in the previous section for the smallest IR cutoff $\hat{\Lambda}_{\rm IR} = 10^{-8}$. We find the alignment of the data points, indicating the universal nature of $\gamma_m = 0.60$. For the case of $m_p \ll \Lambda_{\rm IR} \ll \Lambda$, we adopt the ansatz (5). The fit works well (black-solid line) and gives $\gamma_m \simeq 0.56 =: \gamma_2$, which is somewhat smaller than $\gamma_1 = 0.60$ but the strong suppression has disappeared. Thus, the right FSS formula in the right place recovers the approximate universality $\gamma_1 \simeq \gamma_2$, or equivalently, the approximate data alignments in the whole mass region including both $\Lambda_{\rm IR} \ll m_p \ll \Lambda$ and $m_p \ll \Lambda_{\rm IR} \ll \Lambda$. Then, two scaling variables, $X(\gamma_1)$ and $Y(\gamma_2)$, are responsible for two slopes of the alignments. In lattice studies, indeed, the formula (4) is widely used in lattice study, but the formula (5) is rarely used[6]. However, these result are of the case without paying attention to the scope of application, and opposed to each other[2].

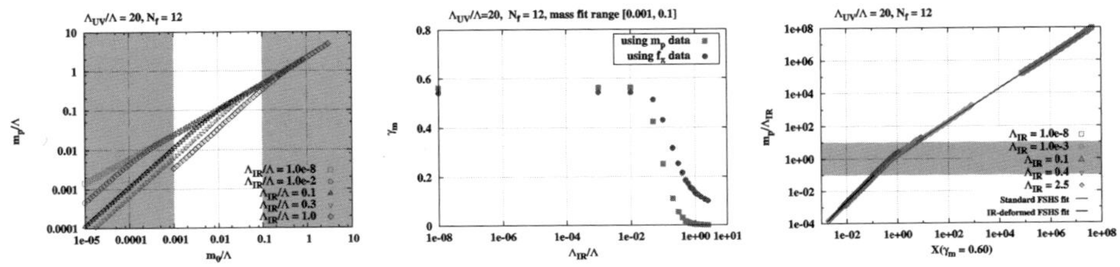

Fig. 2. Left: The pole mass $\hat{m}_p = m_p/\Lambda$ vs the fermion mass $\hat{m}_0 = m_0/\Lambda$ for various IR cutoffs $\hat{\Lambda}_{\rm IR} = \Lambda_{\rm IR}/\Lambda$ in $N_f = 12$. Middle: The mass anomalous dimension γ_m obtained by fitting the non-shaded region data of the left panel with the conformal ansatz. Right: The pole mass $\hat{m}_p$ vs $X = \hat{m}_0^{1/(1+\gamma_m)}/\hat{\Lambda}_{\rm IR}|_{\gamma_m=0.60}$ in the conformal phase $N_f = 12$. The data upper and lower than the shaded region are used for the fits, giving the blue- and black-solid lines.

Acknowledgement

This work is supported by the JSPS Grant-in-Aid for Scientific Research (S) No.22224003. We thank Robert Shrock and Seyong Kim for fruitful discussions during the conference.

References

1. For a recent review, see R. Contino, arXiv:1005.4269 [hep-ph].
2. The lattice values of γ_m for $N_f = 12$ are summarized in M. P. Lombardo *et al.*, JHEP **1412**, 183 (2014); E. Itou, PTEP **2015**, no. 4, 043B08 (2015).
3. Y. Aoki *et al.*, Phys. Rev. D **85**, 074502 (2012).
4. T. Appelquist, A. Ratnaweera, J. Terning and L. C. R. Wijewardhana, Phys. Rev. D **58**, 105017 (1998).
5. L. Del Debbio and R. Zwicky, Phys. Rev. D **82**, 014502 (2010).
6. T. DeGrand, Phys.Rev. D **84** (2011) 116901.
7. K. Miura, K-i. Nagai and A. Shibata, in preparation.

Sizable D-Term Contributions as a Signature of $E_6 \times SU(2)_F \times U(1)_A$ SUSY GUT model

Yoshihiro Shigekami

Department of Physics, Nagoya University,
Nagoya 464-8602, Japan
E-mail: sigekami@eken.phys.nagoya-u.ac.jp

We show that the sizable D-term contributions to the sfermion mass spectrum can be signatures of a certain grand unified theory (GUT), $E_6 \times SU(2)_F \times U(1)_A$ GUT. Note that these D-term contributions destroy the degeneracy of sfermion masses among different generations in this model. Such D-terms are strongly constrained by the FCNC processes if the SUSY breaking scale is the weak scale. However, in $E_6 \times SU(2)_F \times U(1)_A$, natural SUSY type sfermion mass spectrum is obtained, and if the masses of $\mathbf{10}_3$ sfermions are larger than $O(1)$ TeV to realize 125 GeV Higgs and the other sfermion masses are $O(10)$ TeV, then sizable D-term contribution is allowed. If the deviations by these D-terms can be observed in future experiments like 100 TeV proton collider or muon collider, we may confirm the $E_6 \times SU(2)_F \times U(1)_A$ GUT.

Keywords: E_6 GUT, SUSY GUT, D-term.

1. Introduction

Grand unified theory (GUT) can unify not only three Standard Model (SM) gauge couplings into a single one but also matter fields into a few multiplets. Furthermore, there are experimental supports for both unifications. One is quantitative consistency of three gauge couplings between experimental and theoretical value, and the other is qualitative explanation of various hierarchies of matters and mixings.

If the E_6 supersymmetric (SUSY) GUT is considered with $SU(2)_F$ family symmetry[1] and the anomalous $U(1)_A$ gauge symmetry, we can obtain more attractive GUT model.[2] This model can solve several problems caused in SUSY GUT model by imposing the CP symmetry and considering the spontaneous CP violation. Moreover, in this model, we can obtain the natural SUSY-type sfermion mass spectrum. We call this spectrum modified universal sfermion masses (MUSM). The degeneracy of this spectrum can suppress the SUSY flavor-changing neutral current (FCNC) and stabilize the weak scale at the same time.

However, there are non-vanishing D-term contributions to sfermion mass in this model, and in fact, these contributions destroy the degeneracy of MUSM. Then, D-terms were constrained to be small by the various FCNC processes[3] before the observation of 125 GeV Higgs. But now, in order to realize the 125 GeV Higgs mass, the stop mass must be larger than 1 TeV, and the other sfermion masses can be $\mathcal{O}(10)$ TeV. This leads to weaken the FCNC constraints, and therefore D-terms may be allowed to be sizable.

We compute the D-term contributions of this model. We will conclude that sizable D-term contribution is possible even if we consider the strongest FCNC constraint, ϵ of $K^0\bar{K}^0$ mixing. If the D-term contributions are sufficiently large and observed by future experiments, then we can obtain the precious evidence of the GUT scenario. This talk is based on our work.[4]

2. $E_6\times SU(2)_F\times U(1)_A$ SUSY GUT model

2.1. *Matter fields and $\bar{\mathbf{5}}$-mixing*

In this model, we introduce three $\mathbf{27}$ fundamental fields of E_6 as matters. This is decomposed in the $E_6 \supset SO(10)\times U(1)_{V'}$ notation (and in the $[SO(10) \supset SU(5)\times U(1)_V]$ notation) as

$$\mathbf{27} = \mathbf{16_1}[\mathbf{10_1} + \bar{\mathbf{5}}_{-\mathbf{3}} + \mathbf{1_5}] + \mathbf{10}_{-\mathbf{2}}[\mathbf{5}_{-\mathbf{2}} + \bar{\mathbf{5}}'_{\mathbf{2}}] + \mathbf{1}'_{\mathbf{4}}[\mathbf{1}'_{\mathbf{0}}]. \tag{1}$$

This decomposition says that three generation of $\mathbf{27}$s of E_6 contains six $\bar{\mathbf{5}}$s of $SU(5)$. Three of these $\bar{\mathbf{5}}$s become superheavy through the superpotential, and the remaining three of $\bar{\mathbf{5}}$s become the SM mode. This mode, denote $\bar{\mathbf{5}}^0_i$, mainly come from the first two generations of $\mathbf{27}$, as like $(\bar{\mathbf{5}}^0_1, \bar{\mathbf{5}}^0_2, \bar{\mathbf{5}}^0_3) \sim (\bar{\mathbf{5}}_1, \bar{\mathbf{5}}'_1, \bar{\mathbf{5}}_2)$. We call this $\bar{\mathbf{5}}$-mixing. This plays an important role in calculating the D-term contributions to the sfermion masses. Thanks to $\bar{\mathbf{5}}$-mixing and SUSY breaking potential $V_{SB} \ni m_0^2|\Psi_a|^2 + m_3^2|\Psi_3|^2$, where Ψ_a and Ψ_3 denote the matter fields of doublet and singlet of $SU(2)_F$, respectively, we can obtain following natural SUSY-type sfermion mass spectrum

$$\tilde{m}^2_{10} \sim \begin{pmatrix} m_0^2 & & \\ & m_0^2 & \\ & & m_3^2 \end{pmatrix}, \quad \tilde{m}^2_{\bar{5}^0} \sim \begin{pmatrix} m_0^2 & & \\ & m_0^2 & \\ & & m_0^2 \end{pmatrix}. \tag{2}$$

This structure that we call MUSM, is good for weak scale stabilization and SUSY FCNC. However, the degeneracy of MUSM is destroyed by D-term contributions.

2.2. *D-term contributions and predictions*

D-term contributions to the sfermion masses are written as

$$\Delta\tilde{m}^2_\psi = \sum_I Q_I(\psi)D_I, \tag{3}$$

where $Q_I(\psi)$ is the $U(1)$ charge of the field ψ and D_I is the squared gauge coupling times D-term of symmetry of the model. In this model, there are four D-terms described as D_6 (for $U(1)_{V'}$), D_{10} (for $U(1)_V$), D_F (for cartan part of $SU(2)_F$) and D_A (for $U(1)_A$). From Eq. (2), Eq. (3) and $\bar{\mathbf{5}}$-mixing, one can write down the sfermion mass matrices for $\bar{\mathbf{5}}^0$ and $\mathbf{10}$ of $SU(5)$. These matrices can be written as

$$\tilde{m}_{10}^2 \sim (m_0^2 + D_6 + D_{10} + D_F + 4D_A)\mathbf{1}_{3\times 3} + \begin{pmatrix} 0 & & \\ & -2D_F & \\ & & -D_F - \frac{5}{2}D_A + m_3^2 - m_0^2 \end{pmatrix}$$

$$\equiv (\tilde{m}_{10}^2)_{11}\mathbf{1}_{3\times 3} + \begin{pmatrix} 0 & & \\ & (\tilde{m}_{10}^2)_{22} - (\tilde{m}_{10}^2)_{11} & \\ & & (\tilde{m}_{10}^2)_{33} - (\tilde{m}_{10}^2)_{11} \end{pmatrix}, \tag{4}$$

$$\tilde{m}_{\bar{5}^0}^2 \sim (m_0^2 + D_6 - 3D_{10} + D_F + 4D_A)\mathbf{1}_{3\times 3} + \begin{pmatrix} 0 & & \\ & -3D_6 + 5D_{10} & \\ & & -2D_F \end{pmatrix}$$

$$\equiv (\tilde{m}_{\bar{5}^0}^2)_{11}\mathbf{1}_{3\times 3} + \begin{pmatrix} 0 & & \\ & (\tilde{m}_{\bar{5}^0}^2)_{22} - (\tilde{m}_{\bar{5}^0}^2)_{11} & \\ & & (\tilde{m}_{\bar{5}^0}^2)_{33} - (\tilde{m}_{\bar{5}^0}^2)_{11} \end{pmatrix}, \tag{5}$$

where $\mathbf{1}_{3\times 3}$ is a 3×3 unit matrix. This leads to a non-trivial prediction of this model, $(\tilde{m}_{10}^2)_{22} - (\tilde{m}_{10}^2)_{11} = (\tilde{m}_{\bar{5}^0}^2)_{33} - (\tilde{m}_{\bar{5}^0}^2)_{11}$. This prediction is specific to $E_6 \times SU(2)_F \times U(1)_F$ SUSY GUT model.

3. FCNC constraints to D-terms

Since large sfermion mass makes FCNC constraints weak, we focus on the natural SUSY-type sfermion masses, i.e., $m_0 \gg m_3$, in this section. Then, $(\tilde{m}_{10}^2)_{33} - (\tilde{m}_{10}^2)_{11}$ can be fixed m_0^2. In order to obtain the 125 GeV Higgs and weak FCNC constraints, we assume that $m_3 \sim \mathcal{O}(1)$TeV and $m_0 \sim \mathcal{O}(10)$TeV.

Here, we consider the constraints from the CP-violating parameter ϵ in $K^0\bar{K}^0$ mixing, which is the strongest constraint in FCNC. Usually, this constrain the mass insertion parameters as follow:[5]

$$\sqrt{|\mathrm{Im}(\delta_{12}^d)_{LL,RR}^2|} < 2.9\times 10^{-3}\left(\frac{m_0}{500\mathrm{GeV}}\right), \tag{6}$$

$$\sqrt{|\mathrm{Im}(\delta_{12}^d)_{LL}(\delta_{12}^d)_{RR}|} < 1.1\times 10^{-4}\left(\frac{m_0}{500\mathrm{GeV}}\right). \tag{7}$$

We can calculate the mass insertion parameters $(\delta_{12}^d)_{LL}$ and $(\delta_{12}^d)_{RR}$ of this model from Eq. (4), Eq. (5) and the diagonalizing matrices of Yukawa matrices obtained from the superpotential as

$$(\delta_{12}^d)_{LL} \sim \left(\frac{2}{3} + i\frac{4}{27}\right)\left(\lambda\frac{(\tilde{m}_{10}^2)_{22} - (\tilde{m}_{10}^2)_{11}}{m_0^2} + \lambda^5\frac{(\tilde{m}_{10}^2)_{33} - (\tilde{m}_{10}^2)_{11}}{m_0^2}\right) \tag{8}$$

$$(\delta_{12}^d)_{RR} \sim \frac{2}{3}(1+i)\left(\lambda^{0.5}\frac{(\tilde{m}_{\bar{5}^0}^2)_{22} - (\tilde{m}_{\bar{5}^0}^2)_{11}}{m_0^2} + \lambda^{1.5}\frac{(\tilde{m}_{\bar{5}^0}^2)_{33} - (\tilde{m}_{\bar{5}^0}^2)_{11}}{m_0^2}\right). \tag{9}$$

Because of our assumption and non-trivial prediction of this model, these mass insertion parameters can be parameterized by two variables, $\sqrt{|(\tilde{m}_{10}^2)_{22} - (\tilde{m}_{10}^2)_{11}|}/m_0$ and $\sqrt{|(\tilde{m}_{\bar{5}^0}^2)_{22} - (\tilde{m}_{\bar{5}^0}^2)_{11}|}/m_0$. We show the allowed region in these variable space

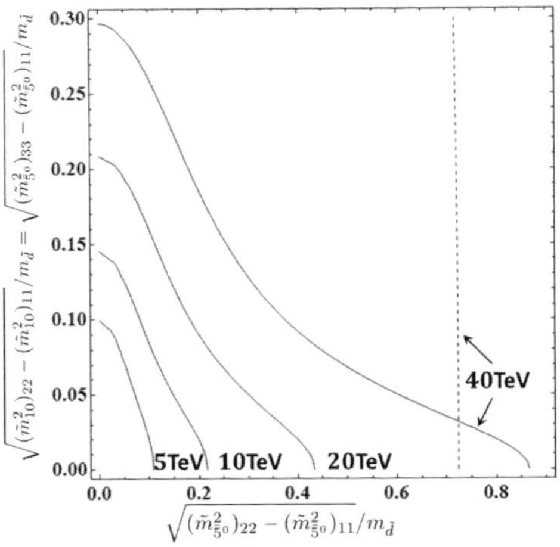

Fig. 1. Allowed region in $(\sqrt{|(\tilde{m}^2_{\bar{5}0})_{22} - (\tilde{m}^2_{\bar{5}0})_{11}|}/m_{\tilde{d}}, \sqrt{|(\tilde{m}^2_{10})_{22} - (\tilde{m}^2_{10})_{11}|}/m_{\tilde{d}}$ space, where $m_{\tilde{d}}$ means m_0. The allowed region for the condition of $\sqrt{|\mathrm{Im}(\delta^d_{12})_{LL}(\delta^d_{12})_{RR}|}$ is obtained below the solid lines for various $m_0 = 5$ TeV, 10 TeV, 20 TeV and 40 TeV. The allowed region for $\sqrt{|\mathrm{Im}(\delta^d_{12})^2_{RR}|}$ is the left side of the dotted line for $m_0 = 40$ TeV. The other conditions are satisfied in the allowed region for $\sqrt{|\mathrm{Im}(\delta^d_{12})_{LL}(\delta^d_{12})_{RR}|}$.

in Fig. 1. This figure means that if m_0 is $\mathcal{O}(10)$ TeV, D-term contribution is allowed to be $\mathcal{O}(1)$ TeV, which is the same scale of m_3.

4. Summary

We have shown that in $E_6 \times SU(2)_F \times U(1)_A$ SUSY GUT model, D-term contributions to the sfermion masses are allowed to be $\mathcal{O}(1)$ TeV when m_0 is $\mathcal{O}(10)$ TeV and this can be one of the signatures of this model. Of course, a non-trivial prediction $(\tilde{m}^2_{10})_{22} - (\tilde{m}^2_{10})_{11} = (\tilde{m}^2_{\bar{5}0})_{33} - (\tilde{m}^2_{\bar{5}0})_{11}$ is derived from the $\bar{\mathbf{5}}$-mixing and the D-term contributions to MUSM, and is important to check this model. If these D-terms can be observed in future experiments like 100 TeV proton collider or muon collider, $E_6 \times SU(2)_F \times U(1)_A$ SUSY GUT model may be confirmed.

References

1. K. S. Babu and S. M. Barr, *Phys. Lett. B* **387**, 87 (1996).
2. N. Maekawa, *Phys. Lett. B* **561**, 273 (2003).
3. K. Inoue, K. Kojima and K. Yoshioka, *JHEP* **0707**, 027 (2007).
4. N. Maekawa, Y. Muramatsu and Y. Shigekami, *Prog. Theor. Exp. Phys.* **113B02** (2014).
5. M. Ciuchini, V. Lubicz, L. Conti, A. Vladikas, A. Donini, E. Franco, G. Martinelli and I. Scimemi *et al.*, *JHEP* **9810**, 008 (1998).

New Approach to the Dirac Spectral Density in Lattice Gauge Theory Applications

Zoltan Fodor

University of Wuppertal, Department of Physics, Wuppertal D-42097, Germany
E-mail: fodor@bodri.elte.hu

Kieran Holland

University of the Pacific, 3601 Pacific Ave, Stockton CA 95211, USA
E-mail: kholland@pacific.edu

Julius Kuti*

University of California, San Diego, 9500 Gilman Drive, La Jolla, CA 92093, USA
E-mail: jkuti@ucsd.edu

Santanu Mondal

Eotvos University, MTA-ELTE Lendulet Lattice Gauge Theory Group, Budapest, Hungary
E-mail: santanu@bodri.elte.hu

Daniel Nogradi

Eotvos University, MTA-ELTE Lendulet Lattice Gauge Theory Group, Budapest, Hungary
E-mail: nogradi@bodri.elte.hu

Chik Him Wong*

University of Wuppertal, Department of Physics, Wuppertal D-42097, Germany
E-mail: cwong@uni-wuppertal.de

We report tests and results from a new approach to the spectral density and the mode number distribution of the Dirac operator in lattice gauge theories. The algorithm generates the spectral density of the lattice Dirac operator as a continuous function over all scales of the complete eigenvalue spectrum. This is distinct from an earlier method where the integrated spectral density (mode number) was calculated efficiently for some preselected fixed range of the integration. The new algorithm allows global studies like the chiral condensate from the Dirac spectrum at any scale including the cutoff-dependent IR and UV range of the spectrum. Physics applications include the scale-dependent mass anomalous dimension, spectral representation of composite fermion operators, and the crossover transition from the ϵ-regime of Random Matrix Theory to the p-regime in chiral perturbation theory. We present thorough tests of the algorithm in the 2-flavor sextet SU(3) gauge theory that we continue to pursue for its potential as a minimal realization of the composite Higgs scenario.

*Poster contributors

1. Introduction and brief history

We introduce a new approach to the Dirac spectral density and mode number distribution in lattice gauge theories. The algorithm effectively generates the spectral density of the lattice Dirac operator as a continuous function over the entire range of the eigenvalue spectrum in large lattice volumes. This is distinct from an earlier method[1] where the integrated spectral density (mode number) was calculated efficiently for some preselected fixed range of the integration and averaged over gauge configurations. Motivated by[1] we set the goal to calculate efficiently the spectral density over the entire Dirac spectrum which can be integrated over any range to generate the mode number distribution on arbitrary scales in a single application to the gauge configuration before the average is taken over the gauge ensemble. Just like the method introduced in[1] for gauge theory applications, our method is also rooted in known applications of the Chebyshev expansion from approximation theory when combined with stochastic evaluation of operator traces in large vector spaces.

We have been developing and testing the reported lattice gauge theory algorithm over the last few years with results appearing in our earlier publications including[2,3]. Our implementation of the algorithm itself was only presented for the first time at this conference[4] with similar material to the one presented at the follow-up Lattice 2015 conference. We hope to motivate new work by our successful and thorough tests of this new lattice gauge theory application, implemented with staggered lattice fermions in our case. As an example, for extension to other type of lattice fermions, interesting new results were presented at the Lattice 2015 conference on the chiral condensate of the Dirac operator with domain wall lattice fermions using the same approach[5].

Based on the poster we presented at this conference, we demonstrate the effectiveness of the new algorithm. The plots in this short report are updated from the poster for better illustration of the algorithm and for pedagogical purposes. Section 2 is a brief summary of the method. In Section 3 we present some tests and implementations as applied to the spectral density of the chiral condensate and the mode number distribution. Section 4 is a brief illustration of physics applications where we present our first tests of the GMOR relation and the scale dependent mass anomalous dimension of the chiral condensate in the 2-flavor sextet SU(3) gauge theory that we continue to pursue for its viability as a minimal realization of the composite Higgs scenario. The method we present projects interesting new applications for future studies.

2. Resolution of spectral lines from a stochastic Chebyshev expansion

In[1] the projector operator was used for the determination of the mode number of the Dirac spectrum. A rational approximation to the projector operator was estimated by a Chebyshev polynomial expansion. Here we will approximate the δ-function of the Dirac operator, with its Chebyshev-Jackson polynomial approximation which will determine the moments of the spectral density to a preset high order. Similar methods have been found and referenced for the curious reader in[5] from recent history of approximation theory. Adding to the historical perspective, we note some very early work applying Chebyshev approximation to moments of spectral densities for Hamiltonian spectra[6], curiously with co-authors from the same institutional affiliation as one of us, but from an earlier era. Our method can be viewed as a similar approximation but has broader scope and general applicability to a large class of Hamiltonian problems with stochastic implementation from the modern computer era.

We will describe the new algorithm for the spectrum of the staggered Dirac operator in a finite lattice volume, but the generalization to the spectrum of a large class of other operators under some general spectral conditions is straightforward. For a finite lattice four-volume V, with periodic or antiperiodic boundary conditions for fermions, the euclidean Dirac operator D on any given gauge field configuration has purely imaginary eigenvalues $i\lambda_1, i\lambda_2, ...$, with the associated average spectral density $\rho(\lambda, m)$,

$$\rho(\lambda, m) = \frac{1}{N_{eig}} \sum_{i=1}^{N_{eig}} \langle \delta(\lambda - \lambda_i) \rangle. \tag{1}$$

In Eq. (1) the sum over N_{eig} individual λ_i eigenvalues is averaged over gauge configurations which depend on the bare fermion mass m. The spectral density is a renormalizable quantity in gauge theories and the entire function in λ can be computed at fixed m on the lattice using the Chebyshev-Jackson expansion we will introduce. It is convenient to consider the mode number $\nu(\Lambda, m)$ of the positive definite hermitian operator $D^\dagger D + m^2$ in the integrated form of the spectral density as given by Eq. (2),

$$\nu(M, m) = V \int_{-\Lambda}^{\Lambda} d\lambda \rho(\lambda, m), \qquad \Lambda = \sqrt{M^2 - m^2}. \tag{2}$$

The important role of the mode number distribution in the analysis of the chiral condensate was emphasized in[1] with a demonstration of its renormalization group invariance $\nu_{\mathrm{R}}(M_{\mathrm{R}}, m_{\mathrm{R}}) = \nu(M, m)$. After rescaling the spectrum of the Dirac operator D and its equivalent $D^\dagger D$ quadratic form, the spectral density $\rho(t)$ depends on scaled eigenvalues $\bar{\lambda}_i$ with

$$\rho(t) = \frac{1}{N_{eig}} \sum_{i=1}^{N_{eig}} \langle \delta(t - \bar{\lambda}_i) \rangle_{gauge\ ensemble} \tag{3}$$

where the variable t is restricted to the [-1,1] interval with rescaled eigenvalues $\bar{\lambda}_i$ in the [-1,1] interval. The implicit dependence on the bare fermion mass m often will not be shown for convenience. The density function $\rho(t)$ can be expanded into a series of $T_k(t)$ Chebyshev polynomials of the first kind,

$$\rho(t) = \frac{1}{\sqrt{1-t^2}} \sum_{k=0}^{\infty} c_k T_k(t), \tag{4}$$

with Chebyshev expansion coefficients

$$c_k = \begin{cases} \frac{2}{\pi} \int_{-1}^{1} T_k(t)\rho(t)dt & k = 0 \\ \\ \frac{1}{\pi} \int_{-1}^{1} T_k(t)\rho(t)dt & k \neq 0. \end{cases} \tag{5}$$

In practical implementations D will be replaced by the positive definite $D^\dagger D$ operator and the Chebyshev coefficients can be expressed in terms of $D^\dagger D$ eigenvalues,

$$c_k = \begin{cases} \frac{2}{N_{eig}\pi} \sum_{i=1}^{N_{eig}} T_k(\bar{\lambda}_i^2) & k = 0 \\ \\ \frac{1}{N_{eig}\pi} \sum_{i=1}^{N_{eig}} T_k(\bar{\lambda}_i^2) & k \neq 0. \end{cases} \tag{6}$$

Based on Eq.(6), the evaluation of the operator traces $\mathrm{Tr}(T_k(D^\dagger D))$ is needed to calculate the spectral density function. We use the well-known stochastic evaluation with Z(2) noise vectors ξ,

$$\mathrm{Tr}(T_k(D^\dagger D)) \approx \frac{1}{N_{noise}} \sum_{n=1}^{N_{noise}} \xi_n^T \cdot T_k(D^\dagger D) \cdot \xi_n. \tag{7}$$

Recursion relations for Chebyshev polynomials of the operators,

$$\begin{aligned} T_{k+1}(D^\dagger D)) \cdot \xi &= 2D^\dagger D \cdot T_k(D^\dagger D) \cdot \xi - T_{k-1}(D^\dagger D) \cdot \xi, \\ \xi^{(n)} = T_n(D^\dagger D)) \cdot \xi^{(0)} &\Leftrightarrow \xi^{(i+1)} = 2D^\dagger D \cdot \xi^{(i)} - \xi^{(i-1)}, \end{aligned} \tag{8}$$

are used in averages over noise vectors ξ in repeated recursions. The series has to be truncated at some finite order which will provide the Dirac δ-function and the Heaviside step function Θ at finite resolution. Figure 1 shows the spectral resolution of $\delta(\lambda)$ for a spectral line at $\lambda = 0.005$ and $\Theta(t)$ at step $t = 1$ for two different orders of the Chebyshev expansion. On the left panel, the resolution of the spectral function for a sharp δ-function spectral line at $\lambda = 0.005$ is color coded. The width of the finite resolution is scaling with $1/N$ in the order N of the Chebyshev expansion. In red color the resolution of the spectral density at truncation $N = 8000$ from the Chebyshev coefficients of Eq. (5) is shown for the sharp δ-function spectral line. The truncation, as the insert of the left panel shows, introduces well-known Gibbs oscillations in the truncated spectral function at finite resolution. The Gibbs oscillations can be damped by the modified expansion

$$\rho(t) = \frac{1}{\sqrt{1-t^2}} \sum_{k=0}^{\infty} c_k T_k(t) \Rightarrow \frac{1}{\sqrt{1-t^2}} \sum_{k=0}^{\infty} c_k g_k T_k(t), \tag{9}$$

with Jackson coefficients g_k which are well-known in approximation theory[7] for damping the Gibbs oscillations with slight loss in the resolution. Blue color on the

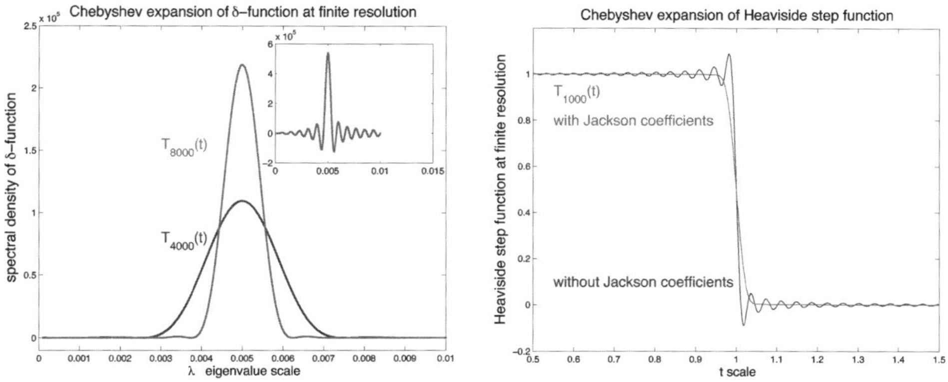

Fig. 1. The left panel shows the sharp δ-function spectral line in finite resolution with $1/N$ scaling and Gibbs oscillations (red insert) damped with Jackson coefficients in the main part of the left panel (red). The right panel shows the finite resolution of the Heaviside step function.

left panel of Figure 1 shows the Chebyshev-Jackson expansion for the same spectral line at lower resolution consistent with $1/N$ scaling. Color coding on the right panel shows the resolution of the Heaviside step function comparing the Gibbs oscillation and its Jackson damping at the same Chebyshev order.

3. Spectral density and Mode number

To illustrate the efficiency of the method, Figure 2 shows results from the calculation of the spectral density and the related mode number distribution on all scales using the Chebyshev-Jackson expansion of Eq. (9). The calculation targets here an important BSM gauge theory with a fermion doublet in the two-index symmetric (sextet) representation of the SU(3) BSM color gauge group as reviewed in talks at this conference[8]. The upper left panel of the figure shows the spectral density of the staggered $D^\dagger D$ Dirac operator using ten independent gauge configurations on the lattice volume with size $64^3 \times 96$ at bare gauge coupling set by $\beta = 3.25$ in the lattice action[8] and the fermion mass set at $m = 0.001$. Chebyshev polynomials up to order 8000 were used in the expansion with 20 noise vectors defined in Eq. (7). The size of the Jackknife errors in the spectral density is not visible on the scale of the upper left panel of the plot. The upper right panel shows the mode number counting the eigenmodes from the integral of the spectral density. It converges to the correct total count which is half of what is shown in the upper left panel from the implementation of $D^\dagger D$ on the staggered lattice counting only one of two degenerate eigenvalues of each pair in the spectrum. The lower left panel magnifies the far infrared (IR) scale of the spectral density illustrating the convergence of the Chebyshev-Jackson expansion as the polynomial order is increased. This panel also illustrates that the convergence rate is the slowest in the IR region and reached at polynomial order $N = 8000$ with the magenta line which is practically identical to

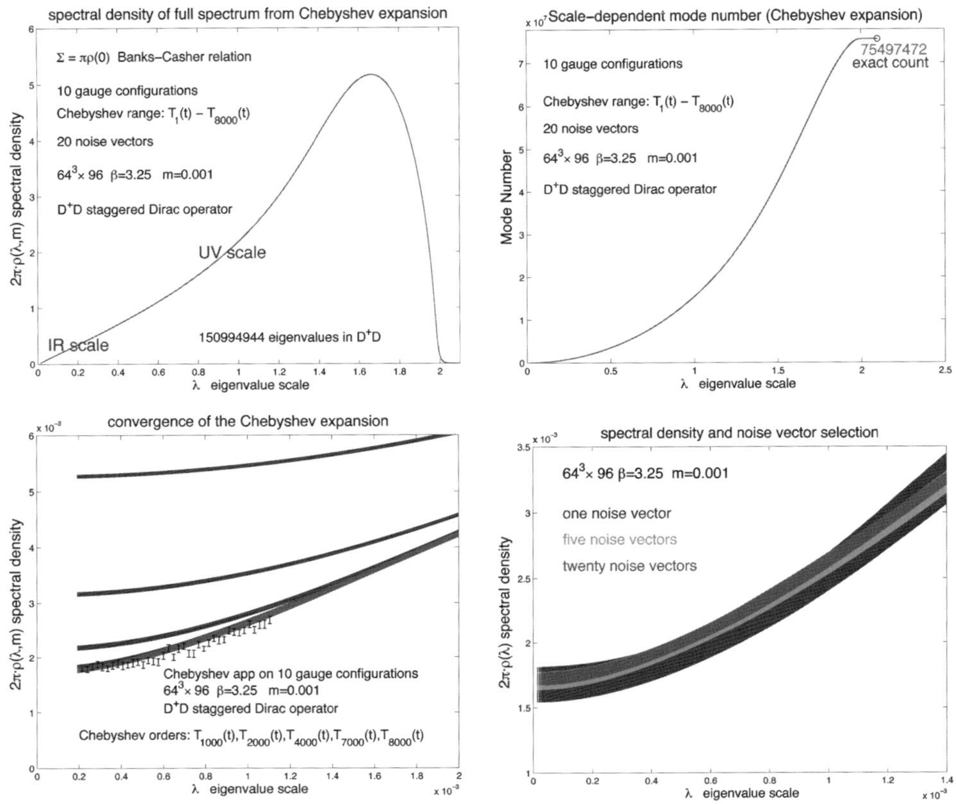

Fig. 2. Tests with a fermion doublet in the two-index symmetric (sextet) representation of the SU(3) BSM color gauge group in the composite figure are discussed in the text. The black data points in the lower left panel come from direct diagonalization which is an important part of the testing procedure. The magenta line of the panel shows the result with $N = 8000$ Chebyshev order in perfect match to the direct diagonalization data.

the lowest blue line at $N = 7000$ even in the lowest eigenvalue range. Bands in the magnified IR part of the plot show the visible Jackknife errors of the spectral density. All tests were made with 20 noise vectors. Data points from the direct diagonalization of $D^\dagger D$ are in excellent agreement with the expansion even in the low IR region. The lower right panel shows the test when the number of noise vectors is varied at fixed expansion order which is kept lower. The results show no bias as a function of the noise vector number and with twenty (or fewer) noise vectors the variance is dominated by the fluctuations over the gauge ensemble. The increased size of the error band simply comes from the larger magnification.

4. Physics applications: GMOR and mass anomalous dimension

Excellent agreement between low eigenvalues from direct diagonalization of $D^\dagger D$ and the full spectral density function from the Chebyshev-Jackson expansion, as shown in Figure 2, provides checks on the GMOR relation for the chiral condensate as discussed in[2,3]. Chiral perturbation theory of the effective chiral condensate $\Sigma_{\rm eff}$,

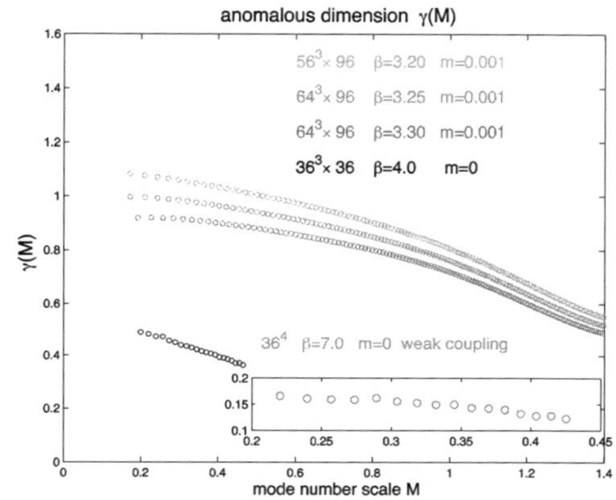

Fig. 3. Results for the mass anomalous dimension $\gamma(M)$, as defined in the text, are shown for the model with a fermion doublet in the two-index symmetric (sextet) representation of the SU(3) BSM color gauge group. There are results at five different lattice spacings, one of them at exactly zero fermion mass. With the renormalization constants Z_p and Z_m determined in separate calculations, the continuum scale dependent mass anomalous dimension and its role in fermion mass generation is left for a future publication in preparation.

aided by the Chebyshev-Jackson expansion was used in our tests of the GMOR relation for added evidence of chiral symmetry breaking in the sextet BSM theory of the composite Higgs. The effective chiral condensate $\Sigma_{\rm eff}$ was analyzed based on Eq. (10) as derived in[9],

$$\frac{\Sigma_{\rm eff}}{\Sigma}=1+\frac{\Sigma}{32\pi^3 N_F F^4}\left[2N_F^2|\Lambda|\arctan\frac{|\Lambda|}{m}-4\pi|\Lambda|-N_F^2 m\log\frac{\Lambda^2+m^2}{\mu^2}-4m\log\frac{|\Lambda|}{\mu}\right]. \tag{10}$$

Additional probing of chiral symmetry breaking includes the study of the mass anomalous dimension $\gamma(M)$ as shown in Figure 3. The anomalous dimension of the chiral condensate can be determined from access to the mass anomalous dimension $\gamma(M)$ in the eigenmode function[10–12],

$$\nu_{\rm R}(M_{\rm R}, m_{\rm R}) = \nu(M,m) \approx {\rm const}\cdot M^{\frac{4}{1+\gamma_m(M)}} \tag{11}$$

The results are reported in Figure 3 with details left for a future publication in preparation.

Acknowledgements

Julius Kuti and Ricky Wong are grateful to Koichi Yamawaki and the organizing committee for their kind hospitality and for the most interesting SCGT 2015 workshop at KMI. We acknowledge support by the DOE under grant DE-SC0009919, by the NSF under grants 0970137 and 1318220, by the DOE ALCC award for the BG/Q Mira platform of Argonne National Laboratory, by OTKA under the grant OTKA-NF-104034, and by the Deutsche Forschungsgemeinschaft grant SFB-TR 55. Computational resources were provided by the Argonne Leadership Computing Facility under an ALCC award, by USQCD at Fermilab, by the University of Wuppertal, by The Juelich Supercomputing Center on Juqueen and by the Institute for Theoretical Physics, Eotvos University. We are grateful to Szabolcs Borsanyi for his code development for the BG/Q platform. We are also grateful to Sandor Katz and Kalman Szabo for their code development for the CUDA platform[13]. KH wishes to thank the Institute for Theoretical Physics and the Albert Einstein Center for Fundamental Physics at the University of Bern for their support.

References

1. L. Giusti and M. Luscher, *Chiral symmetry breaking and the Banks-Casher relation in lattice QCD with Wilson quarks*, JHEP **0903**, 013 (2009) [arXiv:0812.3638 [hep-lat]].
2. Z. Fodor, K. Holland, J. Kuti, D. Nogradi and C. H. Wong,*The chiral condensate from the Dirac spectrum in BSM gauge theories*, PoS LATTICE **2013**, 089 (2014) [arXiv:1402.6029 [hep-lat]].
3. Z. Fodor, K. Holland, J. Kuti, S. Mondal, D. Nogradi and C. H. Wong, *Toward the minimal realization of a light composite Higgs*, PoS LATTICE **2014**, 244 (2015) [arXiv:1502.00028 [hep-lat]].
4. Z. Fodor, K. Holland, J. Kuti, S. Mondal, D. Nogradi and C. H. Wong, *New approach to the Dirac spectral density with applications*, Contribution to Sakata Memorial KMI Workshop on "Origin of Mass and Strong Coupling Gauge Theories (SCGT15)", 3-6 March 2015, Nagoya University. To be published by World Scientific.
5. G. Cossu, H. Fukaya, S. Hashimoto, T. Kaneko and J. Noaki, *Stochastic calculation of the QCD Dirac operator spectrum with Mobius domain-wall fermion*, [arXiv:1601.00744 [hep-lat]].
6. J. C. Wheeler, M. G. Prais, C. Blumstein,*Analysis of spectral densities using modified moments*, Phys. Rev. B **10**, 2429 (1974).
7. D. Jackson, *The theory of approximation* , Am. Math. Soc. Coll Publ., Vol. XI. Am. Math. Soc., New York (1930) 178.
8. Z. Fodor, K. Holland, J. Kuti, S. Mondal, D. Nogradi and C. H. Wong, *Status of a minimal composite Higgs theory*, PoS LATTICE **2015**, 219 (2016).
9. J. C. Osborn, D. Toublan and J. J. M. Verbaarschot, *From chiral random matrix theory to chiral perturbation theory*, Nucl. Phys. B **540**, 317 (1999).

10. L. Del Debbio and R. Zwicky, *Hyperscaling relations in mass-deformed conformal gauge theories*, Phys. Rev. D **82**, 014502 (2010).
11. A. Patella, *A precise determination of the psibar-psi anomalous dimension in conformal gauge theories*, Phys. Rev. D **86**, 025006 (2012).
12. A. Cheng, A. Hasenfratz, G. Petropoulos and D. Schaich, *Scale-dependent mass anomalous dimension from Dirac eigenmodes*,JHEP **1307**, 061 (2013).
13. G. I. Egri, Z. Fodor, C. Hoelbling, S. D. Katz, D. Nogradi and K. K. Szabo, *Lattice QCD as a video game*, Comput. Phys. Commun. **177**, 631 (2007) [hep-lat/0611022].

Workshop Timetable

March 3, 2015

	Chair: H. Georgi	
9:00 - 9:45	Y. Iwasaki (KEK/U. Tsukuba)	IR Fixed Points in SU(3) Guage Theories
9:45 - 10:30	J. Kuti (U. C. San Diego)	Toward the Minimal Realization of a Light Composite Higgs
	Chair: A. Hasenfratz	
10:50 - 11:15	Y. Aoki (KMI)	Spectral Properties of $N_f = 8$ SU(3) Gauge Theory
11:15 - 12:00	E. Rinaldi (LLNL)	Investigation of the scalar spectrum in SU(3) with 8 flavors
12:00 - 13:30	Group Photo and Lunch	
	Chair: J. Kuti	
13:30 - 14:15	A. Hasenfratz (U. Colorado Boulder)	Strongly coupled gauge theories : What can lattice calculations teach us?
14:15 - 15:00	S. Matsuzaki (Nagoya)	One-family walking technicolor in light of LHC-Run II
	Chair: F. Sannino	
15:30 - 15:55	K.-i. Nagai (KMI)	Topological insights in many flavor QCD on the lattice
15:55 - 16:20	E. Bennett (Swansea U.)	The infrared regime of SU(2) with one adjoint Dirac Fermion
16:20 - 16:45	C.-J.D. Lin (NCTU)	Lattice study of the Higgs-Yukawa model with a dimension-6 operator
16:45 - 17:10	H. Matsufuru (KEK)	SU(2) lattice gauge theory with many flavors of domain-wall fermions
18:00 - 20:00	Poster Session and Welcome Party	

March 4, 2015

	Chair: R.S. Chivukula	
9:00 - 9:45	R.E. Shrock (Stony Brook)	Some Recent Results on Strongly Coupled Gauge Theories
9:45 - 10:30	F. Sannino (CP3-Oeigins/DIAS)	Rethinking Naturalness: Can the Higgs be elementary?
	Chair: E.H. Simmons	
10:50 - 11:35	K. Tuominen (Helsinki)	Dynamical origin of the electroweak scale and a 125 GeV scalar
11:35 - 12:00	H.S. Fukano (KMI)	Top-Mode Pseudo-Nambu-Goldstone Boson Higgs Model
	Chair: R. Kitano	
13:30 - 14:15	E. Rinaldi (subst. G. Fleming)	Composite Dark Matter
14:15 - 14:40	H. Ohki (KMI)	Lattice study of the scalar and baryon spectra in many flavor QCD
14:40 - 15:05	C.H. Wong (Wuppertal)	Baryon Spectroscopy in the Sextet Gauge Model
	Chair: K.-I. Kondo	
15:35 - 16:20	K-I. Aoki (Kanazawa)	Weak Renormalization Group Approach for Dynamical Chiral Symmetry Breaking
16:20 - 16:45	K. Itoh (Niigata)	Gauge symmetry and Functional RG
19:00 - 21:00	Banquet	

March 5, 2015

	Chair: T. Kugo	
9:00 - 9:45	M. Rho (CEA Saclay)	Hidden and Mended Symmetries and Compact Stars
9:45 - 10:10	D-K. Hong (Pusan)	Holographic Estimate of Isospin splitting in hadron mass
10:10 - 10:35	C. Nonaka (KMI/Nagoya)	Scalar mesons in lattice QCD
	Chair: S. Nojiri	
10:55 - 11:20	K.-I. Kondo (Chiba)	Magnetic monopole versus vortex as gauge-invariant topological objects for quark confinement
11:20 - 11:45	A. Shibata (KEK)	Confinement/deconfinement phase transition in SU(3) Yang-Mills theory in view of dual superconductivity
11:45 - 12:10	N. Yamada (KEK)	Linking $U(2) \times U(2)$ to $O(4)$ via decoupling
	Chair: T. Sakai	
13:30 - 14:15	D. Elander (Purdue)	Walking dynamics from gauge/gravity duality
14:15 - 14:40	M. Jarvinen (ENS, Paris)	Holography and the conformal window in the Veneziano limit
	Chair: T. Yamazaki	
15:30 - 15:55	D. Kadoh (KEK)	Gauge/gravity duality - From lattice gauge theory to black hole
15:55 - 16:20	D. Schaich (Syracuse)	Maximally supersymmetric Yang–Mills on the lattice
16:20 - 16:45	N. Maru (Osaka City U.)	Higgs Mass in D-term triggered Dynamical SUSY Breaking
16:45 - 17:10	K. Bamba (Ochanomizu)	Large-scale magnetic fields, non-Gaussianity, and gravitational waves from inflation
18:00 - 21:00	Poster Session and Group Dinner	

March 6, 2015		
	Chair: M. Tanabashi	
9:00 - 9:45	R.S. Chivukula (Michigan State)	The Jet Energy Profile: A BSM Analysis Tool
9:45 - 10:30	E.H. Simmons (Michigan State)	Separating Di-jet Resonances using the Color Discriminant Variable
	Chair: N. Maekawa	
10:50 - 11:35	R. Kitano (KEK)	Hidden Local Symmetry as magnetic gauge theory
11:35 - 12:00	M.J.S. Yang (KEK)	A symmetry breaking mechanism by parity assignment in the noncommutative Higgs model
	Chair: Y. Kikukawa	
13:30 - 14:15	E. Pallante (Groningen)	Around and across the endpoint of the conformal window
14:15 - 14:40	S.Y. Kim (Sejon)	Strong coupling limit of lattice QCD with many staggered quarks
	Chair: R. Shrock	
15:10 - 15:55	K. Yamawaki (KMI)	Old Wine in a New Wine Bottle – Technidilaton as the 125 GeV Higgs
	impromptu talk filling cancelation of	
	V.A. Miransky (Western)	Who is afraid of the Landau Pole?
15:55 - 16:30	Best Poster Award and Break	
	Chair: K. Yamawaki	
16:30 - 17:15	H. Georgi (Harvard)	SCGT15 – Summary Talk
17:30 - 19:30	Free Discussions and Refreshments	

Photo Gallery

running (walking) coupling in progress
chiral condensate, large γ(Λ)
spectroscopy
dark matter
Outlook: is it worth the big effort?
chances of sextet model ~ ε
would be significance ~ 1/ε
it makes sense to work on it ~ O(ε/ε = 1)
and we learn more about SCGT!

pionic tensor force, with the
sor strongly suppressed
over at high den
nsed crystal ma

名古屋大学
NAGOYA UNIVERSITY

SCGT15
Sakata Memorial KMI Workshop on
Origin of Mass and Strong Coupling Gauge Theories
Old Wine in a New Wine Bottle----
Technidilaton as the 125 GeV Higgs

Energy is Eternal Delight!

List of Participants

Yuki Adachi	Matsue College of Technology
Satoshi Akagi	Nagoya University
Yukinao Akamatsu	KMI, Nagoya University
Yasumichi Aoki	KMI, Nagoya University
Ken-Ichi Aoki	Kanazawa University
Tatsumi Aoyama	KMI, Nagoya University
Yuhma Asano	Kyoto University
Kazuharu Bamba	Ochanomizu University
Ed Bennett	Swansea University
R. Sekhar Chivukula	Michigan State University
Daniel Elander	Purdue University
Mitsutoshi Fujita	YITP, Kyoto University
Takanori Fujiwara	Faculty of Science, Ibaraki University
Hidenori Fukano	KMI, Nagoya University
Howard Georgi	Harvard University
Anna Hasenfratz	University of Colorado Boulder
BingRan He	Nagoya University
Takayuki Hirayama	Chubu University
Deog-Ki Hong	Pusan National University
Katsumi Itoh	Niigata University
Masaya Iwanaga	Nagoya University
Yoichi Iwasaki	University of Tsukuba and KEK

Matti Jarvinen	Ecole Normale Superieure, Paris
Daisuke Kadoh	KEK
Eu Siang Kam	Nagoya University
Taishi Katsuragawa	Nagoya University
Mamiya Kawaguchi	Nagoya University
Takeo Kawaguchi	Nagoya University
Ken Kikuchi	Nagoya University
Yoshio Kikukawa	Institute of Physics, the University of Tokyo
Seyong Kim	Sejong University
Ryuichiro Kitano	KEK
Meguru Komada	Nagoya University
Kei-Ichi Kondo	Chiba University
Taichiro Kugo	Maskawa Institute, Kyoto Sangyo Univ.
Shin-Ichiro Kumamoto	Kanazawa University
Masafumi Kurachi	KMI, Nagoya University
Wataru Kuramoto	Nagoya University
Tsunehide Kuroki	KMI, Nagoya University
Julius Kuti	UC San Diego
Takumi Kuwahara	Nagoya Univ.
David Lin	National Chiao-Tung University, Taiwan
Toshihide Maskawa	KMI, Nagoya University
Nobuhiro Maekawa	KMI / Nagoya University
Nobuhito Maru	Osaka City University
Hideo Matsufuru	KEK

Mamoru Matsunaga	Mie University
Yuto Matsushima	Nagoya University
Shinya Matsuzaki	Nagoya University
Kohtaroh Miura	KMI, Nagoya University
Yu Muramatsu	Nagoya University
Kei-ichi Nagai	KMI, Nagoya Univ.
Ryo Nagai	Nagoya University
Katsumasa Nakayama	Nagoya University
Jotaro Nakane	Nagoya University
Keita Nii	Nagoya University
Hiroki Nishihara	Nagoya University
Shin'ichi Nojiri	KMI / Nagoya University
Chiho Nonaka	KMI / Nagoya University
Hiroshi Ohki	KMI, Nagoya University
Shohei Okawa	Nagoya University
Yuji Omura	Nagoya Univ.
Kazuma Ono	Nagoya University
Elisabetta Pallante	University of Groningen
Mannque Rho	CEA Saclay
Enrico Rinaldi	Lawrence Livermore National Laboratory
Tadakatsu Sakai	KMI / Nagoya University
Francesco Sannino	CP3-Origins, University of Southern Denmark
David Schaich	Syracuse University
Akihiro Shibata	KEK

Yoshihiro Shigekami	Nagoya University
Robert Shrock	Stony Brook University
Elizabeth H. Simmons	Michigan State University
Masaharu Tanabashi	KMI / Nagoya University
Kazuhiro Tobe	KMI / Nagoya University
Koji Tsumura	Kyoto University
Kimmo Tuominen	University of Helsinki
Naoki Watamura	Nagoya University
Chik Him Wong	University of Wuppertal
Norikazu Yamada	KEK
Koichi Yamawaki	KMI, Nagoya University
Takeshi Yamazaki	University of Tsukuba
Masaki J.S. Yang	KEK

Author Index